IMMUNOLOGY

IMMUNOLOGY

Editor

Norbert Oursler

scitus
academics

Immunology

Edited by **Norbert Oursler**

This Book is Printed on Acid-Free Paper. ♾

ISBN: 9781681179261

Library of Congress Control Number: 2017948955

SCITUS Academics LLC,

616, Corporate Way, Suite 2, 4766,

Valley Cottage, NY 10989

www.scitusacademics.com

Printed in United States of America.

Preface

The science of Immunology encompasses the study of the development, anatomy functions and malfunctions of the immune system, all of which are of fundamental importance to the understanding of disease. The immune system protects animals from infection through various lines of defense. If the immune system is not functioning as it should, it can result in disease, such as autoimmunity, allergy and cancer. It is also now becoming clear that immune responses contribute to the development of many common disorders not traditionally viewed as immunologic, including metabolic, cardiovascular, and neurodegenerative conditions such as Alzheimer's. Veterinary immunology deals the study of all aspects of immune system in animals under biomedical science and related to zoology and veterinary sciences including malfunctions and disorders, but also health, of the immune system in animals. It is interested in how the immune system works, how vaccines prevent disease and why vaccines sometimes do not work and cause adverse reactions.

This volume 'Immunology' reports basic, comparative and clinical immunology as they pertain to the animal species designated here: livestock, poultry, and fish species that are major food animals and companion animals such as cats, dogs, horses and camels, and wildlife species that act as reservoirs for food, companion or human infectious diseases, or as models for human disease. It brings a clinical framework complete with real-world examples to integrate the theory and practice of veterinary medicine. It also covers fundamental aspects of immunology; methods for the demostration of the immune state, and their application to the diagnosis of disease; vaccines and antisera; physiological and pathological aspects of the specific immune responses. This book provides crucial information not only for poultry health professionals and avian biologists, but also for comparative and veterinary immunologists, graduate students and veterinary students with an interest in veterinary immunology.

Table of Contents

CHAPTER 1

Corynebacterium pseudotuberculosis: Immunological Responses in Animal Models and Zoonotic Potential

Bruno lopes bastos[1], Ricardo Wagner Dias Portela[1], Fernanda Alves Dorella[2], Dayana Ribeiro[2], Núbia Seyffert[2], Thiago Luiz de Paula Castro[2], Anderson Miyoshi[2], Sérgio Costa Oliveira[3], Roberto Meyer[1] and Vasco Azevedo[2*]

[1]*Instituto de Ciências da Saúde, Universidade Federal da Bahia, Salvador – BA, Brazil*

[2]*Departamento de Biologia Geral, Instituto de Ciências Biológicas, Universidade Federal de Minas Gerais, Belo Horizonte – MG, Brazil*

[3]*Departamento de Bioquímica e Imunologia, Instituto de Ciências Biológicas, Universidade Federal de Minas Gerais, Belo Horizonte – MG, Brazil*

ABSTRACT

Corynebacterium pseudotuberculosis is a member of the Corynebacterium, Mycobacterium and Nocardia (CMN) group that comprises species of medical, veterinary and biotechnological interest. This pathogen mainly affects small ruminants, causing caseous lymphadenitis (CLA), but it also infects bovines, equines, pigs, deer, camels and humans, showing its zoonotic relevance. Phospholipase D (PLD) and the toxic lipid cell wall are the two most wellstudied virulence factors of this bacterium. They are responsible, in part, for the establishment of disease in the host. Current knowledge on the immunity induced by *C. pseudotuberculosis* indicates that the resistance to infection is a complex process involving components of both the non-specific and specific host responses, in which humoral and cellular immune responses are both operative. Despite this knowledge and the importance of the disease, a satisfactory vaccine model for sheep and goats has not been developed. Moreover, a control program that includes an efficient diagnostic method in addition to vaccination is crucial for avoiding the spread of bacteria inside flocks. Further, because of its zoonotic potential, *C. pseudotuberculosis* infection of animals can contaminate meat and milk, putting consumers at risk. The ability of *C. pseudotuberculosis* to infect both animals and humans makes studies on prevention and diagnosis of this pathogen important.

Keywords: *Corynebacterium pseudotuberculosis;* Caseous lymphadenitis; Immunology; Zoonotic potential; Vaccine; Phospholipase D; Diagnosis

1. INTRODUCTION

The genus Corynebacterium is classified into the suborder Corynebacterineae (Actinobacteria: Actinomycetales), which includes the families Corynebacteriaceae, Mycobacteriaceae and Nocardiaceae (http://www.ncbi.nlm.nih.gov/Taxonomy), commonly designated as the CMN group. Some characteristics common to this group include the organization of the cell wall, which is composed mainly of peptidoglycan, arabinogalactan and mycolic acids, as well as the high guanine-cytosine (G + C) content of the genome (47-74%) [1].

The CMN group comprises species of significant medical, veterinary and biotechnological interest, and thus has an important socioeconomic role. For example, the CMN group includes the species Mycobacterium tuberculosis and M. leprae, the etiologic agents of tuberculosis and leprosy, respectively. Within the genus Corynebacterium, there has been a significant expansion in the number of described species in the last decade due to increased concern regarding their potential pathogenic significance. The species that are of major clinical relevance are Corynebacterium diphtheriae, C. jeikeium and C. pseudotuberculosis, which are the causative agents of diphtheria, nosocomial infections in humans and caseous lymphadenitis (CLA) in goats and sheep, respectively [1-3]. C. glutamicum is used in the production of amino acids such as L-aspartate and L-lysine [4], and C. ulcerans causes pathologies in both humans and animals [5].

Most studies mainly report the incidence of C. pseudotuberculosis infection in small ruminants. Affected animals have characteristic abscesses in the lymph nodes, mainly in the parotid or retropharyngeal and/or viscera [6]. However, some infected sheep may have only internal abscesses, which are often present in the lungs or mediastinal lymph nodes, and show no major outward signs of infection [7]. This pathogen has a broad spectrum of hosts and causes clinical disease in cattle, horses, pigs, deer, camels and laboratory animals [8]. According to the World Animal Health Organization, among the 201 countries that described their sanitary conditions, 64 reported animals with CLA within their borders in 2009 [9]. Although human cases of infection are relatively rare, some case reports note the potential risks of infection of veterinarians and farm practitioners [10]. In this review, we present the general characteristics and virulence factors of C. pseudotuberculosis related to its immunopathogenesis, the response of the immune system, aspects associated with resistance to infection in animal models and diagnostics. We also discuss the zoonotic potential of the bacteria, presenting an update on human infections by C. pseudotuberculosis. Finally, we suggest guidelines for future research.

2. GENERAL CHARACTERISTICS OF *CORYNEBACTERIUM PSEUDOTUBERCULOSIS*

Corynebacterium pseudotuberculosis is the etiological agent of CLA, which mainly affects populations of small ruminants throughout the world and generates significant economic losses [11]. This bacterium was first described in 1888 by a French veterinarian, Edmound Nocard, in a case of lymphangitis in cattle. In 1891, Hugo von Preisz isolated a similar bacterium in a renal abscess in sheep. Hence, the pathogen was named "Preisz-Nocard" bacillus, the name with which it was linked for decades thereafter. In Australia, which is currently the country with the second largest sheep flock in the world, CLA was firstly described in 1934 by Churchward [8,12,13].

The current nomenclature was adopted in 1948 in the sixth edition of Bergey's Manual; however, the designation C. ovis was used synonymously. C. pseudotuberculosis is characterized as a Grampositive facultative intracellular pathogen that displays pleomorphic forms, ranging from coccoid to filamentous rods and measures approximately 0.5-0.6 μm in width and 1.0-3.0 μm in length. These bacteria do not have a capsule or flagella and they do not sporulate; however, fimbriae are present. The cell-wall peptidoglycan appears to be based on meso-diaminopimelic acid (meso-DAP), and the sugars arabinose and galactose are present in large quantities. Short chains of mycolic acid have also been found [14-16].

C. pseudotuberculosis is a mesophilic facultative anaerobic organism, and its optimum growth conditions are at 37ºC with pH values ranging between 7.0 and 7.2 [12,17]. However, according to Batey [18], C. pseudotuberculosis can grow well in a pH range of 7.0 to 8.0. C. pseudotuberculosis is a demanding organism from a nutritional standpoint, growing well on enriched media such as blood agar, brain heart infusion (BHI) agar or broth or enriched medium with animal serum. Its cultivation improves when BHI is complemented with yeast extract, tryptone or lactalbumin [19]. Its growth in a solid culture medium is initially sparse on an agar surface and then becomes organized in clumps or in palisades, taking on a cream to yellowish coloration. Colonies are dry, opaque and concentrically ringed, and after several days of incubation, colonies can reach 3 mm in diameter [20]. When grown in blood agar, it is possible to detect beta-hemolysis after a period of 48 to 72 hours of incubation [19,20]. Growth in liquid medium occurs as a biofilm on the surface, without medium clouding. This film is dismantled by agitation, forming flakes that precipitate [17,21]. The film formation is attributed to surface lipids, and the amount of these lipids in the cell membrane is directly correlated with the thickness of the biofilm and the virulence of the bacterial strain [22]. Some authors have reported better growth in an atmosphere of 5% CO2 [12,21]. Recently, a chemically defined medium (CDM) for C. pseudotuberculosis cultivation composed of dibasic phosphate, vitamins and amino acids was developed. This medium is free of macromolecules in its composition, allowing researchers to obtain the complex antigenic solution composed only of secreted/ excreted bacterial proteins, which would aid studies on this bacterium [23]. This CDM has been recently used with success to obtain pure secreted antigens for an exoproteome analysis of C. pseudotuberculosis [24].

Biochemically, C. pseudotuberculosis is characterized by the production of catalase, phospholipase D, and urease and by the fermentation of carbohydrates such as maltose, mannose, glucose, and galactose (the latter is used only occasionally) [17]. It does not perform lactose fermentation or gas production [21,25-27], has no proteolytic activity, does not have the ability to hydrolyze gelatin or digest casein and is oxidase negative [17,20]. It has been suggested that there are two biotypes of these bacteria based on the production of nitrate reductase, demonstrated by analysis with restriction enzymes. The reduction of nitrate to nitrite characterizes strains that infect mainly horses (biovar equi) and present sensitivity to streptomycin, while the strains that infect goats and sheep (biovar ovis) are mainly nitrate negative and resistant to streptomycin [27,28]. Cattle are infected by both reducing and non-reducing nitrate strains [25,26].

3. ANTIGENS AND VIRULENCE FACTORS

Compared to other pathogenic bacteria, few studies have investigated the virulence determinants of C. pseudotuberculosis. However, an understanding of the molecular mechanisms and the genetic basis of virulence in C. pseudotuberculosis has rapidly improved. Two virulence determinants of C. pseudotuberculosis have been well characterized: the exotoxin PLD and the toxic lipid cell wall. PLD hydrolyzes phosphatidylcholine and sphingomyelin from mammalian cell membranes, resulting in the formation of choline and ceramide phosphate [29-31]. Biologically, this exotoxin works as a permeability factor that promotes the dissemination of the pathogen from the initial site of infection to all the body tissues of the host. Additionally, PLD causes the dermonecrosis of endothelial cells, which contributes to the passage of C. pseudotuberculosis from the dermis to small blood vessels, thereby gaining access to lymphatic vessels [32,33]. Further, PLD is considered a cytotoxic exotoxin for white blood cells, as it promotes the destruction of goat macrophages during experimental infection [34].

Studies using specific anti-PLD antibodies have demonstrated that this exotoxin is essential to the spread of C. pseudotuberculosis, as the use of this antibody prevents its spread. In addition, the vaccination of goats with inactivated exotoxin is able to prevent the spread of the bacteria after experimental challenge [35]. The role of PLD as a virulence determinant of C. pseudotuberculosis was demonstrated by the generation of pld-mutants. The mutant was unable to spread but did induce an immune response. However, these results suggest that the use of this mutant as a live vaccine model for CLA would be insufficient [36]. The spread of bacteria by the action of PLD can be explained by its induction of dermonecrosis, which causes direct damage to endothelial cells. During hydrolysis, PLD is able to work synergistically with cholesterol oxidase and phospholipase C, which are both, produced by Rhodococcus equi, and causes the β-hemolysis of sheep erythrocytes when cultured in blood agar. The phosphate ceramide generated, in turn, is hydrolyzed by phospholipase C of R. equi, producing ceramide [14,30]. In addition to the degradation of sphingomyelin to ceramide, PLD is also able to activate the complement system through the alternative pathway, although the exact mechanism is not well understood [37].

A 40-kDa protein from C. pseudotuberculosis that was previously identified as a protective antigen against ovine CLA [38] was further investigated. Molecular and biochemical characterization revealed that this protein is a 379 amino acid protein encoded by an open reading frame of 1,137 bp. No significant homology with previously published DNA or amino acid sequences in the databases was found, suggesting that this is a novel protein. Thus, the recombinant 40-kDa protein was over-expressed in E. coli for the biochemical analysis of the native and the recombinant 40-kDa proteins. The protein was found to be a serine protease with proteolytic activity and was designated corynebacterial protease 40 (CP40) [39].

Other antigens from both the secreted and somatic fractions were identified by an immunoblotting analysis. One study used the serum of naturally infected sheep for immunoblotting to identify antigens from sonicated C. pseudotuberculosis and found eleven antigens with the following molecular weights: 20 kDa, 22.4 kDa, 31.6 kDa, 35.5 kDa, 36.3 kDa, 39.8 kDa, 45.7 kDa, 56.2 kDa, 63.1 kDa, 79.4 kDa and 100 kDa. When the analyses were performed with secreted antigens from the culture supernatant, antigens with the following molecular weights were found: 20 kDa, 25.1 kDa, 31.6 kDa, 39.8 kDa and 63.1 kDa [40]. A previous immunoblotting analysis with sera from experimentally infected goats on secreted antigens from bacterial culture supernatants revealed antigens with the following molecular weights: 16, 20, 27, 30, 36, 40, 43, 58, 64, 68 and 125 kDa [41].

The introduction of more advanced technologies such as mass spectrometry (MS) has brought new perspectives to the study of bacterial secreted/excreted proteins. A comparative exoproteome analysis of C. pseudotuberculosis was recently performed using a highthroughput proteomic strategy based on liquid chromatography-mass spectrometry (LC-MS). This study investigated the proteins that are exported from the bacteria, which might represent key components of the host-pathogen interplay. Ninety-three different extracellular proteins of C. pseudotuberculosis were identified with high confidence using this strategy; 44 proteins were commonly identified in two different strains that were isolated from distinct hosts, thus composing the core of the C. pseudotuberculosis exoproteome. This study generated a catalog of exoproteins, in which novel targets for future work on C. pseudotuberculosis molecular determinants of virulence can be identified [24].

As C. pseudotuberculosis is classified in the Actinomycetales order, it shares some common characteristics with this group. C. pseudotuberculosis has a significant amount of lipids in its cell wall called corynomycolic acids, and they are similar to the mycolic acids in M. tuberculosis cell walls. These surface lipids have long been described as important factors for the pathogenesis of the disease, as virulent strains have more lipids than attenuated strains, and there is a direct relationship between the percentage of surface lipids and chronic abscesses [21]. This lipid layer protects against proteolytic enzymes present in phagolysosomes, allowing the microorganism to adhere, thereby promoting local cytotoxicity. The toxicity of the extracted lipid material was demonstrated by the induction of hemorrhagic necrosis after intradermal injection into guinea pigs [22,42,43].

Recently, four genes within an operon that is involved in iron acquisition, designated fag A, B, C and D, were shown to have an important role in the virulence of C. pseudotuberculosis [44]. Recent advances in the understanding of pathogenicity pathways used by C. pseudotuberculosis in the infection process have been achieved through complete genome sequencing [45]. A C. pseudotuberculosis strain (FRC41) isolated from a 12-year-old girl with necrotizing lymphadenitis was completely sequenced and assembled, yielding the identification of specific gene sets associated with a variety of metabolic and pathogenic bacterial functions. Two gene clusters encoding proteins involved in the sortase-mediated polymerization of adhesive pili were found, and these proteins likely mediate the adherence to host tissue, thereby facilitating additional ligand-receptor interactions and the delivery of virulence factors. In addition, the prominent virulence factors PLD and CP40 were found in the genome of this human isolate. The genome also revealed additional serine proteases, neuraminidase H, nitric oxide reductase, an invasion-associated protein, and acyl-CoA carboxylase subunits involved in mycolic acid biosynthesis as potential virulence factors [46].

The genetic characterization of two strains of C. pseudotuberculosis from goat (Cp1002) and sheep (CpC231) using a predicted genome analysis contributed new information on genome evolution and lateral acquisition of virulence functions. C. pseudotuberculosis was compared with other Corynebacterium species, revealing that this pathogenic species has lost numerous genes, resulting in one of the smallest genomes in the genus. Other differences found were a lower GC content (approximately 52%) as well as a reduced gene repertoire. These characteristics suggest adaptations that give the bacteria its pathogenesis. In addition, seven putative pathogenicity islands were found in its genome that contain several classical virulence factors, including genes for fimbrial subunits, adhesion factors, iron uptake and secreted toxins [9].

4. IMMUNOPATHOGENESIS OF CLA

The progression of CLA in sheep and goats starts as primary wound infection, with lymphatic and hematogenous dissemination, followed by secondary infection of the lymph nodes and various visceral organs. This is followed by the elimination or containment of infection, the latter presenting as characteristic caseous lesions. The steps of infection have been separated into the following phases: an initial phase (day 1–4 p.i.), characterized by the recruitment of neutrophils to the inoculation site and the draining of the lymph nodes; an amplification phase (day 5–10 p.i.), characterized by the development of pyogranuloma; and a stabilization phase, characterized by the maturation and persistence of the pyogranuloma [47]. Bacterial factors, including PLD and cytotoxic lipids, contribute to pathogenesis at a local level but have little effect on the systemic disease. After C. pseudotuberculosis is captured by phagocytic cells such as neutrophils and macrophages, the phagosome fuses with the lysosome, forming the phagolysosome [28]. However, C. pseudotuberculosis is a facultative intracellular pathogen that is capable of surviving within macrophages for more than 48 hours. During that time, bacteria are released as a result of a process that leads to phagocyte death, although this property varies among different strains. The specific mechanisms of cell death caused by C. pseudotuberculosis are still unclear, as it does not induce the autophagy or apoptosis of macrophages. This has been demonstrated in murine macrophage cell lines, as evidenced by stable levels of microtubule-associated protein I light chain 3 (MAP-I LC3) activity and caspase-3 activity and an absence of nuclear fragmentation in infected macrophages [48]. The bacteria survive within macrophages because some macrophages cannot produce nitric oxide in response to C. pseudotuberculosis in vivo, which results in ineffective clearing of the organism [49]. These effects might be associated with the outer lipid layer in the cell wall of C. pseudotuberculosis and other antigenic components that attenuate the production of nitric oxide by macrophages.

As a result of the uncontrolled bacterial growth within macrophages, the host attempts to restrain and limit the infection through the formation of pyogranulomas, which are characterized by the encapsulation of the C. pseudotuberculosis infected cells. The formation of pyogranulomas is dependent on adaptive immunity, which is a complex process in the case of infection by C. pseudotuberculosis that involves both humoral and cell-mediated immunity [28,50]. Immunohistochemical studies on the cellular composition of the pulmonary lesions in sheep infected by C. pseudotuberculosis have revealed a predominance of large macrophages that express major histocompatibility complex (MHC) class II molecules on their surfaces in the abscess walls and surrounding lung parenchyma. T lymphocytes were prominent in the same areas within the naturally occurring lesions, with a CD4+ T cell to CD8+ T cell ratio of 3.5:1. B lymphocytes and granulocytes comprised a minor portion of the infiltrating cells. These data revealed the participation of macrophages and MHC class II-restricted T lymphocytes in the pathogenesis of CLA [51].

Another immunohistochemical study revealed more features of the pyogranuloma cellular makeup. Lymphocytes localized to the inside of the collagen capsule were found to be in close contact with necrotic tissue and were organized into three layers. The innermost layer, which is immediately adjacent to the central necrotic tissue, consists of a narrow band of MHC class II-macrophages. CD4+, CD8+ and gamma delta T cells were unevenly distributed throughout the lymphoid layer, tending to be more numerous immediately peripheral to the macrophage layer. The intracapsular lymphoid tissue contained a high proportion of CD8+ lymphocytes (CD4:CD8, 1.5:1) and gamma delta lymphocytes (CD4:CD8: gamma delta, 1:0.7:0.8). However, in contrast to the previous immunohistochemical study, CD8+ and gamma delta+ T cells were found to be more abundant than CD4+ T cells [52].

Recent studies have clarified the cellular makeup of pyogranulomas. In an experimental pyogranuloma induced in lambs infected with C. pseudotuberculosis, three different monoclonal antibodies were used to characterize the presence of macrophages in situ and demonstrated that macrophages are the predominant cells in pyogranulomas [53]. Further immunohistochemical experiments on the cellular composition of 46 experimentally induced pyogranulomas in sheep also demonstrated that lesions localized to inoculation sites or draining lymph nodes consisted of macrophage and lymphocyte layers distributed around the necrotic center, surrounded by a fibrous capsule. However, the most significant observation from this work was that in immature lesions, CD4+ T cells were most predominant, whereas in mature lesions, the proportions of CD8+ T lymphocytes and gamma/delta receptor-expressing cells increased. In addition, this study found that many pyogranuloma cells expressed the interleukin-2 receptor, and a large individual variability in the proportions of macrophage and T cell subsets was observed in lesions at the same maturation point, with particular discrepancies in the number of epithelioid macrophages. This heterogeneity suggests that there are different cellular patterns based on the persistence or elimination of bacteria by the host [54].

A recent study aiming to evaluate the outcomes in mice after inoculation with four equine-origin C. pseudotuberculosis strains (slow and rapid-growing strains) demonstrated that the tropism for the liver, spleen, lungs, and mesenteric lymph nodes is distinct for each strain. Specifically for the liver, the histological lesions associated with rapidly growing strains included focally extensive unencapsulated areas of acute, massive coagulative necrosis of hepatocytes. These areas had intralesional colonies of bacteria and variable portal hepatitis characterized by the accumulation of mononuclear and polymorphonuclear inflammatory cells. In contrast, the livers from mice inoculated with slow-growing strains had multiple discrete, randomly distributed foci of hepatocellular necrosis and neutrophilic hepatitis that were considerably less severe than the lesions in the mice inoculated with the rapidly growing strains [55].

In addition to the formation of pyogranulomas, infection with C. pseudotuberculosis leads to the formation of subcutaneously located abscesses that do not originate from primary lymph node lesions. However, some abscesses can be closely association with lymph nodes, and as a result of the tissue destruction associated with abscesses and the expanding nature of these lesions, micro-abscesses that are composed of multiple caseous lesions of 0.5–1.0 cm in diameter can occur. Microscopically, these encapsulated abscesses contain partially calcified pus arranged in a concentric lamellar structure [56].

5. RESISTANCE TO C. *PSEUDOTUBERCULOSIS*

Resistance to infection by C. pseudotuberculosis is a complex process involving components of both the non-specific and specific host response, in which both the humoral and cellular immune responses are operative [28,57,58]. The importance of humoral mechanisms has been demonstrated by assessing numerous commercial and experimental vaccine trials, which include preparations based on inactivated cellculture supernatant or toxoids [59,60], bacterial cell-wall fractions [61,62], attenuated and killed bacteria [63,64], exotoxins and their subunits [34,65] or genetically modified pathogens [66,67]. Advances regarding immunoprophylaxis have been made, and the use of these preparations reduces the number and size of granulomas in challenged animals. However, to adequately control infection, the activation of macrophages needs to be improved; indeed, many researchers are now focusing on the gamma IFN (IFN-γ) response.

Early studies in mice evaluated the ability of levamisole to promote nonspecific immunity to C. pseudotuberculosis infection. The enhanced nonspecific resistance was demonstrated with a quantitative reduction in immunoglobulin levels; the levels of IgG2 and IgA were lower in mice pretreated with levamisole. This result suggested that cell-mediated immunity might play a more important role than humoral immunity in resistance to C. pseudotuberculosis infection [68]. The role of IFN-γ was first demonstrated in experimental infections of mice deficient for the IFN-γ receptor with attenuated C. pseudotuberculosis mutants. The study found that the production of IFN-γ and the presence of its receptors were directly associated with the control of primary infection in mice [63]. Another important cytokine that controls primarily infections by C. pseudotuberculosis is tumor necrosis factor alpha (TNF-α) [63,69,70]. The administration of anti-TNF-α and anti-IFN- γ-monoclonal antibodies (mAbs) increased bacterial proliferation in infected mouse organs, leading to the death of the animals. Further, the injection of anti-CD4 or anti-CD8 mAbs also resulted in significantly increased mortality and a marked suppression of IFN-γ production but had no effect on TNF-α production. Therefore, endogenous TNF-α and IFN-γ may have addictive effects on the anti-corynebacterial defense in the early stage of infection and could be critical for the generation of resistance [69].

The same group also investigated the role of these cytokines during secondary C. pseudotuberculosis infection in mice. They found that both TNF-α and IFN-γ are required for survival and the development of protective immunity. During secondary infection, the mice recovered 50% faster, suggesting the importance of these cytokines in activating macrophages. In this trial, the administration of anti-CD4 mAb alone or anti-CD4 plus anti-CD8 mAbs increased mortality, bacterial proliferation and reduced the production of these cytokines, while treatment with anti-CD8 mAb alone showed no effect on either the resistance to infection or cytokine production. In conclusion, these data suggest that CD4+ T cells, likely Th1 T cells, play an important role in immunity against secondary C. pseudotuberculosis infection [71].

Another study on the innate immune response during the early course of infection revealed the participation of the complement system in the defense against C. pseudotuberculosis. Experiments demonstrated that the type 3 complement receptor (CR3) plays a key role in inflammatory cell recruitment during the course of infection. Indeed, the treatment of mice with an anti-CR3 mAb resulted in the unrestricted proliferation of bacteria in the spleen and livers and dramatically increased the mortality of the infected mice within 3 days of infection. A histological examination also revealed that mononuclear phagocytes did not migrate to the sites of bacterial multiplication, as evidenced by the inhibition of inflammatory cell migration and large numbers of bacteria in their organs. These results revealed the importance of CR3 in the resistance against primary as well as secondary C. pseudotuberculosis infection in mice [72].

Although the previous results highlight the importance of T cellmediated immunity against the infection, some evidence has also indicated that the production of antitoxin may protect the host during secondary exposures to the pathogen [14,57]. In mice previously immunized with washed bacterial suspensions, the antitoxin was unable to prevent the formation of pus after inoculation, despite the fact that the antitoxin was able to prevent the spread of infection from the site of inoculation to internal organs [57]. These results were later reproduced in several experiments demonstrating that anti-PLD antibodies present in host blood before infection exert a protective effect by neutralizing the permeability induced by PLD, thereby hindering the dissemination of C. pseudotuberculosis to draining lymph nodes. Further, the antibodies reduced secondary spread and prevented lesion development [73].

As CLA affects sheep and goats, a vast number of studies have been performed in these species since the 1980s [28,50-54,62,74]. A groundbreaking study was conducted in sheep inoculated with attenuated and wild-type strains of C. pseudotuberculosis; the authors assessed the expression of cytokine mRNA during infection. This study revealed that at the site of experimental inoculation, the levels of the inflammatory cytokines TNF-α and interleukin-1β (IL-1β) were increased, and the expression of interleukin-4 (IL-4) was decreased. Further, the cytokines IL-2 and IL-4 were up-regulated. The highest expression of TNF-α, IL-2 and IFN-γ occurred at the seventh day post-inoculation, and the highest levels of IL-1β and IL-6 mRNA were detected 28 days after inoculation. The levels of mRNA for IL- 1β, IL-6, IL-8, TNF-α and MCP-1 (monocyte chemoattractant protein 1) were elevated in animals with granulomas in the draining lymph nodes. These results suggested that the development of granulomas could be a consequence of the presence of Th1 and Th2 cells as well as the elevated production of cytokines such as IFN-γ, IL-2, IL-4, TNF-α and MCP-1. These data suggest that granulomas constitute an important factor for limiting C. pseudotuberculosis infection [47].

Another study assessed the kinetics of IgG and IFN-γ production in Corynebacterium pseudotuberculosis primary infection in goats. The authors demonstrated that the patterns of IgG production varied between animals, but the maximum titers were detected between days 11 and 21 post infection (PI). The levels further declined until 140 days PI. Serological positivity was detected from days 6 to 11 PI, but not all individuals remained positive throughout the 20 weeks of follow-up, indicating that humoral immunity is not long lasting. The same work demonstrated the existence of two patterns of the IFN-γ response, one considered high and another considered medium/low. In general, IFN-γ production was observed as a short-lived primary response on day 5 PI for the animals of both groups and a strong secondary response starting on day 16 and declining from days 42 to 56 after infection in the high response group. The "IFN-γ low producer" group exhibited only a short-lived peak on day 5 after infection and thereafter displayed no further significant antigen-specific cell-mediated stimulation [50]. This wide variation in the profiles of IFN-γ production was recently confirmed in a large study of sheep and goats. These differences have necessitated the use of assays with low sensitivity for measuring IFN-γ produced by peripheral leukocytes after stimulus with C. pseudotuberculosis; nonetheless, the production of IFN-γ remains a highly specific indicator of CLA [75].

The role of the innate immune system in C. pseudotuberculosis infection is currently under investigation, including the response proinflammatory cytokines and the acute phase protein. Previous studies in sheep have demonstrated an acute phase response during the initial phase of infection, with a significant increase in the concentrations of haptoglobin approximately 5 days PI [76]. A recent study confirmed the acute phase reaction of haptoglobin in CLA but also revealed that serum amyloid A (SAA) and α1 acid glycoprotein (AGP) were increased in an experimental model of CLA in sheep. The study also revealed the continued production of AGP during the transition to the chronic phase of the disease. These data suggest that AGP may be a valuable quantitative marker of the level of pathogenesis in experimental studies of CLA as well as a useful biomarker of natural infection and of chronic conditions in sheep [74]. Another recent study demonstrated that sheep that did not develop clinical signs of CLA presented significantly elevated levels of haptoglobin during the acute phase of the disease compared to sheep that developed superficial abscesses. Although the exact participation of this protein in the defense against infection by C. pseudotuberculosis requires further investigation, this suggests that innate immune mechanisms contribute to the resolution of infection or resistance to the development of CLA pyogranulomas. Further, haptoglobin has been suggested as a promising candidate marker for the clinical progression of C. pseudotuberculosis infection in sheep [77].

6. VACCINE MODELS TESTED FOR CLA

Studies have shown that primary infection with viable bacteria induces strong protection against subsequent exposures. Indeed, ewes with primary infection did not develop lesions as a result of further exposure, whereas immune-naïve ewes developed numerous pyogranulomas at different sites. However, ewes with primary infection remained carriers of the bacterium as a result of primary inoculation [78,79]. A study conducted in alpacas confirmed this observation, showing that primary infection with a low dose of 1.1×103 CFU protected animals against a significantly higher dose of 9×108 CFU of C. pseudotuberculosis. The primary infected alpacas had a febrile response and abscesses at the inoculation site and regional lymph nodes; however, after the challenge, the primary infected animals showed no superficial lesions, in contrast to the immune-naïve alpacas that developed severe disease characterized by fever and abscesses in regional lymph nodes. In addition, primary infected alpacas had a robust antibody response against a C. pseudotuberculosis cell wall antigen [61]. These studies have encouraged scientists to create a vaccine model based on the use of killed bacterial cells, called bacterin. Killed C. pseudotuberculosis cells can theoretically mimic the primary infection, promoting immunoprotection without causing disease. A summary of data from studies published in the last 40 years on vaccine models is presented in Table 1, including the country of origin of the research, the animal model, the composition of the immunogens, the type of adjuvants used and the route of administration of the preparations. Killed bacteria have been used by many scientists as potential vaccines [59,62,80-84]. The degrees of protection have varied among these studies, but a common feature is that the protective effects were only partial. The immunogens did not completely prevent the disease, but the clinical course was milder, with significantly lower numbers of granulomas in immunized animals compared to unvaccinated control animals.

Another vaccine model uses the secreted exotoxins of C. pseudotuberculosis. Specifically, PLD has been widely studied, as it is one of the few virulence factors known and is the best characterized protein of C. pseudotuberculosis. PLD has been purified, cloned and expressed in E. coli [32,85-87]. In some studies, the exotoxins are treated with various concentrations of formaldehyde, producing toxoids. These toxoid-based vaccines have been tested and increase antibody levels against the exotoxins as shown by ELISAs, which decreases CLA spread in sheep [35,60,88,89]. This boost in humoral immunity still only partially protected animals because vaccinated animals only had reduced numbers of lesions after challenge compared to the unvaccinated sheep. In a recent study, animals inoculated with secreted antigen associated with Freund's incomplete adjuvant and oligodeoxynucleotide containing unmethylated CpG dinucleotides (CpG ODN) showed a strong humoral response, but this inoculation could not prevent the spread of infection [90].

Table 1: Summary of data from scientific works published in the last 40 years related to vaccine models against infection caused by *Corynebacterium pseudotuberculosis* in different animal models.

Reference	Year of publication	Country of origin	Animal model	Composition of the immunogens	Type of adjuvants	Route of administration
[81]	1972	Australia	Sheep	Bacterins	AP	-----
[62]	1984	USA	Sheep	Bacterin or CW	WOE	IM
[119]	1985	USA	Mice	Bacterin or CW	MDP, TDM, BCG and *Cpar*	IP
[35]	1986	USA	Goat	Toxoid	FIA	SC
[120]	1987	USA	Sheep	Bacterin	-----	SC
[82]	1988	Brazil	Goat	Bactenn	AP	-----
[121]	1989	Norway	Goat	Bacterin or toxoid	Levamisole	-----
[122]	1989	Norway	Goat	Bacterin + toxoid	-----	-----
[123]	1990	USA	Sheep / mice	Bactenn	MDP	IM / IP
[83]	1991	USA	Sheep / goat	Dried whole *Cp* cells	MO + AA	IM
[40]	1991	USA	Sheep	CST	BPA	-----
[60]	1991	Australia	Sheep	Toxoid or bacterin + toxoid	AH	SC
[124]	1991	Australia	Sheep	Toxoid or toxoid *Cp* + 5 Clostridial toxoids	AH / Sodium selenate	SC
[125]	1991	Australia	Sheep	Varying concentrations of toxoid *Cp* + 5 Clostridial toxoids	AH	SC
[96]	1991	Brazil	Goat	Bacterin or live attenuated *Cp*	AP (bacterin)	ID
[36]	1992	Australia	Sheep	Toxminus *Cp*	-----	SC
[98]	1994	Australia	Sheep	Toxminus *Cp* or recombinant Toxminus *Cp*	-----	OR
[38]	1994	Australia	Sheep	40-kDa antigen	AH	SC
[84]	1996	USA	Sheep / goat	Bacterin	MDP + MO	IM
[3]	1997	Australia	Mice	Live *Cp aroQ* mutant or live *Cp pld* mutant	-----	IP
[64]	1998	Australia	Sheep	Live *Cp aroQ* mutant or live *Cp pld* mutant	-----	SC
[59]	1998	Canada	Sheep	Glanvac® 6, Case-Vac® or bacterin	MDP + MO (bacterin)	SC / IM
[91]	1998	USA	Sheep	Bacterin + toxoid	AH	SC
[65]	1999	Australia	Sheep	Glanvac 6 or recombinant Glanvac® 6	MA	SC

[126]	2000	German	Goat	CW	-----	-----
[127]	2000	Australia	Sheep	Toxoid Cp + 5 Clostridial toxoids	AH / Moxidectin	SC
[97]	2002	Brazil	Goat	Liophilized live attenuated *Cp*	no	ID
[128]	2003	Perù	Mice	CST	AH	SC
[58]	2005	Egypt	Mice	Bacterin, toxoid or bacterin + toxoid	MO	SC
[99]	2006	United Kingdon	Sheep	Bacterin, rPLD, bacterin + rPLD, or Glanvac® 3	AH	SC
[61]	2007	Peru	Alpacas	Live Cp	-----	SC
[92]	2007	Peru	Alpacas	CW or CST	MDP	SC
[129]	2007	Egypt	Mice	Bacterin, rPLD, or bacterin + rPLD	MO	SC
[90]	2008	Brazil	Goat	Crude CST, concentrated CST + oligodeoxynucleotide CpG, or live attenuated *Cp*	FIA	SC
[93]	2009	Brazil	Mice	Recombinant Heat-shock protein 60 (rHsp60)	FCA / FIA	SC
[130]	2010	Egypt	Sheep	Bacterin, bacterin + rPLD, Gamma-irradiated Cp + rPLD or BCG + rPLD	MO (except BCG)	SC
[131]	2010	USA	Mice	Bacterin + toxoid	-----	-----
[132]	2010	Turkey	Sheep	Bacterin + toxoid	FCA	SC
[133]	2011	Saudi Arabia	Sheep	Glanvac® 6	Olive oil	SC

Abbreviations: Immunogens - *Cp* = *Corynebacterium pseudotuberculosis*; Bacterin = *Cp* killed whole cells; CW = sonicated *Cp* cell wall; CST = filtrated culture supernatant exotoxins; Toxoid = inactivated exotoxins or phospholipase D (PLD); rPLD = recombinant PLD; Glanvac® and Case-Vac® = commercial vaccines; Toxminus = Cp with deleted *pld* gene. **Adjuvants** - WOE = Water-in-oil emulsion; MO = mineral oil; AA = Arlacel A or monooleate of manitol; AH = aluminum hydroxide; AP = Aluminum phosphate, MDP = muramyl dipeptide; TDM = trehalose dimycolate; BCG = heat-killed *Mycobacterium bovis* BCG; *Cpar* = heat killed *Corynebacterium parvum*; FCA = complete Freund's adjuvant; FIA = incomplete Freund's adjuvant; BPA = block polymer adjuvant; MA= multicomponent adjuvant. **Route of administration** - SC = subcutaneous; IM = intramuscular; IP = intraperitoneal; ID = intradermal; OR = oral.

Some studies have used a combination of bacterins and toxoids. A product composed of killed C. pseudotuberculosis and PLD inactivated by formaldehyde was tested in field trials, and the serological results demonstrated the presence of antibodies against PLD as well as against cellular antigens of C. pseudotuberculosis. A progressive significant increase in humoral response was observed in the eighth week, but antibody titers decreased by the thirty-second week. After the challenge, the number of granulomas was significantly lower in vaccinated animals compared to controls, supporting the use of a vaccine combining exotoxin with somatic antigens. However, the immune protection was still not adequate [91].

Subunit vaccine models have also been developed in an attempt to improve the specific immunological response and promote greater rates of protection. The CP40 antigen protein from C. pseudotuberculosis was identified by a strategy that employs locally derived antibody-secreting cells (ASCs). ASC probes generated by culture of ASCs obtained from lymph nodes draining at the site of infection showed specificity for CP40. Sheep vaccinated twice with 100 μg per dose of CP40 in aluminum hydroxide adjuvant were protected against infection with C. pseudotuberculosis, with an 82% reduction in the number of infected sheep and a 98% reduction in lung lesions, suggesting that the 40-kDa antigen plays a major role in immunity to CLA [38]. Another study used an antigen of the C. pseudotuberculosis cell wall with muramyl dipeptide as an adjuvant and observed a reduction in CLA pyogranulomas [92]. C. pseudotuberculosis heat-shock protein (Hsp60) has also been tested as a vaccine candidate against CLA. The immunization of BALB/c mice with recombinant Hsp60 (rHsp60) that was expressed in E. coli and purified induced a significant anti-Hsp60 IgG response, with greater production of IgG1 rather than IgG2a. Cell-mediated immune responses induced by immunization were characterized by an elevated production of IFN-γ and IL-10, while IL-4 concentrations were not significantly increased. However, the subcutaneous administration of rHsp60 did not induce effective protection against intraperitoneal infection with C. pseudotuberculosis [93]. Other studies have identified immunodominant antigens by immunoblotting using the secreted/ excreted proteins of C. pseudotuberculosis with immune sera from infected goats and sheep [50,94,95], but these antigens have not yet been investigated in the context of vaccines.

Experiments with live attenuated bacterial strains as vaccine models have also been performed, but the same pattern of results have been observed: evident humoral induction, varied degrees of immune protection and reduction only in the number of CLA lesions [90,96,97]. To improve the results of vaccine models based on live strains, researchers genetically engineered a mutant strain of C. pseudotuberculosis in which the gene for PLD was deleted by sitespecific mutagenesis, named "Toxminus". When inoculated into sheep, this mutant was less virulent, causing no local granulomas (108 CFU / dose). With a dose of 1010 CFU of mutant Toxminus, small granulomas were observed in draining lymph nodes, and the mutant bacterium was isolated in only one of these granulomas. These results confirmed that PLD is essential for C. pseudotuberculosis survival in vivo. The humoral and cellular responses to this mutant were assessed and were found to be less severe than those induced by wild type strains. When animals vaccinated with Toxminus were challenged with a wild type strain (107 CFU), the number of granulomas and their extension were much smaller when compared with control animals. Thus, the Toxminus strain that encodes a genetically modified PLD is non-toxic and confers immunoprotection [36].

In further studies, the Toxminus group developed a new mutant, Toxminus with an additional gene encoding an inactive form of PLD. This inactive form was obtained by substituting a histidine for a tyrosine at the active site of the enzyme. Animals orally vaccinated with this mutant were 100% protected against a wild type strain. These results confirmed the importance of PLD as a protective antigen and demonstrated both the potential for developing an oral CLA vaccine and C. pseudotuberculosis Toxminus as a live vaccine vector [98].

Another mutant strain of C. pseudotuberculosis, designated aroQ, was constructed by allelic exchange [63] and tested in sheep models for its ability to act as a vaccine against a homologous challenge. The results demonstrated that aroQ mutants failed to elicit detectable specific IFN-γ-secreting lymphocytes and induced only low levels of antibodies against C. pseudotuberculosis culture supernatant antigens. Subcutaneous vaccination with aroQ did not protect sheep from infection with the wild-type strain, but the clinical severity of disease resulting from challenge did appear to be lower. Attempts to improve the Toxminus strain have been made by replacing a histidine residue at position 20 in the active site of PLD with a serine residue. In vaccination trials, all the immunized sheep showed evidence of low antibody titers to PLD; however, after the challenge, the titers increased significantly, indicating that animals had been sensitized by the vaccine. Nevertheless, the protection rate of the immunized animals was only 44% [65].

Recombinant DNA vaccine strategies have been pursued further. A model containing a DNA sequence for a genetically attenuated PLD and a model for the sequence that encodes the protein CTLA-4 were constructed. The goal of this study was to direct the toxin to the antigenpresenting cells (APCs) because CTLA-4 binds to B7-1 (CD80) or B7-2 (CD86) on APCs. Sheep vaccinated with the DNA vaccine for PLD protected sheep against infection, and this protection was even more pronounced in animals vaccinated with DNA for PLD associated with CTLA-4. The immune protection rates conferred by these vaccines in sheep were 56% and 70%, respectively [66]. The same DNA vaccine was tested for different routes of administration. In sheep, the vaccine was administered intramuscularly and subcutaneously. The route of administration significantly influenced the effectiveness of the vaccine, and the intramuscular route was the most efficient for this vaccine [67]. These studies opened new possibilities for improvement and innovation in veterinary immunoprophylaxis for CLA.

In a recent study, sheep were immunized with a recombinant derivative of PLD, a formalin-killed bacterin, and a bacterin supplemented with recombinant PLD. Following homologous experimental challenge, the PLD and bacterin vaccines conferred statistically significant protection against infection and appeared to restrict the dissemination of bacteria beyond the inoculation site in the majority of animals. The most interesting observation was that the combined vaccine provided absolute protection against infection, whereby challenge bacteria were eradicated from all the vaccinated sheep. In this experiment, in addition to the experimental vaccines, a commercially available CLA vaccine that is not licensed for use in the European Union was assessed in a heterologous challenge. The vaccine conferred significant protection, although the dissemination of infection beyond the inoculation site was not restricted as it was with the previous vaccines. The authors of this study highlighted the fact that if a single commercial vaccine is chosen, it must have the capacity to protect against infection by a variety of different isolates, irrespective of their geographical origin; this must be considered when developing and proposing any vaccine model [99].

Commercial vaccines are available in various regions of the world. For example, Fort Dodge Animal Health (now Pfizer, www.pfizer.com.br) produces a vaccine called Biodectin™ that is marketed in many countries. The combined vaccines of the Glanvac™ series, which are also marketed by Pfizer, are available in Australia. Caseous D - T™ and Case - Bac™ are produced in the United States by the Colorado Serum Company (www.colorado-serum.com). A live attenuated strain of C. pseudotuberculosis (strain 1002) was licensed in the year 2000 for use as a vaccine in Brazil and was developed by the Empresa Baiana de Desenvolvimento Agrícola (www.ebda.ba.gov.br) in collaboration with the Health Sciences Institute of the Federal University of Bahia, Brazil. Another attenuated live vaccine, LinfoVac (Laboratórios Vencofarma do Brasil Ltda; www.vencofarma.com.br), which is licensed for use in sheep and goats, is also currently available in Brazil. Despite the fact that these vaccines have been commercially available for a long time, CLA remains prevalent.

There have been great advances in understanding the immune response against C. pseudotuberculosis, but improvements are still needed to develop a vaccine model for sheep and goats that promotes 100% protection. The partial protection provided by the immunization of goats and sheep with commercial vaccines may be associated with the type of immune response elicited. Protection against C. pseudotuberculosis is mainly dependent on an immune response that involves INF-γ production and cytotoxic T cells. A humoral response alone is insufficient to protect the animal, and a good cellular response is not achieved with inactivated vaccines.

Once CLA is a long-term chronic disease, the success of a vaccination program can to be attributed to the correct use of vaccines and good practices of animal management. These measures would allow younger animals that are already vaccinated to replace the older infected ones, thus eradicating the disease in the flock. In other words, vaccines should only be a part, though an essential part, of CLA control programs.

7. SEROLOGICAL DIAGNOSIS OF CLA

Currently, CLA diagnosis in small ruminants is based on characteristic clinical symptoms and the microbiological identification of C. pseudotuberculosis in material collected from abscesses. However, efficient control requires a serological diagnosis because infected animals that have no apparent symptoms are a source of infection for healthy animals [100]. To date, the isolation and identification of C. pseudotuberculosis by microbiological culture and biochemical testing are still the most reliable methods of CLA diagnosis [101]. However, the puncture of the abscess for the removal of purulent material for culture infects the animal skin and the environment, representing a high risk for pathogen transmission within a herd. As such, the development of less invasive immunodiagnostic tests based on the detection of specific antibodies to C. pseudotuberculosis in the serum of animals has been widely promoted throughout the world [6].

Several diagnostic methods have been proposed for the diagnosis of CLA, including indirect hemagglutination [102], complement fixation [100], immunodiffusion [103], the synergistic inhibition of hemolysis [35] and PCR [104]. Numerous serological tests have been developed to detect antibodies against C. pseudotuberculosis in small ruminants [105]. ELISAs have been frequently used to control CLA in flocks around the world, as they detect subclinical infections in a highly specific manner [106-109]. However, no test has been found to be satisfactory alone [88]. Various antigen preparations have been assayed in ELISAs, including bacterial culture supernatants [108], cell wall antigens [7], PLD [106] and recombinant exotoxins [107]. Typically, the tests work well for goats, but they have reduced sensitivity in sheep, especially in sheep with subclinical infection or that only have internal abscesses. Most of these ELISAs are still not commercially available, and those that are have a relatively high cost [7,105,106].

As C. pseudotuberculosis is a facultative intracellular pathogen, cell-mediated immunity is an important component of the protective immune response [50]. Thus, a whole-blood IFN-γ assay is a promising detection tool for CLA in small ruminant flocks [75,107,110]. This method has been optimized, and its ability to correctly detect infected animals has been compared to ELISAs. Indeed, the IFN-γ test accurately detected C. pseudotuberculosis experimentally infected goats over a 363 day period with a reliability of 89.2% and in non-infected goats with a reliability of 97.1%, while a recombinant PLD-based ELISA detected C. pseudotuberculosis in experimentally infected goats over the same period with a reliability of 81.0% and in non-infected goats with a reliability of 97.0% [107]. A recent study developed an assay for the in vitro quantification of IFN-γ that is secreted in the supernatant of antigen-stimulated cells and concentrated. The samples are simple and fast to prepare, and the test is inexpensive and convenient for IFN-γ quantification. Although it is promising, the sensitivity of this method needs improvement, including efforts to improve the blood-cell culture conditions, antigen extraction procedures and stimulation protocols [75]. Sensitivity and specificity are important factors that should be considered when selecting a diagnostic assay(s) for a screening program. A test with a lower specificity can lead to false positives, whereas reduced sensitivity can lead to false-negative results. An alternative approach to the diagnosis of CLA might be the combined assessment of the in vitro cellular immune response to C.

pseudotuberculosis antigens combined with ELISAs, which would fully evaluate the humoral response.

8. ZOONOTIC POTENTIAL OF C. PSEUDOTUBERCULOSIS

Sheep and goats are the most common animals infected within the broad spectrum of hosts in which C. pseudotuberculosis causes clinical disease. However, C. pseudotuberculosis is considered an emergent public health problem. This bacterium can produce diphtheria toxin as a consequence of lysogenization with a diphtheria phage. This has been clearly demonstrated in vitro [111] and has led authorities from many countries to urge their health services to be aware of diseases caused by all the diphtheria toxin-producing species of Corynebacterium, including C. pseudotuberculosis and C. ulcerans, in the differential diagnosis of diphtheria. For many years, the identification of any toxigenic C. pseudotuberculosis required notification to communicable disease control agencies [112]. Recently, a case of toxigenic C. pseudotuberculosis was reported. This is the only report in the literature of the isolation of toxigenic C. pseudotuberculosis from a human, and this isolation occurred in the UK. The organism was isolated from the aortic root vegetation of an intravenous drug user with endocarditis; this patient had no history of animal contact, and no possible source of infection was identified [113]. In addition to the production of diphtheria toxin by lysogenic C. pseudotuberculosis, common strains can infect humans and cause suppurative lymphadenopathy; these events are traditionally associated with farm animal contact and contaminated dairy products. Pathologically, the infected lymph node tissue is replaced by necrotizing granulomas, while the general histological appearance resembles tuberculous lymphadenitis, catscratch disease and lymphogranuloma venereum, which must be considered in the differential diagnosis. A summary of the human cases of infection is presented in Table 2, including the country of origin of the report, the sex and age of the patients, the source of exposure to the pathogen (professional or occasional), the main clinical presentations of the disease and treatment.

The first human case was described in 1966; a 37-year old man from Panama that worked as a grass cutter developed an inguinal lymphadenopathy. Initially, the clinical presentation was thought to be a lymphogranuloma venereum, and the patient was treated with tetracycline for a period of 3 weeks. The lymph node was excised and presented a characteristic histological appearance of the CLA pyogranulomas that normally occur in sheep and goats. The bacterial strain was isolated and was found to produce acid from glucose, fructose, galactose, sucrose, mannose, maltose, dextrin and xylose, as well as from hydrogen sulfide, but it did not reduce nitrate to nitrite and was negative for the urease reaction [114]. Since that time, 32 new clinical cases have been reported, and the 19 human cases reported from Australia is higher than that from all other countries, with reports varying between 1 and 3 cases per country. A profile including the age, sex and source of exposure for the patients infected with C. pseudotuberculosis found in the reviewed literature is presented in Figure 1. Of the 33 human cases, most cases presented as axillary or epitroclear lymphadenopathy, and the main group of infected patients was men aged 21 to 40 years old that were exposed to farm animals, mostly sheep, at work. These data strongly indicate that human infection by C. pseudotuberculosis constitutes a bona fide zoonosis.

In most cases, lymphadenopathy has a prolonged course, and the formation of relapsing abscesses is frequent. Treatment with antibiotics is prolonged, normally lasting for more than two weeks, and typically comprises administration of intracellularly active antibiotics, such as tetracycline and macrolides. Although antibiotic therapy alone has been successful in a few cases, C. pseudotuberculosis is a facultative intracellular pathogen, and it is very difficult for antibiotics to reach the bacteria inside the pyogranuloma macrophages. Thus, surgical interventions including drainage or excision were often required to clear the purulent content of the thick collagen capsule. The combination of antimicrobial therapy with excision and drainage is the most successful course of treatment.

It should be noted that two specific cases of human infection due to C. pseudotuberculosis have involved clinical presentations other than the characteristic lymphadenopathy. In the first case, a veterinary student who worked with equines was infected by C. pseudotuberculosis and presented with an eosinophilic pneumonia [115]. In the second case, a 63-year old man presented an ocular infection by C. pseudotuberculosis, involving a scleral buckle after retinal reattachment intervention; this case may be the first human ocular C. pseudotuberculosis infection [116].

The prevalence of the human disease caused by C. pseudotuberculosis is likely underestimated, as only 33 cases have been reported over a period of 42 years (from 1966 to 2008). In addition to this low prevalence, some cases of human exposure to C. pseudotuberculosis that are found in the literature were written in Russian, making them difficult to utilize. In one report, a microbiological study of 69 Russian patients with allergic annual rhinitis (AAR) and infectious rhinitis (IR) was performed and demonstrated that among the species isolated in IR, C. pseudotuberculosis was the predominant species associated with Streptococci [117]. In another study from Russia, the bacteriologic examination of 1589 patients showed that, aside from C. diphtheriae, 11% of acute upper respiratory tract infections were caused by Corynebacterium species. The disease processes varied significantly, presenting as bronchitis, pyelonephritis, urethritis, colpitis, dermatitis and arthritis. Although C. pseudodiphtheriticum and C. xerosis were isolated more frequently from clinical specimens, C. diphtheriae and C. pseudotuberculosis were identified as the most virulent species. Corynebacterium species (but not C. diphtheriae) were frequently isolated from clinical specimens with Staphylococci and Streptococci, and in those cases, pathogenicity and resistance to antibiotics were more pronounced; the strains isolated with other bacteria were resistant to tetracycline, penicillin, and erythromycin [118], which are the most common active antibiotics used for the treatment of intracellular bacteria. These two Russian studies provide evidence demonstrating that C. pseudotuberculosis is a much more widespread pathogen than suggested by the 33 published cases. Additionally, because the published human cases involved zoonotic transmission due to work with farm animals, the proper training and education of these workers is needed.

9. FUTURE DIRECTIONS

The two virulence determinants of C. pseudotuberculosis, the toxic lipid cell wall and the exotoxin PLD, have proved to be highly associated with dissemination in tissues and development of granulomatous lesions, respectively. However, knowledge of the molecular mechanisms and genetic basis of virulence of C. pseudotuberculosis is central to understanding the host-pathogen associations.

Because C. pseudotuberculosis is an intracellular pathogen, its immunopathogenesis is characterized by pyogranulomas; biologically, this represents a form of containment in the host's tissues. Data suggest that antibodies might help to protect animals against infection, but full protection by any vaccine model must provide better stimulation of cellular immunity, such as the activation of CD8+ cells and the secretion of IFN-γ, to control the bacteria in an early phase of the infection process. Thus, the molecular mechanisms of infection by C. pseudotuberculosis need to be more thoroughly understood so that specific stimulation of cellular immune responses can be accomplished through immunization. A promising area of research is the acute phase response of the infection, as innate immunity is responsible for the early events in the control the bacterial spread; vaccines could be improved to better stimulate innate immunity.

For the diagnosis of C. pseudotuberculosis infection, ELISA and IFN-γ quantification are promising techniques that provide high levels of reliability. With some improvement, a great variety of tests will be available to flock owners and veterinarians. However, these methodologies require a minimal laboratory structure for implementation, including microplate readers and well-trained professionals. These requirements would not allow these tests to be applied in field situations during the routine care of small ruminant flocks or in the inspection of slaughterhouses. Thus, scientists are also encouraged to develop simpler technologies that could be used in field conditions, such as a rapid immune-chromatographic test.

Table 2: Summary of data from 33 previously published cases of human infection by *Corynebacterium pseudotuberculosis.*

Case n°	Year	Reference	Country of origin	Sex, age of patients	Occupation / Source of exposition	Main clinical presentation	Treatment
1	1966	[114]	Panama	M, 37	Grass cutter	Inguinal lymphadenopathy	EX and TET
2	1967	[134]	Australia	M, 28	Manager of a sheep and cattle farm	Lymphadenopathy in the inguinal, leg and thigh lymph nodes	EX, DR and PEN
3	1968	[135]	Australia	M, 24	Sheep shearer	Asymptomatic axillary lymphadenopathy	EX, PEN and TET
4	1968	[136]	Australia	M, 23	Butcher	Axillary lymphadenopathy	EX and TET
5	1974	[137]	Australia	M, 20	Sheep rancher	Axillary lymphadenopathy	EX and DR
6	1974	[137]	Australia	M, 40	Rural worker	Inguinal lymphadenopathy	EX and ERY
7	1974	[137]	Australia	F, 50	Housewife of rural worker	Asymptomatic cervical lymphadenopathy	EX
8	1979	[138]	Australia	M, 21	Abbatoir worker	Axillary lymphadenopathy	EX and CLO
9	1979	[115]	USA	M, 28	Veterinary student	Eosinophilic pneumonia	ERY
10	1980	[139]	France	M, 27	Shepherd	Axillary lymphadenopathy	EX, TET and CHL
11	1981	[140]	USA	M, 30	Raw milk ingestion	Cervical lymphadenopathy	EX, DR, PEN and ERY
12	1985	[141]	Australia	M, 18	Butcher	Epitrochlear and axillary lymphadenopathy	EX, DR, PEN, FLU, TET and ERY
13	1985	[10]	Australia	M, 41	Farm worker	Axillary lymphadenopathy	EX
14	1985	[10]	Australia	F, 29	Farm worker	Inguinal lymphadenopathy	EX, PEN and FLU
15	1986	[142]	New Zeland	M, 29	Sheep rancher	Asymptomatic inguinal lymphadenopathy	EX, PEN, CLO, FLU and ERY
16	1986	[10]	Australia	M, 29	Meat inspector	Axillary lymphadenopathy	EX and DR
17	1988	[10]	Australia	M, 22	Butcher	Axillary lymphadenopathy	EX and ERY
18	1988	[10]	Australia	M, 20	Slaughterman	Axillary lymphadenopathy	EX, DR and ERY
19	1988	[10]	Australia	F, 53	Unknown exposure	Supraclavicular lymphadenopathy	EX
20	1989	[10]	Australia	M, 40	Abbatoir worker	Epitrochlear and axillary lymphadenopathy	EX and ERY
21	1991	[143]	Belgium	NF	NF	Axillary lymphadenopathy	NF
22	1992	[10]	Australia	M, 27	Worker of a sheep saleyard	Axillary lymphadenopathy	EX, FLU and AMOX-CLAV
23	1992	[10]	Australia	M, 26	Abbatoir worker	Axillary lymphadenopathy	EX and PEN
24	1992	[10]	Australia	M, 40	Contact with sheep	Axillary lymphadenopathy	EX, ERY and FLU
25	1995	[144]	Spain	M, 34	Shepherd	Inguinal lymphadenopathy	EX and ERY
26	1997	[145]	Switzerland	M, 30	Sheep rancher and butcher (Turkey)	Epitrochlear and axillary lymphadenopathy	EX and CLA
27	1997	[146]	Australia	M, 17	Contact with farm animals	Suppurative lymphadenopathy	NF
28	1997	[147]	New Zeland	M, 22	Shepherd	Axillary lymphadenopathy	EX, DR and ATB
29	2004	[148]	Spain	F, 33	History of contact with a white rat	Cervical lymphadenopathy	AMOX-CLAC
30	2004	[149]	France	F, 63	Living in a rural area	Axillary lymphadenopathy	EX, DR, CLO, GEN, CIP and PRI
31	2005	[117]	China	M, 63	Contact only with a dog maintained as pet	Ocular infection and mucopurulent discharge	AMP, CLO, PEN and VAN
32	2006	[150]	France	F, 12	Contact with sheep during vacation in a rural area	Asymptomatic lymphadenopathy	EX, DR, IMI-CIL, RIF and OFL
33	2008	[113]	United Kingdon	NF	Injecting drug user, unknown exposure	Endocarditis	NR

Legend: EX = excision of the lymph node; DR = drainage of the lymph node content; ATB = antibiotics; TET = tetracycline; PEN = penicillin; CLO = cloxacillin; FLU = flucloxacillin; ERY = erythromycin; CHL = chloramphenicol; AMOX-CLAV = amoxicillin; CLA = clarithromycin; GEN = gentamicin; CIP = ciprofloxacin; PRI = pristinamycin; AMP = ampicillin; VAN = vancomycin; IMI-CIL = imipenem-cilastatin; RIF = rifampin; OFL = ofloxacin; NR = not reported; NF = not found.

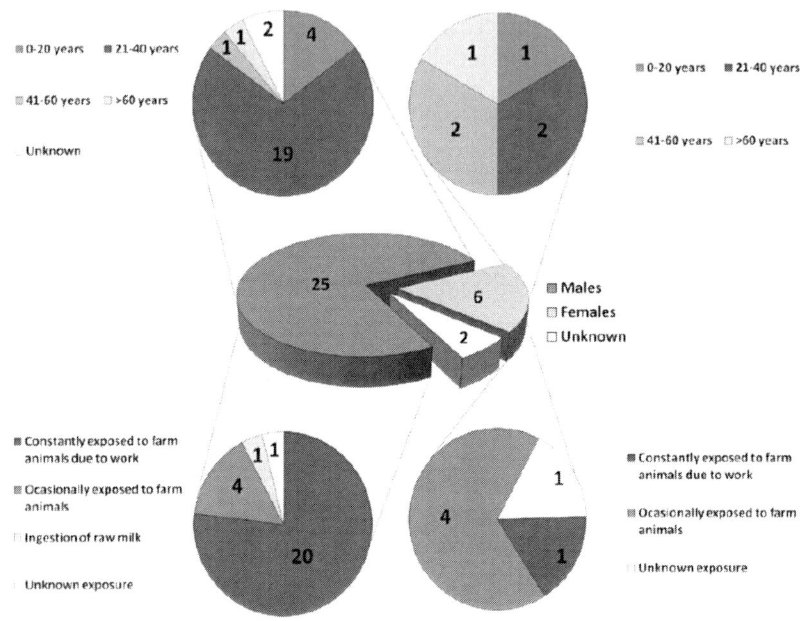

Figure 1: Profile of patients infected with *C. pseudotuberculosis* in 33 reviewed literature cases, according to sex, age and exposition to the pathogen.

Because of the zoonotic potential of C. pseudotuberculosis, studies should be performed to characterize the presence of the bacteria in carcasses and the rates of condemnation due to C. pseudotuberculosis. Further, the risks to consumers who come into contact with or ingest contaminated meat or milk need to be quantified.

ACKNOWLEDGEMENTS

The authors are grateful for the financial support from CAPES, CNPq, FAPEMIG, FAPEX, FUNDECE and RENORBIO. R Meyer, S C Oliveira, A Miyoshi and V Azevedo are research fellows of CNPq. BL Bastos and D Ribeiro are Ph.D. students with scholarships from CNPq and CAPES, respectively.

REFERENCES

1. Dorella FA, Pacheco LGC, Oliveira SC, Miyoshi A, Azevedo V (2006) Corynebacterium pseudotuberculosis: microbiology, biochemical properties, pathogenesis and molecular studies of virulence. Vet Res 37: 201-218.
2. Mattos-Guaraldi AL, Moreira LO, Damasco PV, Hirata Junior R (2003) Diphtheria remains a threat to health in the developing world - an overview. Mem Inst Oswaldo Cruz 98: 987-993.
3. Tauch A, Kaiser O, Hain T, Goesmann A, Weisshaar B, et al. (2005) A Complete genome sequence and analysis of the multiresistant nosocomial pathogen *Corynebacterium jeikeium* K411, a lipid-requiring bacterium of the human skin flora. J Bacteriol 187: 4671-4682.
4. Koffas M, Stephanopoulos G (2005) Strain improvement by metabolic engineering: lysine production as a case study for systems biology. Curr Opin Biotechnol 16: 361-366.
5. Taylor J, Saavedra-Campos M, Harwood D, Pritchard G, Raphaely N, et al. (2010) Toxigenic *Corynebacterium ulcerans* infection in a veterinary student in London, United Kingdom, May 2010. Euro Surveillance 15: 19634.
6. Baird GJ, Fontaine MC (2007) Corynebacterium pseudotuberculosis and its role in ovine caseous lymphadenitis. J Comp Pathol 137: 179-210.
7. Binns SH, Green LE, Bailey M (2007) Development and validation of an ELISA to detect antibodies to Corynebacterium pseudotuberculosis in ovine sera. Vet Microbiol 20: 169-179.
8. Moore R, Miyoshi A, Pacheco LGC, Seyffert N, Azevedo V (2010) Corynebacterium and Arcanobacterium In: Pathogenesis of bacterial infections in animals. (4th ed.), Blackwell Publishing, Iowa.
9. Ruiz JC, D'Afonseca V, Silva A, Ali A, Pinto AC, et al. (2011) Evidence for reductive genome evolution and lateral acquisition of virulence functions in two Corynebacterium pseudotuberculosis strains. PLoS ONE 6: e18551.
10. Peel MM, Palmer GG, Stacpoole AM, Kerr TK (1997) Human lymphadenitis due to Corynebacterium pseudotuberculosis: report of ten cases from Australia and review. Am J Med 24: 185-191.
11. Arsenault J, Girard C, Dubreuil P, Daignault D, Galarneau JR, et al. (2003) Prevalence of and carcass condemnation from maedi-visna , paratuberculosis and caseous lymphadenitis in culled sheep from Quebec , Canada. Prev Vet Med 59: 67-81.
12. Benham CL, Seaman A, Woodbine M (1962) Corynebacterium pseudotuberculosis *and its role in diseases of animals. Vet Bull 32: 645-657.*
13. Collett MG, Bath GF, Cameron CM (1994) *Corynebacterium pseudotuberculosis* infections In: Infectious diseases of livestock with special reference to Southern Africa. (2nd ed.), Oxford University Press, Cape Town.
14. Jolly RD (1965) The pathogenesis of experimental *Corynebacterium ovis* infection in mice. N Z Vet J 13: 141-147.
15. Puech V, Chami M, Lemassu A, Lanéelle M-A, Schiffler B, et al. (2001) Structure of the cell envelope of corynebacteria: importance of the non covalently bound lipids in the formation of the cell wall permeability barrier and fracture plane. Microbiology 147: 1365-1382.
16. Selim SA (2001) Oedematous skin disease of buffalo in Egypt. J Vet Med B Infect Dis Vet Public Health 48: 241-258.
17. Merchant IA, Packer RA (1975) The Genus *Corynebacterium* In: Veterinary bacteriology and virology. The Iowa State University Press, Iowa.
18. Batey RG (1986) Factors affecting the yield of viable cells of Corynebacterium pseudotuberculosis in a liquid medium. Vet Microbiol 11: 145-152.
19. Cameron CM, Swart CF (1965) A new liquid medium for the cultivation of *Corynebacterium pseudotuberculosis.* J S Afr Vet Med Ass 36: 185-188.

20. Quinn PJ, Carter ME, Markey B, Carter GR (1994) *Corynebacterium* species and *Rhodococcus equi* In: Clinical Veterinary Microbiology. Wolfe Publishing Company, London.
21. Muckle CA, Gyles CL (1982) Characterization of strains of Corynebacterium pseudotuberculosis. Can J Comp Med 46: 206-208.
22. Jolly RD (1966) Some observations on surface lipids of virulent and attenuated strains of *Corynebacterium ovis.* J Appl Bact 29: 189-196.
23. Moura-Costa LF, Paule BJA, Azevedo V, Freire SM, Nascimento I, et al. (2002) Chemically defined synthetic medium for Corynebacterium pseudotuberculosis culture. Braz J Anim Health Prod 3: 1-9.
24. Pacheco LGC, Slade SE, Seyffert N, Santos AR, Castro TLP, et al. (2011) A combined approach for comparative exoproteome analysis of Corynebacterium pseudotuberculosis. *BMC Microbiology* 11: 12.
25. Songer JG, Beckenbach K, Marshall MM, Olson GB, Kelley L (1988) Biochemical and genetic characterization of Corynebacterium pseudotuberculosis. Am J Vet Res 49: 223-226.
26. Sutherland SS, Hart RA, Buller NB (1996) Genetic differences between nitratenegative and nitrate-positive Corynebacterium pseudotuberculosis strains using restriction fragment length polymorphisms. Vet Microbiol 49: 1-9.
27. Costa LRR, Spier SJ, Hirsh DC (1998) Comparative molecular characterization of Corynebacterium pseudotuberculosis of different origin. Vet Microbiol 62: 135-143.
28. Batey RG (1986) Pathogenesis of caseous lymphadenitis in sheep and goats. Aust Vet J 63: 269-272.
29. Carne HRA (1939) Bacteriological study of 134 strains of *Corynebacterium ovis.* J Pathol Bacteriol *49: 313-328.*
30. Brown CC, Olander HJ, Alves SF (1987) Synergistic hemolysis-inhibition titers associated with caseous lymphadenitis in a slaughterhouse survey of goats and sheep in Northeastern Brazil. Can J Vet Res 51: 46-49.
31. Onon EO (1979) Purification and partial characterization of the exotoxin of *Corynebacterium ovis. Biochem J 177: 181-186.*
32. Egen NB, Cuevas W, Mcnamara PJ, Sammons DW, Humphreys R, et al. (1989) Purification of the phospholipase D of Corynebacterium pseudotuberculosis by recycling isoelectric focusing. Am J Vet Res 50: 1319-1322.
33. Songer JG (1997) Bacterial phospholipases and their role in virulence. Trends Microbiol 5: 156-160.
34. Tashjian JJ, Campbell SG (1983) Interaction between caprine macrophages and Corynebacterium pseudotuberculosis*:* an electron microscopy study. Am J Vet Res 44: 690-693.
35. Brown CC, Olander HJ, Biberstein EL, Morse SM (1986) Use of a toxoid vaccine to protect goats against intradermal challenge exposure to Corynebacterium pseudotuberculosis. Am J Vet Res 47: 1116-1119.
36. Hodgson ALM, Krywult J, Corner LA, Rothel JS, Radford AJ (1992) Rational attenuation of Corynebacterium pseudotuberculosis*:* potential cheesy gland vaccine and live delivery vehicle. Infect Immun 60: 2900-2905.
37. Muckle CA, Gyles, CL (1986) Exotoxic activities of Corynebacterium pseudotuberculosis. Curr Microbiol 13: 57-60.
38. Walker J, Jackson IHJ, Eggleton DG, Meeusen ENT, Wilson MJ (1994) Identification of a novel antigen from Corynebacterium pseudotuberculosis that protects sheep against caseous lymphadenitis. Infect Immun 62: 2562-2567.
39. Wilson M J, Brandon MR, Walker J (1995) Molecular and biochemical characterization of a protective 40- kilodalton antigen from Corynebacterium pseudotuberculosis. Infect Immun 63: 206-211.
40. Ellis JA, Hawk DA, Mills KW, Pratt DL (1991) Antigen specificity and activity of ovine antibodies induced by immunization with Corynebacterium pseudotuberculosis culture filtrate. Vet Immunol Immunopathol 28: 303-316.

41. Paule BJA, Azevedo V, Moura-Costa LF, Freire SM, Regis LF, et al. (2004) SDS-PAGE and Western blot analysis of somatic and extracellular antigens of Corynebacterium pseudotuberculosis. Med Biol Sci J 3: 44-52.
42. Hard GC (1972) Examination by electron microscopy of the interaction between peritoneal phagocytes and *Corynebacterium ovis.* J Med Microbiol 5: 483-491.
43. Alves FSF, Pinheiro RR, Pires PC (1997) Linfadenite caseosa: patogenia, diagnóstico, controle. Artigo Técnico Documento n° 27, Embrapa Sobral-CE, Ministério da Agricultura, Pecuária e Abastecimento.
44. Billington SJ, Esmay PA, Songer JG, Jost BH (2002) Identification and role in virulence of putative iron acquisition genes from Corynebacterium pseudotuberculosis. FEMS Microbiol Lett 208: 41-45.
45. D'Afonseca V, Prosdocimi F, Dorella FA, Pacheco LGC, Moraes PM, et al. (2010) Survey of genome organization and gene content of Corynebacterium pseudotuberculosis. Microbiol Res 165: 312-320.
46. Trost E, Ott L, Schneider J, Schröder J, Jaenicke S, et al. (2010) The complete genome sequence of Corynebacterium pseudotuberculosis FRC41 isolated from a 12-year-old girl with necrotizing lymphadenitis reveals insights into gene- regulatory networks contributing to virulence. BMC Genomics 11: 728.
47. Pépin M, Seow HF, Corner L, Rothel JS, Hodgson AL, et al. (1997) Cytokine gene expression in sheep following experimental infection with various strains of Corynebacterium pseudotuberculosis differing in virulence. Vet Res 28: 149- 163.
48. Stefańska I, Gieryńska M, Rzewuska M, Binek M (2010) Survival of Corynebacterium pseudotuberculosis within macrophages and induction of phagocytes death. Pol J Vet Sci 13: 143-149.
49. Bogdan JR, Newlands-Monteith CF, Ellis JA (1997) Nitric oxide production following in vitro stimulation of ovine pulmonary alveolar macrophages. Vet Immunol Immunopathol 56: 299-310.
50. Paule BJA, Azevedo V, Regis LF, Carminati R, Bahia R, et al. (2003) Experimental Corynebacterium pseudotuberculosis primary infection in goats: kinetics of IgG and interferon-γ production, IgG avidity and antigen recognition by Western blotting. Vet Immunol Immunopathol 96: 129-139.
51. Ellis JA (1988) Immunophenotype of pulmonary cellular infiltrates in sheep with visceral caseous lymphadenitis. Vet Pathol 23: 362-368.
52. Walker J, Jackson H, Brandon MR, Meeusen E (1991) Lymphocyte subpopulations in pyogranulomas of caseous lymphadenitis. Clin Exp Immunol 86: 13-18.
53. Pépin M, Cannella D, Fontaine JJ, Pittet JC, Le Pape A (1992) Ovine mononuclear phagocytes in situ: identification by monoclonal antibodies and involvement in experimental pyogranulomas. J Leukoc Biol 51: 188-198.
54. Pépin M,Pittet JC,Olivier M,Gohin I (1994) Cellular composition of Corynebacterium pseudotuberculosis pyogranulomas in sheep. J Leukoc Biol 56: 666-670.
55. Nieto NC,Foley JE,MacLachlan NJ,Yuan T,Spier SJ (2009) Evaluation of hepatic disease in mice following intradermal inoculation with Corynebacterium pseudotuberculosis. Am J Vet Res 70: 257-262.
56. Moller K, Agerholm JS, Ahrens P, Jensen NE, Nielsen TK (2000) Abscess disease, caseous lymphadenitis, and pulmonary adenomatosis in imported sheep. J Vet Med B 62: 55-62.
57. Zaki MM (1976) Relation between the toxogenicity and pyogenicity of Corynebacterium ovis in experimentally infects mice. Res Vet Sci 20: 197-200.
58. El-Enbaawy MI,Saad MM,Selim SA (2005) Humoral and cellular immune responses of a murine model against Corynebacterium pseudotuberculosis antigens. Egypt J Immunol 12:13-19.
59. Stanford K, Brogden KA, McClelland LA, Kozub, GC, Audibert F (1998) The incidence of caseous lymphadenitis in Alberta sheep and assessment of impact by vaccination with commercial and experimental vaccines. Can J Vet Res 1: 38-43.
60. Eggleton DG, Middleton HD, Doidge CV, Minty DW (1991) Immunisation against ovine caseous lymphadenitis: comparison of Corynebacterium pseudotuberculosisvaccines with and without bacterial cells. Aust Vet J 68: 317–319.

61. Braga, W, Schul S, Nu A, Pezo D, Franco E (2007) A primary Corynebacterium pseudotuberculosis low dose infection in alpacas (Lama pacos) protects against a lethal challenge exposure. Small Rumin Res 72: 81-86.
62. Brogden KA,Cutlip RC,Lehmkuhl HD (1984) Experimental Corynebacterium pseudotuberculosis infection in lambs. Am J Vet Res 45:1532-1534.
63. Simmons CP, Hodgson ALM, Strugnell RA (1997) Attenuation and vaccine potential of aroQ mutants of Corynebacterium pseudotuberculosis. Infect Immun 65: 3048-3056.
64. Simmons CP, Dunstan SJ, Tachedjian M, Krywult J, Hodgson ALM, et al. (1998) Vaccine potential of attenuated mutants of Corynebacterium pseudotuberculosis in sheep. Infect Immun 66: 474-479.
65. Hodgson ALM, Carter K, Tachedjian M, Krywult J, Corner LA, et al. (1999) A Efficacy of an ovine caseous lymphadenitis vaccine formulated using a genetically inactive form of the Corynebacterium pseudotuberculosis phospholipase D. J Bacteriol 17: 802-808.
66. Chaplin PJ,De Rose R,Boyle JS,McWaters P,Kelly J,et al. (1999) Targeting improves the efficacy of a DNA vaccine against Corynebacterium pseudotuberculosis in sheep. Infect Immun 67: 6434-6438.
67. De Rose R, Tennent J, McWaters P, Chaplin PJ, Wood PR, et al. (2002) Efficacy of DNA vaccination by different routes of immunisation in sheep. Vet Immunol Immunopathol 90: 55-63.
68. Irwin MR, Knight HD (1975) Enhanced resistance to Corynebacterium pseudotuberculosis infections associated with reduced serum immunoglobulin levels in levamisole-treated mice. Infect Immun 12: 1098-1103.
69. Lan DT,Taniguchi S,Makino S,Shirahata T,Nakane A (1998) Role of endogenous tumor necrosis factor alpha and gamma interferon in resistance to Corynebacterium pseudotuberculosis infection in mice. Microbiol Immunol 42: 863-870.
70. Ellis JA, Campos M, Snyder M, Chelak B, Haines DM (1995) Local production of tumor necrosis factor-alpha in corynebacterial pulmonary lesions in sheep. Vet Pathol 32: 68-71.
71. Lan DTB, Makino S, Shirahata T, Yamada M, Nakane A (1999) Tumor necrosis factor alpha and gamma interferon are required for the development of protective immunity to secondary Corynebacterium pseudotuberculosis infection in mice. J Vet Med Sci 61: 1203-1208.
72. Lan DT,Makino S,Shirahata T,Yamada M,Nakane A (1999) Complement receptor type 3 plays an important role in development of protective immunity to primary and secondary Corynebacterium pseudotuberculosis infection in mice. Microbiol Immunol43: 1103-1106.
73. Yozwiak ML, Songer JG (1993) Effect of Corynebacterium pseudotuberculosis phospholipase D on viability and chemotac- tic responses of ovine neutrophils. Am J Vet Res 54: 392-397.
74. Eckersall PD, Lawson FP, Bence L, Waterston MM, Lang TL, et al. (2007) Acute phase protein response in an experimental model of ovine caseous lymphadenitis. BMC Vet Res 3:35.
75. Rebouças MF, Portela RW, Lima DD, Loureiro D, Bastos BL, et al. (2011) Corynebacterium pseudotuberculosis secreted antigen-induced specific gamma-interferon production by peripheral blood leukocytes: potential diagnostic marker for caseous lymphadenitis in sheep and goats. J Vet Diag Invest 23:213-220.
76. Pépin M, Pardon P, Lantier F, Marly J, Levieux D, et al. (1991) Experimental Corynebacterium pseudotuberculosis infection in lambs – kinetics of bacterial dissemination and inflammation. Vet Microbiol 26: 381-392.
77. Bastos BL, Meyer R, Guimarães JE, Ayres MCC, Guedes MT, et al. (2011) Haptoglobin, fibrinogen concentrartions and Leukocyte counts in the clinical investigation of caseous lymphadenitis in sheep. Vet Clin Pathol 40: 496-503.
78. Pépin M,Pardon P,Marly J,Lantier F (1988) Corynebacterium pseudotuberculosis infection in adult ewes by inoculation in the external ear. Am J Vet Res 49: 459-463.
79. Pépin M,Pardon P,Marly J,Lantier F,Arrigo JL (1993) Acquired immunity after primary caseous lymphadenitis in sheep. Am J Vet Res 54: 873-877.

80. Quevedo JM, Rodriguez-Lostau JA, Risso HR, Mettler, EAJ (1957) Evaluación de unavacuna contra linfadenite caseosa en la oveja. Revista de Investigaciones Ganaderas 1: 47.
81. Cameron CM,Minnaar JL,Engelbrecht MM,Purdom MR (1972) Immune response of merino sheep to inactivated Corynebacterium pseudotuberculosis vaccine. Onderstepoort J Vet Res 39:11-24.
82. Ribeiro OC, Silva JAH, Maia PCC, Campos WG (1988) Avaliação de vacina contra linfadenite caseosa em caprinos mantidos em regime extensivo. Pesq Vet Bras 8: 27-29.
83. Menzies PI, Muckle CA, Brogden KA, Robinson L (1991) A field trial to evaluate a whole cell vaccine for the prevention of caseous lymphadenitis in sheep and goat flocks. Onderstepoort J Vet Res 55: 362-366.
84. Brogden KA, Glenn JS, East N, Audibert FA (1996) Corynebacterium pseudotuberculosis bacterin with muramyl dipeptide induces antibody titers, increases the time of onset, and decreases naturally occurring external abscesses in sheep and goats. Small Rumin Res 19: 161-168.
85. Hodgson AL,Bird P,Nisbet IT (1990) Cloning, nucleotide sequence, and expression in Escherichia coli of the phospholipase D gene from Corynebacterium pseudotuberculosis. J Bacteriol 172: 1256-1261.
86. Songer JG,Libby SJ,Iandolo JJ,Cuevas WA (1990) Cloning and expression of the phospholipase D gene from Corynebacterium pseudotuberculosis in Escherichia coli. Infect Immun 58:131-136.
87. McNamara PJ, Cuevas WA, Songer JG (1995) Toxic phospholipases D of Corynebacterium pseudotuberculosis, C. ulcerans and Arcanobacterium haemolyticum: cloning and sequence homology. Gene 156: 113-118.
88. Paton MW,Sutherland SS,Rose IR,Hart RA,Mercy AR,et al. (1995) The spread of Corynebacterium pseudotuberculosis infection to unvaccinated and vaccinated sheep. Aust Vet J 72: 266-269.
89. Alves FSF, Olander H (1999) Uso de vacina toxóide no controle da linfadenite caseosa em caprinos. Veterinária Notícias 5: 69-75.
90. Moura-Costa LF,Bahia RC,Carminati R,Vale VL,Paule BJ,et al. (2008) Evaluation of the humoral and cellular immune response to different antigens of Corynebacterium pseudotuberculosis in Canindé goats and their potential protection against caseous lymphadenitis. Vet Immunol Immunopathol 126: 131-141.
91. Piontkowski MD,Shivvers DW (1998) Evaluation of a commercially available vaccine against Corynebacterium pseudotuberculosis for use in sheep. J Am Vet Med Assoc 212: 1765-1768.
92. Braga WU (2007) Protection in alpacas against Corynebacterium pseudotuberculosis using different bacterial components. Vet Microbiol 119: 297-303.
93. Pinho JMR, Dorella FA, Coelho S, Fonseca CT, Cardoso FC, et al. (2009) Heat-shock protein (Hsp)-60 is able to induce an immune response in mice, but fails to confer protection against infection. Open Vet Sci J 3: 22-27.
94. Muckle CA,Menzies PI,Li Y,Hwang YT,van Wesenbeeck M (1992) Analysis of the immunodominant antigens of Corynebacterium pseudotuberculosis. Vet Microbiol 30: 47-58.
95. Braithwaite CE,Smith EE,Songer JG,Reine AH (1993) Characterization of detergent-soluble proteins of Corynebacterium pseudotuberculosis. Vet Microbiol 38: 59-70.
96. Ribeiro OC, Silva JAH, Oliveira SC, Meyer R, Fernandes GB (1991) Dados preliminares sobre uma vacina viva contra a linfadenite caseosa. Pesq Agropec Bras 26: 461-465.
97. Meyer R, Carminati R, Bahia RC, Vale VL, Viegas S, et al. (2002) Evaluation of the goats humoral immune response induced by the Corynebacterium pseudotuberculosis lyophilized live vaccine. Med Biol Sci J 1: 42-48.
98. Hodgson ALM, Tachedjian M, Corner LA, Radfordt AJ (1994) Protection of sheep against caseous lymphadenitis by use of a single oral dose of live recombinant Corynebacterium pseudotuberculosis. Vaccine 62: 5275-5280.

99. Fontaine MC, Baird G, Connor KM, Rudge K, Sales J, et al. (2006) Vaccination confers significant protection of sheep against infection with a virulent United Kingdom strain of Corynebacterium pseudotuberculosis. Vaccine 24: 5986-5996.
100. Shigidi MT (1978) An indirect haemagglutination test for the sero-diagnosis of Corynebacterium ovis infection in sheep. Res Vet Sci 24: 57-60.
101. Ribeiro MG, Júnior JGD, Paes AC, Barbosa PG, Júnior GN, et al. (2001) Punção aspirativa com agulha fina no diagnóstico de Corynebacterium pseudotuberculosis na linfadenite caseosa caprina. Arq Inst Biol 68: 23-28.
102. Shigidi MT (1979) A comparison of five serological tests for the diagnosis of experimental Corynebacterium ovis infection in sheep. Br Vet 135: 172-177.
103. Burrel DH (1980) A simplified double immunodiffusion technique for detection of Corynebacterium ovis antitoxin. Res Vet Sci 28: 234-237.
104. Pacheco LGC, Pena RR, Castro TLP, Dorella FA, Bahia RC, et al. (2007) Multiplex PCR assay for identification of Corynebacterium pseudotuberculosis from pure cultures and for rapid detection of this pathogen in clinical samples. J Med Microbiol 56: 480-486.
105. Kaba J, Kutschke L, Gerlach GF (2001) Development of an ELISA for the diagnosis of Corynebacterium pseudotuberculosis infections in goats. Vet Microbiol 78: 155-163.
106. Dercksen DP, Brinkhof JMA, Dekker-Nooren T, Maanen K, Bode CF, et al. (2000) A comparison of four serological tests for the diagnosis of caseous lymphadenitis in sheep and goats. Vet Microbiol 75: 167-175.
107. Menzies PI, Hwang Y, Prescott JF (2004) Comparison of an interferon-? to a phospholipase D enzyme-linked immunosorbent assay for diagnosis of Corynebacterium pseudotuberculosis infection in experimentally infected goats. Vet Microbiol 100: 129-137.
108. Seyffert N, Guimarães AS, Pacheco LGC, Portela RW, Bastos BL, et al. (2010) High seroprevalence of caseous lymphadenitis in Brazilian goat herds revealed by Corynebacterium pseudotuberculosis secreted proteins-based ELISA. Res Vet Sci 88: 50-55.
109. Guimarães AS, Seyffert N, Bastos BL, Portela RWD, Meyer R, et al. (2009) Caseous lymphadenitis in sheep flocks of the state of Minas Gerais, Brazil: Prevalence and management surveys. Small Rumin Res 87: 86-91
110. Prescott JF, Menzies PI, Hwang YT (2002) An interferon-gamma assay for diagnosis of Corynebacterium pseudotuberculosis infection in adult sheep from a research flock. Vet Microbiol 88: 287-297.
111. Maximescu P, Oprisan A, Pop A, Potorac E (1974) Further studies on Corynebacterium species capable of producing diphtheria toxin (C. diphtheriae, C. ulcerans, C. ovis). J General Microbiol 82: 49-56.
112. DeWinter LM, Bernard KA, Romney MG (2005) Human clinical isolates of Corynebacterium diphtheriae and Corynebacterium ulcerans collected in Canada from 1999 to 2003 but not fitting reporting criteria for cases of diphtheria. J Clin Microbiol 43: 3447-3449.
113. Wagner KS, White JM, Crowcroft NS, Mann G, Efstratiou A (2010) Diphtheria in the United Kingdom, 1986-2008: the increasing role of Corynebacterium ulcerans. Epidemiol Infect 138: 1519-1530.
114. Lopez JF, Wonc FM, Quesada J (1966) Corynebacterium pseudotuberculosis First case of human infection. Am J Clin Path 46:562.
115. Keslin MH, McCoy EL, McCusker JJ, Lutch JS (1979) Corynebacterium pseudotuberculosis: A new cause of infectious and eosinophilic pneumonia. Am J Med 67: 228-231.
116. Liu DTL, Chan W-M, Fan DSP, Lam DSC (2005) An infected hydrogel buckle with Corynebacterium pseudotuberculosis. Br J Ophthalmol 89: 238-251.
117. Romanenko EE,Baturo AP,Mokronosova MA,Tarasova GD,Sergeev AV (2003) Microflora of the nasal mucosa in allergic perennial and infectious rhinitis. Zh Mikrobiol Epidemiol Immunobiol 3: 66-71.
118. Kraeva LA,Manina ZN,Tseneva GY,Radchenko AG (2007) Etiologic role of Corynebacterium non diphtheriae in patients with different pathology. Zh Mikrobiol Epidemiol Immunobiol 5: 3-7.

119. Brogden KA,Cutlip RC,Lehmkuhl HD (1985) Immunogenicity of Corynebacterium pseudotuberculosis and the effect of adjuvants in mice. J Comp Pathol 95: 167-173.
120. LeaMaster BR,Shen DT,Gorham JR,Leathers CW,Wells HD (1987) Efficacy of Corynebacterium pseudotuberculosis bacterin for the immunologic protection of sheep against development of caseous lymphadenitis. Am J Vet Res 48: 869-872.
121. Holstad G,Teige JJr,Larsen HJ (1989) Corynebacterium pseudotuberculosis infection in goats VIII. The effect of vaccination against experimental infection. Acta Vet Scand 30: 275-283.
122. Holstad G (1989) Corynebacterium pseudotuberculosis infection in goats IX. The effect of vaccination against natural infection. Acta Vet Scand30: 285-293.
123. Brogden KA,Chedid L,Cutlip RC,Lehmkuhl HD,Sacks J (1990) Effect of muramyl dipeptide on immunogenicity of Corynebacterium pseudotuberculosis whole-cell vaccines in mice and lambs. Am J Vet Res 51: 200-202.
124. Eggleton DG,Doidge CV,Middleton HD,Minty DW (1991) Immunisation against ovine caseous lymphadenitis: efficacy of monocomponent Corynebacterium pseudotuberculosis toxoid vaccine and combined clostridial-corynebacterial vaccines. Aust Vet J 68: 320-321.
125. Eggleton DG,Haynes JA,Middleton HD,Cox JC (1991) Immunisation against ovine caseous lymphadenitis: correlation between Corynebacterium pseudotuberculosis toxoid content and protective efficacy in combined clostridial-corynebacterial vaccines. Aust Vet J 68: 322-325.
126. Kutschke L,Ganter M,Kaba J (2000) Efficacy of a flock-specific pseudotuberculosis vaccine in goats. Dtsch Tierarztl Wochenschr 107: 495-500.
127. Schwartzkoff CL, Cobb RM (2000) Ensayos clínicos con una formulación que contiene Moxidectina y una vacuna hexavalente para el ganado ovino. Patologia Animal 15: 437-440
128. Medrano G, Hung CA, Alvarado SA, Li EO (2003) Evaluación de una vacuna contra Corynebacterium pseudotuberculosis en ratones albinos. Revista de Investigacion Veterinaria del Peru 14: 61-67.
129. Ibrahim AK, El-Seedy FR, Hassan WH (2007) The role of corynebacterial phospholipase D vaccine in activation of macrophages. Beni-Suef Vet Med J 17: 50-55.
130. Selim SA, Ghoneim ME, Mohamed KhF (2010) Vaccinal efficacy of genetically inactivated phospholipase D against caseous lymphadenitis in small ruminants. Int J Microbiol Res 1: 129-136.
131. Gohrman J,Gabriel M,Maclachlan NJ,Nieto N,Foley J,et al. (2010) Pilot immunization of mice infected with an equine strain of Corynebacterium pseudotuberculosis. Vet Ther 11: E1-8.
132. Izgür M, Akan M, Ilhan Z, Yazicioglu N (2010) Studies on vaccine development for ovine caseous lymphadenitis. Vet J Ankara University 57: 161-165.
133. Alharbi KBA (2011) Control of abscess disease of sheep by concurrent vaccination, zinc injection and antiseptic washing. Res J Vet Sci 4: 9-13.
134. Rountree PM, Carne HR (1967) Human infection with an unusual Corynebacterium. J Pathol Bacteriol 94: 19-27.
135. Hamilton NT, Perceval A, Aarons BJ, Goodyear JE (1968) Pseudotuberculous axillary lymphadenitis caused by Corynebacterium pseudotuberculosis. Med J Aust 2: 356-361.
136. Batey YM, Tonge JI, Horsfall WR, McDonald IR (1968) Human infection with Corynebacterium ovis. Med J Aust 2: 540-543.
137. Blackwell JB, Smith FH, Joyce PR (1974) Granulomatous lymphadenitis caused by Corynebacterium ovis. Pathology 6: 243-249.
138. Henderson A (1979) Pseudotuberculous adenitis caused by Corynebacterium pseudotuberculosis. J Med Microbiol 12: 147-149.
139. Peloux Y, Maresca C, Oddou JH (1980) Suppurative lymphadenitis caused by Corynebacterium pseudotuberculosis: Report of a case observed in an alpine shepherd. Nouv Presse Med 9: 3182-3184.
140. Goldberger AC,Lipsky BA,Plorde JJ (1981) Suppurative granulomatous lymphadenitis caused by Corynebacterium ovis (pseudotuberculosis). Am J Clin Pathol 76: 486-490.

141. Richards M, Hurse A (1985) Corynebacterium pseudotuberculosis abscesses in a young butcher. Aust NZ J Med 15: 85-86.
142. House RW, Schousboe M, Allen JP (1986) Corynebacterium ovis (pseudotuberculosis) lymphadenitis in a sheep farmer: a new occupational disease in New Zealand. NZ Med J 99: 659-662.
143. Prospert F, Lemoine O, Serruys E, Sternon J (1991) Anatomo-clinical conference of Erasmus Hospital A case of febrile axillary adenopathy. Acta Clin Belg 46:183-188.
144. Bartolomé J,Roca MJ,Marcote E,Moreno R (1995) Corynebacterium pseudotuberculosis adenitis in a shepherd. Med Clin (Barc) 104: 699-701.
145. Bregenzer T, Frei R, Ohnacker H, Zimmerli W (1997) Corynebacterium pseudotuberculosis infection in a butcher. Clin Microbiol Infect 3: 696-698.
146. Mills AE,Mitchell RD,Lim EK (1997) Corynebacterium pseudotuberculosis is a cause of human necrotising granulomatous lymphadenitis. Pathology 29: 231-233.
147. Lester WA, Schousboe M, Chambers S, Patton WN (1997) A second case of Corynebacterium pseudotuberculosis in New Zealand. NZ Med J 110: 469-470.
148. Romero-Pérez JC, Suñer-Machado M, Batista-Díaz N (2004) Linfadenitis porCorynebacterium pseudotuberculosisen una paciente joven. Ver Clin Esp 204: 388-389.
149. Hémond V, Rosenstingl S, Auriault ML, Galanti MJ, Gatfosse M (2009) Lymphadénite axillaire à Corynebacterium pseudotuberculosis chez une patiente de 63 ans. Médecine et maladies infectieuses 39:136-139.
150. Join-Lambert OF, Ouache M, Canioni D, Beretti JL, Blanche S, et al. (2006) Corynebacterium pseudotuberculosis necrotizing lymphadenitis in a twelve-year-old patient. Pediatr Infect Dis J 25: 848-851.

CHAPTER 2

Marek's disease in chickens: A review with focus on immunology

Nitish Boodhoo[1], Angila Gurung[1], Shayan Sharif[2] and Shahriar Behboudi[1*]

1 The Pirbright Institute, Ash Road, Pirbright, Woking, Surrey GU24 0NF, UK.

2 Department of Pathobiology, Ontario Veterinary College, University of Guelph, Guelph, ON, Canada.

ABSTRACT

Marek's disease (MD), caused by Marek's disease virus (MDV), is a commercially important neoplastic disease of poultry which is only controlled by mass vaccination. Importantly, vaccines that can provide sterile immunity and inhibit virus transmission are lacking; such that vaccines are only capable of preventing neuropathy, oncogenic disease and immunosuppression, but are unable to prevent MDV transmission or infection, leading to emergence of increasingly virulent pathotypes. Hence, to address these issues, developing more efficacious vaccines that induce sterile immunity have become one of the important research goals for avian immunologists today. MDV shares very close genomic functional and structural characteristics to most mammalian herpes viruses such as herpes simplex virus (HSV). MD also provides an excellent T cell lymphoma model for gaining insights into other herpesvirus-induced oncogenesis in mammals and birds. For these reasons, we need to develop an in-depth knowledge and understanding of the host-viral interaction and host immunity against MD. Similarly, the underlying genetic variation within different chicken lines has a major impact on the outcome of infection. In this review article, we aim to investigate the pathogenesis of MDV infection, host immunity to MD and discuss areas of research that need to be further explored.

1 INTRODUCTION

Characterized after its human orthologue (Herpes Simplex Virus; HSV a DNA containing virus), Marek's disease virus (MDV), or Gallid herpesvirus 2 (GaHV-2), the etiologic agent for Marek's disease (MD) is an alpha-herpes virus that targets avian species (Gallus gallus domesticus) where it establishes chronic infection. Notably recognized as a multifaceted disease, MD is characterized based on immunosuppression, neurological disorders and neoplastic transformation of CD4+ T cells, localised around peripheral nerves and visceral organs of the host [1]. Since the late 1960s, both large and small scale poultry production systems have been dependent on vaccine use for disease control. Although not a notifiable disease according to the World Organization for animal health (OIE), disease distribution is acknowledged as worldwide although precise estimates of morbidity, annual economic loses and report of disease distribution on each continent are lacking. Effective global surveillance for MDV requires accuracy of reporting source and comprehensiveness. Current data from OIE estimates that about half of the world countries have reported cases of MDV infection (Figure 1). As global requirement increases, so does our dependence on intensive poultry production facilities. The control of MDV infection is arguably very challenging due to its ubiquitous presence at the expense of already pre-established biosecurity programs.

Several MDV pathotypes have been characterised over the years based on morbidity and mortality rates [2]. These pathotypes can be distinguished into three subfamilies (GaHV-2, GaHV-3 and MeHV-1) based on biological and genomic similarities: GaHV-2 (RB-1B, Md5 and CVI988/RISPENS), GaHV-3 (SB-1) and MeHV-1 (HVT; FC-126). Like all herpes viruses, MDV is strictly cell associated. MeHV-1 also known as the Herpes Virus of Turkey (HVT/FC-126) is non-pathogenic in chickens and turkeys [3], but can induce viremia, which is associated with the induction of protective immunity against MD. Chickens infected with HVT become persistently infected and maintain long-lasting immunity. Comparative genomic analysis identified sequence similarities, features and structures to that of infectious MDV (GaHV-2) antigens [4]. Live HVT (FC-126 strain) in combination with SB-1 and 301B/1 are widely used in bivalent and polyvalent vaccine formulation to provide enhanced protection against GaHV-2, but they became less effective with the emergence of highly virulent MDV.

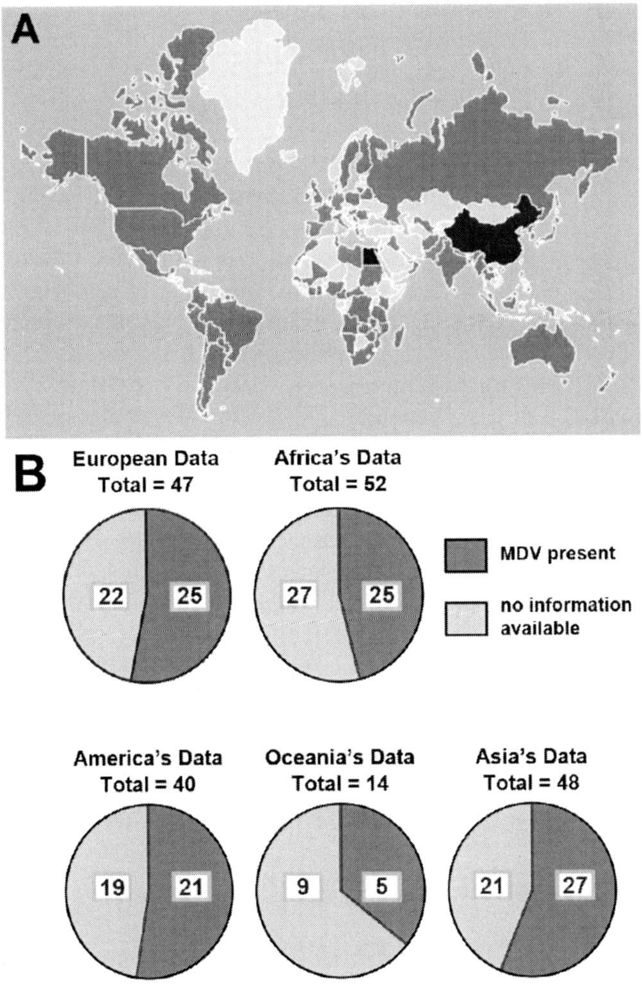

Figure 1 Worldwide map depicting distribution of MDV whereby evidence for presence has been submitted to the OIE. Distribution data was obtained from the World Organization for Animal Health (OIE) distributed through the World Animal Health Information Database (WAHIS) interface and summarized above based on absence (blue) and presence of disease with reported cases (red) before and after 2009. A World map depicting countries positive for MDV was constructed using the imapbuilder software. Both China and Egypt (black) are endemic areas for MDV with outbreaks reported on a yearly basis. B Pie chart demonstrates the number of countries that have reported MDV cases to the OIE based on their geographical location (continent). Data is summarized based on absence (blue), and presence of disease with reported cases (red).

The most effective vaccines developed to date is CVI988/RISPENS (GaHV-2) vaccines [5] which provides protective immunity against disease progression (oncogenesis and neuropathy; transient and acute paralysis) but not sterile immunity (establishment of primary infection and shedding of mature virus particle) which is complicated by reactivation of virus particles from latency. Since its introduction, global vaccination efforts have been efficacious in controlling disease outbreaks thus increasing poultry operations productivity and reduced losses associated with large scale culling of diseased chickens. MD is the first oncogenic disease for which an effective vaccine has been developed. New surging evidence suggests that current vaccination protocol, whether imperfect [6] may be acting in synergy with a plethora of environmental factors resulting in MDV genetic drifts [7, 8, 9]. The human alpha-herpes virus, HSV and Varicella Zoster Virus (VZV) [10] are well documented with similar pathogenesis to MDV. The Oka vaccine provides highly protective immunity against VZV primary infection (chicken pox) and controls viral dissemination (virus reactivation from latency in nervous system cells), unlike the CVI988/RISPENS strain. In MD, although clinical onset of disease has been controlled with implementation of vaccination programs, MDV continues to infect and replicate into fully infectious virus particles in vaccinated chickens; followed by the shedding of these highly contagious, cell-free mature virions from the feather follicle epithelium into skin dander and poultry dust. All these make MDV environmentally persistent as well as a highly infectious and contagious pathogen of chickens. In this review article, we discuss recent progress made in understanding MDV pathogenesis and immunity, aim to scrutinize the pathogenesis of MDV infection, host's immunity to MD and outline areas of research that need to be further explored.

2 PATHOGENESIS OF MAREK'S DISEASE

2.1 Marek's disease

MDV, the causative agent of MD in chickens, result in transformation of CD4+ T cells. The natural route of infection is defined by inhalation of airborne cell-free virus particles within the contaminated dust and dander (Figure 2), shed from infected host produced in terminally differentiated feather follicle epithelium [11], into a naïve respiratory track [12]. MD pathogenesis has four phases in the susceptible birds; an early cytolytic phase within 2–7 days post-infection (dpi) which delineates as semi productive lytic viral replication in lymphocytes. This is followed by a latency phase that occurs between 7 and 10 dpi in CD4+ T cell subset that result in systemic viral dissemination. Cutaneous viral infection can occurs as early as 4 dpi and eventually results in fully productive viral replication and shedding [13]. MDV reactivation in CD4+ T cells initiates a late cytolytic and immunosuppressive phase starting around 18 dpi. Finally a proliferative phase around 28 dpi [14, 15] is characterized by formation of visceral tumours that originate from CD4+ T cells lymphoma. There is no transmission from chicken to eggs (vertical transmission) but the birds are usually infected in early stage after hatching (horizontal transmission). The presence of maternal antibody against MD can protect the neonatal chicks, and with the development of a functional immune system a degree of resistant to MD is developed [5]. However, husbandry-related stress or concurrent infection with other immunosuppressive pathogens such as chicken infectious anaemia virus (CIAV), reovirus and infectious bursal disease virus (IBDV), significantly enhances susceptibility to MDV [5]. Another important factor in the susceptibility to MD is the genetic background of chicken lines which is to some extent associated with the major histocompatibility complex (MHC). However, it should be noted that MHC does not necessarily play a critical role in resistance or susceptibility to MD. Chicken lines 6-3 and 7-2 harbour the same MHC haplotype (B2/B2) while line 6-3 is relatively resistant and line 7-2 is highly susceptible to MD. It has been demonstrated that some very virulent (vv+) strains of MDV induces tumours even in the resistant line 6-3. MDV-associated lymphoma can only be developed in genetically susceptible chickens; however the virus can replicate and shedding still occurs in both susceptible and resistant chicken lines.

2.2 Establishment of primary infection

It is speculated that lung epithelial cells are one of the primary target cells for MDV infection. MDV antigens, with well-defined expression during cytolytic and latent phase of replication, have been detected at significant levels at various time points in lung epithelial cells in ovo [16], and in vivo [17] suggesting an establishment of successful infection. The later was performed via an aerosol method which simulates natural infection as a respiratory disease [12]. Viral replication in the lungs could be detected as early as 1 dpi. Purchase et al. [18] were among the first to demonstrate a novel route for high replication kinetics of infectious MDV antigens in lungs epithelial cells of chicks inoculated via intra-abdominal route. However when they repeated the experiment, a lower immunofluorescence was detected at 5 dpi compared to 7 dpi. The route of administration, whether intra-abdominal or intra-tracheal might affect viral replication as well as systemic dissemination that results in MD [19].

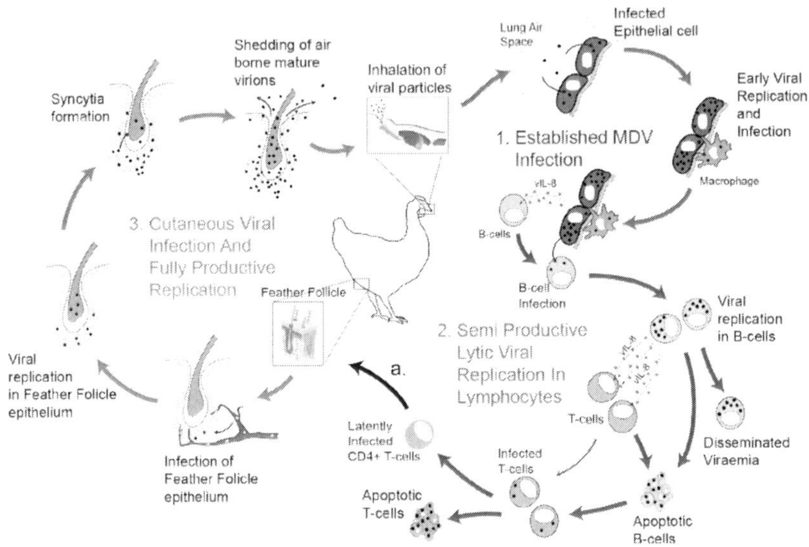

Figure 2 Model of MDV infectious life cycle in resistant birds. MDV infection of naive host occurs via inhalation of dust or skin dander encapsulated viral particles into the respiratory tract. 1 Primary infection occurs when virus particle breaks mucosal tolerance in the lungs, site of entry into the epithelial cells. Local viral replication establishes infection and initiates viral immediate-early gene, viral Interleukin-8 (vIL-8), transcription and translation. Inflammatory responses in the underlying tissue recruit innate immune system cells which result in uptake of infectious virus particle by macrophages. Infiltration of lymphocytes via action of vIL-8 follows resulting in MDV infection of B-cells. 2 Viral replication in B cells initiates Semi Production Lytic Viral Infection and disease progression. MDV infected B cells secret vIL-8 that acts as a chemotactic factor for and gains access to T-cells. This specific lymphotropism (B cells and T cells) enables systemic disseminated viraemia. Viral replication causes apoptosis of B and T lymphocytes in a hallmark of immunosuppression. MDV integrates specifically into the genome of CD4+ T cells enabling escape from immune detection and initiates Latent Viral Infection. Early latently infected and activated CD4+ T cells have not been phenotypically characterised by cell surface markers. a Early latently infected and activated CD4+ T cells migrate to cutaneous sites of replication namely feather follicle. 3 Infection of feather follicle epithelium enables fully productive viral replication. Viral replication results in syncytia formation. Infection of feather epithelium leads to secretion of mature virion in skin danders and dust that act as the major source of infectious materials. Horizontal transmission is the only recognized form for environmental persistence and infection in field conditions. Systemic infection and neoplastic transformation of CD4+ T cells in susceptible birds is further discussed (Figure 3).

In addition, infection of lung resident antigen presenting cells (APCs), such as macrophages, is thought to result in subsequent transport to primary and secondary lymphoid organs such as thymus, bursa of fabricius, and spleen [20]. Although it is unclear whether macrophages and lung epithelial cells get infected simultaneously or rather infected lung epithelial cells may play a role in transmitting viral particles to macrophages. It is evident that post MDV infection, immune responsiveness leads to macrophage infiltration although viral replication is unaffected [17]. It is also believed that presence of MDV particles in the lung, during the earliest infection, stimulates secretion of cytokines and chemotactic factors that help in attracting B cells to site of infection [21, 22]. One of the defined chemokine is a viral IL-8 which is similar to CXCL13 and is involved in recruiting immune systems cells to site of viral replication [23] and is defined as a homologue to the host IL-8 gene. IL-8 has a well-defined role as a chemotactic molecule for T cell [22, 24] and B cells [22]. Immune cells recruited to the lung such as B cells can be detected as early as 2 dpi [25].

2.3 Semi productive lytic viral replication

MDV has a specific tropism for immune system cells and preferentially infects lymphocytes; B cells and T cells (αβ). Infection of B cells may occur in the lung and viral replication in B cells is defined as semi-productive lytic viral replication. Lytic activity due to viral replication has been linked to phosphoprotein 38 (pp38) activities [26, 27]. PP38 role as an early immediate gene is defined only in lymphocytes, specifically B cells and T cells [28, 29]. It has been shown that an MDV rMd5Deltapp38 deletion mutant for pp38 lacked the ability to induce cytolytic activity characterized by B cell apoptosis [30]. This is in accordance with the notion that recruitment of lymphocytes such as B cells to site of viral replication is a key step for transmission of virus and dissemination. Deletion mutants of vIL-8 (RB1BvIL-8ΔsmGFP) when tested in vivo showed a reduced capacity to successfully infect lymphocytes and induce lytic infection [23]. A lack of IL-8 therefore result in impaired ability to recruit B cells and as well as an observable reduction in cytolytic activity due to reduced viral titer and dissemination by lymphocytes. Infection with wild-type MDV restored lytic activity and viral dissemination to primary lymphoid organs [28]. Viral lytic activity in B cells results in a drastic downfall in the overall antibody production and ability to fight against an infection. Consequently, the infected chicken suffers from immunosuppression, making it more susceptible to MD and other infections [31]. During the early cytolytic phase, transcriptional modification and epigenetic changes (DNA methylation, histone post-translational modifications and non-coding RNAs), along with post-transcriptional and post-translational modifications, regulate viral replication cycle and subsequent expression of cellular and viral genes. Either way, it's been postulated that disseminated viremia to various organs in a cell associated manner results in systemic infection. Infected B cells may also be able to produce vIL-8 mRNA which would have chemotactic activity for T cells. Infected B cells therefore are able to recruit T cells which would allow for transmission of MDV virus particle from infected B cells to activated T cells. Viral replication in T cells also lead to cytolysis associated with pp38 activity. The RB1BvIL-8ΔsmGFP deletion mutant lacked the ability to attract B cells and subsequently impaired its ability to infect T cells albeit at a lower frequency [23]. Furthermore, MDV preferentially targets CD4+ T cell subsets and infection results in viral latency and immune evasion [22].

2.4 Immune evasion and latency

Hereafter, MDV enters the latency phase (Figure 2) such that it can no longer be detected by the host immune system while it continues to replicate inside the lymphocytes [15]. One of the important immune evasion strategies include down regulation of MHC I molecules on infected lymphocytes. A specific gene, encoding for viral RNA telomerase (vTR) subunit, has been reported to be present only in the oncogenic MDV pathotypes, but not in their non-oncogenic counterparts. In addition, further study has confirmed about 88% identity of the vTR gene with the chicken terminal repeat (cTR) gene of the host. This would mean that vTR gene is also complement to host Telomerase reverse transcriptase (TERT) for cTR. Consequently, MDV and other herpes virus integrate at the ends of host's chromosomes, preferably at TR sequences. It is, therefore, thought that vTR subunit might, in fact, have a significant role in maintaining the viral oncogenicity through telomeric elongations at the host's chromosomal ends thereby inhibiting programmed cell death through telomere shortening [32]. However, it has recently been shown that both oncogenic and vaccinal strains of MDV have the ability to integrate into DNA, suggesting that integration alone is not sufficient for MDV-induced transformation [33]. Chromosomal insertions of alpha-herpes virus DNA segments, including those from HSV and equine herpes virus types 1 and 3, have been associated with their oncogenesis, because many of the cells carrying integrated viral DNA displayed a transformed phenotype [32]. Consistent with this, Delecluse et al. [34] has demonstrated evidences of MDV genome integration into host's chromosomes for all lymphoma established, MDV cell lines. Additionally, expression of MDV-Meq gene is essential for MDV-induced neoplastic transformation of latently infected cells (Figure 3). Meq expressed specifically by oncogenic GaHV-2 has transcriptional activities that lends to modulation of host genomic activity.

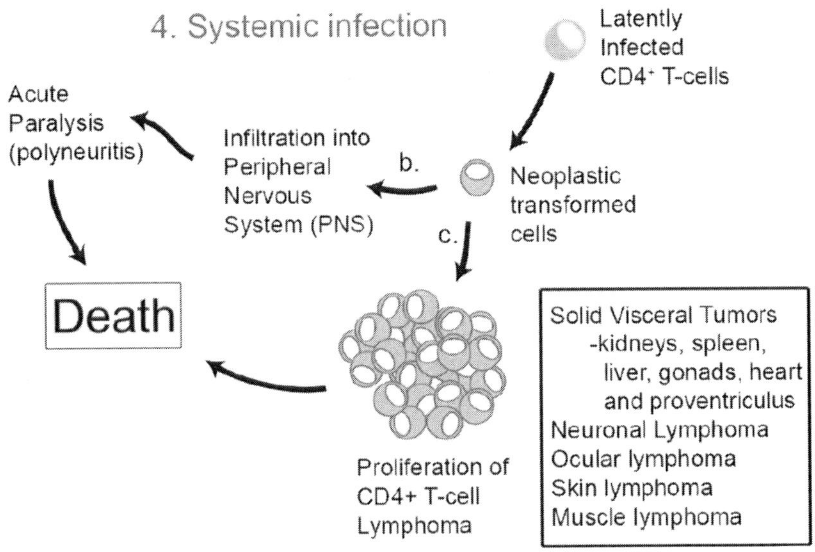

Figure 3 Model of MDV neoplastic transformed CD4+ T cells and subsequent disseminated Systemic Infection in susceptible birds. 4 Fully latent MDV-Transformed CD4+ T-cells proliferate in all sites where immune systems cells are involved in primary and secondary line of defence. Early latently activated CD4+ T cells undergo neoplastic transformation due to transcriptional and transrepressional activity of viral oncogenic proteins such as Meq. Pathogenesis arise in cases where vaccination failure is suspected (suboptimal dosage/titre, damage to vaccine or confirmed immunological vaccine failure), vaccination is not performed (backyard birds). b Fully latent neoplastic MDV-Transformed CD4+ T cells infiltrate and establish a reservoir of MDV genome in peripheral nerve fibres interspace. These cells have a CD4+ CD25+ Treg phenotype although additional cell surface markers have yet to be determined. Expression of viral neurovirulence factor, phosphoprotein 14 (pp14), promotes neuropathy and cell survival. Neuropathy (Polyneuritis) is presented as transient or acute paralysis of legs, wings, neck, with vision impairment and weight loss depending on MDV-1 virulence factor. Birds infected with serotype-1 eventually succumb to death from paralysis. c Reactivation from latency enables a second phase of replication whereby viral oncogenic protein Meq acts on T cell signaling pathways causing uncontrolled cellular proliferation leading to disseminated lymphoma formation in visceral organs, peripheral and central nervous system, musculoskeletal systems, skin and eyes. Severe lymphoma eventually causes death in birds. Highly pathogenic viruses (serotype-1, vv+ MDV) kill birds before they reach the lymphoproliferative phase of the disease.

In relation to this, Lupiani et al. [35] conducted a study using a recombinant Md5 pathotype, attenuated by deletion of Meq-gene. As expected, MDV did infect and replicate in the lymphoid organs and feather follicles; but there was no tumour induction, implicating that the integrating viral genome requires a Meq gene for induction and maintenance of oncogenic properties. Reactivation of MDV from latency and tumorigenic transformation of latently infected lymphocytes mainly occurs in CD4+ T cells [36].

Another important factor in the susceptibility to MD is the genetic background of established chicken lines. MDV-associated CD4+ T cell lymphoma can only be developed in genetically susceptible chickens; however, the virus can replicate and be shed from both susceptible and resistant chicken lines. It should be noted that highly pathogenic strains of MDV can induce T cell lymphoma in the resistant chicken lines. It is believed that only a few subsets of CD4+ T lymphocytes undergo transformation, and thus are the origin of lymphoma [1]. This may explain why in most cases of MD tumour cells obtained from different anatomical sites, such as liver, kidneys, gonads, skin and muscles, all have similar CDR3 length profile suggesting that tumour cells are monoclonal [37]. In most lymphoma cells, the virus is in the latent phase and does not produce viral particles and only 0.1% of tumour cells are in the lytic phase [38]. Omar and Schat [26] have revealed an association between the genetic background of chickens and their resistance to MDV; in relation to slight variations on MDV associated-MHC presentations between resistant (B21) and susceptible (B19) chicken lines. These studies showed evidences of T cell mediated immune responses against the MDV antigen, *ICP4*, only in the resistant chicken lines; but not in the susceptible ones. In contrast, both B21 and B19 chicken lines have been reported to show T cell immune responses against the other three MDV antigens: *gB, Meq* and *pp38*. Nevertheless, these studies suggest that T cell mediated responses in combination with MHC play an important role in genetic resistance against MD.

2.5 Cutaneous infection, replication and shedding

Viral genomes are detectable by quantitative PCR in blood cells and feather tips of birds infected with oncogenic or vaccinal strains [13]. Similar to other herpes viruses, MDV has a tendency to be transported towards cutaneous sites such as skin, and feather follicles. MDV is shed into the environment via scales and feather debris, which is a major source of contamination [12, 39]. It is possible that T cells transport the virus to feather follicles, but the role of other immune cells in transporting the virus has not yet been ruled out. Infiltration of CD4+ and CD8+ T cells and expression of pro-inflammatory cytokines into the skin of birds infected with a highly virulent virus (like RB-1B) or by a vaccine strain of the virus (like Rispens or HVT) suggest that the immunity against the virus is ineffective at blocking virus replication and shedding [40, 41]. Cell-free MDV is only produced in feather follicle epithelial cells, and it is believed that MDV relies exclusively on cell-to-cell transmission [42, 43]. However, in a recent report, this notion has been challenged by demonstrating that in a cell-blebbing phenomenon and cell apoptotic corps clearance mechanism, MDV can be transmitted in a cell-free condition [44]. Meq expressing tumour tissues can also be found in the skin of infected birds and neoplastic cutaneous lesions in the scaleless chickens indicates that feather follicles are not necessary for skin tumour development. Finally, the data indicate that inoculation with supernatant fluid from homogenized and sonicated skin samples of MDV-infected scaleless chickens induces MD in susceptible birds, suggesting that skin epithelial cells not associated with feather follicles also harbour infectious viral particles [45]. The process of apoptotic corps clearance may well be related to its specific tropism for adaptive immune cells which leads to immunosuppression. It is still unclear whether either phenomenon also occurs in vivo during the early hours of infection or vaccination as well as the respective cells involved. Furthermore, vaccinated challenge birds don't show clinical signs of immunosuppression therefore an inherent resistance to cellular apoptosis is observed. This may have important implications in our understanding of MDV pathogenesis and development of next generation MDV vaccines. MDV can be detected in the feather follicles at around 11–14 dpi, using standard biochemical methods, and 6–7 dpi, using sensitive methods such as qPCR. In cutaneous sites, fully productive infection and replication is re-activated in feather-follicle epithelium and enveloped infectious viral particles in a cell free form are released. MDV can be found in the epithelium of feather follicles infected chickens more frequently than other tissues, both, in terms of incidence and levels of viral antigen expression [46]. The virus replicates as enveloped, cell free MDVs in the feather follicles epithelium of infected chickens. These cell-free MDVs are highly infectious and are easily released into the poultry dust or litter. They have a very protective envelope, allowing them to survive for months in poultry thus facilitating horizontal transmission and infection of naïve animals in following production cycles [47].

MDV infection and its global presence could also be a product of natural reservoirs located in backyard and migratory birds (Figure 4) such as Common teal, White-fronted goose, Pintail, European wigeon, and Mallard [48] and more recently in Roulroul partridges [49]. The presence of MDV in wild and migratory birds has been well documented since the early 1980. So it is of no great surprise that new reports emerge as these viruses cause disease in newly identified host. Further monitoring is required to understand the importance of wild birds as reservoirs along migratory routes for pathogenic serotypes.

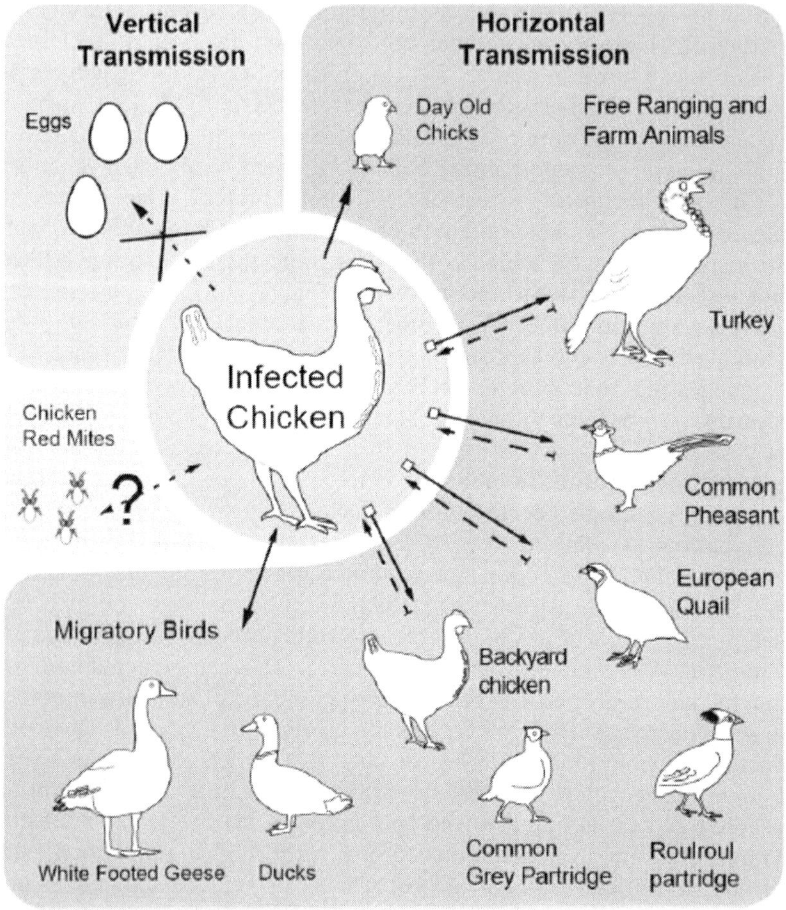

Figure 4 Model for horizontal and vertical transmission of MDV between avian species. MDV serotypes can infect several different avian species globally. It has been confirmed that MDV cannot be transmitted horizontally from layer hens to eggs but day old chicks become infected in broiler production housing systems from shedders. Several avian species have been grouped based on literature: Free ranging and farm animals and Migratory birds. Chicken red mites have also been identified as carriers of avian pathogens. Free ranging and farm animals consist of turkeys, common pheasant, common grey partridge and European quail. Migratory birds that have been confirmed positive are ducks and white footed geese. Although these birds become infected and are carriers, what has not been identified is their role in viral virulence factors as well as ability to infect broiler production system. Furthermore, the presence of chicken red mites has yet to be confirmed in broiler production system which could contribute to viral persistence.

2.6 Infection of immune cells

Based on the current model of MDV pathogenesis, phagocytic cells such as macrophages and dendritic cells in the respiratory tract become infected either directly or after an initial round of viral replication. The virus can also replicate in macrophages and induce cytolysis in the infected phagocytic cells as demonstrated by high levels of cell death in splenic macrophages expressing three herpesvirus kinetic classes: ICP4 (immediate early), pp38 (early) and gB (late). The level of infection in macrophages isolated from MDV-infected birds depends on the virulence of MDV. The in vivo results demonstrate that, at 4–6 dpi, more virulent viruses (C12/130) induce 3–10 times higher percentages of pp38+ macrophages compared to that infected with less virulent MDV. The results also show that pp38+ macrophages are prone to cell death. Interestingly, MDV antigens could only be detected in MDV infected macrophages, but not in uninfected macrophages that have phagocytosed MDV infected cells [20]. Consistent with this view, Abdul-Careem et al. [17] have shown an up-regulation in nitric oxide synthase (NOS) gene in lung macrophages of MDV infected chickens. This results in production of nitric oxide (NO) reported to have anti-microbial activity against many viruses including MDV for both in vivo and in vitro conditions [20]. Likewise, a similar distribution profiles of MDV antigen have been observed in MDV infected B lymphocytes during the early cytolytic phase. Replication of MDV in the lung induces host innate immune response as demonstrated by up-regulation of Toll-like receptors (e.g. TLR 3 and TLR 7), pro-inflammatory cytokines (e.g. IL-1β and IL-8) and iNOS genes as well as infiltration of macrophages [17]. However, mucosal immune response to MDV in the lungs is ineffective in controlling virus replication [17, 20, 25].

Generally, the infection of mammalian immune cells such as macrophages and B cells with herpes viruses down-regulates the expression of both MHC class I and II molecules consequently, evading hosts' cell mediated immunity [50]. It has been shown that MDV infection down-regulates surface expression of MHC (B complex) class I (BF) glycoproteins during active but not latent infection of chicken cells [51]. Further studies demonstrated that MDV012 and MDV pUL.49.5 genes (Table 1) are involved in down-regulation of MHC class I molecules by interfering with transporter associated with antigen processing (TAP) function. However, the effects of MHC class I down-regulation in the pathogenesis of the disease is unknown and recombinant viruses lacking the cytoplasmic tail of pUL49.5 exhibit almost similar pathogenicity as wild type virus in both the resistant and susceptible chicken lines [52, 53]. NK cells and more importantly cytotoxic T lymphocytes (CTL) which monitor cell surface MHC class I molecules and viral peptide complex respectively may play a major role in host defence against infection. Contrary to MHC class I, MDV infection up-regulates MHC class II molecules on chicken APC's such as macrophages [54, 55], and this may contribute to virus spread within the infected host by increasing the interaction between infected macrophages and activated CD4+ T cells. Since MDV requires cell-to-cell contact for virus spread and productive infection of T cells and other immune cells, this up-regulation of MHC class II molecules might, in fact, be essential in the pathogenesis of MD within the infected host by increasing the interaction between infected macrophages and activated CD4+ T cells [51, 54]. HIV infection can also induce activation of innate response and up-regulation of MHC class II molecules which can lead to immuno-pathology [56, 57]. It is still unclear whether MHC class II up-regulation observed after MDV infection can induce immunopathology which is clearly manifested in MDV pathogenesis. In the infected cells, MDV virus expresses a viral antigen with high amino acid sequence homology to CXCL13, termed viral IL-8 (vIL-8). CXCL13 exerts its chemotactic effects by interacting with chemokine receptor CXCR5 and is a major regulator trafficking for B cells and subsets of T cells. This viral chemokine has the ability to recruit immune cells to the site of infection [23]. The deletion of vIL-8 from open reading frames severely affects MDV pathogenesis and tumour incidence [22, 23, 58]. vIL-8 induces chemotaxis of B cells and regulatory T cells (CD4+ CD25+ T cells), and these cells are targets for both lytic and latently infected cells, thus demonstrating a virus specific tropism [22]. Within 24 h after infection, the virus can be detected in the bursa of Fabricius, spleen, and thymus [14, 15]. It is believed that macrophages and dendritic cells can disseminate the virus from the lungs to B cells and CD4+ T cells in these lymphoid organs. Splenic B cells play an important role in the replication of MDV, and the high titre of circulating virus is due to the replication of the virus within B cells, but in the absence of B cells, the virus replicates in other immune cells and MD can still be developed. In the cytolytic phase of the disease, large number of splenic B cells undergoes apoptosis and cell death. Depletion of B cells and CD4+ T lymphocytes in lymphoid tissues, such as cecal tonsils (CTs), of susceptible chicken lines within 5 dpi may contribute to immunosuppression observed in late cytolytic phase of the disease. The depletion of B and T cells in the CTs of the resistant line was minimal at 5 dpi, which also recovered by 21 dpi [59]. Virus replication in B and T cells reaches its peak between 3 and 7 dpi. B lymphocytes constitute the majority (around 90%) of cytolytically infected (MDV-antigen positive) cells, while CD4+ and CD8+ T lymphocytes represent only 3 and 8%, respectively [60, 61]. MDV can be transmitted directly to T cells from the infected macrophages or dendritic cells; however, the transfer of virus from B cells to T cells is also conceivable.

Table 1 MDV genes and their respective products involved in immune modulation and pathogenesis

MDV gene (protein)	Function	Antigenic potential	Infection stage	Reference
MDV003/078 (vIL-8)	Viral chemotactic (CXCL-) protein involved in recruiting immune systems cells to site of viral replication	–	Lytic replication	[23, 58]
MDV010 (vLIP)	Shares homology to host lipase enzyme and forms covalent bonds with lipids	–	Lytic replication	[136]
MDV011/012	Immune evasion protein that down regulates cell surface MHC I expression	–	Lytic replication	[52]
MDV012 ORF012	Phosphoprotein required for viral growth both in vivo and in vitro	–		[137]
MDV040 (gB)	*Hypothetical function*: Virion membrane protein that heterodimerizes with other glycoproteins facilitating viral fusion with host membrane	Yes	Lytic replication	[26, 27]
MDV052/053 UL39/40 (RR)	Viral Ribonucleotide Reductase (RR) is essential for replication both in vivo and in vitro	–	Lytic replication	[138]
MDV064 (U_L49.5)	Partially reduces MHC I expression by interacting with TAP protein	–	Lytic replication	[53]
MDV073 (pp38)	Early protein expressed during cytolytic infection and phosphorylated by U_S3p	Yes	Lytic replication	[26–29]
MDV092 (U_S3p)	Serine/threonine protein kinase that phosphorylates pp38	–	Lytic replication	[139]

MDV084/100 (ICP4)	Viral gene transactivation function	Yes	Lytic replication	[26, 27]
MDV001a (vTR)	Required for integration of viral genome into host DNA for immune evasion, neoplastic transformation and viraemia	–	Latency	[140]
MDV006 (pp14)	Neurovirulence factor required for PNS neuropathy (acute or transient paralysis)	–	Latency	[141]
MDV062 (VP22)	Tegument protein essential for viral replication and modulates host cell cycle	–	Latency	[142]
MDV057 (gC/UL44)	Type 1 transmembrane protein required for horizontal transmission/shedding from feather follicle epithelium	–	Feather follicle shedding	[143]
MDV005/076 (MEQ)	Viral oncogenic protein involved in T cell neoplastic transformation which forms homodimers and heterodimers with specific intracellular signalling proteins that modulates host cell cycle	Yes	Neoplastic transformation	[26, 27, 35]
MDV029 (pUL17)	Co-localizes with VP5 and VP13/14 tegument protein and essential for in vivo viral growth, capsid maturation and DNA packaging	–	Neoplastic transformation	[144]

Cytolytic infection of B and T cells is semi-productive, which is defined by their inability to express certain viral structural components. No cell-free virus is produced by the infected T and B lymphocytes and only non-enveloped intra-nuclear particles are detected [5]. The precise mechanism as to how in vivo MDV spreads from cell to cell has not been elucidated. However, it is assumed that MDV glycoproteins (g) B, gC and gD, similarly observed in most herpes viruses, are likely to interact with host's cell surface receptors; thereby, forming an intracellular bridge between infected and uninfected cells which might, in turn, contribute to cell associated viral spread in MD. Similarly, co-existence of MDV gH and gL have been reported to be vital for GaHV-2 cell to cell viral spread. The primary peptide of gL has a high affinity for specific region of gH, and therefore binds to it giving rise to a complex, hetero-oligomer structure that anchors itself onto the cell surface of infected host cell, promoting MDV proteins (gp) cell surface expression [62]. In support, Schumacher et al. [63] demonstrated that deletions of gE (20DeltagE) or gI (20DeltagI) were essential in restricting viral spread and plaque formation, although viral replication was not abolished. Transfecting gE or gI did not support viral spread indicating that gE and gI could work in synergistic manner to aid viral spread. Furthermore, it is possible that MDV mainly replicates within B and T cells by mitosis of infected cells rather than production of virions. Thus, MDV does not require expressing all the viral genes to replicate in vivo. By not producing all the MDV viral antigens that may be highly immunogenic, the virus can escape immune control. This notion is supported by the fact that the resting T cells, with low proliferative abilities, are less susceptible to MDV infections than the activated lymphocytes. T cell activation may increase the expression of surface molecules that is engaged in virus entry, thus enhancing the occurrence of MDV infection in these cells.

Similar to MDV, the infection of human CD4+ T cells by human T cell leukaemia virus type 1 (HTLV-1) is increased in activated CD4+ T cells. This is reflected on up-regulation of heparan sulfate proteoglycans (HSPGs), a receptor for HTLV-1, on activated CD4+ T cells [64]. No specific receptor for MDV entry into chicken CD4+ T cells has yet been identified. It is conceivable that MDV infection induces cytolysis/inflammatory responses and T cell activation, which leads to T cell infection and viral replication via mitosis by passing the virus to daughter lymphocytes.

The underlying genetic variation within different chicken lines may play a key role in the pathogenesis and prognosis of the MDV infection. The percentages of T cells that become cytolytically infected in the lymphoid organs are less than 2% in MD-susceptible chickens and 0.2% in the resistant chickens [5, 60]. In support, Omar et al. [26] have demonstrated evidences of T cell mediated immune responses against the MDV antigen, *ICP4*, only in resistant (B21) chicken lines; but not in susceptible (B19) ones. In contrast, both B21 and B19 chicken lines have been reported to show T cell immune responses against MDV antigens: *gB, Meq* and *pp38*. Nevertheless, these studies suggest that the role of T cell mediated immunity, and differences in T cell receptor repertoire generation in conjuncture with other inherent genetic resistance mechanism against MD cannot be ruled out. In latent phase of infection, MDV antigens cannot be detected in the lymphoid tissues and there is no production of infectious viruses. It is difficult to distinguish latent infection from transformation phase, as both represent non-productive infections. Following in vitro reactivation of MDV from virus antigen-negative lymphocytes, the majority of latently infected lymphocytes were identified as T cells with only 3% being B cells [5]. *Meq* antigen from MDV plays a crucial role in maintaining latency by blocking apoptosis of latently infected CD4+ T cells. One of the pronounced differences in gene expression profile between MDV-resistant and susceptible chicken lines after MDV infection are genes that are associated with apoptosis [65]. In addition to *Meq*, it has been shown that microRNA miR-M3, an MDV-encoded miRNA, abrogates apoptosis by directly targeting Smad2, a critical component in the transforming growth factor (TGF)-β [66], providing an environment beneficial for latency and oncogenesis. miR-M4, an ortholog of the oncogenic miR-155, was shown to have a direct effect on inducing MDV-induced T cell lymphoma, as viruses deleted in miR-M4 or having mutations in the seed region failed to induce lymphoma [67]. Viral miR-M4 exerts its effects by reducing the levels of latent TGF-β binding protein-1, which is involved in the maturation of TGF-β. This leads to a reduction in the levels of active Smad2/3 and release of the inhibition of the c-Myc promoter, resulting in a rise in c-Myc transcription. The production of viral protein *Meq* allows the formation of complexes with c-Myc [68], which is associated with transformation. In the susceptible chickens, a second wave of semi-productive infection and cytolysis are observed between 14 and 21 dpi [14]. This late cytolytic phase is associated with immunosuppression, atrophy of lymphoid tissues such as thymus, bursa of fabricius, cecal tonsils and infiltration of mononuclear cells and heterophils [14, 69]. Virus is probably transferred to the skin by latently infected CD4+ T cells and infects skin and feather follicles in a yet unknown mechanism. Syncytia formation in skin epithelium may be a product of viral protein which facilitates and pools greater resources for viral replication. The involvement of other immune system cells such as macrophages and dendritic cells to transport the virus to the skin cannot be excluded. Replication of MDV starts at 7 dpi, well before tumour development. Therefore, it is possible that early infected CD4+ T cells, but not necessarily latent or transformed CD4+ T cells, transport the virus into the skin at this time.

3 THE IMMUNE RESPONSE TO MDV

3.1 Innate immunity

While effective immunity against human alphaherpesviruses relies on both innate and adaptive mechanisms, the innate immune response has been shown to be of paramount importance [70]. Less is known about the role of innate immunity in the control of MDV in chickens.

3.1.1 Interferons

Type I IFNs belong to a family of cytokines that attracted much attention owing to their protective role against viral infection. IFNs are widely expressed cytokines that possess strong antiviral and immunomodulatory properties. The IFN family can be classified into three main types of cytokines—type I, type II and type III IFNs. IFN-α and IFN-β belongs to type I IFN family, while the type II IFN family includes only one cytokine: IFN-γ, which also exhibits antiviral activities [71]. The third type of IFNs is the IFN-λ family. In mammals, plasmacytoid dendritic cells (pDCs), monocytes, epithelial cells and fibroblasts are the main producers of type I IFNs [72], while type II IFNs are predominantly produced by NK cells and activated T cells. In spite of the fact that chicken type I IFNs are shown to inhibit viral infection both in vivo and in vitro, chicken pDCs have not been identified.

Chickens become infected with MDV via the respiratory system by inhaling infected dust. MDV is taken up by phagocytic cells such as macrophages or dendritic cells within the respiratory system. Chicken lung has a different anatomical structure than the mammalian counterpart consisting of air sacs; and due to narrower pulmonary capillaries than in mammals there are fewer airway resident macrophages [73]. Therefore, it is likely that MDV has to cross lung epithelial lining before being transported by phagocytic cells to lymphoid tissues. In the respiratory system, the virus can be recognized by TLRs, such as TLR21 (recognizing unmethylated CpG DNA), leading to the initiation of protein signalling cascaded which stimulates the expression of type I interferons (α and β), shown to be involved in antiviral defence. In fact, an increase in the expression of TLR receptors in the lungs of MDV-infected birds [36], IFN-α expression in the blood of susceptible chickens [74] and interferon regulatory factors (IRF)-1 and IRF-3 in MDV-infected chicken embryonic fibroblasts cells (CEF) have been reported [75, 76]. The role of IFNs in the control of MDV replication has been confirmed in an in vitro model showing that IFNs reduces plaque formation and expression of *PP38* and *gB* in the infected cells [77]. The protective role of IFNs is also indicated by the results demonstrating the differential expression patters of IRF-3 and IFN-β genes in resistant and susceptible chicken lines [78]. Similarly, oral administration of IFN-α are shown to reduce MDV viral replication in vivo [79].

In addition to their direct effects on viral replication, type I interferons may also activate other immune system cells such as natural killer (NK) cells and increases their cytotoxic function [80]. However, it has been suggested that chicken NK cell cytotoxicity is not increased after oral administration of recombinant chicken IFN-α or inoculation of recombinant MDV expressing chicken IFN-α (rMDV-cIFN-α). In chickens inoculated with rMDV-cIFN-α, NK cell cytotoxicity was not enhanced over control chickens at 4 and 7 dpi. Furthermore, at 4 dpi, chickens inoculated with R2/23 actually had decreased NK cell cytotoxic activity. Therefore, it is concluded that ChIFN-α, given at high doses orally in the drinking water or via expression in a recombinant MDV vector, does not increase NK cell cytotoxicity as originally hypothesized [79]. The suppression of immune responses by oral administration of IFN-α is not a novel phenomenon as similar observations are reported in murine model where, bone marrow functions were suppressed by administration of murine IFN-α orally or subcutaneously [81]. In fact, it has been shown that type I IFNs modulate the function of both innate and adaptive immune cells including DCs and T cells in the gut and suppress the intestinal inflammation. Taken together, it is believed that type I IFN response by MDV-infected cells promotes the activation of immune cells and inhibits MDV replication and dissemination. Considering the complexity of MDV and co-evolution of the virus with the host's type I IFNs response, it is very likely that there is a complex relationship between MDV and host response. The exact role of type I IFNs in the pathogenesis of MDV in chickens is poorly understood due to lack of immunological reagents, complexity of MDV infection and the cell-type specific effect of type I IFNs. However, it is possible that type I IFNs may be involved in promoting latency infection of MDV as has been observed in other *alphaherpesviruses* [82].

The role of type II IFNs, IFN-γ, is discussed in more detail in adaptive immunity section. IFN-γ is induced during MDV infection and shown to have inhibitory effects on MDV replication by inducing nitric oxide production [77] and the administration of IFN-γ with MDV vaccine positively influenced vaccine-induced protective immunity in vivo [83]. There is no information on the role of type III IFNs on viral infections in chickens.

3.1.2 Macrophages and dendritic cells

There is very little information on the type and function of antigen presenting cells (APCs) involved in the initiation of immune responses against MDV in the respiratory system of chickens. However, it is believed that chicken professional APC's such as macrophages and dendritic cells play an important role in the development of adaptive immunity against MDV. In mammals, it has been shown that dendritic cells play a crucial role in linking innate to adaptive immunity. The type of dendritic cells and the type of maturation induced by different stimuli influences the immunological outcome, such as differentiation of Th1 vs. Th2 type T cells. There is very limited information on the role of dendritic cells in MDV immunity, however, it is postulated that these cells could be involved in the initiation of innate and adaptive immunity against MDV. In addition to their ability to present MDV antigens in association with MHC class I and II molecules to initiate adaptive immunity, macrophages can also be directly involved in inhibition of MDV replication and development of MD. Macrophages isolated from B21 chickens have higher phagocytic activity than B19 chickens to MDV [84]. Moreover, macrophages obtained from MDV-infected chickens inhibited viral replication in vitro more efficiently than macrophages isolated from non-infected chickens [85]. Further confirmation on the role of macrophages in MDV infection is obtained from the results demonstrating that depletion of macrophages from splenocytes increases MDV replication [84], while stimulating macrophages in vivo reduces the incidence of MD [86]. Taken together, the evidence presented here from several studies confirms that macrophages play a pivotal role in control of MDV replication and MDV-derived tumour incidence. One of the mechanisms involved in the inhibitory function of macrophages on MD is their ability to produce inducible nitric oxide (iNOS). The production of NO has been reported in the spleen, brain and lungs of MDV-infected chickens [13, 77, 87, 88] and NO is shown to inhibit MDV replication with highest level of NO production detected in serum and spleen of resistant chickens compared to susceptible chickens [77, 89]. Further experiments demonstrated that the inhibition of iNOS in chickens increases viral load, suggesting that NO plays an important role in the control of MDV replication in vivo [77]. Another function of macrophages is their ability to kill tumour cells and it is believed that activated chicken macrophages have the ability to lyse MDV-derived tumour cells in vitro [90]. In contrast to activated and fully functional macrophages, tumour associated macrophages (TAMS) represent key regulators of the complex interplay between the immune response and cancer. TAMS produce tumour growth promoting factors and induce immunosuppression by releasing immuno-modulatory factors [91]. Macrophages isolated from tumour tissues of MDV-infected chickens demonstrate similar functional abilities as TAMS and have been shown to suppress T cell proliferation in vitro. The development of immuno-regulatory macrophages in MDV-infected chickens correlates with transient immunosuppression observed during primary cytolytic phase of infection [92], suggesting that TAMS may be involved in MDV-induced immuno-suppression.

3.1.3 Natural killer (NK) cells

NK cells represent important effectors of the innate immunity and can respond to stimuli and produce anti-viral cytokines such as IFN-γ. In addition, these cells have ability to recognize virus- infected cells/tumour cells via ligation of cell death receptors and the release of granules. NK cells from MDV-resistant chickens have higher cytotoxic capability than the MDV-susceptible chickens, suggesting that these cells may be involved in determining resistance to MDV. This hypothesis is also supported by the results demonstrating that MDV-infected chickens have higher NK cell activity than the cells isolated from the non-infected birds and; this activity lasted longer in the resistant chickens than the susceptible chickens [93]. The exact role of NK cells in providing vaccine-induced protection against MDV is still unknown, however, there are some evidences demonstrating that vaccination against MDV increases the functional abilities of NK cells. This may explain how MDV vaccine can provide protection in vaccinated chicks as early as 3 days post-vaccination [93]. NK cells also play a fundamental role as antitumor senses through downregulation of cell surface markers such as MHC I. A notable characteristic of herpes viral infection and specifically MDV is down regulation of MHC I cell surface translocation. Further to that, anti-tumor activity has not yet been demonstrated in an MDV resistance model independent of genetic factors that predispose resistance to neoplastic transformation of CD4+ T cells. Taken together, it has been suggested that NK cells may play an important role in controlling MDV infection [93, 94, 95]. Using recently identified markers such as CD56 and CHIR-AB1 [96] for identification of chicken NK cells, researchers will be able to elucidate the role of these important cells in providing protection against MDV.

3.2 Adaptive immunity

The key components of adaptive immunity are B and T lymphocytes which specifically recognize antigens and generate memory response. B cells are involved in humoral immune response, whereas T cells are involved in cell-mediated immune response.

3.2.1 Humoral immunity

As MDV is a cell-associated herpes virus and is strictly intracellular, antibodies should not have a major role in the protective immunity against MDV infection. However, antibodies against several MDV glycoproteins including gB, gE, gI have been detected in MDV-infected birds [97, 98]. The role of these antibodies in providing protective immunity against MDV has not been clarified. However, there is some evidence suggesting that anti-gB neutralizing antibody may have a protective role against MDV, perhaps via blocking virus entry into the host cells or antibody dependent cell-mediated cytotoxicity (ADCC) of infected cells. The role of antibody response in the control of MDV infection is also confirmed by the fact that the presence of maternal antibody delays the development of clinical signs and tumour. However, the presence of maternal antibody can interfere with live replicating vaccines against MDV by neutralizing the vaccinal virus [5]. Therefore vertical transmission of immunological factors such as maternal antibodies can be a limiting factor for both generations of protective immunity as well as delaying immune responsiveness to infection challenge models of offspring from vaccinated layers. Maternal antibody and interference with vaccination has been reported in both human and veterinary medicine regardless of the types of vaccine formulations used. It has been shown that while maternal antibody can interfere with antibody response, vaccine-induced cell-mediated immune responses are largely unaffected. This has been confirmed in humans, murine models and farmed animals [99, 100, 101, 102, 103, 104]. With regard to the inhibitory effect of maternal antibody response, new strategies have been developed to overcome the inhibitory effects of maternal antibody on vaccine-induced antibody response. For example, it has been shown that the induction of type I interferon in vivo strongly stimulates B cell responses and restores antibody levels after immunization in the presence of maternal antibodies. One way of inducing high levels of type I interferon is the combined use of TLR-3 and TLR-9 agonists as adjuvants for immunization [102]. Therefore, it should be possible to overcome the neutralizing effects of maternal antibody against MDV in novel vaccination strategies.

3.2.2 Cell-mediated immunity

Antibody and cell mediated immunity is involved in the control of infection with highly cell-associated human alphaherpesviruses such as VZV. Both CD4+ and CD8+ effector and memory T cells are shown to be essential for recovery from VZV and maintaining the latent stage of infection in the subclinical state. Generally, it is believed that (a) broad (response to several epitopes), (b) durable (memory response), and (c) multi-functional (capable of producing several Th1 type cytokines/kill virus infected cells) T cell response is associated with the control and resolution of viral infections. MDV is also a highly cell-associated alphaherpesvirus, and thus it is postulated that cell-mediated immunity is crucial for the vaccine induced protection [105, 106]. Since MDV exists in cell associated forms inside the host, except in feather follicles, T cell mediated immunity is thought to be more important than the antibody mediated immunity in the control of the disease in chickens [107]. Studies have shown evidences demonstrating the presence of CD8+ T cells against MDV antigens such as: *gB, Meq, pp38* and *ICP4*. However, the role of CTL in conferring long term immunity, generation of memory cells, in genetically resistant chickens is unknown

[36] and the role of cell-mediated immunity in vaccine-induced protection has not been determined. No cytotoxic response against ICP22 and a weak cytotoxicity against Meq were detected in MDV-infected chickens [5, 26]. Other studies confirmed the presence of anti-*gB* and anti-*gI* TCRαβ1+ CD8+ T cells with cytotoxic abilities in MDV-infected birds [108].

The presence of herpes virus specific CD8+ T cell response coincides with protection from cytomegalovirus (CMV) infection and adoptive cell transfer of CD8+ T cells provide protection in animal models. The importance of CD4+ T cells in providing protection against herpes virus infection was also demonstrated by CD4+ T cell depletion experiments in animal models. There was an inverse correlation between the number of virus-specific CD4+ T cells and prolonged shedding [105, 106]. Therefore, it is very likely that both CD8+ and CD4+ T cells are involved in the control of MDV replication. This notion is confirmed with the early studies showing that T cells are crucial for the control of tumour growth in HVT-vaccinated chickens. The importance of T cells in the control of tumour growth was confirmed by demonstrating that HVT vaccinated birds treated with cyclosporine, a drug shown to inhibit T cell function, develop MDV-lymphoma [109, 110]. However, later studies suggested that T cells are only involved in the control of viral replication but not essential for the control of tumour growth [111]. Further studies are required to confirm the exact role of CD4+ and CD8+ T cells in the control of MDV replication and tumour growth. The availability of CD4+ and CD8+ knockout chicken will provide valuable tools to study the role of these cells in vaccine-induced protective immunity against viral replication and tumour growth. MHC-determined resistance to MDV in MHC:B21/B21 birds indirectly confirms the role of T cells in control of tumour development. For MHC class I molecules, the relative level of expression varies between different MHC haplotypes and reflects the consensus hierarchy of response by different MHC haplotypes with the most susceptible chickens (MHC B19/B19) having the highest expression while the most resistant chickens (MHC: B21/B21) expressing the lowest level [112]. It is believed that the differences in cell surface expression level ensure the development of optimal peripheral T cell responses against MDV. The detection of MDV-specific CD8+ T cells with cytotoxic ability to recognize target cells expressing MDV antigen is technically challenging. However, Schat and Markowski-Grimsrud developed a heterologous system to investigate cytotoxic T cell response to MDV using reticuloendothelial virus-transformed cells stably transfected with specific MDV genes [99]. The cell lines derived from resistant (MHC: B21/B21) and susceptible (MHC: B19/B19) birds were generated and antigen-specific cytotoxicity were analysed in vitro. The cytotoxicity was low compared to mammalian cytotoxic T lymphocyte (CTL) assays [5]. In these studies, relatively a more potent cytotoxicity was detected against target cells expressing pp38 and gB antigens. No cytotoxic response against

ICP22 and a weak cytotoxicity against Meq were detected in MDV-infected chickens [5, 26]. Other studies confirmed the presence of anti-gB and gI TCRαβ1+ CD8+ T cells with cytotoxic abilities in MDV-infected birds [108]. Vaccination of chickens with a recombinant Fowl Pox Virus (FPV) expressing these two glycoproteins have been reported to induce protective immunity when challenged with MDV and shown to induce anti-MDV neutralizing antibodies [113, 114]. Recombinant FPV expressing gB was later shown to induce cytotoxic T cell response recognizing target cells expressing gB [115]. These data support the importance of anti-MDV CD8+ T cells responses in the control of the disease, however, the role of MDV-specific CD8+ T cell responses in the control of viral infection or tumour growth in birds immunized with HVT or CVI988-Rispens is still unknown. In addition, there is very little information on the magnitude and quality of MDV-specific T cells responses in resistant vs. susceptible chickens. The only available information suggest that anti-ICP4 CD8+ T cell responses is only detected in the resistant chickens [26], while anti-gE CD8+ T cell response is only detected in the susceptible chickens. These results suggest that the response to ICP4, an immediate early MDV gene product, but not to gE may be an important factor in genetic resistant to MDV. It has been shown that CTL responses characterized by release perforin and granzymes can be induced against MDV antigens such as PP38, Meq and ICP4 [36]. Moreover, there is still no information on the presence of MDV-specific CD4+ T cell responses and whether transformed CD4+ T cells have antigen specificity to MDV antigens. Further studies are required to identify CD4+ and CD8+ T cell epitopes within MDV antigens and determine their MHC restriction. Moreover, it is important to examine the quality, broadness, durability and magnitude of the CD4+ and CD8+ T responses in MDV infected or vaccinated chickens and determined the correlates between T cell responses and protection.

In addition to T cell responses against MDV viral antigens, it is possible that T cells may also recognize self-antigens expressed by tumour cells. However, the nature of these self-antigens and their relevance to immunity and inhibition of tumour growth in birds is still unknown. A few decades ago, Marek's disease tumour-associated surface antigens (MATSA) was identified in the MDV-transformed T cells [116]. However, it was later discovered that activated T cells can also express MATSA antigens, and this molecule is not solely expressed on transformed T cells [117]. Several of these MATSA are found to be activation associated lymphocyte antigens, and one of them is identified as CD30 molecule [47]. Meanwhile CD30, a co-stimulatory molecule, has been shown to have pleiotropic effects on human T cell activation, apoptosis, effector function (cytotoxicity) and regulating T cell trafficking/migration [118]. It has a fundamental role across all T cell lineages such as CD4+ , CD8+ and Th17 cells [119] and may influence T cell interaction; suppressive action in the tissue microenvironment.

Immunomodulatory processes that influence CTL functional abilities in relation to persistent viral infection and tumorigenesis may be mediated via interaction of inhibitory receptors such as CTLA4, programmed death-1 (PD-1) and its respective ligand programmed death ligand-1 (PD-L1). The evidence suggests that MDV infection up-regulate the expression of CTLA-4, PD1 and PD-L1 in chicken immune system cells. The expression level of PD-1 was increased in chickens at the early cytolytic phase of the MDV infection, while the PD-L1 expression level was increased at the latent phase. In addition, the expression levels of PD-1 and PD-L1 were increased at tumor lesions found in MDV-infected chickens [120, 121]. Furthermore, T cell responsiveness may be affected by chronic antigen stimulation leading to exhaustion thus altering phenotypic characteristics such as PD-1 expression also observed during in vitro MDV infection. A combination of reduced MHC-I and increased PD-1 cell surface translocation provides a platform for highly efficacious immune evasion tactic. CTLA-4, a potent inhibitory receptor, expressed by CD4+ T cells has also be reported during the early cytolytic phase in MDV infected birds [120]. Significant differences in CTLA-4 and PD-1 expression levels, which could result in a delayed immune responsiveness (anergy) and cell death respectively, were reported between resistant and susceptible lines of chickens challenged with MDV. A similar phenotype has been reported in patients infected with human alpha-herpes viruses such as VZV whereby CD4+ T cell predominantly express CTLA-4 and PD-1 [122]. Induction of an early immune unresponsiveness (anergy) and cytolysis (4–7 dpi) could well be a hallmark, for alpha-herpes viruses, to establish early semi-productive viral replication and disseminated viraemia. Blocking or down regulating CTLA-4 or PD-L1 could be therapeutically significant during early infection. Furthermore, PD-L1 expression results in immune suppression thus can be used as a marker for CD4+ T cell lymphoma. PD-L1 expression in MDV challenged chickens, could limit the quality of T cells immune responsiveness to *Meq*, *pp38*, *ICP4* or *gB* at the immunological synapse (Figure 5). PD-1 and CTLA-4 ligation with its cognate receptor results in cross-pathway interference resulting in inhibition of RAS, ERK 1/2, AKT, JNK and PLCγ phosphorylation altering cell fate; growth/proliferation, effector function and survival. Expression of inhibitory molecules on both antigen presenting cells can lead to the generation of chicken regulatory CD4+ T cells (Treg cells). The role of naturally occurring and peripheral derived Treg cells in modulation of anti-MDV immunity is still unknown. However, it is known that MDV infection up-regulates the expression of inhibitory molecules such as CTLA-4 and expression of inhibitory cytokines such as IL-10. The expression of regulatory molecules such as CTLA-4 and IL-10 on both CD4+ and CD8+ T cells are at 10 and 21 dpi and this effect was more pronounced in the MDV-susceptible chicken lines [123].

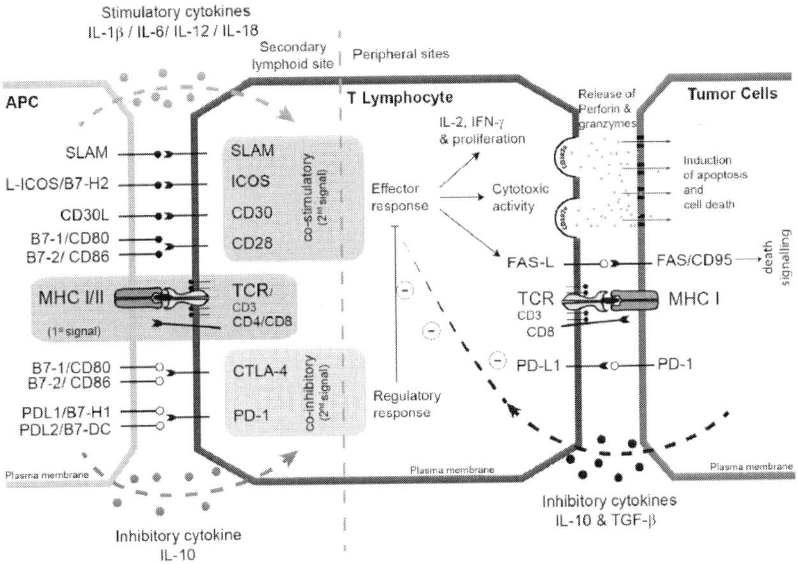

Figure 5 T cell activation and tolerance by tumour cells/antigen presenting cells. Inhibitory or stimulatory molecules expressed on the surface of antigen presenting cells (APC) or tumour cells regulate T cell function. Moreover, stimulatory or inhibitory cytokines may drive the generation of different T cell populations (e.g. Th1, Th17, Treg etc.) with diverse functional properties. Signal one is provided to the T cell receptor of T cells by presentation of antigens via MHC class I or II molecules expressed by APCs or tumour cells, while the signal 2 is provided by co-stimulatory molecules such as B7 family. Co-stimulatory signals induce the generation of effector T cells, which can recognize and lyse target cells or produce cytokines such as IFN-γ, involved in the control of tumour growth. In contrast, inhibitory molecules deliver negative signals and suppressing the effector T cell function, and induce T cell anergy or exhaustion.

4 PERSISTENCE IN THE FACE OF VACCINATION

Vaccination against MD started in the late 1960s using turkey herpesvirus (HVT) which does not induce disease in chickens. The vaccination reduced the incidence of MD by 99% and was the first successful vaccine against naturally occurring virus-induced cancer. Since then MDV has evolved, perhaps due to vaccine induced immune response, and the escape of new mutants from immune pressure. Currently only attenuated MDV strain; CVI988-Rispens is effective in providing protection against the very virulent MDV. Live vaccines are administered either to day-old chicks or to the 18-day-old embryo. MDV vaccine inhibits the development of MDV-induced lymphoma but does not prevent infection and replication of pathogenic strains of MDV. Both vaccination and maternal antibodies against MDV increases viral shedding and onward transmission of hyper-virulent strains of MDV due to the survival of the host without controlling the virus shedding [124]. The failure of current vaccines to induce sterile immunity can be attributed to MDV inducing latency with minimal viral replication and viral protein expression. The ideal MDV vaccine is to control both the disease and viral shedding in the infected birds. Revaccination (prime-boost) with the current cell-associated MDV vaccine (CVI988-Rispens) improve protection against the disease and increases the magnitude of anti-MDV T cell responses as demonstrated by enhancement of anti-MDV neutralizing antibody and proliferation of CD4+ and CD8+ T cells [125]. However, the type and quality of immunity after even revaccination with the current MDV vaccines cannot control virus shedding. Further research is required to understand immune responses required to control MDV shedding as well as the control of the disease. To generate more effective vaccines against MDV, we need to (a) have a better understanding of the type of immunity required for reducing viral shedding/inducing sterile immunity (b) design a vaccine that can induce protective immunity against the virus. It is believed that the quality, broadness, durability and homing of both CD4 and CD8 T cell responses may explain why vaccines against some alpha-herpes viruses (VZV) are more effective than vaccines against other alpha-herpes viruses (HSV) [126]. As yet there is little information as to whether cell-mediated immunity is induced by the MDV vaccine. The lack of established methodology to assess cell-mediated immunity to MDV vaccine is the major obstacle to have a better understanding of how MDV vaccine actually works. It is very likely that the induction of both innate and adaptive immune responses by vaccine strain is similar to those induced by pathogenic MDV [93, 127]. Cytokine response following oncogenic and vaccine strains of MDV has also been reported and the results demonstrate that very virulent virus (vvMDV), compared with vaccine strain (CVI988/Rispens), induced similar levels of the typical Th1-type cytokine IFN-γ in microglia in vitro, while vvMDV induces higher expression of IL-12 (p40), IL-8, and MIP-1β [128].

It is postulated that vaccine-induced adaptive immunity plays a major role in providing protection against the disease. There are some evidences demonstrating that a number of MDV antigens, including gB when administered as recombinant vaccine in a fowl pox vector, are immunogenic and immunity to these antigens confers protection [115]. However, the types and magnitude of vaccine-induced protective immune responses to these antigens are still unknown. Several factors including genetic background of chickens [129], presence or absence of maternal antibodies, virulence of MDV, vaccine dose [130] and concurrent infections with other immunosuppressive pathogens such as CIAV [108] can influence the efficacy of MDV vaccines.

A medley of growth factors and cytokines such as myelomonocytic growth factor (MGF) [131] and IFN-γ [83] have been considered as immuno-modulators and vaccine adjuvants against MDV. The treatment of MDV-susceptible chickens with MGF, a growth and activation factor for monocytes/macrophages, reduced viral load, increased survival rate and reduced tumour incidence after challenge [131]. It is postulated that MGF increases the number of macrophages and their response to stimuli as shown by an increase in NO production by activated macrophages. Similarly, we have recently shown that administering HVT vaccine with a plasmid expressing recombinant chicken IFN-γ enhanced the protective efficacy of the vaccine against MDV and reduced viral load and tumour incidence [83]. IL-18 has been shown to stimulate IFN-γ production from CD4+ T cells and can also indirectly stimulate CD8+ T cell proliferation [132]. In addition, IFN-γ can enhance the development of Th1 type responses which is known to be important in the control of viral infections as well as activating innate immune cells such as macrophages and NK cells. Although not explored in chickens, classical activation of IFN-γ results in interferon stimulatory gene (ISG)-1 up regulation, which could have an additive antiviral effect. Both type I and type II interferons have been tested in various clinical settings but the whole gamete of IFN-regulated signalling pathways have yet to be elucidated. Ligands for TLRs are used as adjuvants that stimulate immune responses, leading to protection against infectious diseases. We have demonstrated that the administration of TLR3 [133], TLR4 and TLR21 ligands [134] reduces the incidence of tumours and MDV genome copy numbers in the infected birds. The importance of TLR in providing protection against tumour development could be extrapolated from the results demonstrating significantly higher basal expression levels of TLR3 and TLR7 in uninfected chicken cells isolated from resistant chicken lines compared to the susceptible lines [135].

5 CONCLUSIONS

Our understanding of innate and adaptive immunity to MDV infection is continuously improving. This growing knowledge can be further enhanced by our understanding of immunity to human alphaherpesviruses and host-pathogen interaction. However, there are many differences between the human and avian immune system as well as different characteristics between MDV and human alphaherpesviruses (VZV, HSV1 and 2). For example, MDV is the only alphaherpesviruses that can modulate the chemokine network by molecular mimicry of a host protein and thus recruit, infect and transform CD4+ T cells. In terms of vaccine development, the developments of potent cell-free vaccines that can inhibit infection as well as the disease are paramount to research in this area.

ACKNOWLEDGEMENTS

We are grateful to Sue Baigent for critical review of the figures. This work was supported by funding from the Biotechnology and Biological Sciences Research Council (award no. BBS/E/I/00001825) to SB.

REFERENCES

1. Morrow C, Fehler F (2004) Marek's disease: an evolving problem. In: Davison F, Nair V (eds) Marek's disease: a worldwide problem. Elsevier Academic Press, London, pp 49–61
2. Gimeno IM, Witter RL, Reed WM (1999) Four distinct neurologic syndromes in Marek's disease: effect of viral strain and pathotype. Avian Dis 43:721–737
3. Witter RL, Solomon JJ (1972) Experimental infection of turkeys and chickens with a herpesvirus of turkeys (HVT). Avian Dis 16:34–44
4. Afonso CL, Tulman ER, Lu Z, Zsak L, Rock DL, Kutish GF (2001) The genome of turkey herpesvirus. J Virol 75:971–978
5. Davidson F (2004) Marek's disease: an evolving problem. Elsevier Academic Press, London
6. Gimeno IM, Cortes AL, Montiel ER, Lemiere S, Pandiri AK (2011) Effect of diluting Marek's disease vaccines on the outcomes of Marek's disease virus infection when challenged with highly virulent Marek's disease viruses. Avian Dis 55:263–272
7. Davison F, Nair V (2005) Use of Marek's disease vaccines: could they be driving the virus to increasing virulence? Expert Rev Vaccines 4:77–88
8. Gimeno IM, Cortes AL, Silva RF (2008) Load of challenge Marek's disease virus DNA in blood as a criterion for early diagnosis of Marek's disease tumors. Avian Dis 52:203–208
9. Read AF, Baigent SJ, Powers C, Kgosana LB, Blackwell L, Smith LP, Kennedy DA, Walkden-Brown SW, Nair VK (2015) Imperfect vaccination can enhance the transmission of highly virulent pathogens. PLoS Biol 13:e1002198
10. Zerboni L, Sen N, Oliver SL, Arvin AM (2014) Molecular mechanisms of varicella zoster virus pathogenesis. Nat Rev Microbiol 12:197–210
11. Nazerian K, Witter RL (1970) Cell-free transmission and in vivo replication of Marek's disease virus. J Virol 5:388–397
12. Abdul-Careem MF, Javaheri-Vayeghan A, Shanmuganathan S, Haghighi HR, Read LR, Haq K, Hunter DB, Schat KA, Heidari M, Sharif S (2009) Establishment of an aerosol-based Marek's disease virus infection model. Avian Dis 53:387–391
13. Abdul-Careem MF, Read LR, Parvizi P, Thanthrige-Don N, Sharif S (2009) Marek's disease virus-induced expression of cytokine genes in feathers of genetically defined chickens. Dev Comp Immunol 33:618–623
14. Calnek BW (1986) Marek's disease–a model for herpesvirus oncology. Crit Rev Microbiol 12:293–320
15. Calnek BW (2001) Pathogenesis of Marek's disease virus infection. Curr Top Microbiol Immunol 255:25–55

16. St Hill CA, Silva RF, Sharma JM (2004) Detection and localization of avian alphaherpesviruses in embryonic tissues following in ovo exposure. Virus Res 100:243–248
17. Abdul-Careem MF, Haq K, Shanmuganathan S, Read LR, Schat KA, Heidari M, Sharif S (2009) Induction of innate host responses in the lungs of chickens following infection with a very virulent strain of Marek's disease virus. Virology 393:250–257
18. Purchase HG (1970) Virus-specific immunofluorescent and precipitin antigens and cell-free virus in the tissues of birds infected with Marek's disease. Cancer Res 30:1898–1908
19. Butter C, Staines K, Baaten B, Smith L, Davison TF (2007) Route of challenge is critical in determining the clinical outcome of infection with a very virulent oncogenic herpesvirus, Marek's disease virus. Avian Pathol 36:93–99
20. Barrow AD, Burgess SC, Baigent SJ, Howes K, Nair VK (2003) Infection of macrophages by a lymphotropic herpesvirus: a new tropism for Marek's disease virus. J Gen Virol 84:2635–2645
21. Witter RL, Calnek BW, Buscaglia C, Gimeno IM, Schat KA (2005) Classification of Marek's disease viruses according to pathotype: philosophy and methodology. Avian Pathol 34:75–90
22. Engel AT, Selvaraj RK, Kamil JP, Osterrieder N, Kaufer BB (2012) Marek's disease viral interleukin-8 promotes lymphoma formation through targeted recruitment of B cells and CD4+ CD25+ T cells. J Virol 86:8536–8545
23. Parcells MS, Lin SF, Dienglewicz RL, Majerciak V, Robinson DR, Chen HC, Wu Z, Dubyak GR, Brunovskis P, Hunt HD, Lee LF, Kung HJ (2001) Marek's disease virus (MDV) encodes an interleukin-8 homolog (vIL-8): characterization of the vIL-8 protein and a vIL-8 deletion mutant MDV. J Virol 75:5159–5173
24. Gesser B, Lund M, Lohse N, Vestergaad C, Matsushima K, Sindet-Pedersen S, Jensen SL, Thestrup-Pedersen K, Larsen CG (1996) IL-8 induces T cell chemotaxis, suppresses IL-4, and up-regulates IL-8 production by CD4+ T cells. J Leukoc Biol 59:407–411
25. Baaten BJ, Staines KA, Smith LP, Skinner H, Davison TF, Butter C (2009) Early replication in pulmonary B cells after infection with Marek's disease herpesvirus by the respiratory route. Viral Immunol 22:431–444
26. Omar AR, Schat KA (1996) Syngeneic Marek's disease virus (MDV)-specific cell-mediated immune responses against immediate early, late, and unique MDV proteins. Virology 222:87–99
27. Omar AR, Schat KA (1997) Characterization of Marek's disease herpesvirus-specific cytotoxic T lymphocytes in chickens inoculated with a non-oncogenic vaccine strain of MDV. Immunology 90:579–585
28. Reddy SM, Lupiani B, Gimeno IM, Silva RF, Lee LF, Witter RL (2002) Rescue of a pathogenic Marek's disease virus with overlapping cosmid DNAs: use of a pp38 mutant to validate the technology for the study of gene function. Proc Natl Acad Sci USA 99:7054–7059
29. Lee LF, Cui X, Cui Z, Gimeno I, Lupiani B, Reddy SM (2005) Characterization of a very virulent Marek's disease virus mutant expressing the pp38 protein from the serotype 1 vaccine strain CVI988/Rispens. Virus Genes 31:73–80
30. Gimeno IM, Witter RL, Hunt HD, Reddy SM, Lee LF, Silva RF (2005) The pp38 gene of Marek's disease virus (MDV) is necessary for cytolytic infection of B cells and maintenance of the transformed state but not for cytolytic infection of the feather follicle epithelium and horizontal spread of MDV. J Virol 79:4545–4549
31. Biggs PM, Nair V (2012) The long view: 40 years of Marek's disease research and Avian pathology. Avian Pathol 41:3–9
32. Morissette G, Flamand L (2010) Herpesviruses and chromosomal integration. J Virol 84:12100–12109
33. Robinson CM, Cheng HH, Delany ME (2014) Temporal kinetics of Marek's disease herpesvirus: integration occurs early after infection in both B and T cells. Cytogenet Genome Res 144:142–154
34. Delecluse HJ, Hammerschmidt W (1993) Status of Marek's disease virus in established lymphoma cell lines: herpesvirus integration is common. J Virol 67:82–92

35. Lupiani B, Lee LF, Cui X, Gimeno I, Anderson A, Morgan RW, Silva RF, Witter RL, Kung HJ, Reddy SM (2004) Marek's disease virus-encoded Meq gene is involved in transformation of lymphocytes but is dispensable for replication. Proc Natl Acad Sci USA 101:11815–11820
36. Haq K, Schat KA, Sharif S (2013) Immunity to Marek's disease: where are we now? Dev Comp Immunol 41:439–446
37. Mwangi WN, Smith LP, Baigent SJ, Beal RK, Nair V, Smith AL (2011) Clonal structure of rapid-onset MDV-driven CD4+ lymphomas and responding CD8+ T cells. PLoS Pathog 7:e1001337
38. Remy S, Blondeau C, Le Vern Y, Lemesle M, Vautherot JF, Denesvre C (2013) Fluorescent tagging of VP22 in N-terminus reveals that VP22 favors Marek's disease virus (MDV) virulence in chickens and allows morphogenesis study in MD tumor cells. Vet Res 44:125
Jarosinski KW, Arndt S, Kaufer BB, Osterrieder N (2012) Fluorescently tagged pUL47 of Marek's disease virus reveals differential tissue expression of the tegument protein in vivo. J Virol 86:2428–2436
39. Abdul-Careem MF, Hunter BD, Sarson AJ, Parvizi P, Haghighi HR, Read L, Heidari M, Sharif S (2008) Host responses are induced in feathers of chickens infected with Marek's disease virus. Virology 370:323–332
40. Abdul-Careem MF, Hunter DB, Shanmuganathan S, Haghighi HR, Read L, Heidari M, Sharif S (2008) Cellular and cytokine responses in feathers of chickens vaccinated against Marek's disease. Vet Immunol Immunopathol 126:362–366
41. Ahmed M, Schidlovsky G (1968) Electron microscopic localization of herpesvirus-type particles in Marek's disease. J Virol 2:1443–1457
42. Churchill AE, Biggs PM (1967) Agent of Marek's disease in tissue culture. Nature 215:528–530
43. Harmache A (2014) A virulent bioluminescent and fluorescent dual-reporter Marek's disease virus unveils an alternative spreading pathway in addition to cell-to-cell contact. J Virol 88:11617–11623
44. Heidari M, Fitzgerald SD, Zhang HM, Silva RF, Lee LF, Dunn JR (2007) Marek's disease virus-induced skin leukosis in scaleless chickens: tumor development in the absence of feather follicles. Avian Dis 51:713–718
45. Couteaudier M, Denesvre C (2014) Marek's disease virus and skin interactions. Vet Res 45:36
46. Burgess SC, Young JR, Baaten BJ, Hunt L, Ross LN, Parcells MS, Kumar PM, Tregaskes CA, Lee LF, Davison TF (2004) Marek's disease is a natural model for lymphomas overexpressing Hodgkin's disease antigen (CD30). Proc Natl Acad Sci USA 101:13879–13884
47. Murata S, Hayashi Y, Kato A, Isezaki M, Takasaki S, Onuma M, Osa Y, Asakawa M, Konnai S, Ohashi K (2012) Surveillance of Marek's disease virus in migratory and sedentary birds in Hokkaido, Japan. Vet J 192:538–540
48. Haesendonck R, Garmyn A, Dorrestein GM, Hellebuyck T, Antonissen G, Pasmans F, Ducatelle R, Martel A (2015) Marek's disease virus associated ocular lymphoma in Roulroul partridges (Rollulus rouloul). Avian Pathol 44:347–351
49. Miller DM, Sedmak DD (1999) Viral effects on antigen processing. Curr Opin Immunol 11:94–99
50. Hunt HD, Lupiani B, Miller MM, Gimeno I, Lee LF, Parcells MS (2001) Marek's disease virus down-regulates surface expression of MHC (B Complex) Class I (BF) glycoproteins during active but not latent infection of chicken cells. Virology 282:198–205
51. Hearn C, Preeyanon L, Hunt HD, York IA (2015) An MHC class I immune evasion gene of Marek's disease virus. Virology 475:88–95
52. Jarosinski KW, Hunt HD, Osterrieder N (2010) Down-regulation of MHC class I by the Marek's disease virus (MDV) UL49.5 gene product mildly affects virulence in a haplotype-specific fashion. Virology 405:457–463
53. Niikura M, Kim T, Hunt HD, Burnside J, Morgan RW, Dodgson JB, Cheng HH (2007) Marek's disease virus up-regulates major histocompatibility complex class II cell surface expression in infected cells. Virology 359:212–219

54. Gimeno IM, Witter RL, Hunt HD, Lee LF, Reddy SM, Neumann U (2001) Marek's disease virus infection in the brain: virus replication, cellular infiltration, and major histocompatibility complex antigen expression. Vet Pathol 38:491–503
55. Stevanovic S, van Bergen CA, van Luxemburg-Heijs SA, van der Zouwen B, Jordanova ES, Kruisselbrink AB, van de Meent M, Harskamp JC, Claas FH, Marijt EW, Zwaginga JJ, Halkes CJ, Jedema I, Griffioen M, Falkenburg JH (2013) HLA class II upregulation during viral infection leads to HLA-DP-directed graft-versus-host disease after CD4+ donor lymphocyte infusion. Blood 122:1963–1973
56. Kimmel PL, Cohen DJ, Abraham AA, Bodi I, Schwartz AM, Phillips TM (2003) Upregulation of MHC class II, interferon-alpha and interferon-gamma receptor protein expression in HIV-associated nephropathy. Nephrol Dial Transplant 18:285–292
57. Cui X, Lee LF, Reed WM, Kung HJ, Reddy SM (2004) Marek's disease virus-encoded vIL-8 gene is involved in early cytolytic infection but dispensable for establishment of latency. J Virol 78:4753–4760
58. Heidari M, Fitzgerald SD, Zhang H (2015) Immune responses in cecal tonsils of Marek's disease virus-infected chickens. Avian Dis 59:213–226
59. Baigent SJ, Ross LJ, Davison TF (1998) Differential susceptibility to Marek's disease is associated with differences in number, but not phenotype or location, of pp38+ lymphocytes. J Gen Virol 79:2795–2802
60. Calnek BW, Schat KA, Ross LJ, Chen CL (1984) Further characterization of Marek's disease virus-infected lymphocytes. II In vitro infection. Int J Cancer 33:399–406
61. Wu P, Reed WM, Lee LF (2001) Glycoproteins H and L of Marek's disease virus form a hetero-oligomer essential for translocation and cell surface expression. Arch Virol 146:983–992
62. Schumacher D, Tischer BK, Reddy SM, Osterrieder N (2001) Glycoproteins E and I of Marek's disease virus serotype 1 are essential for virus growth in cultured cells. J Virol 75:11307–11318
63. Jones KS, Fugo K, Petrow-Sadowski C, Huang Y, Bertolette DC, Lisinski I, Cushman SW, Jacobson S, Ruscetti FW (2006) Human T-cell leukemia virus type 1 (HTLV-1) and HTLV-2 use different receptor complexes to enter T cells. J Virol 80:8291–8302
64. Subramaniam S, Preeyanon L, Cheng HH (2013) Transcriptional profiling of mEq-dependent genes in Marek's disease resistant and susceptible inbred chicken lines. PLoS One 8:e78171
65. Xu S, Xue C, Li J, Bi Y, Cao Y (2011) Marek's disease virus type 1 microRNA miR-M3 suppresses cisplatin-induced apoptosis by targeting Smad2 of the transforming growth factor beta signal pathway. J Virol 85:276–285
66. Zhao Y, Xu H, Yao Y, Smith LP, Kgosana L, Green J, Petherbridge L, Baigent SJ, Nair V (2011) Critical role of the virus-encoded microRNA-155 ortholog in the induction of Marek's disease lymphomas. PLoS Pathog 7:e1001305
67. Chi JQ, Teng M, Yu ZH, Xu H, Su JW, Zhao P, Xing GX, Liang HD, Deng RG, Qu LH, Zhang GP, Luo J (2015) Marek's disease virus-encoded analog of microRNA-155 activates the oncogene c-Myc by targeting LTBP1 and suppressing the TGF-beta signaling pathway. Virology 476:72–84
68. Heidari M, Fitzgerald SD, Zhang H (2014) Marek's disease virus-induced transient cecal tonsil atrophy. Avian Dis 58:262–270
69. Weller SK (2011) Alphaherpesviruses: molecular virology. Caister Academic, Norfolk
70. Gonzalez-Navajas JM, Lee J, David M, Raz E (2012) Immunomodulatory functions of type I interferons. Nat Rev Immunol 12:125–135
71. Fitzgerald-Bocarsly P, Dai J, Singh S (2008) Plasmacytoid dendritic cells and type I IFN: 50 years of convergent history. Cytokine Growth Factor Rev 19:3–19
72. Toth TE (2000) Nonspecific cellular defense of the avian respiratory system: a review. Dev Comp Immunol 24:121–139
73. Quere P, Rivas C, Ester K, Novak R, Ragland WL (2005) Abundance of IFN-alpha and IFN-gamma mRNA in blood of resistant and susceptible chickens infected with Marek's disease virus (MDV) or vaccinated with turkey herpesvirus; and MDV inhibition of subsequent induction of IFN gene transcription. Arch Virol 150:507–519

74. Karaca G, Anobile J, Downs D, Burnside J, Schmidt CJ (2004) Herpesvirus of turkeys: microarray analysis of host gene responses to infection. Virology 318:102–111
75. Morgan RW, Sofer L, Anderson AS, Bernberg EL, Cui J, Burnside J (2001) Induction of host gene expression following infection of chicken embryo fibroblasts with oncogenic Marek's disease virus. J Virol 75:533–539
76. Xing Z, Schat KA (2000) Inhibitory effects of nitric oxide and gamma interferon on in vitro and in vivo replication of Marek's disease virus. J Virol 74:3605–3612
77. Feng ZQ, Lian T, Huang Y, Zhu Q, Liu YP (2013) Expression pattern of genes of RLR-mediated antiviral pathway in different-breed chicken response to Marek's disease virus infection. Biomed Res Int 2013:419256
78. Jarosinski KW, Jia W, Sekellick MJ, Marcus PI, Schat KA (2001) Cellular responses in chickens treated with IFN-alpha orally or inoculated with recombinant Marek's disease virus expressing IFN-alpha. J Interferon Cytokine Res 21:287–296
79. Biron CA (2012) Yet another role for natural killer cells: cytotoxicity in immune regulation and viral persistence. Proc Natl Acad Sci USA 109:1814–1815
80. Koren S, Fleischmann WR Jr (1993) Orally administered interferons suppress bone marrow function. Proc Soc Exp Biol Med 204:155–164
81. De Regge N, Van Opdenbosch N, Nauwynck HJ, Efstathiou S, Favoreel HW (2010) Interferon alpha induces establishment of alphaherpesvirus latency in sensory neurons in vitro. PLoS One 5:e13076
82. Haq K, Elawadli I, Parvizi P, Mallick AI, Behboudi S, Sharif S (2011) Interferon-gamma influences immunity elicited by vaccines against very virulent Marek's disease virus. Antiviral Res 90:218–226
83. Powell PC, Hartley KJ, Mustill BM, Rennie M (1983) Studies on the role of macrophages in Marek's disease of the chicken. J Reticuloendothel Soc 34:289–297
84. Kodama H, Mikami T, Inoue M, Izawa H (1979) Inhibitory effects of macrophages against Marek's disease virus plaque formation in chicken kidney cell cultures. J Natl Cancer Inst 63:1267–1271
85. Gupta MK, Chauhan HV, Jha GJ, Singh KK (1989) The role of the reticuloendothelial system in the immunopathology of Marek's disease. Vet Microbiol 20:223–234
86. Djeraba A, Bernardet N, Dambrine G, Quéré P (2000) Nitric oxide inhibits Marek's disease virus replication but is not the single decisive factor in interferon-gamma-mediated viral inhibition. Virology 277:58–65
87. Jarosinski KW, Njaa BL, O'Connell PH, Schat KA (2005) Pro-inflammatory responses in chicken spleen and brain tissues after infection with very virulent plus Marek's disease virus. Viral Immunol 18:148–161
88. Djeraba A, Musset E, Bernardet N, Le Vern Y, Quéré P (2002) Similar pattern of iNOS expression, NO production and cytokine response in genetic and vaccination-acquired resistance to Marek's disease. Vet Immunol Immunopathol 85:63–75
89. Qureshi MA, Miller L (1991) Signal requirements for the acquisition of tumoricidal competence by chicken peritoneal macrophages. Poult Sci 70:530–538
90. Ostuni R, Kratochvill F, Murray PJ, Natoli G (2015) Macrophages and cancer: from mechanisms to therapeutic implications. Trends Immunol 36:229–239
91. Lee LF, Sharma JM, Nazerian K, Witter RL (1978) Suppression of mitogen-induced proliferation of normal spleen cells by macrophages from chickens inoculated with Marek's disease virus. J Immunol 120:1554–1559
92. Garcia-Camacho L, Schat KA, Brooks R, Bounous D (2003) Early cell-mediated immune responses to Marek's disease virus in two chicken lines with defined major histocompatibility complex antigens. Vet Immunol Immunopathol 95:145–153
93. Sarson AJ, Abdul-Careem MF, Read LR, Brisbin JT, Sharif S (2008) Expression of cytotoxicity-associated genes in Marek's disease virus-infected chickens. Viral Immunol 21:267–272
94. Queré P, Dambrine G (1988) Development of anti-tumoral cell-mediated cyto-toxicity during the course of Marek's disease in chickens. Ann Rech Vet 19:193–201
95. Neulen ML, Viertlboeck BC, Straub C, Gobel TW (2015) Identification of novel chicken CD4(+) CD3(−) blood population with NK cell like features. Dev Comp Immunol 49:72–78

96. Churchil AE, Payne LN, Chubb RC (1969) Immunization against Marek's disease using a live attenuated virus. Nature 221:744–747
97. Schat KA, Markowski-Grimsrud CJ (2001) Immune responses to Marek's disease virus infection. Curr Top Microbiol Immunol 255:91–120
98. Edwards KM (2015) Maternal antibodies and infant immune responses to vaccines. Vaccine 33:6469–6472
99. Endsley JJ, Roth JA, Ridpath J, Neill J (2003) Maternal antibody blocks humoral but not T cell responses to BVDV. Biologicals 31:123–125
100. Siegrist CA, Barrios C, Martinez X, Brandt C, Berney M, Cordova M, Kovarik J, Lambert PH (1998) Influence of maternal antibodies on vaccine responses: inhibition of antibody but not T cell responses allows successful early prime-boost strategies in mice. Eur J Immunol 28:4138–4148
101. Niewiesk S (2014) Maternal antibodies: clinical significance, mechanism of interference with immune responses, and possible vaccination strategies. Front Immunol 5:446
102. Bandrick M, Theis K, Molitor TW (2014) Maternal immunity enhances Mycoplasma hyopneumoniae vaccination induced cell-mediated immune responses in piglets. BMC Vet Res 10:124
103. Gans H, DeHovitz R, Forghani B, Beeler J, Maldonado Y, Arvin AM (2003) Measles and mumps vaccination as a model to investigate the developing immune system: passive and active immunity during the first year of life. Vaccine 21:3398–3405
104. White DW, Suzanne Beard R, Barton ES (2012) Immune modulation during latent herpesvirus infection. Immunol Rev 245:189–208
105. Hanley PJ, Bollard CM (2014) Controlling cytomegalovirus: helping the immune system take the lead. Viruses 6:2242–2258
106. Sharma JM, Witter RL (1975) The effect of B-cell immunosuppression on age-related resistance of chickens to Marek's disease. Cancer Res 35:711–717
107. Markowski-Grimsrud CJ, Schat KA (2002) Cytotoxic T lymphocyte responses to Marek's disease herpesvirus-encoded glycoproteins. Vet Immunol Immunopathol 90:133–144
108. Gupta SK, Kharole MU, Kalra DS (1982) Role of thymus-dependent immune system in HVT protection against Marek's disease. Avian Dis 26:7–13
109. Purchase HG, Sharma JM (1974) Amelioration of Marek's disease and absence of vaccine protection in immunologically deficient chickens. Nature 248:419–421
110. Morimura T, Ohashi K, Sugimoto C, Onuma M (1998) Pathogenesis of Marek's disease (MD) and possible mechanisms of immunity induced by MD vaccine. J Vet Med Sci 60:1–8
111. Dalgaard T, Boving MK, Handberg K, Jensen KH, Norup LR, Juul-Madsen HR (2009) MHC expression on spleen lymphocyte subsets in genetically resistant and susceptible chickens infected with Marek's disease virus. Viral Immunol 22:321–327
112. Nazerian K, Lee LF, Yanagida N, Ogawa R (1992) Protection against Marek's disease by a fowlpox virus recombinant expressing the glycoprotein B of Marek's disease virus. J Virol 66:1409–1413
113. Nazerian K, Witter RL, Lee LF, Yanagida N (1996) Protection and synergism by recombinant fowl pox vaccines expressing genes from Marek's disease virus. Avian Dis 40:368–376
114. Omar AR, Schat KA, Lee LF, Hunt HD (1998) Cytotoxic T lymphocyte response in chickens immunized with a recombinant fowlpox virus expressing Marek's disease herpesvirus glycoprotein B. Vet Immunol Immunopathol 62:73–82
115. Murthy KK, Calnek BW (1979) Marek's disease tumor-associated surface-antigen (matsa) in resistant versus susceptible chickens. Avian Dis 23:831–837
116. McColl KA, Calnek BW, Harris WV, Schat KA, Lee LF (1987) Expression of a putative tumor-associated surface antigen on normal versus Marek's disease virus-transformed lymphocytes. J Natl Cancer Inst 79:991–1000
117. Podack ER, Strbo N, Sotosec V, Muta H (2002) CD30-governor of memory T cells? Ann N Y Acad Sci 975:101–113
118. Sun X, Yamada H, Shibata K, Muta H, Tani K, Podack ER, Iwakura Y, Yoshikai Y (2010) CD30 ligand is a target for a novel biological therapy against colitis associated with Th17 responses. J Immunol 185:7671–7680

119. Parvizi P, Andrzejewski K, Read LR, Behboudi S, Sharif S (2010) Expression profiling of genes associated with regulatory functions of T-cell subsets in Marek's disease virus-infected chickens. Avian Pathol 39:367–373
120. Matsuyama-Kato A, Murata S, Isezaki M, Kano R, Takasaki S, Ichii O, Konnai S, Ohashi K (2012) Molecular characterization of immunoinhibitory factors PD-1/PD-L1 in chickens infected with Marek's disease virus. Virol J 9:94
121. Schub D, Janssen E, Leyking S, Sester U, Assmann G, Hennes P, Smola S, Vogt T, Rohrer T, Sester M, Schmidt T (2015) Altered phenotype and functionality of varicella zoster virus-specific cellular immunity in individuals with active infection. J Infect Dis 211:600–612
122. Parvizi P, Read LR, Abdul-Careem MF, Sarson AJ, Lusty C, Lambourne M, Thanthrige-Don N, Burgess SC, Sharif S (2009) Cytokine gene expression in splenic CD4+ and CD8+ T cell subsets of genetically resistant and susceptible chickens infected with Marek's disease virus. Vet Immunol Immunopathol 132:209–217
123. Gimeno IM (2008) Marek's disease vaccines: a solution for today but a worry for tomorrow? Vaccine 26(Suppl 3):C31–41
124. Wu C, Gan J, Jin Q, Chen C, Liang P, Wu Y, Liu X, Ma L, Davison F (2009) Revaccination with Marek's disease vaccines induces productive infection and superior immunity. Clin Vaccine Immunol 16:184–193
125. Rouse BT, Kaistha SD (2006) A tale of 2 alpha-herpesviruses: lessons for vaccinologists. Clin Infect Dis 42:810–817
126. Levy AM, Heller ED, Leitner G, Davidson I (1999) Effect of native chicken interferon on MDV replication. Acta Virol 43:121–127
127. Yang Q, Wei P, Chen H (2011) Cytokine responses and inducible nitrous oxide synthase expression patterns in neonatal chicken brain microglia infected with very virulent Marek's disease virus strain YL040920. Vet Immunol Immunopathol 142:14–24
128. Bacon LD, Hunt HD, Cheng HH (2001) Genetic resistance to Marek's disease. Curr Top Microbiol Immunol 255:121–141
129. Witter RL (2001) Protective efficacy of Marek's disease vaccines. Curr Top Microbiol Immunol 255:57–90
130. Djeraba A, Kut E, Rasschaert D, Quéré P (2002) Antiviral and antitumoral effects of recombinant chicken myelomonocytic growth factor in virally induced lymphoma. Int Immunopharmacol 2:1557–1566
131. Gobel TW, Schneider K, Schaerer B, Mejri I, Puehler F, Weigend S, Staeheli P, Kaspers B (2003) IL-18 stimulates the proliferation and IFN-gamma release of CD4+ T cells in the chicken: conservation of a Th1-like system in a nonmammalian species. J Immunol 171:1809–1815
132. Parvizi P, Mallick AI, Haq K, Haghighi HR, Orouji S, Thanthrige-Don N, St Paul M, Brisbin JT, Read LR, Behboudi S, Sharif S (2012) A toll-like receptor 3 ligand enhances protective effects of vaccination against Marek's disease virus and hinders tumor development in chickens. Viral Immunol 25:394–401
133. Parvizi P, Abdul-Careem MF, Mallick AI, Haq K, Haghighi HR, Orouji S, Heidari M, Behboudi S, Sharif S (2014) The effects of administration of ligands for Toll-like receptor 4 and 21 against Marek's disease in chickens. Vaccine 32:1932–1938
134. Haunshi S, Cheng HH (2014) Differential expression of Toll-like receptor pathway genes in chicken embryo fibroblasts from chickens resistant and susceptible to Marek's disease. Poult Sci 93:550–555
135. Kamil JP, Tischer BK, Trapp S, Nair VK, Osterrieder N, Kung HJ (2005) vLIP, a viral lipase homologue, is a virulence factor of Marek's disease virus. J Virol 79:6984–6996
136. Schippers T, Jarosinski K, Osterrieder N (2015) The ORF012 gene of Marek's disease virus type 1 produces a spliced transcript and encodes a novel nuclear phosphoprotein essential for virus growth. J Virol 89:1348–1363
137. Lee LF, Heidari M, Sun A, Zhang H, Lupiani B, Reddy S (2013) Identification and in vitro characterization of a Marek's disease virus-encoded Ribonucleotide reductase. Avian Dis 57:178–187

138. Schumacher D, Tischer BK, Trapp S, Osterrieder N (2005) The protein encoded by the US3 orthologue of Marek's disease virus is required for efficient de-envelopment of perinuclear virions and involved in actin stress fiber breakdown. J Virol 79:3987–3997
139. Chbab N, Egerer A, Veiga I, Jarosinski KW, Osterrieder N (2010) Viral control of vTR expression is critical for efficient formation and dissemination of lymphoma induced by Marek's disease virus (MDV). Vet Res 41:56
140. Tahiri-Alaoui A, Smith LP, Kgosana L, Petherbridge LJ, Nair V (2013) Identification of a neurovirulence factor from Marek's disease virus. Avian Dis 57:387–394
141. Trapp-Fragnet L, Bencherit D, Chabanne-Vautherot D, Le Vern Y, Remy S, Boutet-Robinet E, Mirey G, Vautherot JF, Denesvre C (2014) Cell cycle modulation by Marek's disease virus: the tegument protein VP22 triggers S-phase arrest and DNA damage in proliferating cells. PLoS One 9:e100004
142. Tischer BK, Schumacher D, Chabanne-Vautherot D, Zelnik V, Vautherot JF, Osterrieder N (2005) High-level expression of Marek's disease virus glycoprotein C is detrimental to virus growth in vitro. J Virol 79:5889–5899
143. Chbab N, Chabanne-Vautherot D, Francineau A, Osterrieder N, Denesvre C, Vautherot JF (2009) The Marek's disease virus (MDV) protein encoded by the UL17 ortholog is essential for virus growth. Vet Res 40:28

CHAPTERT 3

Fish Cytokines and Immune Response

Sebastián Reyes-Cerpa, Kevin Maisey, Felipe Reyes-López, Daniela Toro-Ascuy, Ana María Sandino and Mónica Imarai

[1] *Dept. Biology, Laboratory of Virology, University of Santiago of Chile, Santiago, Chile*

[2] *Dept. Biology, Laboratory of Immunology, University of Santiago of Chile, Santiago, Chile*

[3] *Dept. Cell Biology, Physiology & Immunology, Unit of Animal Physiology, Autonomous University of Barcelona, Barcelona, Spain*

[4] *Dept. Biology, Laboratory of Virology, University of Santiago of Chile, Santiago, Chile; Activaq S.A., Santiago, Chile*

1. INTRODUCTION

The immune system can be defined as a complex system that protects the organism against organisms or substances that might cause infection or disease. One of the most fascinating characteristics of the immune system is its capability to recognize and respond to pathogens with significant specificity. Innate and adaptive immune responses are able to recognize foreign structures and trigger different molecular and cellular mechanisms for antigen elimination. The immune response is critical to all individuals; therefore numerous changes have taken place during evolution to generate variability and specialization, although the immune system has conserved some important features over millions of years of evolution that are common for all species. The emergence of new taxonomic categories coincided with the diversification of the immune response. Most notably, the emergence of vertebrates coincided with the development of a novel type of immune response. Apparently, vertebrates inherited innate immunity from their invertebrate ancestors [1].

In higher vertebrates, the immune system consists of primary and secondary lymphoid organs with distinct compartments and morphology located in anatomically distinct sites. The thymus and bone marrow constitute the primary lymphoid organs, while the spleen, lymph nodes, and mucosal associated lymphoid tissue (MALT) comprise the secondary lymphoid organs [2].

Fish are a heterogeneous group divided into three classes: Agnatha (jawless fish such as the hagfish and lampreys), Chrondrichthyes (cartilaginous fish such as sharks, rays and skates) and Osteichthyes (bony fish) [3]. As in all vertebrates, fish have cellular and humoral immune responses and organs, the main function of which is immune defence. Most generative and secondary lymphoid organs in mammals are also found in fish, except for lymphatic nodules and bone marrow [3].

The head kidney or pronephros has hematopoietic functions [3, 4], and unlike in higher vertebrates, it is the immune organ involved in phagocytosis [5], antigen processing, production of IgM [6, 7] and immune memory through melanomacrophagic centres [8, 9]. The thymus, another lymphoid organ situated near the opercular cavity in teleosts, produces T lymphocytes involved in allograft rejection, stimulation of phagocytosis and antibody production by B cells [10, 11]. The spleen is a large, blood-filtering organ that undergoes increasing structural complexity in order to augment its efficiency in trapping and processing antigens [12-15]. Melanomacrophage centres are present for clearance of ingested material and can be surrounded by immunoglobulin-positive cells, especially after immunization [8]. Proliferation of granular cells has also been observed in association with ellipsoids and melanomacrophage centres after immunization [16].

1.1. Innate and adaptive immune response

The development of an immune system is essential for the survival of living organisms. In vertebrates, immunity can be divided into two components, the innate immune response and the adaptive immune response. The innate immune response is the initial line of defence against infection, which includes physical barriers and cellular response. The adaptive immune response is capable of specific antigen recognition and is responsible for the secondary immune response.

The innate immune system recognizes conserved molecular structures common to pathogenic microorganisms such as polysaccharides, lipopolysaccharides (LPS), peptidoglycans, bacterial DNA, and double-strand viral RNA, among others, through their interaction with specific receptors like toll receptors (TLRs). These mechanisms of recognition may lead directly to successful removal of pathogens, for instance by phagocytosis, or may trigger additional protective responses through induction of adaptive immune responses [17]. Cells of the innate immune system have a diverse array of functions. Some cells are phagocytic, allowing them to engulf and degrade pathogenic particles. Other cells produce and secrete cytokines and chemokines that can stimulate and help guide the migration of cells and further direct the immune response [18].

The adaptive system recognizes foreign structures by means of two cellular receptors, the B cell receptor (BCR) and the T cell receptor (TCR). Adaptive immunity is highly regulated by several mechanisms. It increases with antigen exposure and produces immunological memory, which is the basis of vaccine development and the preventive function of vaccines [19, 20]. The adaptive response generally starts days after infection and is capable of recognizing specific protein motifs of peptides, which leads to a response that increases in both speed and magnitude with each successive exposure [21]. The main effector cells of the adaptive immune response are the lymphocytes, specifically B cells and T cells. When B cells are activated, they are capable of differentiating into plasma cells that can secrete antibodies. Upon activation T cells differentiate into either helper T cells or cytotoxic T cells. Helper T cells are capable of activating other cells of the adaptive immune response such as B cells and macrophages, while cytotoxic T cells upon activiation are able to kill cells that have been infected [22].

1.2. Fish immune response

Immune responses in fish have not been as well characterized as they have in higher vertebrates. Consequently, there is not enough information about the components of the fish immune system and its function and regulation. Key immune mammalian homologous genes have been identified in several fish species, suggesting that the fish immune system shares many features with the mammalian system. For example, the identification of α and β T cell receptor genes (TCR) [23], key T cell markers such as CD3, CD4, CD8, CD28, CD40L, and a great number of cytokines and chemokines [24-26] suggest that T helper (Th)1, Th2 and Th17 and the regulatory subset Treg are present in fish. Some cell subsets have been better studied mainly because their activity can be easily differentiated and measured, as in the case of cytotoxic cells [27] and macrophages [28, 29]. Finally, B cells have been much more studied due to the availability of monoclonal antibodies that have been isolated and identified by a number of techniques [30, 31]. Phenotypic characterization of leukocytes has been hampered mainly by the lack of membrane cell markers [32, 33]. Researchers anticipate developing antibodies for cell lineage markers of fish immunocompetent cells that can be used to isolate and characterize immune cells to obtain insights into their regulation and role in immune response [34-36].

Antibodies in teleosts play a key role in the immune response. In general, IgM is the main immunoglobulin in teleosts that can elicit effective specific humoral responses against various antigens. For IgM, one gene alone can generate as many as six structural isoforms. Therefore, diversity is the result of structural organization rather than genetic variability [37]. Recently, several reports have provided evidence for the existence of IgD/IgZ/IgT in fish [38-41]. Interestingly, B cells from rainbow trout and salmon have high phagocytic capacity, suggesting a transition in B lymphocyte during evolution in which a key cell type of the innate immunity and phagocytosis evolved into a highly specialized component of the adaptive immune response in higher vertebrates [42, 43].

1.3. Fish cytokines

Cytokines are secreted proteins with growth, differentiation, and activation functions that regulate the nature of immune responses. Cytokines are involved in several steps of the immune response, from induction of the innate response to the generation of cytotoxic T cells and the production of antibodies. In higher vertebrates, the combination of cytokines that are secreted in response to an immune stimulation induces the expression of immune-related genes through multiple signalling pathways, which contributes to the initiation of the immune response. Cytokines can modulate immune responses through an autocrine or paracrine manner upon binding to their corresponding receptors [44].

Cytokines have overlapping and sometimes contradictory pleiotropic functions that make their classification difficult. Cytokines are produced by macrophages, lymphocytes, granulocytes, DCs, mast cells, and epithelial cells, and can be divided into interferons (IFNs), interleukins (ILs), tumor necrosis factors (TNFs), colony stimulating factors, and chemokines [45]. They are secreted by activated immune-related cells upon induction by various pathogens, such as parasitic, bacterial, or viral components [46]. Macrophages can secret IL-1, IL-6, IL-12, TNFα, and chemokines such as IL-8 and MCP-1, all of which are indispensable for macrophage, neutrophil, and lymphocyte recruitment to the infected tissues and their activation as pathogen eliminators [47]. Meanwhile, cytokines released by phagocytes in tissues can also induce acute phase proteins, including mannose-binding lectin (MBL) and C-reactive protein (CRP), and promote migration of DCs [48].

Fish appear to possess a repertoire of cytokines similar to those of mammals. To date several cytokine homologues and suppressors have been cloned in fish species [24, 25, 49]. Some cytokines described in fish are TNFα, IL-1β, IL-6 or IFN.

Current knowledge of fish cytokines is based on mammal models of the cytokines network and their complex interactions. In this review we included the pro-inflammatory cytokines associated with innate and adaptive immunity, regulatory cytokines and anti-inflammatory cytokines.

1.4. Pro-inflammatory fish cytokines

1.4.1. Tumour necrosis factor α (tnfα)

TNFα (tumour necrosis factor alpha) is a pro-inflammatory cytokine that plays an important role in diverse host responses, including cell proliferation, differentiation, necrosis, apoptosis, and the induction of other cytokines. TNFα can induce either NF-kB mediated survival or apoptosis, depending on the cellular context [50]. TNFα mediates powerful anti-microbial responses, including inducing apoptosis, killing infected cells, inhibiting intracellular pathogen replication, and up-regulating diverse host response genes. Many viruses have evolved strategies to neutralize TNFα by direct binding and inhibition of the ligand or its receptor or modulation of various downstream signalling events [51].

TNFα has been identified, cloned, and characterized in several bony fish, including Japanese flounder [52], rainbow trout [53, 54], gilthead seabream [55], carp [56] catfish [57], tilapia [58], turbot [59] and goldfish [60]. These studies have revealed the existence of some obvious differences from their mammalian counterpart, such as the presence of multiple isoforms of TNFα in some teleost species [54, 56] the high constitutive expression of this gene in different tissues of healthy fish and its relatively poor up-regulation by immune challenge *in vitro* and *in vivo* [53, 55, 57]. However, the most unexpected and interesting difference between fish and mammal TNFα concerns the weak *in vitro* effects of TNFα on phagocyte activation in goldfish [60], rainbow trout [57], turbot [59] and gilthead seabram [61]. This weak *in vitro* activity of fish TNFα sharply contrasts with the powerful actions exerted by the i.p. injection of recombinant TNFα in gilthead seabream, which includes the recruitment of phagocytes to the injection site, with a concomitant strong increase in their respiratory burst [61]. Apparently endothelial cells are the main target cells of fish TNFα, suggesting that TNFα is mainly involved in the recruitment of leukocytes to the inflammatory foci rather than in their activation [62]. Despite the above, differential expression has been observed in studies with rainbow trout leucocytes, which have shown increased response to different pro-inflammatory stimuli, as human recombinant TNFα [63], LPS [53, 64], zimosan and muramyl dipeptide as a peptidoglycan constituent of both gram-positive and gram-negative bacteria [64]. Moreover, it is known that

Infectious Pancreatic Necrosis Virus (IPNV)-mediated up-regulation of TNFα regulates both the Bad/Bid-mediated apoptotic pathway and the RIP1 (receptor-interacting protein-1)/ROS-mediated secondary necrosis pathway [65].

1.4.2. Interleukin 1 family

In mammals, the 11 members of the Interleukin-1 family include IL-1α (IL-1F1), IL-1β (IL-1F2), IL-1 receptor antagonist (IL-1ra/IL-1F3), IL-18 (IL-1F4), IL-1F5-10 and IL-33 (IL-1F11). These molecules tend to be either pro-inflammatory or act as antagonists that inhibit the activities of particular family members [66]. Despite these semantic issues, to date only two clear homologues of these molecules have been discovered in fish, IL-1β and IL-18 [24].

1.4.2.1. Interleukin 1β

IL-1β is one of the earliest expressed pro-inflammatory cytokines and enables organisms to respond promptly to infection by inducing a cascade of reactions leading to inflammation. Many of the effector roles of IL-1β are mediated through the up- or down-regulation of expression of other cytokines and chemokines [67]. Mammalian IL-1β is produced by a wide variety of cells, but mainly by blood monocytes and tissue macrophages. IL-1β was the first interleukin to be characterized in fish and has since been identified in a number of fish species, such as rainbow trout [68], carp [69], sea bass [70], gilt head seabram [71], haddock [72], tilapia [73]. A second IL-1β gene (IL-1beta2] has been identified in trout [74]

In mammals pro-IL-1β remains cytosolic and requires cellular proteases to release the mature peptide. It is known that the peptide is cleaved by the IL-1β converting enzyme (ICE)[75]. However, the aspartic acid residue for which this enzyme has specificity is not present in all fish genes sequenced to date. Nevertheless, using a combination of multiple alignments and analysis of the N-terminal sequences of known mature peptides, it is possible to predict fish gene cutting sites. In trout, this gives a mature peptide of 166 and 165 aminoacids for IL-1β1 and IL-1β2 [76].

Like its mammalian counterpart, teleost IL-1β has been found to be regulated in response to various stimuli, such as LPS or poly I:C [68, 70-74, 77-81]. The biological activity of recombinant IL-1β (rIL-1β) has been studied in several fish species, indicating that fish IL-1β is involved in the regulation of immune relevant genes, lymphocyte activation, migration of leucocytes, phagocytosis and bactericidal activities [77, 81-84].

1.4.2.2. Interleukin 18

In mammals, IL-18 is mainly produced by activated macrophages. It is an important cytokine with multiple functions in innate and acquired immunity [85-87]. One of its primary biological properties is to induce interferon gamma (IFNγ) synthesis in Th1 and NK cells in synergy with IL-12 [88, 89]. It promotes T and NK cell maturation, activates neutrophils and enhances Fas ligand-mediated cytotoxicity [90-92]. Like IL-1β, it is synthesized as an inactive precursor of approximately 24 kDa and is stored intracellularly. Activation and secretion of IL-18 is mainly effected through specific cleavage of the precursor after D35 by caspase 1, also termed the IL-1β-converting enzyme (ICE), which is believed to be one of the key processes regulating IL-18 bioactivity [93, 94]. Some other enzymes, including caspase 3 and neutrophil proteinase 3, also cleave the IL-18 precursor to generate active or inactive mature molecules [95, 96].

IL-18 was discovered in fish by analysis of sequenced fish genomes (fugu) and EST databases (medaka) [97, 98]. An alternative splicing form of the IL-18 mRNA was discovered in trout that may have an important role in regulating IL-18 expression and processing in this species. This form shows a lower constitutive expression relative to the full length transcript, but unlike the full length transcript, it increases in response to LPS and polyI:C stimulation in the RTG-2 fibroblast cell line [98]. The expression level of the full length transcript can increase in response to LPS plus IL- 1b in head kidney leucocyte cultures, and by IFNγ in RTS-11 cells [99].

1.4.3. Other pro-inflammatory cytokines

1.4.3.1. Interleukin 6

A number of other interleukins are considered pro-inflammatory, some of which are released during the cytokine cascade that follows bacterial infection. Of these IL-6 is one of the best known, and is itself a member of the IL-6 family of cytokines that includes IL-11 and IL-31, as well as cytokines such as mammalian CNTF, LIF, OSM, CT-1 and CT-2 [24]. Whilst the homology of known fish molecules with many of these IL-6 family members is not conclusive [100], true homologues appear to be present in at least in the cases of IL-6 and IL-11 [24]. IL-6 is produced by a diverse group of cells including T lymphocytes, macrophages, fibroblasts, neurons, endothelial and glial cells. The pleiotropic effects of IL-6 are mediated by a 2-subunit receptor [101] and include the regulation of diverse immune and neuro-endocrine processes. IL-6 has been implicated in the control of immunoglobulin production, lymphocyte and monocyte differentiation, chemokine secretion and migration of leukocytes to inflammation sites [102-104].

IL-6 was first discovered in fugu by analysis of the genome sequence [105] and subsequently in other species as part of EST analysis of immune gene-enriched cDNA libraries [106-108]. However, little is known about the function and signalling pathways of IL-6 in fish. Interestingly, trout IL-6 expression in macrophages is reported to be induced by LPS, poly I:C and IL-1β in the macrophage cell line RTS-11, as well as in head kidney macrophages [109]. Moreover, IL-6 induces the expression of itself, so it can act in an autocrine and paracrine fashion to increase its expression, with the potential to both amplify and exacerbate the inflammatory response. However, IL-6 can significantly down-regulate the expression of trout TNFα1, TNFα2, and IL-1β, suggesting a potential role of trout IL-6 in limiting host damage during inflammation [109].

1.4.3.2. Interleukin 11

In mammals, IL-11 is produced by many cell types throughout the body. Basal and inducible IL-11 mRNA expression can be detected in fibroblasts, epithelial cells, chondrocytes, synoviocytes, keratinocytes, endothelial cells, osteoblasts and certain tumour cells and cell lines [110]. Viral [111] and bacterial infection [112] and cytokine stimulation (IL-1, TNFα and TGF-β1] induce IL-11 expression. IL-11 acts on multiple cell types, including hemotopoietic cells, hepatocytes, adipocytes, intestinal epithelial cells, tumour cells, macrophages, and both osteoblasts and osteoclasts. In the hematopoietic compartment IL-11 supports multilineage and committed progenitors, contributing to myeloid, erythroid, megakaryocyte and lymphoid lineages [113]. IL-11 is also an anti-inflammatory cytokine that inhibits the production of pro-inflammatory cytokines from lipopolysaccharde (LPS)-stimulated macrophages [114]. In combination with its trophic effects on the gastrointestinal epithelium, IL-11 plays an important role in the protection and restoration of gastrointestinal mucosa [115, 116].

The teleostean IL-11 orthologue has been found to consist of duplicate IL-11 genes, named IL-11a and IL-11b [117], with expression patterns indicating that both divergent forms of teleostean IL-11 play roles in antibacterial and antiviral defence mechanisms of fish [117-119]. In trout, IL-11 molecule is grouped with IL-11a and is constitutively expressed in intestine and gills and is highly up-regulated at other immune sites (spleen, head kidney, liver) following bacterial infection. *In vitro*, the macrophage-like RTS-11 cell line has shown enhanced IL-11 expression in response to LPS, bacteria, poly I:C and rIL-1β [118]. In carp, IL-11a is modulated by LPS, ConA and peptidoglycan in head kidney macrophages [117, 120] and cortisol has been found to inhibit IL-11 expression on its own and in combination with LPS [117]. In contrast to carp IL-11a, which shows low levels of constitutive expression in blood leucocytes, IL-11b in Japanese flounder shows higher expression at this site, and strong up-regulation was found in response to rhabdovirus infection in kidney cells [119]. This suggests that these paralogues have some complementarity of function related to their differential expression, although study of both forms in a single experiment is still required [24].

1.5. Chemokines

Chemokines are a superfamily of approximately 40 different small secreted cytokines that direct the migration of immune cells to infection sites. Their activity is coordinated by binding to G-protein-linked receptors with seven transmembrane domains. Four distinct subgroups make up the chemokine superfamily. These are designated as CXC (or a), CC (or b), C (or g) and CX_3C (or d), which are defined by the arrangement of the first two cysteine residues within their peptide structure. The CC subfamily can be further subdivided according to the total number of cysteine residues, as some members of this group contain four cysteines whilst the remainder possesses six (and are known as the C6-b group). Similarly, the CXC subfamily contains two subgroups based on whether or not the first two cysteines are preceded by a Glu-Leu-Arg (ELR) motif associated with specificity to neutrophils [76, 121].

1.5.1. Interleukin 8

An important chemokine related to the pro-inflammatory process is CXCL-8, also called interleukin 8, this chemokine is a member of the CXC chemokine subfamily and attracts neutrophils, T lymphocytes and basophils *in vitro,* but not macrophages or monocytes [122]. Many cell-types, including macrophages, produce IL-8 in response to a variety of stimuli (LPS, cytokines and viruses). The neutrophil-attracting ability of IL-8 can be attributed to the presence of the ELR motif adjacent to the CXC motifs at its N-terminus, presumably by affecting its binding to specific receptors [123, 124]. In contrast, CXC chemokines lack an ELR motif and specifically attract lymphocytes but not neutrophils. The biological effects of IL-8 on neutrophils include increased cytosolic calcium levels, respiratory burst, a change in neutrophil shape and chemotaxis[125].

The fish IL-8 has been found in flounder [126], trout [125, 127], catfish [128], and lamprey [129]. *In vitro* stimulation of a trout macrophage cell line (RTS-11) [125] or *in vivo* intraperitoneal challenge [78] with either LPS or poly I:C did result in clear up-regulation of IL-8 expression. Moreover, induction of IL-8 expression in primary cultures of rainbow trout leukocytes stimulated for 24 hours with LPS and TNFα confirms that this fish chemokine is associated with inflammatory response, as has been suggested in mammals [127]. Interestingly, the ELR motif associated with the neutrophil-attracting ability is absent from the lamprey molecule and it is similar in flounder, where CXCL8 also lacks the ELR motif and appears to be regulated by a bacterial mechanism, since its transcript has only been detected in the major immune organs (spleen and head kidney) of an LPS stimulated flounder. The case of the trout is different, although there is also no ELR preceding the CXC motif, it has a very similar motif (DLR) in this position [130]. The human CXCL8 molecule, where the ELR motif has been mutated to DLR, retains neutrophil-attracting ability, albeit at lower potency [123]. Consequently, it is possible that the trout molecule has similar chemotactic activity to that of mammalian CXCL8 [130].

1.6. The interleukin 2 family

The IL-2 subfamily of cytokines signals via the common gamma chain (gC or CD132), a member of the type I cytokine receptor family expressed in most leucocytes.These cytokines in mammals include IL-2, IL-4, IL-7, IL-9, IL-15 and IL-21. IL-2, IL-4, IL-9 and IL-21 are all cytokines released from Th cells, which affect their responses [24], whilst IL-7 and IL-15 are particularly important for the maintenance of T cell memory [131]. To date molecules with homology to all of these have been found in fish, except IL-9 [24].

1.6.1. Interleukin 2

Interleukin-2 (IL-2 is an important immunomodulatory cytokine that primarily promotes proliferation, activation and differentiation of T cells [132]. IL-2, initially known as T-cell growth factor (TCGF), is synthesized and secreted mainly by Th1 cells that have been activated by stimulation by certain mitogens or by interaction of the T-cell receptor with the antigen/MHC complex on the surface of antigen-presenting cells [133-135]. Although CD4 T cells are the major source of IL-2 production in response to TCR stimulation, transient induction of IL-2 mRNA and production of the protein has been detected in murine dendritic cells activated by gram-negative bacteria [136]. IL-2 can also be produced by B cells in certain situations [137, 138]. The produced IL-2 promotes the expansion and survival of activated T cells and is also required for the activation of natural killer (NK) cells [139] and for immunoglobulin (Ig) synthesis by B cells [140].

The IL-2 gene has been detected only recently in fish by analysis of the fugu genome sequence, which also identified IL-21 as a neighbouring gene, as in mammals, providing the first direct evidence for the existence of a true IL-2 homologue in bony fish [141]. The gene has a 4 exon/3 intron organisation, as in mammals, and showed no constitutive expression in a range of tissues examined. However, injection of Fugu with poly I:C induced expression of IL-2 in the gut and gills [141]. Moreover, IL-2 could be induced in head kidney cell cultures stimulated with PHA, and in T-cell enriched cultures isolated from PBL when stimulated with B7-H3 or B7- H4 Ig fusions proteins in the presence of PHA [24, 142]. IL-2 has since been cloned in rainbow trout [143, 144]. The trout IL-2 was significantly up-regulated in head kidney leucocytes by the T cell mitogen PHA and in classical mixed leucocyte reactions and *in vivo* following infection with bacteria (*Y. ruckeri*) or the parasite *Tetracapsuloides bryosalmonae*. More importantly, the recombinant trout IL-2 produced in *Escherichia coli* was shown to induce expression of two transcription factors (STAT5 and Blimp-1) known to be involved in IL-2 signalling in mammals [143], as well as interferon-g (IFNγ) and IL-2 itself, and a CXC chemokine known to be induced by IFNγ, termed a IFNγ-inducible protein (γIP) [145].

1.6.2. Interleukin 4

Interleukin-4 IL-4 is a pleiotropic cytokine produced by T cells, mast cells, and basophils and is known to regulate an array of functions in B cells, T cells, macrophages, hematopoietic and non-hematopoietic cells [146, 147]. IL-4 serves as a key cytokine in driving Th2 differentiation and mediating humoral immunity, allergic responses and certain autoimmune diseases [148]. The IL-4 gene is conserved evolutionally in the animal kingdom and has been isolated from various animals including humans [149], mice [150, 151] and bovines [152], in which the IL-4 locus has been mapped in a region adjacent to those of IL-5 and IL-13 on the same chromosome [153, 154].

Teleost fish have two genes of the IL-4/13 family, IL-4/13A and IL-4/13B, which are situated on separate chromosomes in regions that duplicated during the fish-specific whole genome duplication (FS-WGD) around 350 million years ago [155, 156]. A few IL-4-like genes have been found in fish to date. The first was discovered by searching the *Tetraodon nigroviridis* genome [157]. In this work, IL-4 was constitutively expressed in head kidney, spleen, liver, brain, gill, muscle and heart. The ubiquitous expression of IL-4 is consistent with a postulated role in immune cytokines regulation. Stimulating the fish with a mixed stimulant containing ConA, PHA and PMA significantly up-regulated the expression of IL-4, which suggests that IL-4 is involved in the immune inflammatory responses triggered by mitogens [157], as in mammals, where it has been observed that this mitogen increases IL-4 expression [158]. However, the homology (amino acid identity) of this molecule was very low [12–15%), making it difficult to be sure it is an IL-4 homologue, although clearly related to Th2-type cytokines [24]. In fugu, T cell enriched PBL was found to express more IL-4/13A and IL-4/13B after stimulation with recombinant B7 molecules [142]. In zebrafish a recombinant IL4/13B was shown to increase the number of IgT-positive and CD209-positive cells in blood [159, 160], and in zebrafish spleen the expression of IL-4/13B and transcription factor related to Th2 immune response as GATA-3, and STAT6 was simultaneously enhanced after PHA stimulation [161]. The IL-4/13A gene was identified in trout and salmon [162], where the tissue distribution of salmonid IL-4/13A and GATA-3 expression were compared to the expression of IL-4, IL-13, and GATA-3 in mice. High levels of these transcripts were found in both salmonid and murine thymus, while constitutive IL-4/13A richness of skin and respiratory tissue was found in salmonids but not in mice. Experiments with isolated cells from gill and pronephros (head kidney) indicated that trout IL-4/13A is mainly expressed by surface IgM-negative cells, readily inducible by PHA but not by poly I:C, and regulated differently from the Th1 cytokine IFNγ gene. In mammals, IL-5 is also considered a Th-2 type cytokine and along with IL-3 and GM-CSF it signals through receptors with a common γ-chain (γC). None of these cytokines have been discovered in fish to date [24].

1.6.3. Interleukin 7

The cytokine IL-7 plays several important roles during lymphocyte development, survival, and homeostatic proliferation [163]. It is produced by many different stromal cell types, including epithelial cells of the thymus and the intestine [164-166]. There is only one report on IL-7 in fish, for the fugu molecule that was discovered using a gene synteny approach by searching with the mammalian IL-7 gene neighbours C8orf70 and PKIA. Fugu IL-7 shows constitutive expression in head kidney, spleen, liver, intestine, gill and muscle, with expression shown to increase in head kidney cultures stimulated with LPS, poly I:C or PHA [24, 167].

1.6.4. Interleukin 15

The central action of IL-15 cytokine is on T-cells, dendritic cells and NK cells. IL- 15 is an important regulator of the innate immune response to infection and autoimmune disease conditions. This gene shares activities with IL-2 and utilizes IL-2R β and γ units [45].

Two genes with homology to IL-15 have been discovered in fish. One shows similar gene organisation and synteny to mammalian and chicken IL-15, and has been termed IL-15. The second gene, which has a 4-exon structure and is in a different genome location, has been termed IL-15-like [168-170]. They show differential expression patterns in terms of the tissues where constitutive expression is apparent, and in terms of inducibility in PBL, with IL-15L being refractory to induction [168]. Two alternative splice variants of IL-15L (IL-15La and IL-15Lb) have also been described [170]. Trout IL-15, which has subsequently been cloned and sequenced, was strongly induced by rIFNγ in two trout cell lines (RTS-11 and RTG-2). rIL-15 could up-regulate IFNγ expression in splenic leucocytes, suggesting a positive feedback loop exists in fish between these two cytokines. Interestingly, unstimulated head kidney leucocytes were not responsive to rIL-15, at least in terms of the IFNγ expression level [171].

1.6.5. Interleukin 21

Interleukin 21 (IL-21) is a newly recognized member of IL-2 cytokine family that utilizes the common γ-chain receptor subunit for signal transduction [172-174]. In humans and other mammals, IL-21 is produced by both Th1 and Th2 cells [172, 175, 176]. IL-21 has pleiotropic effects on both innate and adaptive immune responses and can act on CD4+ and CD8+ T cells, B cells, NK cells, dendritic cells (DC), myeloid cells, and other tissue cells. IL-21 enhances the proliferation of anti–CD3-stimulated T cells and acts in concert with other γc cytokines to enhance the growth of CD4+ T cells [177]. IL-21–producing CD4+ T cells exhibit a stable phenotype of IL-21 production in the presence of IL-6 but retain the potential to produce IL-4 under Th2-polarizing conditions and IL-17A under Th17-polarizing conditions [178]. IL-21 stimulates CD8+ T cell proliferation and synergizes with IL-15 in promoting CD8+ T cell expansion in vitro and their antitumor effects in vivo [177, 179]. B cells that encounter IL-21 in the context of Ag-specific (BCR) stimulation and T cell co/stimulation undergo class-switch recombination and differentiate into Ab-producing plasma cells. In contrast, B cells encountering IL-21 during nonspecific TLR stimulation or without proper T cell help undergo apoptosis [180].

Since its discovery in fugu as a gene neighbour of IL-2 [141], IL-21 has been reported in tetraodon [181, 182] and rainbow trout [182]. Fugu IL-21 shows low constitutive expression. However, stimulation of isolated kidney leucocytes with PHA induced IL-21 expression. IL-21 was also up-regulated at mucosal sites as gill and gut when fish were injected with LPS or poly I:C [141]. Similarly, in tetraodon IL/21 expression is low but detectable in the gut, gonad and gills of healthy fish, and is induced in the kidney, spleen and skin following LPS injection [181]. In trout IL-21 expression is highest in gills and intestine, and is induced *in vivo* by bacterial (*Y. ruckeri*) and viral (VHSV) infection [182]. Relative to IL-2, induction of IL-21 expression in head kidney cells appears more rapidly but has shorter duration after stimulation. The trout rIL-21 has also been produced and shown to increase the expression of IL-10, IL-22 and IFNγ, and to a lesser extent IL-21, and to maintain the expression levels of key lymphocyte markers in primary cultures [182]. Thus, IL-21 may act as a survival factor for fish T and B cells [24].

1.7. The interleukin 10 family

Interleukin-IL-10 is an anti-inflammatory cytokine and a member of the class II cytokine family that also includes IL-19, IL-20, IL-22, IL-24, IL-26 and the interferons [183]. Although the predicted helical structure of these homodimeric molecules is conserved, certain receptor-binding residues are variable and define the interaction with specific heterodimers of different type-2 cytokine receptors. This leads to diverse biological effects through the activation of signal transducer and activator of transcription (STAT) factors [184].

1.7.1. Interleukin 10

Interleukin-10 (IL-10) was discovered initially as an inhibitory factor for the production of Th1 cytokines. Subsequently, pleiotropic inhibitory and stimulatory effects on various types of blood cells were described for IL-10, including its role as a survival and differentiation factor for B cells. IL-10, which is produced by activated monocytes, T cells and other cell types like keratinocytes, appears to be a crucial factor for at least some forms of peripheral tolerance and a major suppressor of the immune response and inflammation. The inhibitory function of IL-10 is mediated by the induction of regulatory T cells [185].

IL-10 was discovered in fish by searching the fugu genome. The translation showed 42–45% similarity to mammalian molecules with very low constitutive expression in tissues [186]. IL-10 has since been cloned in several other fish species including carp [187] zebrafish [188], rainbow trout [189], sea bass [190, 191] and cod [79]. Such studies have shown that IL-10 expression can be increased by LPS stimulation, by bacterial infection, by bath administration of immunostimulants [192] and by IPNV infection which may be associated with mechanisms of immune evasion [78].

1.7.2. Interleukin 20 (il-20like)

In mammals, IL-20 was discovered as a new member of the IL-10 family of cytokines. IL-20 shares the highest amino-acid sequence identity with IL-10, IL-24 and IL-19. It is secreted by immune cells and activated epithelial cells like keratinocytes. A high expression of the corresponding IL-20 receptor chains has been detected on epithelial cells. In terms of function, IL-20 might therefore mediate crosstalk between epithelial cells and tissue-infiltrating immune cells under inflammatory conditions [193].

In fish, the gene of IL-20 has been described in putterfish [183], zebrafish [194] and trout [195]. In the latter work, the IL-20 gene, called IL-20-like (IL-20L) has been described as having a high level of expression in immune related tissues and in the brain, suggesting an important role of the fish IL-20L molecule in both the immune and nervous systems. Although the exact cell types expressing IL-20L have yet to be defined, macrophages express IL-20L. Moreover, IL-20L expression in the macrophage cell line RTS-11 is modulated by pro-inflammatory cytokines, signalling pathway activators, microbial mimics and the immuno-suppressor dexamethasone. These data suggest that trout IL-20L plays an important role in the cytokine network. The increased expression of IL-20L was only detected at late stages (4–24 h) of LPS stimulation in RTS-11 cells and in spleen 24–72 h after infection with *Yersinia ruckeri*, which suggests that the increased expression of IL-20L by LPS and infection is via the rapid increase of pro-inflammatory cytokines (e.g., IL-1β) and other factors known to occur [195].

1.7.3. Interleukin 22/26

In mammals, interleukin-22 is secreted by Th17 cells [196], as well as by a subset of NK cells, designated as NK22 [197]; and even by some Th1 cells [198]. Studies have suggested there is a distinct Th22 cell lineage [199, 200]. Many of the same cytokines that induce differentiation and proliferation of IL-17-producing cells also lead to the secretion of IL-22 by Th17 cells, NK22 cells, and putative Th22 cells, including IL-6, IL-23, IL- 1β, TGF-β, and TNFα [201]. IL-17 and IL-22 are therefore frequently produced together in response to infections [202]. Interleukin-22 interacts with a heterodimeric receptor, IL-10R2/IL-22R1 [203], which is expressed on a variety of non-lymphoid cells, especially epithelial cells. Ligation of this receptor leads to both protective and detrimental effects. In synergy with IL-17, IL-22 induces pro-inflammatory cytokines in human bronchial epithelial cells against *Klebsiella pneumoniae* infection [204] and in colonic myofibroblasts [205]. Independently or in synergy with IL-17, IL-22 acts in defence against intestinal infection of mice with *Citrobacter rodentium* [206]. Moreover, IL-22 has been implicated in intestinal homeostasis keeping commensal bacteria contained in anatomical niches, which is key to our symbiotic relationship and normal intestinal physiology. However, the mechanisms that restrict colonization to specific niches are unclear. David Artis and colleagues have described a crucial role for IL-22-producing innate lymphoid cells (ILCs) in preventing lymphoid-resident commensal bacteria from escaping their niche and causing inflammation [207].

IL-26 can be produced by primary T cells, NK cells and T cell clones following stimulation with specific antigen or mitogenic lectins. IL-26 was initially shown by several groups to be co-expressed with IL-22 [208]. IL-26 is co-expressed with IFNγ and IL-22 by human Th1 clones, but not by Th2 clones. It was subsequently found that IL-26 is co-expressed with IL-17 and IL-22 by Th17 cells, an important subset of CD4+ T-helper cells that are distinct from Th1 and Th2 cells [209-211]. More recently, a novel subset of CD56+ NKp44+ NK cells was identified that co-expresses IL-22 and IL-26, especially following treatment with IL-23 [212]. Furthermore, a different subset of immature NK cells was described that do not express CD56 or NKp44 but do express CD117 and CD161 and constitutively express IL-22 and IL-26 [213].

The mechanisms that regulate transcription of the human IL-26 gene are so far largely undefined. It is possible and perhaps likely that expression of the IL-26 gene is induced in an IL-23-dependent manner because IL-23 is known to induce differentiation of Th17 cells, and IL-23 amplifies expression of IL-17 and IL-22 by Th17 cells [214].

In fish, the IFNγ locus was discovered using a gene synteny approach, and was first reported for fugu [215]. It contained a homologue of IL-22/26, that later studies of the zebrafish genome revealed to be two genes, one with clear homology to IL-22 and one with somewhat less clear homology to IL-26 [216]. The IL-22 gene was expressed constitutively in intestine and gills in all the treated and non-treated tissues. The gene was also expressed in kidney and spleen in LPS and PolyI:C-treated tissues, respectively, while IL-26 was expressed only in intestine treated with PolyI:C without expression [216]. IL-22 expression has been correlated with disease resistance in haddock vaccinated against *V. anguillarum,* with a strong constitutive expression in gills in vaccinated fish but not in control fish 24 hours post bath challenge, resulting in complete protection in fish vaccinated [217]. Moreover, IL-22, a cytokine released by Th-17 cells in mammals, is also interesting, and such responses are thought to be crucial for protection against extracellular microbes and at mucosal sites [218]. This coupled with the recent discovery of novel gill-associated immune tissue in fish [219] may provide a clue to a potential mechanism of resistance elicited by the *V. anguillarum* vaccination [24].

1.8. The interleukin 17 family

Interleukin-17 and a related family of genes are known to have pro-inflammatory actions and are associated with diseases [220]. After the discovery of the human IL- 17 gene [221], five cellular paralogs of IL-17 were identified, namely IL-17B, C, D, E and F [222-227]. These paralogs, identified by ESTs, genomics and proteomic databases, share identities of 20–50% with IL-17A gene. Human IL-17 A and F are present in tandem in opposite transcriptional orientation on the same chromosome 6p12, while IL-17B (Chr 5q24), IL-17C (Chr 16q24), IL-17D (Chr 13q11) and IL-17E (Chr 14q11) are dispersed. The structural similarities lead to the classification of IL-17 A, B, C, D, E, and F genes to a larger IL-17 sub-family [45]. Several IL-17 family members have been discovered in teleost fish, but homology to mammalian genes has not always been easy to assign. Two IL-17A or F homologue genes (IL-17A/F) have been found on the same chromosome. However, it has been difficult to determine which gene codes IL-17A and F. This gene in zebrafish was named IL-17A/F1 and 2. Furthermore, another IL-17A or F homologue gene (IL-17A/F3) has been found in zebrafish localized on a chromosome different from that of IL-17A/F1 and 2 [228]. In addition to those in zebrafish, IL-17A or F homologue genes have been found in rainbow trout [229], Atlantic salmon [230], pufferfish (IL-17A/F1, 2 and 3) [231], and medaka (IL-17A/F1, 2 and 3) [232].

The tissue distribution of the fugu IL-17 gene family also differs. In particular, IL-17 family genes are highly expressed in the head kidney and gills. Moreover, expression of IL-17 family genes is significantly up-regulated in the lipopolysaccharide-stimulated head kidney, suggesting that Fugu IL-17 family members are involved in inflammatory responses [231]. In Atlantic salmon IL-17D expression is widely distributed in tissues, with the highest levels of expression in testis, ovary and skin. Infection with *A. salmonicida* by injection increases IL-17D expression levels in the head kidney (but not the spleen) in a time-dependent manner. Skin and kidney showed an increased IL-17D expression level in fish given a cohabitation challenge with *A. salmonicida* [230]. The two trout IL-17C genes show some degree of differential expression within tissues, with IL-17C1 being more dominant in the gills and skin, whilst IL- 17C2 is more dominant in the spleen, head kidney and brain. Expression of both genes increases significantly with bacterial infection, although the increased expression of IL-17C2 is greater in terms of fold change. Similarly, both genes could be up-regulated in the trout RTS-11 cell line by LPS, poly I:C, calcium ionophore and rIL-1β, with IL-17C2 showing higher fold increases in all cases [229].

1.9. Interleukin 12

IL-12 is a heterodimeric cytokine composed of p35 and p40 subunits. It can mediate a number of different activities, including stimulation of IFNγ secretion from resting lymphocytes, NK cell stimulation and cytolytic T cell maturation. Perhaps most crucially, IL-12 also affects the progression of uncommitted T cells to either the Th1 lineage, which in general is characterized by secretion of lymphokines associated with cell-mediated rather than humoral immunity [233].

The p35 and p40 subunits were discovered in fish by analysis of the fugu genome [234]. The p35 locus is quite well conserved, with Schip1 being the immediate neighbour in all cases. This association has allowed p35 to be cloned by gene walking from Schip1 from fish species for which no genome sequence is available [24, 235]. The p40 subunit in fugu is constitutively expressed in all the tissues examined, except muscle, and no increases in expression were seen 3 h after injection with poly I:C or LPS. This constitutive and broad expression distribution of the p40 subunit suggests that it may be expressed in most cell types. The expression of the p35 subunit is more limited in its tissue expression and is induced after injection with poly I:C in the head kidney and the spleen, but not after injection with LPS. These results show that there are differences from the mammalian data in fugu IL-12 subunit expression. Further investigation will be required to show whether this is unique to fugu, if IL-12 is involved more in antiviral defence in fish and if the two subunits are regulated differently from their regulation in the mammalian system [234].

1.10. Transforming growth factor β (tgf-β)

TGF-β is a pleiotropic cytokine that regulates cell development, proliferation, differentiation, migration, and survival in various leukocyte lineages including lymphocytes, dendritic cells, NK cells, macrophages and granulocytes [236, 237]. In the mammalian immune system, TGF-β1 is a well-known suppressive cytokine and its dominant role is to maintain immune tolerance and suppress autoimmunity [238, 239]. The potent immunosuppressive effects of TGF-β1 are mediated predominantly through its multiple effects on T cells: TGF-β1 suppress Th1 and Th2 cell proliferation, while it promotes T regulatory cell generation by inducing Foxp3 expression. On the other hand, TGF-β also promotes immune responses by inducing the generation of Th17 cells [236, 240, 241]. Therefore, the regulatory roles of TGF-β as a positive or negative control device in immunity are widely acknowledged in mammals [238, 240, 241].

In teleost, despite the lack of extensive investigation on the functional role of TGF-β, some recent studies have revealed that TGF-β1 also exterts powerful immune depressing effects on activated leukocytes, as it does in mammals. For instance, TGF-β1 significantly blocks TNFα-induced activation of macrophage in goldfish and common carp, but induces the proliferation of the goldfish fibroblast cell line CCL71 [242, 243]. In grass carp, TGF-β1 down-regulates LPS/PHA-stimulated the proliferation of peripheral blood lymphocyte by contrast with the stimulatory effect of TGF-b1 alone in the same cells [244]. In red sea bream, similar phenomenon was observed during leukocyte migration under TGF-β1 treatment, with or without LPS challenges [245]. These findings not only define TGF-β1 as an immune regulator in teleost, but also indicate that TGF-β1 may have retained similar functions in immunity during the evolution of vertebrates [246].

1.11. Interferons

Interferons genes are involved in mediating cellular resistance against viral pathogens and modulating innate and adaptive immune systems. Broadly, IFNs are classified into two main groups called type I and type II [45]. Type I IFN includes the classical IFNα/β, which is induced by viruses in most cells, whereas type II IFN is only composed of a single gene called IFNγ and is produced by NK cells (NK cells) and T lymphocytes in response to interleukin-12 (IL-12), IL-18, mitogens or antigens [247]. Structurally both IFN types belong to the class II a-helical cytokine family, but have different 3-dimensional structures and bind to different receptors [248].

Two IFNs (IFNα1 and IFNα2) have been cloned from Atlantic salmon and characterized with respect to sequence, gene structure, promoter, antiviral activity and induction of ISGs [249-252]. Salmon IFNα1 induces both Mx and ISG15 proteins in TO cells and thus has properties similar to mammalian IFNα/β and IFNλ [251, 252]. Furthermore, salmon IFNα1 induces potent antiviral activity against the IPNV *in vitro* [251], but this protection has not been observed *in vivo,* despite a high level of expression of IFNα detected in spleen and head kidney of Atlantic salmon challenged intraperitoneally with IPNV [78].

At least three type I IFNs have been discovered in rainbow trout. The IFN1 (rtIFN1) and rtIFN2 show high sequence similar to Atlantic salmon IFNα1 and IFNα2, which contains two cysteines. On the other hand, rtIFN3 contains four cysteines, which further confirms the relationship between mammalian IFNα and fish IFNs. Recombinant rtIFN1 and rtIFN2 have both been shown to up-regulate expression of Mx and inhibit VHSV replication in RTG-2 cells. In contrast, recombinant rtIFN3 has been found to be a poor inducer of Mx and antiviral activity. Interestingly, the three rtIFNs show differential expression in cells and tissues [253]. This suggests that the three trout IFNs have different functions in the immune system of fish, which is an interesting subject for further research [254].

IFNγ has been identified in several fish species, including rainbow trout and Atlantic salmon [215, 216, 248, 255-257]. In contrast to the type I IFNs, fish and mammalian IFNγ are similar in exon/intron structure and display gene synteny. However, some fish species also possess a second IFNγ subtype named IFN gamma rel, which is quite different from the classical IFNγ [258]. Rainbow trout and carp IFNγ have several functional properties in common with mammalian IFNγ, including the ability to enhance respiratory burst activity, nitric oxide production, and phagocytosis of bacteria in macrophages [257-259]. Far less is known about the antiviral properties of fish IFNγ. However, it has been reported that it induces antiviral activity against both IPNV and the Salmon Alpha Virus (SAV) in salmon cell lines [260]

1.12. Tools for fish cytokine analysis

The major strategy of functional genomics is to identify the types of responses to specific pathogens based on cytokines expression as a predictor of profile immune response, which began by using suppressive subtractive hybridization as major tools at the beginning of the immunogenomics and upgrade to platforms of wide screening that allow identify thousands of EST's that are differentially regulated in their expression and that allow identifying potential candidates as biomarkers in the progression of the immune response at differential environmental conditions, not only against pathogens, but also in captivity stress conditions that affect the fisheries production.

1.13. Suppressive subtractive hybridization (ssh)

One of the most important biological processes in higher eukaryotes against external stimuli is the response mediated by differential gene expression. To understand the molecular regulation of these processes, the relevant subsets of differentially expressed genes of interest must be identified, cloned, and studied in detail using specific molecular techniques. In this matter, subtractive cDNA hybridization has been a powerful approach to identify and isolate cDNAs of differentially expressed genes [261-263]. Numerous cDNA subtraction methods have been reported. In general, they involve hybridization of cDNA from one population (tester) to excess of mRNA (cDNA) from another population (driver) and then separation of the unhybridized fraction (target) from the common hybridized sequences. One of these tools is a PCR-based technique called representational difference analysis, which does not require physical separation of single-strand (ss) and double-strand (ds) cDNAs. Representational difference analysis has been applied to enrich genomic fragments that differ in size or representation [264] and to clone differentially expressed cDNAs [265]. However, representational difference analysis has the problem of the wide differences in abundance of individual mRNA species so that multiple rounds of subtraction are needed [265]. Other strategies, such as mRNA differential display [266] and RNA fingerprinting by arbitrary primed PCR [267], are potentially faster methods for identifying differentially expressed genes, but both of these methods have high levels of false positives [268] that bias high-copy-number mRNA [269], which can inappropriate in experiments where only a few genes are expected to vary [268]. One of the techniques most often used to establish differential expression pattern between two conditions is suppression subtractive hybridization (SSH), which selectively amplifies target cDNA fragments (differentially expressed) and simultaneously suppresses non-target DNA amplification. The method is based on the fact that long inverted terminal repeats attached to DNA fragments can selectively suppress amplification of undesirable sequences in PCR procedures [270]. This method overcomes the problem of differences in mRNA abundance by incorporating a hybridization step that normalizes (equalizes) sequence abundance during the course of subtraction by standard hybridization kinetics [271]. Two types of SSH are possible: forward SSH, when the reaction involves the hybridization of cDNA from one population indicated as the evaluated phenotype (tester) to excess of mRNA (cDNA) from a control phenotype (driver); and reverse SSH, when the conditions described above are inverted. Together, the two processes are called reciprocal SSH.

Different works have been done with SSH to evaluate fish immune response at the gene expression level against challenges with bacteria-derived pathogen-associated molecular patterns (PAMP) like LPS [272, 273] and whole bacteria like *Aeromonas salmonicida* [274, 275], *Listonella anguillarum* [276], *Edwardsiella tarda* [277], and *Vibrio parahaemolyticus* [278] (Table 1).

A critical step in any immune response is the recognition of invading organisms. This is mediated by many proteins, including pattern recognition receptors (PRR), which recognize and bind to molecules present on the surface of microorganisms. LPS is an essential cell wall component of gram-negative bacteria and is recognized by PRR, triggering a series of responses that lead to the activation of the host defence system. These PRRs include a number of toll-like receptors, as well as other cell-surface and cytosolic receptors that, upon stimulation, modulate immunity [279, 280]. In LPS-stimulated yellow grouper spleen a subtracted cDNA library was constructed using SSH. The contigs and singlets obtained were analyzed and a low number of immune-related genes were found [272]. In Asian seabass the up-regulation of differentially expressed genes like pro-inflammatory cytokines and related receptors, such as TNF receptor super family member 14 (TNFRSF14), IL-31 receptor A (IL31RA), chemokine receptor-like 1 (CMKLR1), chemokine (C-X-C motif) receptor 3 (CXCR3), chemokine (C-C motif) receptor 7 (CCR7) and chemokine (C-C motif) ligand 25 (CCL25), was identified at 24h post-challenge by bacterial LPS in spleen Complement components were also identified [273]. These genes are a solid basis for a better understanding of immunity in Asian seabass and for developing effective strategies for immune protection against infections in that species.

Infection of Atlantic salmon by *A. salmonicida* was observed to stimulate an acute-phase response (APR) as part of the innate immune defence system to infection, whose gene expression pattern was remarkably observed in liver at 7 days post-infection [275] indicating that the liver appears to be the main source of APPs in fish, as in mammals. Not surprisingly, the liver gene expression pattern observed in other fish species against *L. anguillarum* [276], *E. tarda* [277], and *V. parahaemolyticus* [278]. The APR is characterized by alterations in the levels of plasma proteins referred to as acute-phase proteins (APPs), as well as the secretion of some other innate defence molecules important for innate immunity, such as complementary systems [281-283]. In Atlantic cod stimulated with atypical *A. salmonicida* (formalin-killed) interleukin-1β (IL-1- β), interleukin-8 (IL-8), CC chemokine type 3, interferon regulatory factor 1 (IRF1), ferritin heavy subunit, cathelicidin, and hepcidin were identified in the forward spleen SSH library. Atlantic cod IRF1 was constitutively expressed at low levels, and expression was significantly elevated in spleen and head kidney at 24 h following A. salmonicida stimulation, with the highest levels of induction observed in the spleen [274]. The target IRFI genes, as well as their importance in innate immune responses in fish, have not yet been determined, although the expression of IRF1 in teleost macrophages can be induced by both IFNγ and IL-1β, with IFNγ being a much more potent inducer of IRF1 than IL-1β [99].

SSH has been in several investigations to evaluate fish gene expression patterns against challenges with PAMPs, such as polyriboinosinic polyribocytidylic acid (poly I:C) [284], Infectious Spleen and Kidney Necrosis Virus (ISKNV) [285], Nodavirus [286-288], and Singapore grouper iridovirus (SGIV) [289].

Spleen gene expression in mandarin fish at 4 days post-infection with ISKNV of Mx protein, interferon-inducible protein Gig-2, and viperin (interferon-inducible and antiviral protein) was up-regulated, suggesting IFN pathway stimulation after ISKNV infection [285]. Also, two inflammatory cytokine genes, CC chemokine and IL-8, were found in the forward SSH library, whereas the CD59/Neurotoxin/Ly-6-like protein gene was down-regulated. In mammals, CD59 is a complement regulatory protein, which can inhibit complement activation and membrane attack complex (MAC) formation on autologous cells [290], suggesting that down-regulation in the ISKNV-infected host cells may make these cells more sensitive to complement attack, mounting an anti-virus mechanism of the host [285].

In orange-spotted grouper after 5 days of infection with Singapore grouper iridovirus (SGIV) novel genes were annotated as immune-related, such as C-type lectin, epinecidin, and complement components C3 and C9. Interestingly, the most abundant clone was C-type lectin, and the microarray results at 1, 5 and 9 days post-infection indicated that its expression was up-regulated in liver, spleen and kidney [289]. Lectins are multivalent carbohydrate-binding proteins that function as important pattern-recognition receptors (PRR) and have been isolated and characterized in fish [291-294]. C-type lectin represents a very large family, most members of which are able to bind PAMP and microorganisms themselves through sugar moieties and play important roles in non-self recognition and clearance of invading microorganisms. The up-regulation of C-type lectin in different organs with immunological functions confirmed as SSH as microarrays suggest an important role in the development of control strategies against SGIV infection.

Table 1. Transcriptomics studies on fish after treatments with bacteria or virus in vivo analyzed with SSH. LPS: Lipopolysaccharide; ISKNV: Infectious spleen and kidney necrosis virus; poly I:C: polyriboinosinic polyribocytidylic acid; SBNNV: Sea bass nervous necrosis virus; ACNNV: Atlantic cod nervous necrosis virus; SGIV: Singapure grouper iridovirus; PBL: Peripheral blood leukocytes.

Microorganism	Fish	Pathogen	Tissue/Cell type	Infection route	Reference
Bacteria	Atlantic salmon	*Aeromonas salmonicida* A449	Liver, head kidney and spleen	Intraperitoneal	Tsoi et al., 2004
	Yellow grouper	*LPS (E. coli)*	Spleen	Intraperitoneal	Wang et al., 2007
	Atlantic cod	*Aeromonas salmonicida (formalin-killed)*	Head kidney, Spleen	Intraperitoneal	Feng et al., 2009
	Asian seabass	*LPS (E. coli)*	Spleen	Intraperitoneal	Xia and Yue 2010
	Ayu	*Listonella anguillarum*	Liver	Intraperitoneal	Li et al., 2011
	Japanese flounder	*Edwardsiella tarda* (NE8003)	PBL	Intraperitoneal	Matsuyama et al., 2011
	Marine medaka	*Vibrio parahaemolyticus*	Liver	Intraperitoneal	Bo et al., 2012
Virus	Mandarian fish	ISKNV	Spleen	Intramuscular	He et al., 2006
	Sea bream	Nodavirus (strain 475-9/99)	Brain	Intramuscular	Dios et al., 2007
	Atlantic cod	poly I:C	Spleen	Intraperitoneal	Rise et al., 2008
	Sea bass	SBNNV	Head kidney	Intramuscular	Poisa-Beiro et al., 2009
	Atlantic cod	ACNNV	Brain	Intraperitoneal	Rise et al., 2010
	Orange-spotted grouper	SGIV	Spleen	Intraperitoneal	Xu et al., 2010

The SSH method was used to generate a subtracted cDNA library enriched in gene transcripts differentially expressed after 1 day post-infection in the brains of sea bream infected with nodavirus. Most of the expressed sequence tags (ESTs) differentially expressed in infected tissues fell into gene categories related to cell structure, transcription, cell signalling or different metabolic routes. Other interesting putative homologies corresponded to genes expressed in stress responses, such as heat shock proteins (Hsp-70) and to immune-related genes such as the Fms-interacting protein, TNFα-induced protein, interferon-induced with helicase C domain protein (mda-5), which in mammals play an important role in the synthesis and secretion of IFN type I [295]. Another nodavirus, sea bass nervous necrosis virus (SBNNV) was studied to identify genes potentially involved in antiviral immune defence in sea bass head kidney using the SSH technique [287]. The results of up-regulated EST from sea bass head kidney SSH showed significant similarities with immune genes, such as β-2 microglobulin, heat shock protein 90 (Hsp-90), IgM, MHC class I and class II, and β-galactoside-binding lectin, identified as a member of the galectin family and closely related to the galectin-1 group (Sbgalectin-1). When the recombinant protein (rSbgalectin-1) was produced and functional assays were conducted, a decrease in IL-1β, TNFα, and Mx expression was observed in the brain of sea bass simultaneously injected with nodavirus and rSbgalectin-1 compared to those infected with the nodavirus alone, suggesting a potential anti-inflammatory protective role of Sbgalectin-1 during viral infection. A similar nodavirus, the Atlantic cod nervous necrosis virus (ACNNV), was studied to evaluate the transcript expression responses in the Atlantic cod (*Gadus morhua*) brain to asymptomatic high nodavirus carrier state [288]. In the forward brain SSH library was identified with significant similarity to genes with immune-relevant functional annotations the interferon stimulated gene 15 (ISG15), IL-8 variant 5, DEXH (Asp-Glu-X-His) box polypeptide 58 (DHX58; LGP2), radical Sadenosyl methionine domain-containing 2 (RSAD2; viperin), β-2-microglobulin (B2M), chemokine CXC-like protein, signal transducer and activator of transcription 1 (STAT1), and CC chemokine type 2. Interestingly, ISG15, DHX58, RSAD2, and sacsin (SACS) transcripts are all strongly upregulated by both high nodavirus carriage and intraperitoneal poly I:C stimulation, suggesting a similar host response is significantly induced in the brain by both nodavirus and poly I:C. This expression pattern is corroborated when the response of Atlantic cod spleen is evaluated against poly I:C stimulation, showing the up-regulation of ISG15, RSAD2, LGP2 and other transcripts such as MHC class I, and IRF1, 7, and 10, indicating that Atlantic cod recognize dsRNA and mount a interferon pathway response [284].

1.14. Microarrays

Microarray analysis measures the expression of large numbers of genes in parallel. This methodology, which combines hypotheses-driven and hypotheses-free research strategies, is used to infer molecular mechanisms, classify samples, and diagnose and search for novel biomarkers. With the use of standard platforms, laboratory protocols and procedures for processing of primary data, the results of microarrays analyses are well suited for database management and meta-analysis across multiple experiments, whilst data mining is based on powerful statistical procedures with support from functional and structural annotations of genes [296].

The Atlantic salmon is of particular importance to the global aquaculture industry. Salmonid cDNA microarrays were constructed shortly after large-scale sequencing of salmon and trout cDNA libraries by several research institutes. One of the projects related to salmon sequencing is GRASP (Genomics Research on Atlantic Salmon Project), an initiative funded by Genome Canada that is intended to improve understanding of physiological and evolutionary processes influencing the survival and phenotype of salmonids and other fish in natural and aquaculture environments. The first salmonid GRASP microarray platform (GRASP-1), containing 7356 salmonid elements representing 3557 different cDNAs (3.7K), was obtained from 80,388 ESTs, principally from cDNA libraries [298] of different salmon species such as Atlantic salmon, rainbow trout, Chinook salmon, sockeye salmon, and lake whitefish cDNA libraries. The second version of the GRASP microarray platform (GRASP-2) was developed and contained cDNAs representing 16,006 genes (16K). The genes identified in the array have been stringently selected from Atlantic salmon and rainbow trout EST databases representing a wide variety of different classes of genes [297]. Finally, a new expanded salmonid cDNA microarray (GRASP-3) of 32,000 features (32K) was created where 69% of the total EST collection used was from Atlantic salmon [298]. The Aleksei Krasnov's group designed the rainbow trout microarray (SFA1.0) by identifying a relatively small number of genes (1300 genes; 1.3K) using clones from normalized and subtracted cDNA libraries, as well as genes selected by the functional categories of Gene Ontology for inclusion in a microarray aimed at characterizing transcriptome responses to environmental stressors [299] to maximize the presence of transcripts related directly to immune response in rainbow trout, because of which this platform is also called Immunochip (SFA1.0 immunochip). The updated SFA platform (1.8K; SFA2.0 immunochip) was specially designed for studies of responses to pathogens and stressors and has substantially improved coverage of immune genes [300]. Another cDNA platform in commercial fish species has been designed in Japanese flounder [301] and European flounder [302], turbot [303], and sole [304].However despite impressive achievements, cDNA platforms suffer from limitations and disadvantages. At present most research groups working with salmonids and other aquaculture species do not have full access to clones required for fabrication of cDNA microarrays. Maintenance and PCR amplification of large clone sets is expensive and time consuming, while the risk of errors is high [296]. Probably the most important drawback of cDNA microarrays is their limited ability to discriminate paralogs since long probes cross-hybridize with highly similar transcripts from members of multi-gene families [305]. In salmonids this problem is aggravated by the large number of expressed gene duplicates. These complications can be resolved with oligonucleotide microarrays (ONM) that also provide greater accuracy and reproducibility of analyses. Until recently, the use of ONM platforms was hampered by the cost, but they are now rapidly replacing cDNA platforms. Construction of ONM platforms begins with establishment of mRNA sequence sets for comprehensive coverage of transcriptomes with low redundancy. The next stage is identifying genes by searching protein databases and annotating them according to functions, pathways and structural features. For successful development and use of ONM, it is necessary to define the gene composition and optimum number of spot replicates and to choose criteria for quality assessment [296].

Because of the commercial importance of salmonid species, there is special interest gene expression pattern against different pathogens. Initially salmonid (rainbow trout) ONM contained 1672 elements, representing more than 1400 genes [306]. Currently, one of the most often used ONM platforms to evaluate the response against different conditions and pathogens is the custom salmon ONM (SIQ-3), based on the Agilent Technology system (21K in 4x44K format). Because limited availability of peripheral blood leukocyte (PBL) markers is a well-recognised problem of fish immunology, this platform compares the transcriptomes of PBL and other tissues to search for genes with preferential expression in leukocytes [296], making it a very significant tool to evaluate the response to pathogens in Atlantic salmon. Another ONM platform based on 500K ESTs Atlantic salmon and 250K ESTs rainbow trout [298] is the cGRASP 44K salmonid oligo array (Agilent eArray), although no studies employing this platform have been published yet. Another ONM has been designed in fish model organisms like zebrafish and in commercial fish species such as channel catfish and turbot [307].

Functional genomic studies based on evaluating immune responses, also called immunogenomics, have been conducted *in vivo* to evaluate the response to different pathogens at the systemic level in different organs, especially the liver and head kidney. The functional genomic approach has been used with *Salmo salar* and *Oncorhynchus mykiss*, where PAMPs, whole bacteria and viruses are the most studied pathogens. Here we present different works in fish challenged by bacteria or viruses where differential gene expression profiles were evaluated using microarray platforms with special emphasis on *in vivo* fish immune response (Table 2).

1.15. Studies with bacterial pathogens, pamps and cytokine network interactions

One of the most commonly studied bacterial pathogens is *Aeromonas salmonicida*, a gram-negative bacteria and the causative agent of furunculosis. In fact prior to the development of species-specific cDNA microarrays a preliminary study used a human microarray (GENEFILTERS GF211) to explore the liver response in Atlantic salmon infected using a cohabitation model [275]. Only 4 mRNAs were consistently up-regulated (p

< 0.01) from the 241 positively identified spots with a clearly detectable hybridization signal, none of them related to cytokine expression. This was probably due to the lack of sequence homology, a problem commonly associated with cross-species cDNA hybridization. Thus the creation of species-specific platforms was a key step in fish immunology. Using a custom Atlantic salmon cDNA microarray (NRC-IMB) consisting of over 4,000 different cDNA amplicons, the first results for challenge with Aeromonas salmonicida were reported in 2005 [275]. The study described a cohabitation challenge and identified 16 up-regulated mRNAs in all three tissues studied (spleen, liver and head kidney), whereas 2 and 19 mRNAs were identified as down-regulated in the head kidney and liver, respectively. The authors found that genes related to the acute phase response were up-regulated in spleen and head kidney of infected salmon, indicating that the infected fish underwent a typical acute phase response to infection.

The effects of an *Aeromonas salmonicida* infection were recently reported in turbot, *Scolphtalmus maximus*, [307]. Using a custom designed oligonucleotide-microarray (8x15K), the authors identified a set of 48 differentially regulated mRNAs in the spleen of challenged fish at 3 dpi, mostly related to the acute-phase and the stress/defence immune response. A study using channel and blue catfish explored the effects of a gram-negative bacterial infection on the acute phase response (APR) [308]. The authors showed up-regulation of mRNA transcripts involved in iron homeostasis, transport proteins, complement components and inflammatory and humoral immune response, indicating that conserved APR occurs as part of the innate immune response in both catfish species. Interestingly, a more acute response was observed composed of several immune pathways in the blue but not the channel catfish. More studies are required to elucidate expression patterns resulting from gram-negative bacterial infection of phylogenetically similar and different fish are required to describe common and divergent responses. This could lead to the development of marker systems, consensus on the APR in fish and treatments tailored to certain species, all of which have significant applied interest.

The activity of LPS from gram-negative bacteria, a common membrane-associated PAMP used in immunological studies, has been explored in several fish species. These studies include effects on the spleen in channel catfish [285], rainbow trout head kidney [309], and liver in the Senegalese sole [310]. Using a 19K oligonucleotide microarray (ONM) it was observed that some pro-inflammatory mRNAs in the catfish spleen were up-regulated very quickly, principally between 2 and 4 hours post-injection with LPS, whereas immunoglobulin- (2h post-injection) and antigenic presentation-related mRNA transcripts were repressed 24h post-injection [311]. A similar inhibition was reported in head kidney of rainbow trout, where the suppression of major cellular processes, including immune function and an initial stress reaction, was followed by a proliferative hematopoietic-type/biogenesis response 3 dpi [309].

Table 2. Transcriptomics studies on fish after treatments with bacteria or virus in vivo analyzed with microarrays. FDKC: Formaldehyde –killed cells; LPS: Lipopolysaccharide; FKC: Formalin-killed cells; ISAV: Infectious salmon anemia virus; PD: Pancreas disease; CMS: Cardiomyopathy syndrome; HSMI: Heart and skeletal muscle inflammation; IHNV: Infectious hematopoietic necrosis virus; VHSV: Viral hemorrhagic septicemia virus; HIRRV: Hirame rhabdovirus; poly I:C: polyriboinosinic polyribocytidylic acid; PBL: Peripheral blood leukocytes; ONM: oligonucleotide microarray.

Microorganism	Fish	Pathogen	Tissue/Cell type	Resource	Platform	Reference
Bacteria	Atlantic salmon	*Aeromonas salmonicida* (A449)	Liver	cDNA	GENE-FILTERS GF211	Tsoi et al., 2003
		Aeromonas salmonicida (A449)	Head kidney/liver/spleen	cDNA	NRC-IMB	Ewart et al., 2005
		Aeromonas salmonicida	Liver/spleen	cDNA	SFA-2	Skugor et al., 2009
		Aeromonas salmonicida (Brivax II)	Liver	cDNA	TRAITS/SPG	Martin et al., 2010
		Piscirickettsia salmonis	Head kidney	cDNA	GRASP-1	Rise et al., 2004
	Chinook salmon	*Vibrio anguillarum*	Head kidney	cDNA	GRASP-1	Ching et al., 2010
	Rainbow Trout	*Vibrio anguillarum* (FDKC)	Liver	ONM	OSUrbt	Gerwick et al., 2007
		LPS (*E.coli* 026:B6)	Head kidney	cDNA	SFA-1	MacKenzie et al., 2008
		LPS (*E. coli* 026:B6)	HK macrophages	cDNA	SFA-1	Mackenzie et al., 2008
		LPS, PGN (*E. coli* 0111:B4)	HK macrophages	cDNA	SFA-2	Boltaña et al., 2011
		PGN (*E. coli* 0111:B4, K12)	HK macrophages	cDNA	SFA-2	Boltaña et al., 2011
		LPS, PGN (*E. coli* 0111:B4)/poly I:C	Erythrocytes	cDNA	SFA-2	Morera et al., 2011
	Brook trout	LPS (*E. coli* 026:B6)	HK macrophages	cDNA	SFA-1	Mackenzie et al., 2006
	Channel catfish	LPS (*E.coli* 0127:B8)	Spleen	ONM	UMSMED-1	Li et al., 2006
		Edwardsiella ictaluri (MS-S97-773)	Gills/head kidney/liver/skin/spleen	ONM	UMSMED-2	Peatman et al., 2007

	Blue catfish	*Edwardsiella ictaluri* (MS-S97-773)	Liver	ONM	UMSMED-2	Peatman et al., 2008
	Japanese flounder	*Edwardsiella tarda* (NE9505)	Head kidney	cDNA	Japanese flounder custom-3	Yasuike et al., 2010
		Streptococcus iniae (02; FKC)	Head kidney cells	cDNA	Japanese flounder custom-3	Dumrongphol et al., 2008
		Mycobacterium bovis (TUMSAT-Msp001) FKC	Kidney cells	cDNA	Japanese flounder custom-3	Kato el al., 2010
	Zebrafish	*Mycobacterium marinum* (M, E11)	Whole fish	ONM	MWG - Sigma Genosys - Affimetrix	Meijer et al., 2005
		Mycobacterium marinum (E11; Mma20)	Whole fish	ONM	ZF Agilent	Van der Sar et al., 2009
		Streptococcus suis (HA9801)	Whole fish	ONM	Affimetrix Zebrafish GeneChip	Wu et al., 2010
	Turbot	*Aeromonas salmonicida*	Spleen	ONM	Turbot custom Agilent	Millán et al., 2010
	Solea	LPS (*E. coli* 011:B4)	Liver	cDNA	GENIPOL-1	Osuna-Jimenez et al., 2009
Virus	Atlantic salmon	ISAV (Glesvaer 2/90)	Gills/heart/liver/spleen	cDNA	SFA-2	Jorgensen et al., 2008
		ISAV (Glesvaer2/90)	Heart/PBL	ONM	SIQ-3	Krasnov et al., 2011
	Chinook salmon	ISAV NA-HRP4 (970)	Head kidney	cDNA	GRASP-3	Leblanc et al., 2010
	Rainbow Trout	IHNV (32/87)/attenuated IHNV	Head kidney	cDNA	SFA-1	MacKenzie et al., 2008
		IHNV (strain 220-90)	Head kidney	cDNA	GRASP-2	Purcell et al., 2011
	Japanese flounder	VHSV (KRRV9822)	Head kidney cells	cDNA	Japanese flounder custom-1	Byon et al., 2005
		VHSV (KRRV9822)	Head kidney cells	cDNA	Japanese flounder custom-2	Byon et al., 2006
		HIRRV (8601H)	Kidney cells	cDNA	Japanese flounder custom-1	Yasuike et al., 2007
		HIRRV (8601H)	Kidney cells	cDNA	Japanese flounder custom-3	Yasuike et al., 2010
	Zebrafish	VHSV (07.71)	Fin/head kidney/liver/spleen	ONM	ZF Agilent	Encinas et a., 2010
	Turbot	Nodavirus (AH95)/poly (I:C)	Kidney	cDNA	Turbot custom	Park et al., 2009

However, in the Senegalese sole a clear up-regulation of transcripts related to the immune response was reported 24 hpi in the liver [310]. These results collectively highlight the diversity of responses observed at the tissue level and reflect the nature of the immune system that is diffusely located throughout many organ compartments.

For gram-positive infections in fish at the level of transcriptome analyses, infection of zebrafish with *Streptococcus suis* is the only model reported [272]. *Streptococcus suis* is a pathogen associated with zoonosis reported in several countries [312, 313]. The Affymetrix Zebrafish GeneChip was used to identify 125 up-regulated transcripts where the most significant pathways were antigen processing and presentation, leukocyte trans-endothelial migration and the proteosome. The authors suggested that the target list obtained could serve as infection markers for gram-positive infection in fish.

Undoubtedly, the identification of prognostic biomarkers for disease resistance is a major aim for aquaculture. Functional genomics has the potential to identify such potential tools. Disease resistance is normally measured by challenge with the pathogen of interest and assessing the cumulative mortalities. Surviving fish or non-challenged siblings from the same family are then considered 'resistant'. Because this process is costly there is a need for non-lethal methodologies of measuring resistance, ideally based on molecular determinants of resistance. An initial example of this approach used the GRASP 3.7K cDNA array to identify *in vitro* macrophage and *in vivo* head kidney biomarkers in response to *Piscirickettsia salmonis* infection, yielding a number of 11 regulated genes common to both challenges. The researchers proposed 19 highly regulated transcripts as potential biomarkers to evaluate the efficacy of vaccines against *Piscirickettsia salmonis* [314]. C-type lectin 2-1, a gene whose product is involved in endocytosis and the C/EBP-driven inflammatory response [315] was identified and has been identified in almost all reports in which bacterial preparations have been used to challenge live fish [275, 309, 314, 316, 317]. Another study aimed at identifying biomarkers at the transcriptional level described differences between triploid and diploid Chinook salmon under live Vibrio anguillarum challenge using the GRASP 3.7K cDNA microarray [318]. Twelve annotated mRNAs were identified as showing significant differences between diploid and triploid fish. The authors however were unable to provide a description of the underlying mechanisms to explain the observed reduced immune function of triploid salmon.

Individual variation is a major hurdle for the development of prognostic markers as both genetic and epigenetic factors must be taken into account. The utility of, for example, C-type lectin in salmon, and other potential biomarkers in other species for bacterial disease resistance, requires further development. The future publication of several fish genomes coupled to array platforms with a much increased transcript representation could provide an exciting route to further develop this strategy by combining both functional and structural genomics for species of commercial interest with a sequenced genome. Several studies have attempted to correlate gene expression profiles with the activity of bacterins (killed bacteria preparations) used to vaccinate fish in culture. Most studies have concentrated on the rainbow trout and Japanese flounder [262, 277, 319, 320]. In trout, intraperitoneal administration of killed *V. anguillarum* resulted in identifying 36 differentially expressed transcripts [320]. Most of the identified targets are involved in inflammatory response and respond to a broad range of stimuli. This suggests that these targets have little use as markers for vaccination, contrary to previous descriptions in other studies. Both the second and fourth versions of the Japanese flounder cDNA microarray have been used to address vaccination [262, 277, 319]. The results of experimental infection with Gram-negative *E. tarda* indicated that a formalin-killed preparation reduced mortality in vaccinated fish from 90% to 20% [277]. However, a correlation between the transcriptome and the efficacy of vaccination could not be identified.

The effects of a commercial vaccine for Atlantic salmon (a six-component oil-adjuvant vaccine from PHARMAQ) were evaluated to correlate vaccine protection to high and low resistance to furunculosis. The authors did not find any association between either group and suggested that "outcomes of vaccination depend largely on the ability of host to prevent the negative impacts of immune response and to repair damages" [305]. Although this study did not identify correlations between vaccination and gene expression profiles, the potential of a functional genomics approach to evaluate the efficacy and underlying mechanisms of vaccination is highlighted. In terms of the immune response and the resulting complexity in expression patterns resulting from multiple cell types and different tissue responses, the investigator has the potential to obtain a clearer 'image' of the biological response from global expression data. A key objective is therefore to increase the available genomic resources much facilitated by next generation sequencing technologies to form a more robust representation of the immune system among different fish species. Furthermore, the increasing use of ONM platforms will also improve comparison across species as data sets become more easily comparable.

It remains difficult to compare microarray experiments across distinct platforms. In this respect, Meijer et al. 2005, evaluated host transcriptome profiling to *Mycobacterium marinum* infection of adult zebrafish employing three oligonucleotides platforms (MWG, Sigma Genosys, and Affimetrix). At a significance level of $P < 1.00E\text{-}5$, there were differences among the platforms in the total number of more than 2-fold up-regulated genes, whereas the 2-fold down-regulated genes were in a similar range.

Evaluation of the distribution of infection-induced genes over different categories reveled some divergence in the set from MWG, probably due to the abundance of genes of the same UniGene cluster. As well, from the total overlap of 4,138 UniGene clusters among the three microarrays, only 66 and 93 genes were up- and down-regulated, respectively [321]. With this antecedent, the same group generated a new platform (Agilent 44K) that includes their 22K probes, a 16K set probes similar to the Sigma-Compugen oligonucleotide library, and 6K set of probes for selected genes of interest indentified by previous data mining of zebrafish transcript and genome databases [322], and they evaluated the transcriptome response to acute and chronic infection by *Mycobacterium marinum*. This important effort in combing different platforms makes it clear that not all relevant genes, including immune-related ones, are represented in all platforms. Consequntly, new efforts are necessary to broaden our understanding of the immune response in fish challenged with a pathogen of interest.

1.16. Wide screening in fish challenged with viral pathogens

Two studies have reported host responses to IHNV with the SFA and GRASP platforms [309, 323]. The potential mechanisms responsible for host-specific virulence were assessed in rainbow trout infected with high (M) and low virulence (U) strains of IHNV. A marked down-regulation in biological processes, including the immune response, lymphocyte activation, response to stress, transcription and translation, together with a greater viral load (M), suggest that the higher virulence is due to the ability to suppress the immune response via the transcriptional and translational machinery of cell [323]. Furthermore, in rainbow trout was compared the expression profiles of IHNV and attenuated IHNV were compared in rainbow trout over a short time frame of one and three days post-challenge. At 3 dpi, a significant change in the transcriptional program of head kidney revealed an immunological shift orientated toward the activation of adaptive immunity. This shift was IHNV-dependent as determined by differences between the attenuated and virulent IHNV specific expression profiles. The rapid systemic spreading of IHNV inhibited TNFα, MHC class I, and several macrophage and cell cycle/differentiation markers and favored a MHC class II, immunoglobulin and MMP/TBX4-enhanced immune response [309].

Parallel studies were conducted with a cDNA microarray enriched with 213 immune-related genes to study the immune response and the efficacy of DNA vaccines containing the viral G proteins of VHSV and HIRRV administered intramuscularly in the Japanese flounder [301, 324]. As expected, all DNA vaccines containing the viral G glycoprotein conferred specific protection to the challenged fish one month after vaccination. It is suggested that the protection occurs via the IFN type I system due to the number of IFN-related genes up-regulated in both studies, ISG-15, interferon-stimulated gene 56kDa (ISG56) and the Mx protein. In both studies, VHSV and HIRRV in the Japanese flounder, the majority of differentially up-regulated genes were identified between 3 and 7 days post-vaccination (dpv), including the less effective DNA vaccine containing N protein of HIRRV. Interestingly, Mx, an antiviral protein commonly used as a marker for antiviral activity in animal species, was consistently up-regulated across vaccinations [301]. In a similar observation, IRF-3, Mx, Vig-1 and Vig-8 were up-regulated in trout at the site of DNA vaccination against IHNV at 7 dpv [323]. In turbot challenged with nodavirus, both Mx and IFN-inducible proteins were identified 24 hpi [303]. These and the previously described results suggest that both the host-expressed viral glycoprotein and the virulent rhadovirus induce a systemic anti-viral state indicative of non-specific IFN type1 innate immune response and that this canonical response is conserved among all fish. However, the mechanisms to develop a specific cytotoxic T or B lymphocyte-mediated humoral response in fish vaccinated with plasmid DNA-IHNV G that confers protective immunity have not been identified [323].

In direct relation to the above, a significant increase in transcript markers for adaptive immunity was reported in Atlantic salmon during ISA virus (ISAV) infection [325]. Importantly, a progressive increase was observed in IgZ mRNA parallel to a decrease in IgM expression that peaked > 30 days post-infection. This coordinated increase in a group of genes related to B lymphocyte differentiation and maturation and activation of T lymphocyte-mediated immunity, including CD4, TGF-β, CD8α and IFNγ, provides strong evidence for the coordinated regulation of the two arms of the immune system in response to viral infection. An important technological contribution derived from the above study was the development of an ONM for Atlantic salmon (SIQ-3). The first assessment of the performance of these arrays was carried out in Atlantic salmon for the study of virus-responsive genes from samples infected with ISA, salmonid alpha virus/PD-virus, cardiomyopathy syndrome (CMS) agent, heart and skeletal muscle inflammation (HSMI) and PBL from fish infected with ISAV. Some 95 up-regulated transcripts were identified. Most of the regulated transcripts are related directly to the immune response or associated with antiviral response [296]. As previously mentioned, the creation of species-specific platforms has been a key challenge for the study of the immune response against pathogens. Despite impressive achievements, cDNA microarrays suffer from limitations and disadvantages, the most important drawback being the limited ability to discriminate between paralogs as long cDNA probes cross-hybridize with highly similar transcripts from members of multi-gene families [305]. Furthermore this information needs to be supplemented to establish if the increased level of detected transcripts is consistent with specific protein synthesis. Recently, a study employed a combined proteomic and transcriptomic approach to evaluate the immune response against VHS [326]. In the fins of infected fish a series of mRNA transcripts principally related to complement components, immunoglobulin-related proteins, and macrophages were up-regulated (> 2-fold), whereas in parallel using two dimensional differential gel electrophoresis (2D-DIGE), enzymes of the glycolytic pathway and some proteins related to cytoskeletal remodelling and apoptosis (such as annexin A1a) increased with infection. However, very few proteins related to anti-viral response were identified.

2. CONCLUDING REMARKS

A complex network exists to regulate the innate and adaptive immune responses of fish from the various cytokines that have been reported. The study of the functional activity of these cytokines is in progress and it will be interesting to know whether mammalian Th1, Th2, Th17 and Treg responses are present in fish, regulating specific cell-mediated immunity. The recombinant production of these cytokines and antibodies against them will be the next challenge in understanding the balance of such immune responses and aid in the effective design of therapeutic strategies to manipulate the fish immune system. towards humoral or cellular immunity in response to specific antigen stimulation, vaccine strategies, functional diets to increase the quality of fishery production and predict the health of cultured fish.

The study of functional genomics in fish has provided substantial data on species of commercial interest. The major aim has been to functionally identify the intensity of responses to specific pathogens and their associated molecular components and to identify transcripts in a whole organism or specific tissue that contribute to such responses. However, the complex biology of the immune response, in which different spatial-temporal expression occurs in multiple cell types at distinct body locations, makes complete mapping of a response difficult and expensive. Moreover, considering that arrays are only as good as the transcripts represented upon them. Thus, the representation of transcripts relevant to the immune response is intimately linked to gene discovery efforts through large-scale sequencing projects, where strategies like SSH contribute not only to understanding transcriptomic response against specific pathogens but also to gene discovery. In this area, access to high-throughput NGS technology has increased in recent years and promises to make an important contribution to understanding immune response in fish. The major task now is the meta-analysis of transcriptomic data to delineate responses common among fish species to specific pathogen groups and highly specific responses. This approach will reveal host specific expression profiles and facilitate the identification of prognostic markers for diseases.

ACKNOWLEDGMENTS

The authors are grateful for the support of VRID-USACH, INNOVA-CORFO 07CN13PBT-90, INNOVA-CORFO 09MCSS6691, INNOVA-CORFO 09MCSS 6698 and CONICYT Fellowships to K. Maisey and D. Toro-Ascuy.

REFERENCES

1. Flajnik MF, Du Pasquier L. Evolution of the Immune System. 6 th ed. Paul WE, editor: Lippincott Williams & Wilkins; 2008.
2. Akirav EM, Liao S, Ruddle NH. Lymphoid Tissues and Organs. 6 th ed. Paul WE, editor: Lippincott Williams & Wilkins; 2008.
3. Zapata AG, Chibá A, Varas A. Cells and tissues of the immune system of fish. In: Iwama G, Nakanishi T, editors. The fish immune system Organism, pathogen and environment: Academic Press; 1996. p. 1-62.
4. Meseguer J, Lopez-Ruiz A, Garcia-Ayala A. Reticulo-endothelial stroma of the head-kidney from the seawater teleost gilthead seabream (Sparus aurata L.): an ultrastructural and cytochemical study. Anat Rec. 1995 Mar;241(3):303-9.
5. Danneving BH, Lauve A, Press MC, Landsverk T. Receptor-mediated endocytosis and phagocytosis by rainbow trout head kidney sinusoidal cells. Fish Shellfish Immunol. 1994;4:3-18.
6. Brattgjerd S, Evensen O. A sequential light microscopic and ultrastructural study on the uptake and handling of Vibrio salmonicida in phagocytes of the head kidney in experimentally infected Atlantic salmon (Salmo salar L.). Vet Pathol. 1996 Jan;33(1):55-65.
7. Kaattari SL, Irwin MJ. Salmonid spleen and anterior kidney harbor populations of lymphocytes with different B cell repertoires. Dev Comp Immunol. 1985 Summer; 9(3):433-44.
8. Herraez MP, Zapata AG. Structure and function of the melano-macrophage centres of the goldfish Carassius auratus. Vet Immunol Immunopathol. 1986 Jun;12(1-4):117-26.
9. Tsujii T, Seno S. Melano-macrophage centers in the aglomerular kidney of the sea horse (teleosts): morphologic studies on its formation and possible function. Anat Rec. 1990 Apr;226(4):460-70.
10. Bowden TJ, Cook P, Rombout JH. Development and function of the thymus in teleosts. Fish Shellfish Immunol. 2005 Nov;19(5):413-27.
11. Zapata AG, Amemiya CT. Phylogeny of lower vertebrates and their immunological structures. Curr Top Microbiol Immunol. 1999;248:67-107.
12. Espenes A, Press CM, Dannevig BH, Landsverk T. Immune-complex trapping in the splenic ellipsoids of rainbow trout (Oncorhynchus mykiss). Cell Tissue Res. 1995 Oct;282(1):41-8.
13. Vallejo AN, Miller NW, Harvey NE, Cuchens MA, Warr GW, Clem LW. Cellular pathway(s) of antigen processing and presentation in fish APC: endosomal involvement and cell-free antigen presentation. Dev Immunol. 1992;3(1):51-65.

14. Vallejo AN, Miller NW, Clem LW. Cellular pathway(s) of antigen processing in fish APC: effect of varying in vitro temperatures on antigen catabolism. Dev Comp Immunol. 1992 Sep-Oct;16(5):367-81.
15. Ellis AE. Antigen trapping in the spleen and kidney of the plaice, Pleuronectes platessa. J Fish Dis. 1980;3:413-26.
16. Agius C, Roberts RJ. Melano-macrophage centres and their role in fish pathology. J Fish Dis. 2003 Sep;26(9):499-509.
17. Medzhitov R. The Innate Immune System. 6 th ed. Paul WE, editor: Lippincott Williams & Wilkins; 2008.
18. Janeway CA, Jr. How the immune system protects the host from infection. Microbes Infect. 2001 Nov;3(13):1167-71.
19. McHeyzer-Williams LJ, McHeyzer-Williams MG. Antigen-specific memory B cell development. Annu Rev Immunol. 2005;23:487-513.
20. Sallusto F, Lanzavecchia A. Heterogeneity of CD4+ memory T cells: functional modules for tailored immunity. Eur J Immunol. 2009 Aug;39(8):2076-82.
21. Dixon B, Stet RJ. The relationship between major histocompatibility receptors and innate immunity in teleost fish. Dev Comp Immunol. 2001 Oct-Dec;25(8-9):683-99.
22. Abbas AK, Lichtman AH, Pillai S. Cells and Tissues of the Adaptive Immune System. 6 th ed: Saunders; 2007.
23. Castro R, Bernard D, Lefranc MP, Six A, Benmansour A, Boudinot P. T cell diversity and TcR repertoires in teleost fish. Fish Shellfish Immunol. 2011 Nov;31(5):644-54.
24. Secombes CJ, Wang T, Bird S. The interleukins of fish. Dev Comp Immunol. 2011 Dec;35(12):1336-45.
25. Wang T, Gorgoglione B, Maehr T, Holland JW, Vecino JL, Wadsworth S, et al. Fish Suppressors of Cytokine Signaling (SOCS): Gene Discovery, Modulation of Expression and Function. J Signal Transduct. 2011;2011:905813.
26. Alejo A, Tafalla C. Chemokines in teleost fish species. Dev Comp Immunol. 2011 Dec;35(12):1215-22.
27. Nakanishi T, Toda H, Shibasaki Y, Somamoto T. Cytotoxic T cells in teleost fish. Dev Comp Immunol. 2011 Dec;35(12):1317-23.
28. Boltana S, Donate C, Goetz FW, MacKenzie S, Balasch JC. Characterization and expression of NADPH oxidase in LPS-, poly(I:C)- and zymosan-stimulated trout (Oncorhynchus mykiss W.) macrophages. Fish Shellfish Immunol. 2009 Apr;26(4):651-61.
29. Hanington PC, Hitchen SJ, Beamish LA, Belosevic M. Macrophage colony stimulating factor (CSF-1) is a central growth factor of goldfish macrophages. Fish Shellfish Immunol. 2009 Jan;26(1):1-9.
30. Salinas I, Zhang YA, Sunyer JO. Mucosal immunoglobulins and B cells of teleost fish. Dev Comp Immunol. 2011 Dec;35(12):1346-65.
31. Zhang YA, Salinas I, Oriol Sunyer J. Recent findings on the structure and function of teleost IgT. Fish Shellfish Immunol. 2011 Nov;31(5):627-34.
32. Randelli E, Buonocore F, Scapigliati G. Cell markers and determinants in fish immunology. Fish Shellfish Immunol. 2008 Oct;25(4):326-40.
33. Rombout JH, Huttenhuis HB, Picchietti S, Scapigliati G. Phylogeny and ontogeny of fish leucocytes. Fish Shellfish Immunol. 2005 Nov;19(5):441-55.
34. Araki K, Akatsu K, Suetake H, Kikuchi K, Suzuki Y. Characterization of CD8+ leukocytes in fugu (Takifugu rubripes) with antiserum against fugu CD8alpha. Dev Comp Immunol. 2008;32(7):850-8.
35. Toda H, Saito Y, Koike T, Takizawa F, Araki K, Yabu T, et al. Conservation of characteristics and functions of CD4 positive lymphocytes in a teleost fish. Dev Comp Immunol. 2011 Jun;35(6):650-60.
36. Takizawa F, Dijkstra JM, Kotterba P, Korytar T, Kock H, Kollner B, et al. The expression of CD8alpha discriminates distinct T cell subsets in teleost fish. Dev Comp Immunol. 2011 Jul;35(7):752-63.
37. Kaattari S, Evans D, Klemer J. Varied redox forms of teleost IgM: an alternative to isotypic diversity? Immunol Rev. 1998 Dec;166:133-42.

38. Wilson M, Bengten E, Miller NW, Clem LW, Du Pasquier L, Warr GW. A novel chimeric Ig heavy chain from a teleost fish shares similarities to IgD. Proc Natl Acad Sci U S A. 1997 Apr 29;94(9):4593-7.
39. Stenvik J, Jorgensen TO. Immunoglobulin D (IgD) of Atlantic cod has a unique structure. Immunogenetics. 2000 May;51(6):452-61.
40. Danilova N, Bussmann J, Jekosch K, Steiner LA. The immunoglobulin heavy-chain locus in zebrafish: identification and expression of a previously unknown isotype, immunoglobulin Z. Nat Immunol. 2005 Mar;6(3):295-302.
41. Hansen JD, Landis ED, Phillips RB. Discovery of a unique Ig heavy-chain isotype (IgT) in rainbow trout: Implications for a distinctive B cell developmental pathway in teleost fish. Proc Natl Acad Sci U S A. 2005 May 10;102(19):6919-24.
42. Li J, Barreda DR, Zhang YA, Boshra H, Gelman AE, Lapatra S, et al. B lymphocytes from early vertebrates have potent phagocytic and microbicidal abilities. Nat Immunol. 2006 Oct;7(10):1116-24.
43. Overland HS, Pettersen EF, Ronneseth A, Wergeland HI. Phagocytosis by B-cells and neutrophils in Atlantic salmon (Salmo salar L.) and Atlantic cod (Gadus morhua L.). Fish Shellfish Immunol. 2010 Jan;28(1):193-204.
44. Wang T, Huang W, Costa MM, Secombes CJ. The gamma-chain cytokine/receptor system in fish: more ligands and receptors. Fish Shellfish Immunol. 2011 Nov;31(5):673-87.
45. Savan R, Sakai M. Genomics of fish cytokines. Comp Biochem Physiol Part D Genomics Proteomics. 2006 Mar;1(1):89-101.
46. Salazar-Mather TP, Hokeness KL. Cytokine and chemokine networks: pathways to antiviral defense. Curr Top Microbiol Immunol. 2006;303:29-46.
47. Svanborg C, Godaly G, Hedlund M. Cytokine responses during mucosal infections: role in disease pathogenesis and host defence. Curr Opin Microbiol. 1999 Feb;2(1):99-105.
48. DeVries ME, Ran L, Kelvin DJ. On the edge: the physiological and pathophysiological role of chemokines during inflammatory and immunological responses. Semin Immunol. 1999 Apr;11(2):95-104.
49. Whyte SK. The innate immune response of finfish--a review of current knowledge. Fish Shellfish Immunol. 2007 Dec;23(6):1127-51.
50. Rahman MM, McFadden G. Modulation of tumor necrosis factor by microbial pathogens. PLoS Pathog. 2006 Feb;2(2):e4.
51. Benedict CA, Banks TA, Ware CF. Death and survival: viral regulation of TNF signaling pathways. Curr Opin Immunol. 2003 Feb;15(1):59-65.
52. Hirono I, Nam BH, Kurobe T, Aoki T. Molecular cloning, characterization, and expression of TNF cDNA and gene from Japanese flounder Paralychthys olivaceus. J Immunol. 2000 Oct 15;165(8):4423-7.
53. Laing KJ, Wang T, Zou J, Holland J, Hong S, Bols N, et al. Cloning and expression analysis of rainbow trout Oncorhynchus mykiss tumour necrosis factor-alpha. Eur J Biochem. 2001 Mar;268(5):1315-22.
54. Zou J, Wang T, Hirono I, Aoki T, Inagawa H, Honda T, et al. Differential expression of two tumor necrosis factor genes in rainbow trout, Oncorhynchus mykiss. Dev Comp Immunol. 2002 Mar;26(2):161-72.
55. Garcia-Castillo J, Pelegrin P, Mulero V, Meseguer J. Molecular cloning and expression analysis of tumor necrosis factor alpha from a marine fish reveal its constitutive expression and ubiquitous nature. Immunogenetics. 2002 Jun;54(3):200-7.
56. Saeij JP, Stet RJ, de Vries BJ, van Muiswinkel WB, Wiegertjes GF. Molecular and functional characterization of carp TNF: a link between TNF polymorphism and trypanotolerance? Dev Comp Immunol. 2003 Jan;27(1):29-41.
57. Zou J, Secombes CJ, Long S, Miller N, Clem LW, Chinchar VG. Molecular identification and expression analysis of tumor necrosis factor in channel catfish (Ictalurus punctatus). Dev Comp Immunol. 2003 Dec;27(10):845-58.
58. Praveen K, Evans DL, Jaso-Friedmann L. Constitutive expression of tumor necrosis factor-alpha in cytotoxic cells of teleosts and its role in regulation of cell-mediated cytotoxicity. Mol Immunol. 2006 Feb;43(3):279-91.

59. Ordas MC, Costa MM, Roca FJ, Lopez-Castejon G, Mulero V, Meseguer J, et al. Turbot TNFalpha gene: molecular characterization and biological activity of the recombinant protein. Mol Immunol. 2007 Jan;44(4):389-400.
60. Grayfer L, Walsh JG, Belosevic M. Characterization and functional analysis of goldfish (Carassius auratus L.) tumor necrosis factor-alpha. Dev Comp Immunol. 2008;32(5):532-43.
61. Garcia-Castillo J, Chaves-Pozo E, Olivares P, Pelegrin P, Meseguer J, Mulero V. The tumor necrosis factor alpha of the bony fish seabream exhibits the in vivo proinflammatory and proliferative activities of its mammalian counterparts, yet it functions in a species-specific manner. Cell Mol Life Sci. 2004 Jun;61(11):1331-40.
62. Roca FJ, Mulero I, Lopez-Munoz A, Sepulcre MP, Renshaw SA, Meseguer J, et al. Evolution of the inflammatory response in vertebrates: fish TNF-alpha is a powerful activator of endothelial cells but hardly activates phagocytes. J Immunol. 2008 Oct 1;181(7):5071-81.
63. Hardie LJ, Chappell LH, Secombes CJ. Human tumor necrosis factor alpha influences rainbow trout Oncorhynchus mykiss leucocyte responses. Vet Immunol Immunopathol. 1994 Jan;40(1):73-84.
64. Iliev DB, Liarte CQ, MacKenzie S, Goetz FW. Activation of rainbow trout (Oncorhynchus mykiss) mononuclear phagocytes by different pathogen associated molecular pattern (PAMP) bearing agents. Mol Immunol. 2005 Jun;42(10):1215-23.
65. Wang WL, Hong JR, Lin GH, Liu W, Gong HY, Lu MW, et al. Stage-specific expression of TNFalpha regulates bad/bid-mediated apoptosis and RIP1/ROS-mediated secondary necrosis in Birnavirus-infected fish cells. PLoS One. 2011;6(2):e16740.
66. Dinarello C, Arend W, Sims J, Smith D, Blumberg H, O'Neill L, et al. IL-1 family nomenclature. Nat Immunol. 2010 Nov;11(11):973.
67. Dinarello CA. Interleukin-1. Cytokine Growth Factor Rev. 1997 Dec;8(4):253-65.
68. Zou J, Grabowski PS, Cunningham C, Secombes CJ. Molecular cloning of interleukin 1beta from rainbow trout Oncorhynchus mykiss reveals no evidence of an ice cut site. Cytokine. 1999 Aug;11(8):552-60.
69. Fujiki K, Shin DH, Nakao M, Yano T. Molecular cloning and expression analysis of carp (Cyprinus carpio) interleukin-1 beta, high affinity immunoglobulin E Fc receptor gamma subunit and serum amyloid A. Fish Shellfish Immunol. 2000 Apr;10(3):229-42.
70. Scapigliati G, Buonocore F, Bird S, Zou J, Pelegrin P, Falasca C, et al. Phylogeny of cytokines: molecular cloning and expression analysis of sea bass Dicentrarchus labrax interleukin-1beta. Fish Shellfish Immunol. 2001 Nov;11(8):711-26.
71. Pelegrin P, Garcia-Castillo J, Mulero V, Meseguer J. Interleukin-1beta isolated from a marine fish reveals up-regulated expression in macrophages following activation with lipopolysaccharide and lymphokines. Cytokine. 2001 Oct 21;16(2):67-72.
72. Corripio-Miyar Y, Bird S, Tsamopoulos K, Secombes CJ. Cloning and expression analysis of two pro-inflammatory cytokines, IL-1 beta and IL-8, in haddock (Melanogrammus aeglefinus). Mol Immunol. 2007 Feb;44(6):1361-73.
73. Lee DS, Hong SH, Lee HJ, Jun LJ, Chung JK, Kim KH, et al. Molecular cDNA cloning and analysis of the organization and expression of the IL-1beta gene in the Nile tilapia, Oreochromis niloticus. Comp Biochem Physiol A Mol Integr Physiol. 2006 Mar; 143(3):307-14.
74. Pleguezuelos O, Zou J, Cunningham C, Secombes CJ. Cloning, sequencing, and analysis of expression of a second IL-1beta gene in rainbow trout (Oncorhynchus mykiss). Immunogenetics. 2000 Oct;51(12):1002-11.
75. Cerretti DP, Kozlosky CJ, Mosley B, Nelson N, Van Ness K, Greenstreet TA, et al. Molecular cloning of the interleukin-1 beta converting enzyme. Science. 1992 Apr 3;256(5053):97-100.
76. Secombes CJ, Wang T, Hong S, Peddie S, Crampe M, Laing KJ, et al. Cytokines and innate immunity of fish. Dev Comp Immunol. 2001 Oct-Dec;25(8-9):713-23.
77. Lu DQ, Bei JX, Feng LN, Zhang Y, Liu XC, Wang L, et al. Interleukin-1beta gene in orange-spotted grouper, Epinephelus coioides: molecular cloning, expression, biological activities and signal transduction. Mol Immunol. 2008 Feb;45(4):857-67.

78. Reyes-Cerpa S, Reyes-Lopez FE, Toro-Ascuy D, Ibanez J, Maisey K, Sandino AM, et al. IPNV modulation of pro and anti-inflammatory cytokine expression in Atlantic salmon might help the establishment of infection and persistence. Fish Shellfish Immunol. 2012 Feb;32(2):291-300.
79. Seppola M, Larsen AN, Steiro K, Robertsen B, Jensen I. Characterisation and expression analysis of the interleukin genes, IL-1beta, IL-8 and IL-10, in Atlantic cod (Gadus morhua L.). Mol Immunol. 2008 Feb;45(4):887-97.
80. Zou J, Holland J, Pleguezuelos O, Cunningham C, Secombes CJ. Factors influencing the expression of interleukin-1 beta in cultured rainbow trout (Oncorhynchus mykiss) leucocytes. Dev Comp Immunol. 2000 Sep-Oct;24(6-7):575-82.
81. Buonocore F, Forlenza M, Randelli E, Benedetti S, Bossu P, Meloni S, et al. Biological activity of sea bass (Dicentrarchus labrax L.) recombinant interleukin-1beta. Mar Biotechnol (NY). 2005 Nov-Dec;7(6):609-17.
82. Hong S, Zou J, Crampe M, Peddie S, Scapigliati G, Bols N, et al. The production and bioactivity of rainbow trout (Oncorhynchus mykiss) recombinant IL-1 beta. Vet Immunol Immunopathol. 2001 Aug 30;81(1-2):1-14.
83. Peddie S, Zou J, Cunningham C, Secombes CJ. Rainbow trout (Oncorhynchus mykiss) recombinant IL-1beta and derived peptides induce migration of head-kidney leucocytes in vitro. Fish Shellfish Immunol. 2001 Nov;11(8):697-709.
84. Peddie S, Zou J, Secombes CJ. Immunostimulation in the rainbow trout (Oncorhynchus mykiss) following intraperitoneal administration of Ergosan. Vet Immunol Immunopathol. 2002 May;86(1-2):101-13.
85. Dinarello CA, Fantuzzi G. Interleukin-18 and host defense against infection. J Infect Dis. 2003 Jun 15;187 Suppl 2:S370-84.
86. Gracie JA, Robertson SE, McInnes IB. Interleukin-18. J Leukoc Biol. 2003 Feb;73(2):213-24.
87. Sugawara I. Interleukin-18 (IL-18) and infectious diseases, with special emphasis on diseases induced by intracellular pathogens. Microbes Infect. 2000 Aug;2(10):1257-63.
88. Micallef MJ, Ohtsuki T, Kohno K, Tanabe F, Ushio S, Namba M, et al. Interferon-gamma-inducing factor enhances T helper 1 cytokine production by stimulated human T cells: synergism with interleukin-12 for interferon-gamma production. Eur J Immunol. 1996 Jul;26(7):1647-51.
89. Okamura H, Nagata K, Komatsu T, Tanimoto T, Nukata Y, Tanabe F, et al. A novel costimulatory factor for gamma interferon induction found in the livers of mice causes endotoxic shock. Infect Immun. 1995 Oct;63(10):3966-72.
90. Dao T, Mehal WZ, Crispe IN. IL-18 augments perforin-dependent cytotoxicity of liver NK-T cells. J Immunol. 1998 Sep 1;161(5):2217-22.
91. Leung BP, Culshaw S, Gracie JA, Hunter D, Canetti CA, Campbell C, et al. A role for IL-18 in neutrophil activation. J Immunol. 2001 Sep 1;167(5):2879-86.
92. Xu D, Trajkovic V, Hunter D, Leung BP, Schulz K, Gracie JA, et al. IL-18 induces the differentiation of Th1 or Th2 cells depending upon cytokine milieu and genetic background. Eur J Immunol. 2000 Nov;30(11):3147-56.
93. Ghayur T, Banerjee S, Hugunin M, Butler D, Herzog L, Carter A, et al. Caspase-1 processes IFN-gamma-inducing factor and regulates LPS-induced IFN-gamma production. Nature. 1997 Apr 10;386(6625):619-23.
94. Gu Y, Kuida K, Tsutsui H, Ku G, Hsiao K, Fleming MA, et al. Activation of interferon-gamma inducing factor mediated by interleukin-1beta converting enzyme. Science. 1997 Jan 10;275(5297):206-9.
95. Akita K, Ohtsuki T, Nukada Y, Tanimoto T, Namba M, Okura T, et al. Involvement of caspase-1 and caspase-3 in the production and processing of mature human interleukin 18 in monocytic THP.1 cells. J Biol Chem. 1997 Oct 17;272(42):26595-603.
96. Sugawara S, Uehara A, Nochi T, Yamaguchi T, Ueda H, Sugiyama A, et al. Neutrophil proteinase 3-mediated induction of bioactive IL-18 secretion by human oral epithelial cells. J Immunol. 2001 Dec 1;167(11):6568-75.

97. Huising MO, Stet RJ, Savelkoul HF, Verburg-van Kemenade BM. The molecular evolution of the interleukin-1 family of cytokines; IL-18 in teleost fish. Dev Comp Immunol. 2004 May 3;28(5):395-413.
98. Zou J, Bird S, Truckle J, Bols N, Horne M, Secombes C. Identification and expression analysis of an IL-18 homologue and its alternatively spliced form in rainbow trout (Oncorhynchus mykiss). Eur J Biochem. 2004 May;271(10):1913-23.
99. Martin SA, Zou J, Houlihan DF, Secombes CJ. Directional responses following recombinant cytokine stimulation of rainbow trout (Oncorhynchus mykiss) RTS-11 macrophage cells as revealed by transcriptome profiling. BMC Genomics. 2007;8:150.
100. Wang T, Secombes CJ. Identification and expression analysis of two fish-specific IL-6 cytokine family members, the ciliary neurotrophic factor (CNTF)-like and M17 genes, in rainbow trout Oncorhynchus mykiss. Mol Immunol. 2009 Jul;46(11-12):2290-8.
101. Taga T, Kishimoto T. Gp130 and the interleukin-6 family of cytokines. Annu Rev Immunol. 1997;15:797-819.
102. Hirano T. Interleukin 6 and its receptor: ten years later. Int Rev Immunol. 1998;16(3-4):249-84.
103. Inoue K, Takano H, Shimada A, Morita T, Yanagisawa R, Sakurai M, et al. Cytoprotection by interleukin-6 against liver injury induced by lipopolysaccharide. Int J Mol Med. 2005 Feb;15(2):221-4.
104. Kaplanski G, Marin V, Montero-Julian F, Mantovani A, Farnarier C. IL-6: a regulator of the transition from neutrophil to monocyte recruitment during inflammation. Trends Immunol. 2003 Jan;24(1):25-9.
105. Bird S, Zou J, Savan R, Kono T, Sakai M, Woo J, et al. Characterisation and expression analysis of an interleukin 6 homologue in the Japanese pufferfish, Fugu rubripes. Dev Comp Immunol. 2005;29(9):775-89.
106. Iliev DB, Castellana B, Mackenzie S, Planas JV, Goetz FW. Cloning and expression analysis of an IL-6 homolog in rainbow trout (Oncorhynchus mykiss). Mol Immunol. 2007 Mar;44(7):1803-7.
107. Nam BH, Byon JY, Kim YO, Park EM, Cho YC, Cheong J. Molecular cloning and characterisation of the flounder (Paralichthys olivaceus) interleukin-6 gene. Fish Shellfish Immunol. 2007 Jul;23(1):231-6.
108. Castellana B, Iliev DB, Sepulcre MP, MacKenzie S, Goetz FW, Mulero V, et al. Molecular characterization of interleukin-6 in the gilthead seabream (Sparus aurata). Mol Immunol. 2008 Jul;45(12):3363-70.
109. Costa MM, Maehr T, Diaz-Rosales P, Secombes CJ, Wang T. Bioactivity studies of rainbow trout (Oncorhynchus mykiss) interleukin-6: effects on macrophage growth and antimicrobial peptide gene expression. Mol Immunol. 2011 Sep;48(15-16):1903-16.
110. Czupryn MJ, McCoy JM, Scoble HA. Structure-function relationships in human interleukin-11. Identification of regions involved in activity by chemical modification and site-directed mutagenesis. J Biol Chem. 1995 Jan 13;270(2):978-85.
111. Bartz H, Buning-Pfaue F, Turkel O, Schauer U. Respiratory syncytial virus induces prostaglandin E2, IL-10 and IL-11 generation in antigen presenting cells. Clin Exp Immunol. 2002 Sep;129(3):438-45.
112. Kernacki KA, Goebel DJ, Poosch MS, Hazlett LD. Early cytokine and chemokine gene expression during Pseudomonas aeruginosa corneal infection in mice. Infect Immun. 1998 Jan;66(1):376-9.
113. Nandurkar HH, Robb L, Begley CG. The role of IL-II in hematopoiesis as revealed by a targeted mutation of its receptor. Stem Cells. 1998;16 Suppl 2:53-65.
114. Trepicchio WL, Bozza M, Pedneault G, Dorner AJ. Recombinant human IL-11 attenuates the inflammatory response through down-regulation of proinflammatory cytokine release and nitric oxide production. J Immunol. 1996 Oct 15;157(8):3627-34.
115. Orazi A, Du X, Yang Z, Kashai M, Williams DA. Interleukin-11 prevents apoptosis and accelerates recovery of small intestinal mucosa in mice treated with combined chemotherapy and radiation. Lab Invest. 1996 Jul;75(1):33-42.

116. Liu Q, Du XX, Schindel DT, Yang ZX, Rescorla FJ, Williams DA, et al. Trophic effects of interleukin-11 in rats with experimental short bowel syndrome. J Pediatr Surg. 1996 Aug;31(8):1047-50; discussion 50-1.
117. Huising MO, Kruiswijk CP, van Schijndel JE, Savelkoul HF, Flik G, Verburg-van Kemenade BM. Multiple and highly divergent IL-11 genes in teleost fish. Immunogenetics. 2005 Jul;57(6):432-43.
118. Wang T, Holland JW, Bols N, Secombes CJ. Cloning and expression of the first nonmammalian interleukin-11 gene in rainbow trout Oncorhynchus mykiss. FEBS J. 2005 Mar;272(5):1136-47.
119. Santos MD, Yasuike M, Kondo H, Hirono I, Aoki T. Teleostean IL11b exhibits complementing function to IL11a and expansive involvement in antibacterial and antiviral responses. Mol Immunol. 2008 Jul;45(12):3494-501.
120. Ribeiro CM, Hermsen T, Taverne-Thiele AJ, Savelkoul HF, Wiegertjes GF. Evolution of recognition of ligands from Gram-positive bacteria: similarities and differences in the TLR2-mediated response between mammalian vertebrates and teleost fish. J Immunol. 2010 Mar 1;184(5):2355-68.
121. Murphy PM, Baggiolini M, Charo IF, Hebert CA, Horuk R, Matsushima K, et al. International union of pharmacology. XXII. Nomenclature for chemokine receptors. Pharmacol Rev. 2000 Mar;52(1):145-76.
122. Mukaida N, Harada A, Matsushima K. Interleukin-8 (IL-8) and monocyte chemotactic and activating factor (MCAF/MCP-1), chemokines essentially involved in inflammatory and immune reactions. Cytokine Growth Factor Rev. 1998 Mar;9(1):9-23.
123. Hebert CA, Vitangcol RV, Baker JB. Scanning mutagenesis of interleukin-8 identifies a cluster of residues required for receptor binding. J Biol Chem. 1991 Oct 5;266(28):18989-94.
124. Clark-Lewis I, Schumacher C, Baggiolini M, Moser B. Structure-activity relationships of interleukin-8 determined using chemically synthesized analogs. Critical role of NH2-terminal residues and evidence for uncoupling of neutrophil chemotaxis, exocytosis, and receptor binding activities. J Biol Chem. 1991 Dec 5;266(34):23128-34.
125. Laing KJ, Zou JJ, Wang T, Bols N, Hirono I, Aoki T, et al. Identification and analysis of an interleukin 8-like molecule in rainbow trout Oncorhynchus mykiss. Dev Comp Immunol. 2002 Jun;26(5):433-44.
126. Lee EY, Park HH, Kim YT, Choi TJ. Cloning and sequence analysis of the interleukin-8 gene from flounder (Paralichthys olivaceous). Gene. 2001 Aug 22;274(1-2):237-43.
127. Sangrador-Vegas A, Lennington JB, Smith TJ. Molecular cloning of an IL-8-like CXC chemokine and tissue factor in rainbow trout (Oncorhynchus mykiss) by use of suppression subtractive hybridization. Cytokine. 2002 Jan 21;17(2):66-70.
128. Chen L, He C, Baoprasertkul P, Xu P, Li P, Serapion J, et al. Analysis of a catfish gene resembling interleukin-8: cDNA cloning, gene structure, and expression after infection with Edwardsiella ictaluri. Dev Comp Immunol. 2005;29(2):135-42.
129. Najakshin AM, Mechetina LV, Alabyev BY, Taranin AV. Identification of an IL-8 homolog in lamprey (Lampetra fluviatilis): early evolutionary divergence of chemokines. Eur J Immunol. 1999 Feb;29(2):375-82.
130. Laing KJ, Secombes CJ. Chemokines. Dev Comp Immunol. 2004 May 3;28(5):443-60.
131. Osborne LC, Abraham N. Regulation of memory T cells by gammac cytokines. Cytokine. 2010 May;50(2):105-13.
132. Sogo T, Kawahara M, Ueda H, Otsu M, Onodera M, Nakauchi H, et al. T cell growth control using hapten-specific antibody/interleukin-2 receptor chimera. Cytokine. 2009 Apr;46(1):127-36.
133. Morgan DA, Ruscetti FW, Gallo R. Selective in vitro growth of T lymphocytes from normal human bone marrows. Science. 1976 Sep 10;193(4257):1007-8.
134. Smith KA. T-cell growth factor. Immunol Rev. 1980;51:337-57.
135. Smith KA. Interleukin-2: inception, impact, and implications. Science. 1988 May 27;240(4856):1169-76.

136. Granucci F, Vizzardelli C, Pavelka N, Feau S, Persico M, Virzi E, et al. Inducible IL-2 production by dendritic cells revealed by global gene expression analysis. Nat Immunol. 2001 Sep;2(9):882-8.
137. Walker E, Leemhuis T, Roeder W. Murine B lymphoma cell lines release functionally active interleukin 2 after stimulation with Staphylococcus aureus. J Immunol. 1988 Feb 1;140(3):859-65.
138. Gaffen SL, Wang S, Koshland ME. Expression of the immunoglobulin J chain in a murine B lymphoma is driven by autocrine production of interleukin 2. Cytokine. 1996 Jul;8(7):513-24.
139. Yu TK, Caudell EG, Smid C, Grimm EA. IL-2 activation of NK cells: involvement of MKK1/2/ERK but not p38 kinase pathway. J Immunol. 2000 Jun 15;164(12):6244-51.
140. Gold MR, DeFranco AL. Biochemistry of B lymphocyte activation. Adv Immunol. 1994;55:221-95.
141. Bird S, Zou J, Kono T, Sakai M, Dijkstra JM, Secombes C. Characterisation and expression analysis of interleukin 2 (IL-2) and IL-21 homologues in the Japanese pufferfish, Fugu rubripes, following their discovery by synteny. Immunogenetics. 2005 Mar;56(12):909-23.
142. Sugamata R, Suetake H, Kikuchi K, Suzuki Y. Teleost B7 expressed on monocytes regulates T cell responses. J Immunol. 2009 Jun 1;182(11):6799-806.
143. Diaz-Rosales P, Bird S, Wang TH, Fujiki K, Davidson WS, Zou J, et al. Rainbow trout interleukin-2: cloning, expression and bioactivity analysis. Fish Shellfish Immunol. 2009 Sep;27(3):414-22.
144. Zhang YA, Hikima J, Li J, LaPatra SE, Luo YP, Sunyer JO. Conservation of structural and functional features in a primordial CD80/86 molecule from rainbow trout (Oncorhynchus mykiss), a primitive teleost fish. J Immunol. 2009 Jul 1;183(1):83-96.
145. Laing KJ, Bols N, Secombes CJ. A CXC chemokine sequence isolated from the rainbow trout Oncorhynchus mykiss resembles the closely related interferon-gamma-inducible chemokines CXCL9, CXCL10 and CXCL11. Eur Cytokine Netw. 2002 Oct-Dec;13(4):462-73.
146. Banchereau J, Bidaud C, Fluckiger AC, Galibert L, Garrone P, Malisan F, et al. Effects of interleukin 4 on human B-cell growth and differentiation. Res Immunol. 1993 Oct;144(8):601-5.
147. Mosmann TR, Cherwinski H, Bond MW, Giedlin MA, Coffman RL. Two types of murine helper T cell clone. I. Definition according to profiles of lymphokine activities and secreted proteins. J Immunol. 1986 Apr 1;136(7):2348-57.
148. Finkelman FD, Urban JF, Jr. The other side of the coin: the protective role of the TH2 cytokines. J Allergy Clin Immunol. 2001 May;107(5):772-80.
149. Yokota T, Otsuka T, Mosmann T, Banchereau J, DeFrance T, Blanchard D, et al. Isolation and characterization of a human interleukin cDNA clone, homologous to mouse B-cell stimulatory factor 1, that expresses B-cell- and T-cell-stimulating activities. Proc Natl Acad Sci U S A. 1986 Aug;83(16):5894-8.
150. Noma Y, Sideras P, Naito T, Bergstedt-Lindquist S, Azuma C, Severinson E, et al. Cloning of cDNA encoding the murine IgG1 induction factor by a novel strategy using SP6 promoter. Nature. 1986 Feb 20-26;319(6055):640-6.
151. Lee F, Yokota T, Otsuka T, Meyerson P, Villaret D, Coffman R, et al. Isolation and characterization of a mouse interleukin cDNA clone that expresses B-cell stimulatory factor 1 activities and T-cell- and mast-cell-stimulating activities. Proc Natl Acad Sci U S A. 1986 Apr;83(7):2061-5.
152. Heussler VT, Eichhorn M, Dobbelaere DA. Cloning of a full-length cDNA encoding bovine interleukin 4 by the polymerase chain reaction. Gene. 1992 May 15;114(2):273-8.
153. Li-Weber M, Krammer PH. Regulation of IL4 gene expression by T cells and therapeutic perspectives. Nat Rev Immunol. 2003 Jul;3(7):534-43.
154. Ansel KM, Djuretic I, Tanasa B, Rao A. Regulation of Th2 differentiation and Il4 locus accessibility. Annu Rev Immunol. 2006;24:607-56.
155. Ohtani M, Hayashi N, Hashimoto K, Nakanishi T, Dijkstra JM. Comprehensive clarification of two paralogous interleukin 4/13 loci in teleost fish. Immunogenetics. 2008 Jul;60(7):383-97.

156. Jaillon O, Aury JM, Brunet F, Petit JL, Stange-Thomann N, Mauceli E, et al. Genome duplication in the teleost fish Tetraodon nigroviridis reveals the early vertebrate proto-karyotype. Nature. 2004 Oct 21;431(7011):946-57.
157. Li JH, Shao JZ, Xiang LX, Wen Y. Cloning, characterization and expression analysis of pufferfish interleukin-4 cDNA: the first evidence of Th2-type cytokine in fish. Mol Immunol. 2007 Mar;44(8):2078-86.
158. Secrist H, Egan M, Peters MG. Tissue-specific regulation of IL-4 mRNA expression in human tonsil. J Immunol. 1994 Feb 1;152(3):1120-6.
159. Lin AF, Xiang LX, Wang QL, Dong WR, Gong YF, Shao JZ. The DC-SIGN of zebrafish: insights into the existence of a CD209 homologue in a lower vertebrate and its involvement in adaptive immunity. J Immunol. 2009 Dec 1;183(11):7398-410.
160. Hu YL, Xiang LX, Shao JZ. Identification and characterization of a novel immunoglobulin Z isotype in zebrafish: implications for a distinct B cell receptor in lower vertebrates. Mol Immunol. 2010 Jan;47(4):738-46.
161. Mitra S, Alnabulsi A, Secombes CJ, Bird S. Identification and characterization of the transcription factors involved in T-cell development, t-bet, stat6 and foxp3, within the zebrafish, Danio rerio. FEBS J. 2010 Jan;277(1):128-47.
162. Takizawa F, Koppang EO, Ohtani M, Nakanishi T, Hashimoto K, Fischer U, et al. Constitutive high expression of interleukin-4/13A and GATA-3 in gill and skin of salmonid fishes suggests that these tissues form Th2-skewed immune environments. Mol Immunol. 2011 Jul;48(12-13):1360-8.
163. Fry TJ, Mackall CL. The many faces of IL-7: from lymphopoiesis to peripheral T cell maintenance. J Immunol. 2005 Jun 1;174(11):6571-6.
164. Alves NL, Richard-Le Goff O, Huntington ND, Sousa AP, Ribeiro VS, Bordack A, et al. Characterization of the thymic IL-7 niche in vivo. Proc Natl Acad Sci U S A. 2009 Feb 3;106(5):1512-7.
165. Mazzucchelli RI, Warming S, Lawrence SM, Ishii M, Abshari M, Washington AV, et al. Visualization and identification of IL-7 producing cells in reporter mice. PLoS One. 2009;4(11):e7637.
166. Repass JF, Laurent MN, Carter C, Reizis B, Bedford MT, Cardenas K, et al. IL7-hCD25 and IL7-Cre BAC transgenic mouse lines: new tools for analysis of IL-7 expressing cells. Genesis. 2009 Apr;47(4):281-7.
167. Kono T, Bird S, Sonoda K, Savan R, Secombes CJ, Sakai M. Characterization and expression analysis of an interleukin-7 homologue in the Japanese pufferfish, Takifugu rubripes. FEBS J. 2008 Mar;275(6):1213-26.
168. Bei JX, Suetake H, Araki K, Kikuchi K, Yoshiura Y, Lin HR, et al. Two interleukin (IL)-15 homologues in fish from two distinct origins. Mol Immunol. 2006 Mar;43(7):860-9.
169. Fang W, Xiang LX, Shao JZ, Wen Y, Chen SY. Identification and characterization of an interleukin-15 homologue from Tetraodon nigroviridis. Comp Biochem Physiol B Biochem Mol Biol. 2006 Mar;143(3):335-43.
170. Gunimaladevi I, Savan R, Sato K, Yamaguchi R, Sakai M. Characterization of an interleukin-15 like (IL-15L) gene from zebrafish (Danio rerio). Fish Shellfish Immunol. 2007 Apr;22(4):351-62.
171. Wang T, Holland JW, Carrington A, Zou J, Secombes CJ. Molecular and functional characterization of IL-15 in rainbow trout Oncorhynchus mykiss: a potent inducer of IFN-gamma expression in spleen leukocytes. J Immunol. 2007 Aug 1;179(3):1475-88.
172. Parrish-Novak J, Dillon SR, Nelson A, Hammond A, Sprecher C, Gross JA, et al. Interleukin 21 and its receptor are involved in NK cell expansion and regulation of lymphocyte function. Nature. 2000 Nov 2;408(6808):57-63.
173. Asao H, Okuyama C, Kumaki S, Ishii N, Tsuchiya S, Foster D, et al. Cutting edge: the common gamma-chain is an indispensable subunit of the IL-21 receptor complex. J Immunol. 2001 Jul 1;167(1):1-5.
174. Vosshenrich CA, Di Santo JP. Cytokines: IL-21 joins the gamma(c)-dependent network? Curr Biol. 2001 Mar 6;11(5):R175-7.

175. Chtanova T, Tangye SG, Newton R, Frank N, Hodge MR, Rolph MS, et al. T follicular helper cells express a distinctive transcriptional profile, reflecting their role as non-Th1/Th2 effector cells that provide help for B cells. J Immunol. 2004 Jul 1;173(1):68-78.
176. Wurster AL, Rodgers VL, Satoskar AR, Whitters MJ, Young DA, Collins M, et al. Interleukin 21 is a T helper (Th) cell 2 cytokine that specifically inhibits the differentiation of naive Th cells into interferon gamma-producing Th1 cells. J Exp Med. 2002 Oct 7;196(7):969-77.
177. Leonard WJ, Zeng R, Spolski R. Interleukin 21: a cytokine/cytokine receptor system that has come of age. J Leukoc Biol. 2008 Aug;84(2):348-56.
178. Suto A, Kashiwakuma D, Kagami S, Hirose K, Watanabe N, Yokote K, et al. Development and characterization of IL-21-producing CD4+ T cells. J Exp Med. 2008 Jun 9;205(6):1369-79.
179. Frederiksen KS, Lundsgaard D, Freeman JA, Hughes SD, Holm TL, Skrumsager BK, et al. IL-21 induces in vivo immune activation of NK cells and CD8(+) T cells in patients with metastatic melanoma and renal cell carcinoma. Cancer Immunol Immunother. 2008 Oct;57(10):1439-49.
180. Konforte D, Simard N, Paige CJ. IL-21: an executor of B cell fate. J Immunol. 2009 Feb 15;182(4):1781-7.
181. Wang HJ, Xiang LX, Shao JZ, Jia S. Molecular cloning, characterization and expression analysis of an IL-21 homologue from Tetraodon nigroviridis. Cytokine. 2006 Aug;35(3-4):126-34.
182. Wang T, Diaz-Rosales P, Costa MM, Campbell S, Snow M, Collet B, et al. Functional characterization of a nonmammalian IL-21: rainbow trout Oncorhynchus mykiss IL-21 upregulates the expression of the Th cell signature cytokines IFN-gamma, IL-10, and IL-22. J Immunol. 2011 Jan 15;186(2):708-21.
183. Lutfalla G, Roest Crollius H, Stange-Thomann N, Jaillon O, Mogensen K, Monneron D. Comparative genomic analysis reveals independent expansion of a lineage-specific gene family in vertebrates: the class II cytokine receptors and their ligands in mammals and fish. BMC Genomics. 2003 Jul 17;4(1):29.
184. Fickenscher H, Hor S, Kupers H, Knappe A, Wittmann S, Sticht H. The interleukin-10 family of cytokines. Trends Immunol. 2002 Feb;23(2):89-96.
185. Moore KW, de Waal Malefyt R, Coffman RL, O'Garra A. Interleukin-10 and the interleukin-10 receptor. Annu Rev Immunol. 2001;19:683-765.
186. Zou J, Clark MS, Secombes CJ. Characterisation, expression and promoter analysis of an interleukin 10 homologue in the puffer fish, Fugu rubripes. Immunogenetics. 2003 Aug;55(5):325-35.
187. Savan R, Igawa D, Sakai M. Cloning, characterization and expression analysis of interleukin-10 from the common carp, Cyprinus carpio L. Eur J Biochem. 2003 Dec;270(23):4647-54.
188. Zhang DC, Shao YQ, Huang YQ, Jiang SG. Cloning, characterization and expression analysis of interleukin-10 from the zebrafish (Danio rerion). J Biochem Mol Biol. 2005 Sep 30;38(5):571-6.
189. Inoue Y, Kamota S, Ito K, Yoshiura Y, Ototake M, Moritomo T, et al. Molecular cloning and expression analysis of rainbow trout (Oncorhynchus mykiss) interleukin-10 cDNAs. Fish Shellfish Immunol. 2005 Apr;18(4):335-44.
190. Buonocore F, Randelli E, Bird S, Secombes C, Facchiano A, Constantini S, et al. Interleukin-10 expression by real-time PCR and homology modelling analysis in the European sea bass (Dicentrarchus Labrax L.). Aquaculture. 2007;270(1-4):512 - 22.
191. Pinto RD, Nascimento DS, Reis MI, do Vale A, Dos Santos NM. Molecular characterization, 3D modelling and expression analysis of sea bass (Dicentrarchus labrax L.) interleukin-10. Mol Immunol. 2007 Mar;44(8):2056-65.
192. Zhang Z, Swain T, Bogwald J, Dalmo RA, Kumari J. Bath immunostimulation of rainbow trout (Oncorhynchus mykiss) fry induces enhancement of inflammatory cytokine transcripts, while repeated bath induce no changes. Fish Shellfish Immunol. 2009 May;26(5):677-84.

193. Wegenka UM. IL-20: biological functions mediated through two types of receptor complexes. Cytokine Growth Factor Rev. 2010 Oct;21(5):353-63.
194. Stein C, Caccamo M, Laird G, Leptin M. Conservation and divergence of gene families encoding components of innate immune response systems in zebrafish. Genome Biol. 2007;8(11):R251.
195. Wang T, Diaz-Rosales P, Martin SA, Secombes CJ. Cloning of a novel interleukin (IL)-20-like gene in rainbow trout Oncorhynchus mykiss gives an insight into the evolution of the IL-10 family. Dev Comp Immunol. 2010 Feb;34(2):158-67.
196. Liang SC, Tan XY, Luxenberg DP, Karim R, Dunussi-Joannopoulos K, Collins M, et al. Interleukin (IL)-22 and IL-17 are coexpressed by Th17 cells and cooperatively enhance expression of antimicrobial peptides. J Exp Med. 2006 Oct 2;203(10):2271-9.
197. Cupedo T, Crellin NK, Papazian N, Rombouts EJ, Weijer K, Grogan JL, et al. Human fetal lymphoid tissue-inducer cells are interleukin 17-producing precursors to RORC+ CD127+ natural killer-like cells. Nat Immunol. 2009 Jan;10(1):66-74.
198. Duhen T, Geiger R, Jarrossay D, Lanzavecchia A, Sallusto F. Production of interleukin 22 but not interleukin 17 by a subset of human skin-homing memory T cells. Nat Immunol. 2009 Aug;10(8):857-63.
199. Eyerich S, Eyerich K, Pennino D, Carbone T, Nasorri F, Pallotta S, et al. Th22 cells represent a distinct human T cell subset involved in epidermal immunity and remodeling. J Clin Invest. 2009 Dec;119(12):3573-85.
200. Fujita H, Nograles KE, Kikuchi T, Gonzalez J, Carucci JA, Krueger JG. Human Langerhans cells induce distinct IL-22-producing CD4+ T cells lacking IL-17 production. Proc Natl Acad Sci U S A. 2009 Dec 22;106(51):21795-800.
201. Wolk K, Haugen HS, Xu W, Witte E, Waggie K, Anderson M, et al. IL-22 and IL-20 are key mediators of the epidermal alterations in psoriasis while IL-17 and IFN-gamma are not. J Mol Med (Berl). 2009 May;87(5):523-36.
202. Feinen B, Russell MW. Contrasting Roles of IL-22 and IL-17 in Murine Genital Tract Infection by Neisseria gonorrhoeae. Front Immunol. 2012;3:11.
203. Wolk K, Kunz S, Witte E, Friedrich M, Asadullah K, Sabat R. IL-22 increases the innate immunity of tissues. Immunity. 2004 Aug;21(2):241-54.
204. Aujla SJ, Chan YR, Zheng M, Fei M, Askew DJ, Pociask DA, et al. IL-22 mediates mucosal host defense against Gram-negative bacterial pneumonia. Nat Med. 2008 Mar;14(3):275-81.
205. Andoh A, Zhang Z, Inatomi O, Fujino S, Deguchi Y, Araki Y, et al. Interleukin-22, a member of the IL-10 subfamily, induces inflammatory responses in colonic subepithelial myofibroblasts. Gastroenterology. 2005 Sep;129(3):969-84.
206. Zheng Y, Valdez PA, Danilenko DM, Hu Y, Sa SM, Gong Q, et al. Interleukin-22 mediates early host defense against attaching and effacing bacterial pathogens. Nat Med. 2008 Mar;14(3):282-9.
207. Bird L. Mucosal immunology: IL-22 keeps commensals in their place. Nat Rev Immunol. 2012 Jul 6.
208. Wolk K, Kunz S, Asadullah K, Sabat R. Cutting edge: immune cells as sources and targets of the IL-10 family members? J Immunol. 2002 Jun 1;168(11):5397-402.
209. Wilson NJ, Boniface K, Chan JR, McKenzie BS, Blumenschein WM, Mattson JD, et al. Development, cytokine profile and function of human interleukin 17-producing helper T cells. Nat Immunol. 2007 Sep;8(9):950-7.
210. Manel N, Unutmaz D, Littman DR. The differentiation of human T(H)-17 cells requires transforming growth factor-beta and induction of the nuclear receptor RORgammat. Nat Immunol. 2008 Jun;9(6):641-9.
211. Pene J, Chevalier S, Preisser L, Venereau E, Guilleux MH, Ghannam S, et al. Chronically inflamed human tissues are infiltrated by highly differentiated Th17 lymphocytes. J Immunol. 2008 Jun 1;180(11):7423-30.
212. Cella M, Fuchs A, Vermi W, Facchetti F, Otero K, Lennerz JK, et al. A human natural killer cell subset provides an innate source of IL-22 for mucosal immunity. Nature. 2009 Feb 5;457(7230):722-5.

213. Hughes T, Becknell B, McClory S, Briercheck E, Freud AG, Zhang X, et al. Stage 3 immature human natural killer cells found in secondary lymphoid tissue constitutively and selectively express the TH 17 cytokine interleukin-22. Blood. 2009 Apr 23;113(17):4008-10.
214. Donnelly RP, Sheikh F, Dickensheets H, Savan R, Young HA, Walter MR. Interleukin-26: an IL-10-related cytokine produced by Th17 cells. Cytokine Growth Factor Rev. 2010 Oct;21(5):393-401.
215. Zou J, Yoshiura Y, Dijkstra JM, Sakai M, Ototake M, Secombes C. Identification of an interferon gamma homologue in Fugu, Takifugu rubripes. Fish Shellfish Immunol. 2004 Oct;17(4):403-9.
216. Igawa D, Sakai M, Savan R. An unexpected discovery of two interferon gamma-like genes along with interleukin (IL)-22 and -26 from teleost: IL-22 and -26 genes have been described for the first time outside mammals. Mol Immunol. 2006 Mar;43(7):999-1009.
217. Corripio-Miyar Y, Zou J, Richmond H, Secombes CJ. Identification of interleukin-22 in gadoids and examination of its expression level in vaccinated fish. Mol Immunol. 2009 Jun;46(10):2098-106.
218. Kolls JK. Th17 cells in mucosal immunity and tissue inflammation. Semin Immunopathol. 2010 Mar;32(1):1-2.
219. Haugarvoll E, Bjerkas I, Nowak BF, Hordvik I, Koppang EO. Identification and characterization of a novel intraepithelial lymphoid tissue in the gills of Atlantic salmon. J Anat. 2008 Aug;213(2):202-9.
220. Moseley TA, Haudenschild DR, Rose L, Reddi AH. Interleukin-17 family and IL-17 receptors. Cytokine Growth Factor Rev. 2003 Apr;14(2):155-74.
221. Yao Z, Timour M, Painter S, Fanslow W, Spriggs M. Complete nucleotide sequence of the mouse CTLA8 gene. Gene. 1996 Feb 12;168(2):223-5.
222. Li H, Chen J, Huang A, Stinson J, Heldens S, Foster J, et al. Cloning and characterization of IL-17B and IL-17C, two new members of the IL-17 cytokine family. Proc Natl Acad Sci U S A. 2000 Jan 18;97(2):773-8.
223. Shi Y, Ullrich SJ, Zhang J, Connolly K, Grzegorzewski KJ, Barber MC, et al. A novel cytokine receptor-ligand pair. Identification, molecular characterization, and in vivo immunomodulatory activity. J Biol Chem. 2000 Jun 23;275(25):19167-76.
224. Hymowitz SG, Filvaroff EH, Yin JP, Lee J, Cai L, Risser P, et al. IL-17s adopt a cystine knot fold: structure and activity of a novel cytokine, IL-17F, and implications for receptor binding. EMBO J. 2001 Oct 1;20(19):5332-41.
225. Lee J, Ho WH, Maruoka M, Corpuz RT, Baldwin DT, Foster JS, et al. IL-17E, a novel proinflammatory ligand for the IL-17 receptor homolog IL-17Rh1. J Biol Chem. 2001 Jan 12;276(2):1660-4.
226. Starnes T, Robertson MJ, Sledge G, Kelich S, Nakshatri H, Broxmeyer HE, et al. Cutting edge: IL-17F, a novel cytokine selectively expressed in activated T cells and monocytes, regulates angiogenesis and endothelial cell cytokine production. J Immunol. 2001 Oct 15;167(8):4137-40.
227. Starnes T, Broxmeyer HE, Robertson MJ, Hromas R. Cutting edge: IL-17D, a novel member of the IL-17 family, stimulates cytokine production and inhibits hemopoiesis. J Immunol. 2002 Jul 15;169(2):642-6.
228. Gunimaladevi I, Savan R, Sakai M. Identification, cloning and characterization of interleukin-17 and its family from zebrafish. Fish Shellfish Immunol. 2006 Oct;21(4):393-403.
229. Wang T, Martin SA, Secombes CJ. Two interleukin-17C-like genes exist in rainbow trout Oncorhynchus mykiss that are differentially expressed and modulated. Dev Comp Immunol. 2010 May;34(5):491-500.
230. Kumari J, Larsen AN, Bogwald J, Dalmo RA. Interleukin-17D in Atlantic salmon (Salmo salar): molecular characterization, 3D modelling and promoter analysis. Fish Shellfish Immunol. 2009 Nov;27(5):647-59.
231. Korenaga H, Kono T, Sakai M. Isolation of seven IL-17 family genes from the Japanese pufferfish Takifugu rubripes. Fish Shellfish Immunol. 2010 May-Jun;28(5-6):809-18.

232. Kono T, Korenaga H, Sakai M. Genomics of fish IL-17 ligand and receptors: a review. Fish Shellfish Immunol. 2011 Nov;31(5):635-43.
233. Huang D, Cancilla MR, Morahan G. Complete primary structure, chromosomal localisation, and definition of polymorphisms of the gene encoding the human interleukin-12 p40 subunit. Genes Immun. 2000 Dec;1(8):515-20.
234. Yoshiura Y, Kiryu I, Fujiwara A, Suetake H, Suzuki Y, Nakanishi T, et al. Identification and characterization of Fugu orthologues of mammalian interleukin-12 subunits. Immunogenetics. 2003 Aug;55(5):296-306.
235. Nascimento DS, do Vale A, Tomas AM, Zou J, Secombes CJ, dos Santos NM. Cloning, promoter analysis and expression in response to bacterial exposure of sea bass (Dicentrarchus labrax L.) interleukin-12 p40 and p35 subunits. Mol Immunol. 2007 Mar;44(9):2277-91.
236. Li MO, Flavell RA. TGF-beta: a master of all T cell trades. Cell. 2008 Aug 8;134(3):392-404.
237. Li MO, Wan YY, Sanjabi S, Robertson AK, Flavell RA. Transforming growth factor-beta regulation of immune responses. Annu Rev Immunol. 2006;24:99-146.
238. Saxena V, Lienesch DW, Zhou M, Bommireddy R, Azhar M, Doetschman T, et al. Dual roles of immunoregulatory cytokine TGF-beta in the pathogenesis of autoimmunity-mediated organ damage. J Immunol. 2008 Feb 1;180(3):1903-12.
239. Zhang L, Yi H, Xia XP, Zhao Y. Transforming growth factor-beta: an important role in CD4+CD25+ regulatory T cells and immune tolerance. Autoimmunity. 2006 Jun; 39(4):269-76.
240. Li B, Samanta A, Song X, Furuuchi K, Iacono KT, Kennedy S, et al. FOXP3 ensembles in T-cell regulation. Immunol Rev. 2006 Aug;212:99-113.
241. Wan YY, Flavell RA. 'Yin-Yang' functions of transforming growth factor-beta and T regulatory cells in immune regulation. Immunol Rev. 2007 Dec;220:199-213.
242. Haddad G, Hanington PC, Wilson EC, Grayfer L, Belosevic M. Molecular and functional characterization of goldfish (Carassius auratus L.) transforming growth factor beta. Dev Comp Immunol. 2008;32(6):654-63.
243. Kadowaki T, Yasui Y, Takahashi Y, Kohchi C, Soma G, Inagawa H. Comparative immunological analysis of innate immunity activation after oral administration of wheat fermented extract to teleost fish. Anticancer Res. 2009 Nov;29(11):4871-7.
244. Yang M, Wang Y, Wang X, Chen C, Zhou H. Characterization of grass carp (Ctenopharyngodon idellus) Foxp1a/1b/2: evidence for their involvement in the activation of peripheral blood lymphocyte subpopulations. Fish Shellfish Immunol. 2010 Feb;28(2):289-95.
245. Cai Z, Gao C, Li L, Xing K. Bipolar properties of red seabream (Pagrus major) transforming growth factor-beta in induction of the leucocytes migration. Fish Shellfish Immunol. 2010 Apr;28(4):695-700.
246. Yang M, Wang X, Chen D, Wang Y, Zhang A, Zhou H. TGF-beta1 exerts opposing effects on grass carp leukocytes: implication in teleost immunity, receptor signaling and potential self-regulatory mechanisms. PLoS One. 2012;7(4):e35011.
247. Samuel CE. Antiviral actions of interferons. Clin Microbiol Rev. 2001 Oct;14(4):778-809, table of contents.
248. Robertsen B. The interferon system of teleost fish. Fish Shellfish Immunol. 2006 Feb;20(2):172-91.
249. Bergan V, Steinsvik S, Xu H, Kileng O, Robertsen B. Promoters of type I interferon genes from Atlantic salmon contain two main regulatory regions. FEBS J. 2006 Sep;273(17):3893-906.
250. Kileng O, Brundtland MI, Robertsen B. Infectious salmon anemia virus is a powerful inducer of key genes of the type I interferon system of Atlantic salmon, but is not inhibited by interferon. Fish Shellfish Immunol. 2007 Aug;23(2):378-89.
251. Robertsen B, Bergan V, Rokenes T, Larsen R, Albuquerque A. Atlantic salmon interferon genes: cloning, sequence analysis, expression, and biological activity. J Interferon Cytokine Res. 2003 Oct;23(10):601-12.

252. Rokenes TP, Larsen R, Robertsen B. Atlantic salmon ISG15: Expression and conjugation to cellular proteins in response to interferon, double-stranded RNA and virus infections. Mol Immunol. 2007 Feb;44(5):950-9.
253. Zou J, Tafalla C, Truckle J, Secombes CJ. Identification of a second group of type I IFNs in fish sheds light on IFN evolution in vertebrates. J Immunol. 2007 Sep 15;179(6):3859-71.
254. Robertsen B. Expression of interferon and interferon-induced genes in salmonids in response to virus infection, interferon-inducing compounds and vaccination. Fish Shellfish Immunol. 2008 Oct;25(4):351-7.
255. Milev-Milovanovic I, Long S, Wilson M, Bengten E, Miller NW, Chinchar VG. Identification and expression analysis of interferon gamma genes in channel catfish. Immunogenetics. 2006 Feb;58(1):70-80.
256. Stolte EH, Savelkoul HF, Wiegertjes G, Flik G, Lidy Verburg-van Kemenade BM. Differential expression of two interferon-gamma genes in common carp (Cyprinus carpio L.). Dev Comp Immunol. 2008;32(12):1467-81.
257. Zou J, Carrington A, Collet B, Dijkstra JM, Yoshiura Y, Bols N, et al. Identification and bioactivities of IFN-gamma in rainbow trout Oncorhynchus mykiss: the first Th1-type cytokine characterized functionally in fish. J Immunol. 2005 Aug 15;175(4):2484-94.
258. Grayfer L, Garcia EG, Belosevic M. Comparison of macrophage antimicrobial responses induced by type II interferons of the goldfish (Carassius auratus L.). J Biol Chem. 2010 Jul 30;285(31):23537-47.
259. Arts JA, Tijhaar EJ, Chadzinska M, Savelkoul HF, Verburg-van Kemenade BM. Functional analysis of carp interferon-gamma: evolutionary conservation of classical phagocyte activation. Fish Shellfish Immunol. 2010 Nov;29(5):793-802.
260. Sun B, Skjaeveland I, Svingerud T, Zou J, Jorgensen J, Robertsen B. Antiviral activity of salmonid gamma interferon against infectious pancreatic necrosis virus and salmonid alphavirus and its dependency on type I interferon. J Virol. 2011 Sep;85(17):9188-98.
261. Duguid JR, Dinauer MC. Library subtraction of in vitro cDNA libraries to identify differentially expressed genes in scrapie infection. Nucleic Acids Res. 1990 May 11;18(9):2789-92.
262. Hara E, Kato T, Nakada S, Sekiya S, Oda K. Subtractive cDNA cloning using oligo(dT)30-latex and PCR: isolation of cDNA clones specific to undifferentiated human embryonal carcinoma cells. Nucleic Acids Res. 1991 Dec;19(25):7097-104.
263. Hedrick SM, Cohen DI, Nielsen EA, Davis MM. Isolation of cDNA clones encoding T cell-specific membrane-associated proteins. Nature. 1984 Mar 8-14;308(5955):149-53.
264. Lisitsyn N, Wigler M. Cloning the differences between two complex genomes. Science. 1993 Feb 12;259(5097):946-51.
265. Hubank M, Schatz DG. Identifying differences in mRNA expression by representational difference analysis of cDNA. Nucleic Acids Res. 1994 Dec 25;22(25):5640-8.
266. Liang P, Pardee AB. Differential display of eukaryotic messenger RNA by means of the polymerase chain reaction. Science. 1992 Aug 14;257(5072):967-71.
267. Welsh J, Chada K, Dalal SS, Cheng R, Ralph D, McClelland M. Arbitrarily primed PCR fingerprinting of RNA. Nucleic Acids Res. 1992 Oct 11;20(19):4965-70.
268. Sompayrac L, Jane S, Burn TC, Tenen DG, Danna KJ. Overcoming limitations of the mRNA differential display technique. Nucleic Acids Res. 1995 Nov 25;23(22):4738-9.
269. Bertioli DJ, Schlichter UH, Adams MJ, Burrows PR, Steinbiss HH, Antoniw JF. An analysis of differential display shows a strong bias towards high copy number mRNAs. Nucleic Acids Res. 1995 Nov 11;23(21):4520-3.
270. Siebert PD, Chenchik A, Kellogg DE, Lukyanov KA, Lukyanov SA. An improved PCR method for walking in uncloned genomic DNA. Nucleic Acids Res. 1995 Mar 25;23(6):1087-8.
271. Diatchenko L, Lau YF, Campbell AP, Chenchik A, Moqadam F, Huang B, et al. Suppression subtractive hybridization: a method for generating differentially regulated or tissue-specific cDNA probes and libraries. Proc Natl Acad Sci U S A. 1996 Jun 11;93(12):6025-30.
272. Wang L, Wu X. Identification of differentially expressed genes in lipopolysaccharide-stimulated yellow grouper Epinephelus awoara spleen. Fish Shellfish Immunol. 2007 Aug;23(2):354-63.

273. Xia JH, Yue GH. Identification and analysis of immune-related transcriptome in Asian seabass Lates calcarifer. BMC Genomics. 2010;11:356.
274. Feng CY, Johnson SC, Hori TS, Rise M, Hall JR, Gamperl AK, et al. Identification and analysis of differentially expressed genes in immune tissues of Atlantic cod stimulated with formalin-killed, atypical Aeromonas salmonicida. Physiol Genomics. 2009 May 13;37(3):149-63.
275. Tsoi SC, Ewart KV, Penny S, Melville K, Liebscher RS, Brown LL, et al. Identification of immune-relevant genes from atlantic salmon using suppression subtractive hybridization. Mar Biotechnol (NY). 2004 May-Jun;6(3):199-214.
276. Li CH, Chen J, Shi YH, Lu XJ. Use of suppressive subtractive hybridization to identify differentially expressed genes in ayu (Plecoglossus altivelis) associated with Listonella anguillarum infection. Fish Shellfish Immunol. 2011 Sep;31(3):500-6.
277. Matsuyama T, Fujiwara A, Takano T, Nakayasu C. Suppression subtractive hybridization coupled with microarray analysis to examine differential expression of genes in Japanese flounder Paralichthys olivaceus leucocytes during Edwardsiella tarda and viral hemorrhagic septicemia virus infection. Fish Shellfish Immunol. 2011 Oct;31(4):524-32.
278. Bo J, Giesy JP, Ye R, Wang KJ, Lee JS, Au DW. Identification of differentially expressed genes and quantitative expression of complement genes in the liver of marine medaka Oryzias melastigma challenged with Vibrio parahaemolyticus. Comp Biochem Physiol Part D Genomics Proteomics. 2012 Jun;7(2):191-200.
279. Purcell MK, Smith KD, Hood L, Winton JR, Roach JC. Conservation of Toll-Like Receptor Signaling Pathways in Teleost Fish. Comp Biochem Physiol Part D Genomics Proteomics. 2006 Mar;1(1):77-88.
280. Trinchieri G, Sher A. Cooperation of Toll-like receptor signals in innate immune defence. Nat Rev Immunol. 2007 Mar;7(3):179-90.
281. Bayne CJ, Gerwick L. The acute phase response and innate immunity of fish. Dev Comp Immunol. 2001 Oct-Dec;25(8-9):725-43.
282. Bayne CJ, Gerwick L, Fujiki K, Nakao M, Yano T. Immune-relevant (including acute phase) genes identified in the livers of rainbow trout, Oncorhynchus mykiss, by means of suppression subtractive hybridization. Dev Comp Immunol. 2001 Apr;25(3):205-17.
283. Holland MC, Lambris JD. The complement system in teleosts. Fish Shellfish Immunol. 2002 May;12(5):399-420.
284. Rise ML, Hall J, Rise M, Hori T, Gamperl AK, Kimball J, et al. Functional genomic analysis of the response of Atlantic cod (Gadus morhua) spleen to the viral mimic polyriboinosinic polyribocytidylic acid (pIC). Dev Comp Immunol. 2008;32(8):916-31.
285. He W, Yinlt ZX, Li Y, Huo WL, Guan HJ, Weng SP, et al. Differential gene expression profile in spleen of mandarin fish Siniperca chuatsi infected with ISKNV, derived from suppression subtractive hybridization. Dis Aquat Organ. 2006 Dec 14;73(2):113-22.
286. Dios S, Poisa-Beiro L, Figueras A, Novoa B. Suppression subtraction hybridization (SSH) and macroarray techniques reveal differential gene expression profiles in brain of sea bream infected with nodavirus. Mol Immunol. 2007 Mar;44(9):2195-204.
287. Poisa-Beiro L, Dios S, Ahmed H, Vasta GR, Martinez-Lopez A, Estepa A, et al. Nodavirus infection of sea bass (Dicentrarchus labrax) induces up-regulation of galectin-1 expression with potential anti-inflammatory activity. J Immunol. 2009 Nov 15; 183(10):6600-11.
288. Rise ML, Hall JR, Rise M, Hori TS, Browne MJ, Gamperl AK, et al. Impact of asymptomatic nodavirus carrier state and intraperitoneal viral mimic injection on brain transcript expression in Atlantic cod (Gadus morhua). Physiol Genomics. 2010 Jul 7;42(2):266-80.
289. Xu D, Wei J, Cui H, Gong J, Yan Y, Lai R, et al. Differential profiles of gene expression in grouper Epinephelus coioides, infected with Singapore grouper iridovirus, revealed by suppression subtractive hybridization and DNA microarray. J Fish Biol. 2010 Aug;77(2):341-60.
290. Fisicaro N, Aminian A, Hinchliffe SJ, Morgan BP, Pearse MJ, D'Apice AJ, et al. The pig analogue of CD59 protects transgenic mouse hearts from injury by human complement. Transplantation. 2000 Sep 27;70(6):963-8.

291. Janeway CA, Jr., Medzhitov R. Innate immune recognition. Annu Rev Immunol. 2002;20:197-216.
292. Tasumi S, Ohira T, Kawazoe I, Suetake H, Suzuki Y, Aida K. Primary structure and characteristics of a lectin from skin mucus of the Japanese eel Anguilla japonica. J Biol Chem. 2002 Jul 26;277(30):27305-11.
293. Tsutsui S, Iwamoto K, Nakamura O, Watanabe T. Yeast-binding C-type lectin with opsonic activity from conger eel (Conger myriaster) skin mucus. Mol Immunol. 2007 Feb;44(5):691-702.
294. Zhang H, Robison B, Thorgaard GH, Ristow SS. Cloning, mapping and genomic organization of a fish C-type lectin gene from homozygous clones of rainbow trout (Oncorhynchus mykiss). Biochim Biophys Acta. 2000 Nov 15;1494(1-2):14-22.
295. Honda K, Yanai H, Takaoka A, Taniguchi T. Regulation of the type I IFN induction: a current view. Int Immunol. 2005 Nov;17(11):1367-78.
296. Krasnov A, Timmerhaus G, Afanasyev S, Jorgensen SM. Development and assessment of oligonucleotide microarrays for Atlantic salmon (Salmo salar L.). Comp Biochem Physiol Part D Genomics Proteomics. 2011 Mar;6(1):31-8.
297. von Schalburg KR, Rise ML, Cooper GA, Brown GD, Gibbs AR, Nelson CC, et al. Fish and chips: various methodologies demonstrate utility of a 16,006-gene salmonid microarray. BMC Genomics. 2005;6:126.
298. Koop BF, von Schalburg KR, Leong J, Walker N, Lieph R, Cooper GA, et al. A salmonid EST genomic study: genes, duplications, phylogeny and microarrays. BMC Genomics. 2008;9:545.
299. Krasnov A, Koskinen H, Pehkonen P, Rexroad CE, 3rd, Afanasyev S, Molsa H. Gene expression in the brain and kidney of rainbow trout in response to handling stress. BMC Genomics. 2005;6:3.
300. Schiotz BL, Jorgensen SM, Rexroad C, Gjoen T, Krasnov A. Transcriptomic analysis of responses to infectious salmon anemia virus infection in macrophage-like cells. Virus Res. 2008 Sep;136(1-2):65-74.
301. Kurobe T, Yasuike M, Kimura T, Hirono I, Aoki T. Expression profiling of immune-related genes from Japanese flounder Paralichthys olivaceus kidney cells using cDNA microarrays. Dev Comp Immunol. 2005;29(6):515-23.
302. Williams TD, Diab AM, George SG, Godfrey RE, Sabine V, Conesa A, et al. Development of the GENIPOL European flounder (Platichthys flesus) microarray and determination of temporal transcriptional responses to cadmium at low dose. Environ Sci Technol. 2006 Oct 15;40(20):6479-88.
303. Park KC, Osborne JA, Montes A, Dios S, Nerland AH, Novoa B, et al. Immunological responses of turbot (Psetta maxima) to nodavirus infection or polyriboinosinic polyribocytidylic acid (pIC) stimulation, using expressed sequence tags (ESTs) analysis and cDNA microarrays. Fish Shellfish Immunol. 2009 Jan;26(1):91-108.
304. Cerda J, Mercade J, Lozano JJ, Manchado M, Tingaud-Sequeira A, Astola A, et al. Genomic resources for a commercial flatfish, the Senegalese sole (Solea senegalensis): EST sequencing, oligo microarray design, and development of the Soleamold bioinformatic platform. BMC Genomics. 2008;9:508.
305. Skugor S, Jorgensen SM, Gjerde B, Krasnov A. Hepatic gene expression profiling reveals protective responses in Atlantic salmon vaccinated against furunculosis. BMC Genomics. 2009;10:503.
306. Tilton SC, Gerwick LG, Hendricks JD, Rosato CS, Corley-Smith G, Givan SA, et al. Use of a rainbow trout oligonucleotide microarray to determine transcriptional patterns in aflatoxin B1-induced hepatocellular carcinoma compared to adjacent liver. Toxicol Sci. 2005 Dec;88(2):319-30.
307. Millan A, Gomez-Tato A, Fernandez C, Pardo BG, Alvarez-Dios JA, Calaza M, et al. Design and performance of a turbot (Scophthalmus maximus) oligo-microarray based on ESTs from immune tissues. Mar Biotechnol (NY). 2010 Aug;12(4):452-65.
308. Peatman E, Baoprasertkul P, Terhune J, Xu P, Nandi S, Kucuktas H, et al. Expression analysis of the acute phase response in channel catfish (Ictalurus punctatus) after infection with a Gram-negative bacterium. Dev Comp Immunol. 2007;31(11):1183-96.

309. MacKenzie S, Balasch JC, Novoa B, Ribas L, Roher N, Krasnov A, et al. Comparative analysis of the acute response of the trout, O. mykiss, head kidney to in vivo challenge with virulent and attenuated infectious hematopoietic necrosis virus and LPS-induced inflammation. BMC Genomics. 2008;9:141.
310. Prieto-Alamo MJ, Abril N, Osuna-Jimenez I, Pueyo C. Solea senegalensis genes responding to lipopolysaccharide and copper sulphate challenges: large-scale identification by suppression subtractive hybridization and absolute quantification of transcriptional profiles by real-time RT-PCR. Aquat Toxicol. 2009 Mar 9;91(4):312-9.
311. Li RW, Waldbieser GC. Production and utilization of a high-density oligonucleotide microarray in channel catfish, Ictalurus punctatus. BMC Genomics. 2006;7:134.
312. Fittipaldi N, Gottschalk M, Vanier G, Daigle F, Harel J. Use of selective capture of transcribed sequences to identify genes preferentially expressed by Streptococcus suis upon interaction with porcine brain microvascular endothelial cells. Appl Environ Microbiol. 2007 Jul;73(13):4359-64.
313. Lun ZR, Wang QP, Chen XG, Li AX, Zhu XQ. Streptococcus suis: an emerging zoonotic pathogen. Lancet Infect Dis. 2007 Mar;7(3):201-9.
314. Rise ML, von Schalburg KR, Brown GD, Mawer MA, Devlin RH, Kuipers N, et al. Development and application of a salmonid EST database and cDNA microarray: data mining and interspecific hybridization characteristics. Genome Res. 2004 Mar;14(3):478-90.
315. Matsumoto M, Tanaka T, Kaisho T, Sanjo H, Copeland NG, Gilbert DJ, et al. A novel LPS-inducible C-type lectin is a transcriptional target of NF-IL6 in macrophages. J Immunol. 1999 Nov 1;163(9):5039-48.
316. MacKenzie S, Iliev D, Liarte C, Koskinen H, Planas JV, Goetz FW, et al. Transcriptional analysis of LPS-stimulated activation of trout (Oncorhynchus mykiss) monocyte/macrophage cells in primary culture treated with cortisol. Mol Immunol. 2006 Mar; 43(9):1340-8.
317. Martin SA, Blaney SC, Houlihan DF, Secombes CJ. Transcriptome response following administration of a live bacterial vaccine in Atlantic salmon (Salmo salar). Mol Immunol. 2006 Apr;43(11):1900-11.
318. Ching B, Jamieson S, Heath JW, Heath DD, Hubberstey A. Transcriptional differences between triploid and diploid Chinook salmon (Oncorhynchus tshawytscha) during live Vibrio anguillarum challenge. Heredity (Edinb). 2010 Feb;104(2):224-34.
319. Dumrongphol Y, Hirota T, Kondo H, Aoki T, Hirono I. Identification of novel genes in Japanese flounder (Paralichthys olivaceus) head kidney up-regulated after vaccination with Streptococcus iniae formalin-killed cells. Fish Shellfish Immunol. 2009 Jan;26(1): 197-200.
320. Gerwick L, Corley-Smith G, Bayne CJ. Gene transcript changes in individual rainbow trout livers following an inflammatory stimulus. Fish Shellfish Immunol. 2007 Mar;22(3):157-71.
321. Meijer AH, Verbeek FJ, Salas-Vidal E, Corredor-Adamez M, Bussman J, van der Sar AM, et al. Transcriptome profiling of adult zebrafish at the late stage of chronic tuberculosis due to Mycobacterium marinum infection. Mol Immunol. 2005 Jun;42(10):1185-203.
322. van der Sar AM, Spaink HP, Zakrzewska A, Bitter W, Meijer AH. Specificity of the zebrafish host transcriptome response to acute and chronic mycobacterial infection and the role of innate and adaptive immune components. Mol Immunol. 2009 Jul;46(11-12):2317-32.
323. Purcell MK, Nichols KM, Winton JR, Kurath G, Thorgaard GH, Wheeler P, et al. Comprehensive gene expression profiling following DNA vaccination of rainbow trout against infectious hematopoietic necrosis virus. Mol Immunol. 2006 May;43(13):2089-106.
324. Byon JY, Ohira T, Hirono I, Aoki T. Use of a cDNA microarray to study immunity against viral hemorrhagic septicemia (VHS) in Japanese flounder (Paralichthys olivaceus) following DNA vaccination. Fish Shellfish Immunol. 2005 Feb;18(2):135-47.
325. Jorgensen SM, Afanasyev S, Krasnov A. Gene expression analyses in Atlantic salmon challenged with infectious salmon anemia virus reveal differences between individuals with early, intermediate and late mortality. BMC Genomics. 2008;9:179.

326. Encinas P, Rodriguez-Milla MA, Novoa B, Estepa A, Figueras A, Coll J. Zebrafish fin immune responses during high mortality infections with viral haemorrhagic septicemia rhabdovirus. A proteomic and transcriptomic approach. BMC Genomics. 2010;11:518.

CHAPTER 4

The Humoral and Cellular Immune Response Characteristic of Propolis Flavonoid Used as Adjuvant to Inactivated PPV Vaccine in Sows

Xia Ma[1,2], Zhenhuan Guo[2], Yonglu Liu[1], Yanzhou Zhou[1], Xiaolin Wang[1], Zhiqiang Shen[2*] and Jinliang Wang[2]

[1]*Medicinal Engineering Department of Henan University of Animal Husbandry and Economy, Zhengzhou, P.R. China*

[2]*Binzhou Animal Science and Veterinary Medicine Academy of Shandong Province, P.R. China*

ABSTRACT

The present study was conducted to assess the effects of propolis flavonoid used as adjuvant on the immune response of Landrace-Yorkshire hybrid sows immunized with an inactivated vaccine of porcine parvovirus (PPV). Thirty Landrace-Yorkshire hybrid sows were randomly assigned to one of 3 groups, and the sows in two adjuvant groups were intramuscular injected PPV vaccine with 2.0 mL propolis flavonoid adjuvant (PA) or oilemulsion adjuvant (OA), respectively. After that, serum hemagglutination inhibition antibody titers, specific IgM, IgA and IgG subclasses were measured to estimate the humoral immunity for the adjuvant effects of propolis flavonoid, as well as peripheral lymphopoiesis activity, and lymphocyte factor concentrations of Th1 and Th2 for cellular immune responses. Results indicated in a significant enhancing effect of PA on concentrations of IgM, IgG2, IgG3, IL-2 and IFN-γ comparing to OA. Especially in improving the effect of Th1 cellular immune response, the PA was superior to OA. These findings suggested that PA can significantly enhance the immune responses against PPV vaccine and could be exploited to a new PPV vaccine in sows.

Keywords: Propolis flavonoid; Porcine parvovirus; Immunological adjuvant; Humoral immune response; Cellular immune response.

1. INTRODUCTION

Propolis, applied as a traditional cure in folk medicine empirically for centuries, is a mix ingredients composeed by honeybees with various **plant** resinous and mandibular gland secretion [1]. It is well known for latent health benefit and is reported to own valuable biological activities, including antivirus, immune enhancement, **antibiosis**, antioxidation, hepatoprotection, **anticancer**, anti-fatigue, and so on [2-4]. It is an splendid immune system booster and natural **antibiotic** with no sideeffect. In recent years, propolis has become an issue of increasing interest in the investigators for its versatile biological activities and broadly exchanged by the pharmaceutical industries as the health-food and an alternative medicine in various parts of the world.

Propolis flavonoid (PF) is a kind of ingredients extracted from propolis and has been used as a harmless natural adjuvant in chickens vaccinated with activated vaccine, and the results showed that PF could remarkably improve the immune activity in the cellular and humoral immune response [2,4]. But the adjuvant effects and feature PF on inactivated vaccine to sows had not be considered detailedly in humoral and cellular immunity response.

Recently, the animal viral infectious diseases, such as severe untypical classical swine fever, acute respiratory syndrome, cow transmissible spongiform encephalopathy, swine hyperpyrexia disease and so on are worldwide concerned as they made a stage of comeback with emerge continuously and spread quickly, and constantly cause a enormous loss in domestic animal and poultry industry [5]. PPV could cause reproductive failure in **pregnant** sows, which is characterized by stillbirths, fetal and embryonic death, mummification and delayed return to oestrus [6].

Although the acute infection of postnatal and non-pregnant pigs is often sub-clinical, PPV has been linked to skin lesion occurrence in piglets [7], non-suppurative myocarditis in lactating piglets and interstitial nephritis in slaughter aged pigs [8].

Several research results have pointed out that vaccination is the effective method for controlling viral disease [9,10]. The ideal and successful **vaccination** depends on the association with antigen and potent adjuvant which can increase the immune effect of vaccine. A desired adjuvant could not only start specific effectors of the immune system to strengthen the cellular and/or humoral immune responses against that antigen, but also activate auxiliary or cytotoxic T cells [8,11,12]. On the other hand, better adjuvant should have no or lower toxicity and side effects [13]. But there are many deficiencies in commonly used adjuvant, for instance, oilemulsion adjuvant (OA) can cause induration or necrosis in the local, acute inflammation and disseminated granulomas in lungs, lymph nodes and skeletal muscles in rats or rabbits [14]. Therefore, it is urgent to explore a new adjuvant with high humoral and cellular immunity efficacy and security.

Our previous studies demonstrated that PF possessed a better immunological enhancement on cellular and humoral immunity in model **animal** of guinea pig inactivated PPV vaccine [15]. In this study, the authors determined the effects of PF on serum antibody tiers, immunoglobulin (Ig)M, IgA and IgG subclass for humoral immunity, and peripheral lymphopoiesis and lymphocyte excretion of Th1 cytokine (interleukin (IL)-2 and interferon (IFN)-γ) and Th2 cytokine (IL-4, IL-6, IL-10 and IL-12) for cellular immunity of pigs vaccinated with inactivated PPV vaccine compared with OA. Thus, the aim of this study was to observe the adjuvant effect and characteristic of PF on immune response of immunized sows and offer theoretical evidence for developing PF into PPV vaccine adjuvant.

2. MATERIALS AND METHODS

2.1 Preparation of adjuvant and vaccine

Propolis was purchased from Shandong province Qingkun Beeswax honeycomb processing factory. PF was prepared in our laboratory in a final purity of 925 $mg \cdot g^{-1}$. The PF adjuvant (PA) was prepared as previoulsy described [16]. The PF was dissolved with phosphate buffer solution (PBS, pH 6.2) in a final concentration of 40 $mg \cdot mL^{-1}$. PPV, the virus contained 2048 haemagglutination units $(HAU) \cdot mL^{-1}$, was obtained from China Institute of Veterinary Drug Control. The adjuvant vaccines containing PA or OA were prepared by Lvdu Veterinary Biologicals Co. Ltd., Binzhou, China. Their virus **antigen** contents were the same.

2.2 Main reagents

RPMI-1640 (Gibco) supplemented with streptomycin 100 IU·mL^{-1}, benzylpenicillin 100 IU·mL^{-1} and 10% fetal bovine serum (FBS), was used for re-suspending and washing the cells, culturing and dilutingthe mitogen the cells. The mitogen of concanavalin A (ConA, Sigma) was dissolved with 0.1 mg·mL^{-1} RPMI-1640. Sodium heparin was dissolved with 5 mg·mL^{-1} of PBS. The 3-(4,5-dimethylthiazol-2-yl)-2,5- diphenyltetrazolium bromide (MTT, Amresco Co.) was dissolved into 5 mg·mL^{-1} with calcium and magnesium free (CMF) PBS (pH 7.4). These reagents were all filtered through a 0.22 μm syringe filter. Dimethyl sulfoxide (DMSO) was purchased from Shanghai Lingfeng Chemical Reagent Ltd. Lymphocytes separation medium was purchased from Tianjing Haoyang Biological Manufacture co. Ltd. The enzyme-linked immunosorbent assay (ELISA) kits for IgM, IgA, IgG1, IgG2, IgG3, IgG4, IL-2, IFN-γ, IL-4, IL-6, IL^{-1}0 and IL^{-1}2 were purchased from R&D Systems Inc., Minneapolis, USA.

2.3 Animals, housing and treatment

Thirty Landrace-Yorkshire hybrid sows (aged from 198 to 204, and average weight is 64.5 kg) were randomly assigned to one of 3 groups, receiving intramuscular injection of PPV vaccine with 2.0 mL PA, OA or physiological saline (blank control group, BC group), respectively. The animals were kept in ten pigpens divide equally under standard conditions in the Experimental Animal Center of Shandong province Animal Science and Veterinary **Medicine** Academy of Binzhou, (NO. SYXK (lu) 20110066). They were maintained in an air-conditioned room with light from 06:00 h to 18:00 h. The room temperature (24 ± 3oC) and humidity were controlled automatically. They were fed water and food *ad libitum*. All the experiment animals of procedures and their care conformed to the internationally accepted principles as found by the government of China for the Guidelines for Keeping Experimental Animals issued. The antibody against PPV was negative before the experiment.

On days 7, 14, 21, 28, 35 and 42 after vaccination, the blood of 6 sows randomly from each group were sampled for measuring PPV serum hemagglutination inhibition (HI) antibody titer by micro-method, IgM, IgA and IgG subclasses with method of ELISA kit dynamically and continuously. On days 7, 14, 21, 28, 35 and 42 after vaccination, the peripheral blood of 4 pigs randomly from each group were sampled for examination of lymphocyte proliferation by MTT assay [17] and analyzing lymphocyte cytokine of IL-2, IFN-γ, IL-4, IL-6, IL-10 and IL-12 by ELISA kit [16].

2.4 Statistical analysis

Data are expressed as the mean ± S.D. LSD's and Duncan multiple range test was used to determine the differences among groups. P-values of less than 0.05 were considered statistically significant.

3. RESULTS

3.1 The serum variation antibody titer

The serum dynamic variation HI antibody titers of each group are demonstrated in **Table 1**. The HI antibody titers in PA and OA vaccination groups improved significantly ($p<0.05$) from days 7 to 42 compared with the BC group. On days 7 and 14 after vaccination, the HI antibody titers in PA group were higher than OA group ($p>0.05$). On days 21, 28, 35 and 42, the HI antibody titers in group PA were lower significantly than those of group OA ($p<0.05$).

Table 1: The dynamic variation of HI antibody titer after vaccination (log2).

Group	D_7	D_{14}	D_{21}	D_{28}	D_{35}	D_{42}
PA	4.5 ± 0.4^a	7.4 ± 0.5^a	7.5 ± 0.4^b	7.8 ± 0.7^b	7.3 ± 0.5^b	6.5 ± 0.5^b
OA	4.3 ± 0.5^a	7.3 ± 0.5^a	8.5 ± 0.6^a	8.8 ± 0.5^a	8.3 ± 0.7^a	7.5 ± 0.7^a
BC	0 ± 0^b	0 ± 0^b	0 ± 0^c	0 ± 0^c	0 ± 0^c	0 ± 0^c

PA: Propolis Adjuvant; OA: Oilemulsion Adjuvant; BC: Blank Control; The same as follows.
Data within a column without the same superscripts (a-c) differ significantly ($p<0.05$).

3.2 The serum variation of IgM and IgA content

The serum variation of IgM content of every group is illustrated in **Figure 1A**. On days 7 and 21 after vaccination, IgM content of PA group was higher than that of OA group ($p<0.05$), while on days 14, 28, 35 and 42 after vaccination, IgM content of PA group only was numerically higher ($p>0.05$)

than that of OA group, and higher significantly ($p<0.05$) than that of BC group.

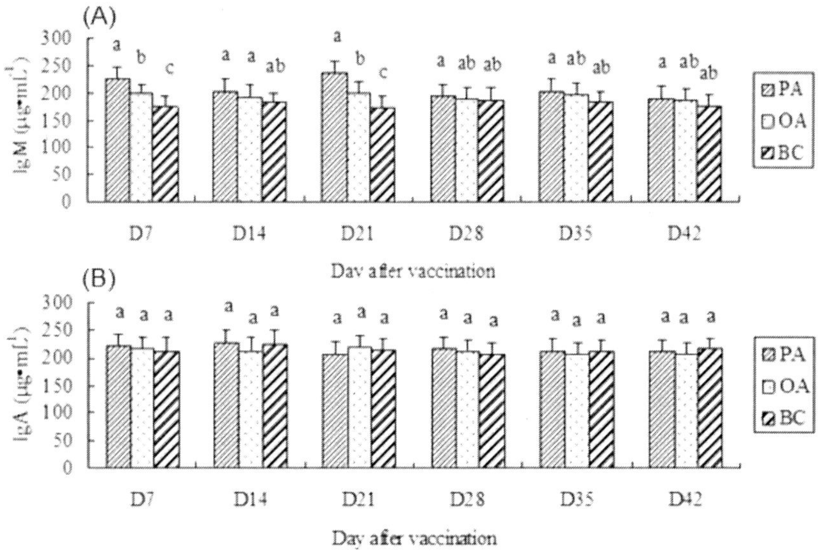

Figure 1: The dynamic changes of IgM and IgA content after vaccination ($\mu g \cdot mL^{-1}$). $^{a-c}$ Bars in the same day without the same superscripts differ significantly ($p<0.05$).

The serum variation of IgA content of every group is shown in **Figure 1B**. From 7 to 42 after immunization, the IgA content in group PA, OA and BC had no significant (($p>0.05$).

3.3 The serum variation of IgG subclasses content

The serum variation of IgG1 content of every group is illustrated in **Figure 2A**. The IgG1 content in PA and OA adjuvant groups was higher significantly ($p<0.05$) than that of BC group from 14 to 35 after vaccination. On day 7 after vaccination, IgG1 content of PA group was higher ($p>0.05$) than that of OA group. But on 21 and 28, the IgG1 content in OA adjuvant group was higher significantly ($p<0.05$) than that of PA group.

The serum variation of IgG2 content of every group is shown in **Figure 2B**. The IgG2 content in PA and OA adjuvant groups was higher significantly ($p<0.05$) than that of BC group from 7 to 28 after vaccination.

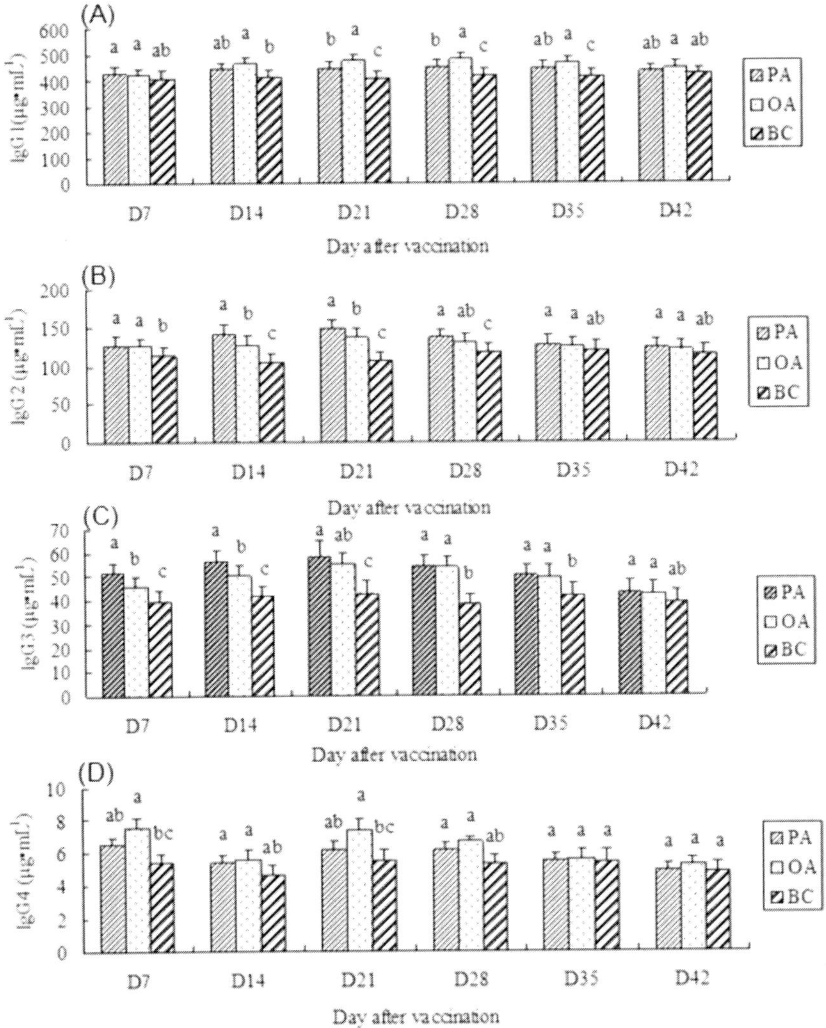

Figure 2: The changes of IgG1, IgG2, IgG3 and IgG4 content after vaccination (μg·mL^{-1}). [a-c] Bars in the same day without the same superscripts differ significantly (p<0.05).

On day 14 and 21 after vaccination, IgG2 content of PA group was higher significantly (p<0.05) than that of OA group. The serum variation of IgG3 content of every group is illustrated in **Figure 2C**. The IgG3 content in PA and OA adjuvant groups was higher significantly (p<0.05) than that of BC group from 7 to 35 after vaccination. On day 7 and 14 after vaccination,

IgG3 content of PA group was higher significantly ($p<0.05$) than that of OA group.

The serum variation of IgG4 content of every group is shown in **Figure 2D**. The IgG4 content in OA adjuvant group was higher significantly ($p<0.05$) than that of BC group on day 7 and 21 after vaccination. From 7 to 42 after vaccination, the IgG4 content of PA group was lower ($p>0.05$) than that of OA group.

3.4 Peripheral lymphocyte proliferation activity

The lymphocyte proliferation activities are shown in **Table 2**. The A_{570} values of PA group were significantly bigger than those of the OA and BC groups from days 7 to 28 after vaccination significantly ($p<0.05$). On day 7 and 21 after vaccination, the A_{570} values in OA group were bigger significantly ($p<0.05$) than that of BC group.

Table 2: The changes in lymphocyte proliferation of blood (A_{570} value).

Group	D_7	D_{14}	D_{21}	D_{28}	D_{35}	D_{42}
PA	0.36 ± 0.02^a	0.33 ± 0.01^a	0.37 ± 0.02^a	0.30 ± 0.01^a	0.29 ± 0.01^a	0.28 ± 0.01^a
OA	0.32 ± 0.01^b	0.29 ± 0.02^b	0.34 ± 0.02^b	0.27 ± 0.02^b	0.28 ± 0.02^a	0.29 ± 0.02^a
BC	0.29 ± 0.02^c	0.28 ± 0.01^b	0.30 ± 0.01^c	0.26 ± 0.02^b	0.27 ± 0.02^a	0.27 ± 0.01^a

Data within a column without the same superscripts a-c differ significantly ($p<0.05$).

3.5 Th1 cytokine level

Th1 cytokine of IL-2 content is listed as **Figure 3A**. The IL-2 content in PA group was significantly higher ($p<0.05$) than those of OA and BC groups from 7 to 42 after vaccination. And on day 7 and 21 after vaccination, IL-2 content of OA group was higher significantly ($p<0.05$) than that of BC group.

Th1 cytokine of IFN-γ content is listed as **Figure 3B**. PA triggered stronger IFN-γ content significantly ($p<0.05$) than the OA group from 7 to 42 after vaccination.

3.6 Th2 cytokine level

Th2 cytokine of IL-4 content is shown in **Figure 4A**. The IL-4 content in PA group was significantly lower ($p<0.05$) than that of OA group from 7 to 42 after vaccination. And on day 7 and 21 after vaccination, IL-4 content in PA group was significantly lower ($p<0.05$) than that of BC group.

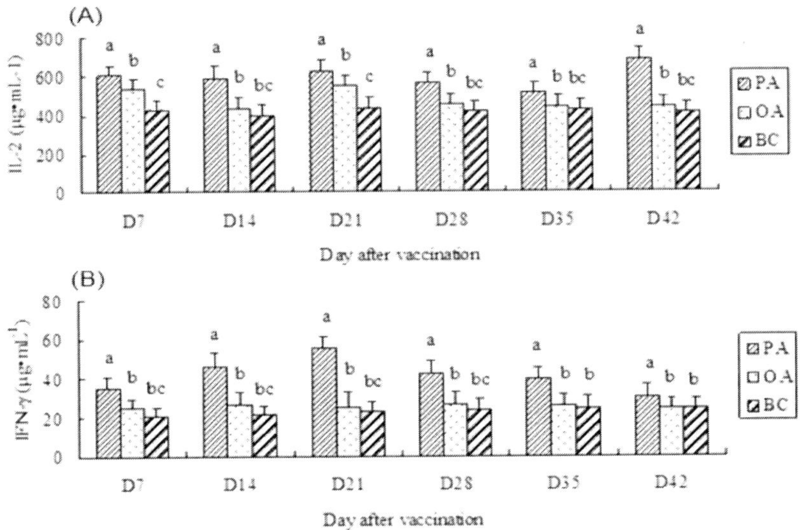

Figure 3: The dynamic changes of IL-2 and IFN-γ content after vaccination ($\mu g \cdot mL^{-1}$). $^{a\text{-}c}$ Bars in the same day without the same superscripts differ significantly ($p<0.05$).

Th2 cytokine of IL-6 content is shown in **Figure 4B**. The IL-6 content in PA group was lower ($p>0.05$) than that of OA group from 7 to 42 after vaccination.

Th2 cytokine of IL-10 content is illustrated in **Figure 4C**. The IL-10 content in PA group was significantly lower ($p<0.05$) than those of OA and BC groups from 7 to 28 after vaccination.

Th2 cytokine of IL-12 content is illustrated in **Figure 4D**. The IL-12 content in PA group was lower ($p>0.05$) than that of OA group from 7 to 42 after vaccination.

4. DISCUSSION

The combinations of adjuvant may combine multiple adjuvant components to achieve the desired mix of immune responses. During early period (7 d) of vaccination, our studies suggested that the serum HI antibody titer against PPV and IgM content of group PA were higher than that of group OA, which suggested that the PA was superior to the OA in the early vaccination. This could get the immuned animals more effectively on resisting wild virus infection. In mice model, Orsatti also verified that propolis could start the initial steps of the immune response [18].

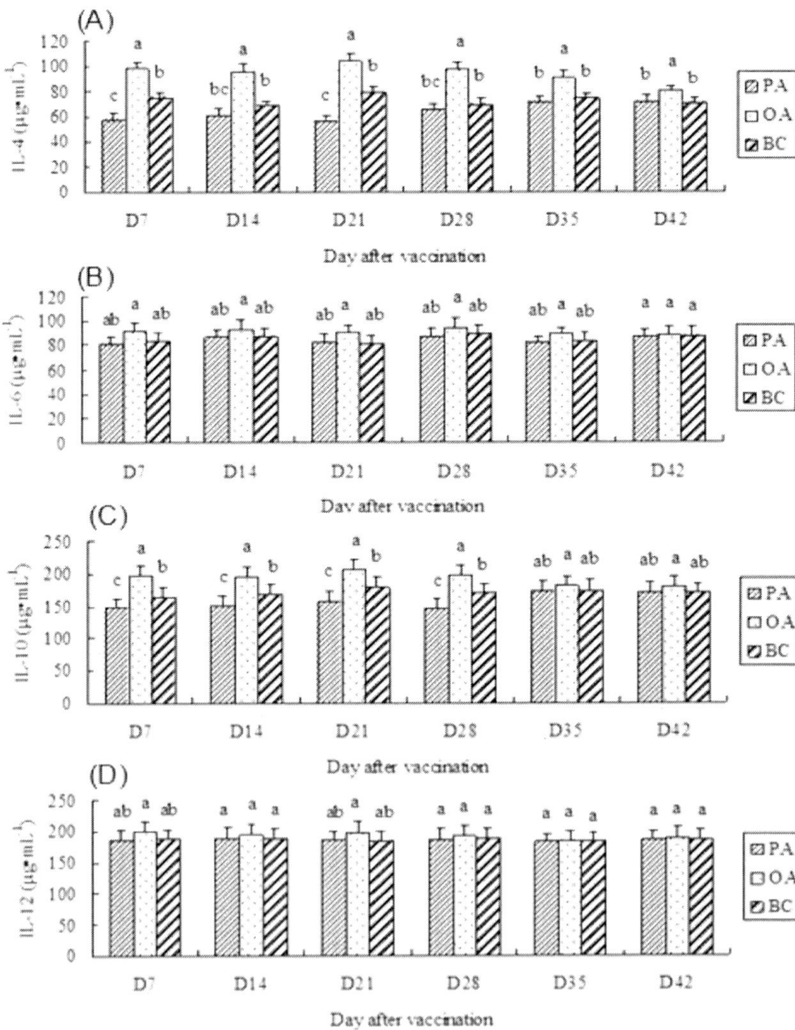

Figure 4: The changes of IL-4, IL-6, IL^{-1}0 and IL^{-1}2 content after vaccination (μg·mL^{-1}). $^{a-c}$ Bars in the same day without the same superscripts differ significantly (p<0.05).

There are different subclasses of IgG in serum such as IgG1, IgG2, IgG3 and IgG4 that afford the bulk of immunity to most infectious **disease**. Raised antibody titers are very important for microbial killing by way of antibody-mediated mechanisms. Our data showed that IgG2 and IgG3 contents were significantly higher in the sows immunized with PA than that of immunized with OA, indicating that PF acts synergistically to promote the IgG subclasses production. Sforcin also concluded that PF stimulates antibody production [19]. And studies in the animal models have shown that antibodies play a significant role in the immune system to defense against PPV [19-22].

PF enhances immune responses in various ways such as formation of the immunostimulating complex, activation of helper and cytotoxic T cells. In addition to activating the humoral immune response, PF also increased cellular immune responses [23,24]. In this study, PF was various in regards to their adjuvant effects on the immune responses stimulated the T lymphocyte cells to secrete cytokines. As shown in **Figure 3A and 3B**, PF could significantly increased the production of Th1 cytokines IL-2 and IFN-γ, which suggested that PF simultaneously elicited a Th1 immune response. These findings are consistent with Blonska's study which demonstrated that PF could improve the activity of T lymphocyte, increase the secretion of some cytokines, such as IL-2 and TNF-α, thus enhance immune function of organism [25].

Th1 immune activation contributes to cell-mediated immunity whereas Th2 immune activation favours the humoral immune response [26]. Th1/Th2 balance is a prerequisite for the functionality of immune system against infections. The immunomodulatory of PF has been widely investigated lately, both *in vivo* and *in vitro* [27]. PF has been advised to be used as a promising adjuvant material on duck inactivated vaccines [28]. Important functional properties of immune cells are their capability to synthesize and secrete soluble polypeptide factors referred to as cytokines. Most secreted cytokines are bind to specific receptors on the surface of target cells. Upon binding they act to regulate differentiation, growth and/or to optimize the immune response [29]. Using a Polish sample, Ansorge reported that propolis had immunoregulatory effects that might be regulated by Erk2 MAPK signals that improve cellular growth [30]. In T cell dependent immune response, the specific antibody produce need a progressive change in the predominant immunoglobulin class, which is adjusted by T cells and their cytokines. IL-4 and IL-10 preferentially switches activated B cells to the IgG1 isotype (Th2 type immune response), IL-2 and IFN-γ enhances IgG2 and IgG3 responses (Th1 type immune response) [31].

5. CONCLUSION

PF significantly increased the serum levels of IgM and IgG2/3, as well as T lymphocyte proliferation when administered in sows with an inactivated vaccine against PPV. The enhanced IgG subclasses levels paralleled the increased cytokine levels of IFN-γ and IL-2. This adjuvant characteristic and activity was evident in both cellular and humoral immune responses, and PF could be exploited to a new PPV vaccine adjuvant in sows.

ACKNOWLEDGEMENTS

The project was jointly supported by Shandong Province National Natural Science Foundation (Grant No.Y2005D10) and Henan Province National Natural Science Foundation of Education Department (Grant No. 13A230383, 14A230012). The authors would like to grateful to all other staff in the experiments for their assistances.

REFERENCES

1. Sforcin JM (2007) Propolis and the immune system: a review.J Ethnopharmacol 113: 1-14.
2. Fan Y, Ma L, Zhang W,Xu Y,Suolangzhaxi et al. (2014) Microemulsion can improve the immune-enhancing activity of propolis flavonoid on immunosuppression and immune response.Int J BiolMacromol 63: 126-132.
3. Yuqing Z,Zhan'ao S, Jiaguo L, Deyun W, Baokang Z, et al. (2012) Flavone ingredients can synergistically inhibit NDV infecting cell and improve ND vaccine's protective rate.Int J BiolMacromol 51: 201-208.
4. Fan YP, Hu YL, Wang DY, Guo ZH, Zhao XN, et al. (2010) Epimedium polysaccharide and propolis flavone can synergistically stimulate lymphocyte proliferation in vitro and enhance the immune responses to ND vaccine in chickens. Int JBiol Macromol 47: 87-92.
5. Kong X Hu Y, Rui R, Wang D, Li X (2004) Effects of Chinese herbal medicinal ingredients on peripheral lymphocyte proliferation and serum antibody titer after vaccination in chicken.IntImmunopharmacol 4: 975-982.
6. Mengeling WL (2006) Porcine parvovirus. In: Straw BE, Zimmerman JJ, D'Allaire S, Taylor DJ (eds.), Diseases of Swine, (9thedn.) Blackwell Publishing, Ames, IA.
7. Lager KM,Mengeling WL (1994) Porcine parvovirus associated with cutaneous lesions in piglets.J Vet Diagn Invest 6: 357-359.
8. Drolet R,D'Allaire S, Larochelle R, Magar R, Ribotta M, et al. (2002) Infectious agents identified in pigs with multifocal interstitial nephritis at slaughter.Vet Rec 150: 139-143.
9. Parke CR, Burgess GW (1993) An economic assessment of porcine parvovirus vaccination.Aust Vet J 70: 177-180.
10. Gardner IA, Carpenter TE, Leontides L, Parsons TD (1996) Financial evaluation of vaccination and testing alternatives for control of parvovirus-induced reproductive failure in swine.J Am Vet Med Assoc 208: 863-869.
11. Leclerc C (2003) New approaches in vaccine development.Comp ImmunolMicrobiol Infect Dis 26: 329-341.

12. Storni T,Kündig TM, Senti G, Johansen P (2005) Immunity in response to particulate antigen-delivery systems.Adv Drug Deliv Rev 57: 333-355.
13. Aguilar JC, Rodríguez EG (2007) Vaccine adjuvants revisited.Vaccine 25: 3752-3762.
14. Holmdahl R,Lorentzen JC, Lu S, Olofsson P, Wester L, et al. (2001) Arthritis induced in rats with nonimmunogenic adjuvants as models for rheumatoid arthritis.Immunol Rev 184: 184-202.
15. Ma X,Guo Z, Shen Z, Wang J, Hu Y, et al. (2011) The immune enhancement of propolis adjuvant on inactivated porcine parvovirus vaccine in guinea pig.Cell Immunol 270: 13-18.
16. Paulino N,Scremin FM, Raichaski LB, Marcucci MC, Scremin A, et al. (2002) Mechanisms involved in the relaxant action of the ethanolic extract of propolis in the guinea-pig trachea in-vitro.J Pharm Pharmacol 54: 845-852.
17. Chen Y, Wang D, Hu Y, Guo Z, Wang J, et al. (2010) Astragalus polysaccharide and oxymatrine can synergistically improve the immune efficacy of Newcastle disease vaccine in chicken.Int J BiolMacromol 46: 425-428.
18. Orsatti CL,Missima F, Pagliarone AC, Bachiega TF, Búfalo MC, et al. (2010) Propolisimmunomodulatory action in vivo on Toll-like receptors 2 and 4 expression and on pro-inflammatory cytokines production in mice.Phytother Res 24: 1141-1146.
19. Sforcin JM,Orsi RO, Bankova V (2005) Effect of propolis, some isolated compounds and its source plant on antibody production.J Ethnopharmacol 98: 301-305.
20. Molitor TW, Joo HS, Thacker BJ (1985) Potentiating effect of adjuvants on humoral immunity to porcine parvovirus vaccines in guinea pigs.Vet Microbiol 10: 209-218.
21. Pye D Bates J, Edwards SJ, Hollingworth J (1990) Development of a vaccine preventing parvovirus-induced reproductive failure in pigs.Aust Vet J 67: 179-182.
22. Paul PS, Mengeling WL (1986) Vaccination of swine with an inactivated porcine parvovirus vaccine in the presence of passive immunity. J Vet Med Assoc 188: 410-413.
23. Fischer G, Paulino N, Marcucci MC, Siedler BS, Munhoz LS (2010)Memorias do InstitutoOswaldo Cruz, Green propolis phenolic compounds act as vaccine adjuvants, improving humoral and cellular responses in mice inoculated with inactivated vaccines. Rio. de Janeiro 105: 908-913.
24. Pagliarone AC,Missima F, Orsatti CL, Bachiega TF, Sforcin JM (2009) Propolis effect on Th1/Th2 cytokines production by acutely stressed mice.J Ethnopharmacol125: 230-233.
25. Blonska M,Bronikowska J, Pietsz G, Czuba ZP, Scheller S, et al. (2004) Effects of ethanol extract of propolis (EEP) and its flavones on inducible gene expression in J774A.1 macrophages.J Ethnopharmacol 91: 25-30.
26. Das J, Chen CH, Yang L, Cohn L, Ray P, et al. (2001) A critical role for NF-kappa B in GATA3 expression and TH2 differentiation in allergic airway inflammation.Nat Immunol 2: 45-50.
27. Fischer G,Cleff MB, Dummer LA, Paulino N, Paulino AS, et al. (2007) Adjuvant effect of green propolis on humoral immune response of bovines immunized with bovine herpesvirus type 5.Vet ImmunolImmunopathol 116: 79-84.
28. Cai JL, Tang XL, Yang LF, Su XY(2001)Propolis inactivated vaccine against infectious serositis in young ducks. Chin J Vet Sci2: 552-553.
29. Sforcin JM,Orsi RO, Bankova V (2005) Effect of propolis, some isolated compounds and its source plant on antibody production.J Ethnopharmacol 98: 301-305.
30. Ansorge S, Reinhold D, Lendeckel U (2003) Propolis and some of its constituents down-regulate DNA synthesis and inflammatory cytokine production but induce TGF-beta1 production of human immune cells.Z Naturforsch C 58: 580-589.
31. Roitt I, Brostoff J, Male D (2001) Immunology, 6th Mosby Publications, London.

CHAPTER 5

Unravelling the Relationship between Animal Growth and Immune Response during Micro-Parasitic Infections

Andrea B. Doeschl-Wilson[1*], Will Brindle[2], Gerry Emmans[2], Ilias Kyriazakis[3,4]

[1]*Sustainable Livestock Systems, Scottish Agricultural College, Edinburgh, United Kingdom*

[2]*Animal Health, Scottish Agricultural College, Edinburgh, United Kingdom*

[3]*School of Agriculture, Food and Rural Development, Newcastle University, Newcastle upon Tyne, United Kingdom, 4 Faculty of Veterinary Medicine, University of Thessaly, Karditsa, Greece*

ABSTRACT

Background

Both host genetic potentials for growth and disease resistance, as well as nutrition are known to affect responses of individuals challenged with micro-parasites, but their interactive effects are difficult to predict from experimental studies alone.

Methodology/Principal Findings

Here, a mathematical model is proposed to explore the hypothesis that a host's response to pathogen challenge largely depends on the interaction between a host's genetic capacities for growth or disease resistance and the nutritional environment. As might be expected, the model predicts that if nutritional availability is high, hosts with higher growth capacities will also grow faster under micro-parasitic challenge, and more resistant animals will exhibit a more effective immune response. Growth capacity has little effect on immune response and resistance capacity has little effect on achieved growth. However, the influence of host genetics on phenotypic performance changes drastically if nutrient availability is scarce. In this case achieved growth and immune response depend simultaneously on both capacities for growth and disease resistance. A higher growth capacity (achieved e.g. through genetic selection) would be detrimental for the animal's ability to cope with pathogens and greater resistance may reduce growth in the short-term.

Significance

Our model can thus explain contradicting outcomes of genetic selection observed in experimental studies and provides the necessary biological background for understanding the influence of selection and/or changes in the nutritional environment on phenotypic growth and immune response.

1. INTRODUCTION

Models that predict phenotypic responses from the interaction between animal genotypes and the environment are desirable both in the context of agricultural systems [1] and evolutionary ecology [2], [3]. Whereas extensive knowledge exists about the effects of animal genotype and nutrition on performance in infection-free environments [1], [4]–[8], understanding of the interactive influence of a host's capacities for growth and disease resistance and of nutrition on phenotypic responses under pathogen challenge is relatively limited.

There is plenty of evidence that nutrient availability can affect the ability of a host to control invading pathogens [9]–[11]. The effect can be attributed to increased nutritional costs associated with the development of an effective immune response [2], [12]. A problem arises if nutrient availability is scarce, in which case a trade-off between mounting an immune response and other body functions (e.g. growth, reproduction) occurs. Trade-offs are central concepts in evolutionary biology [13], [14], and a large body of studies has either indicated trade-offs occurring frequently amongst natural and semi-natural populations [15]–[17], or considered them theoretically [3], [18], [19]. Such populations are frequently confronted with trade-offs due to variation in food resource availability. By contrast, domestic livestock raised in controlled, nutrient rich environments are expected to face the dilemma of appropriate partitioning of nutrients to a lesser extent. However, the situation of sufficient nutrient supply may be compromised when domestic livestock are exposed to pathogens, as a common by-product of infection is a reduction in voluntary food intake, henceforth called anorexia [20]. While there is still speculation about the biological mechanisms underlying anorexia, it is well established that infection induced anorexia occurs across a wide range of pathogen and host species in both natural and domestic populations, and that this causes a trade-off between the immune response and other body functions [10], [21], [22].

As with the role of nutrition, it is well established that host genetics strongly influence how animals allocate nutrients [23]–[25] and thus how they respond to pathogen challenges. The 'resource allocation theory' of Beilharz et al. [26] states that 'when environmental resources are limiting, all major components (e.g. growth, immunity, reproduction) of fitness are selected toward intermediate optimal values'. The theory implies that resource allocation preferences are genetically determined and that such preferences will be tailored to suit the environment in which selection is made. For domestic livestock, changes in host genetics, which are aimed at maximising fitness in natural populations, are often defined by the breeding goal, which primarily targets production traits [27]. Both experimental and simulation approaches have shown that artificial selection for production traits influences the performance-resistance relationship [27], [28], but the outcomes are often contradictory and the underlying mechanisms not fully understood (as reviewed by [27]). Similar conflicting outcomes arise from artificial selection for disease resistance, with some studies reporting improved performance [29], [30], while others report a performance decrease [31], [32]. Previous simulation studies of selection experiments suggest that the relationship between production and immune traits in a population depends strongly on the physiological status of the animals at the time of selection as well as on the genetic relationship (e.g. pleiotropy or linkage) between production and resistance traits [33], [34]. However, these studies have assumed that the performance-resistance relationship is mainly determined by host genetics and the infectious challenge, whereas the potential influence of the nutritional environment has been ignored.

In this study, a different approach to previous theoretical studies is adopted by exploring the hypothesis that the conflicting outcomes of selection experiments may arise through the nutritional environment of the host affecting the relationship between the genetic traits associated with growth and immune response. The underlying assumption of this study is that the host's genetic capacities for growth and immune response determine the animal's nutrient requirements and preference of allocating nutrients to either process, and that nutrient availability stipulates the extent at which the genetic potentials are expressed if the animal is under pathogen challenge. A similar approach had been adopted to model the interactive effect of host genotype and nutrition on gastro-intestinal parasitism in growing lambs [35], [36]. The model developed in this study focuses on micro-parasitic infections, which are characterised by short generation times and high rates of reproduction within the host [37].

2. METHODS

2.1 Assumptions

We consider only two resource demanding biological processes of a host: growth and immune response to pathogen challenge. Resource requirements for all other processes (e.g. maintenance, reproduction, coping with environmental stressors other than pathogen challenges, damage from within-host parasite replication) are assumed to be either negligible (e.g. the animal is assumed to be in a non-reproductive state and that requirements for the development of the reproductive apparatus are encompassed by those for growth) or fully satisfied during the time periods considered here. We further assume that the host has genetically intrinsic capacities for growth and immune response [25], [38] hereafter simply called genetic growth and resistance potential, respectively. Both these biological processes are associated with nutritional costs [10], [12].

Pathogen induced anorexia, i.e. a reduction in the host voluntary food intake, is considered to be the main reason that challenged animals cannot fully cover the nutritional requirements to achieve the above genetic capacities during infection and are required to distribute scarce resources between growth and the immune response [21], [26].

2.2 Model description

2.2.1 Genetically controlled capacity for growth and immune response

We assume that during the typical short time scales of micro-parasitic infection, the animal's intrinsic capacity for change in body weight (dW_p/dt) is constant, i.e.

$$\frac{dW_P}{dt} = GP \tag{1}$$

where GP describes the animal's genetic growth potential in the absence of infection.

The host's genetically determined capacity of the immune response and associated within-host parasite dynamics are described according to the equations of Antia et al. [39], which have two variables, i.e. the (potential) intensity of the immune response (I_p) and the parasite density (P), which interact according to

$$\frac{dI_p}{dt} = \alpha + \rho I_p(\frac{P}{P+\varphi}) - \gamma I_p \tag{2a}$$

$$\frac{dP}{dt} = \lambda P - \kappa P I_p \tag{2b}$$

where λ is the replication rate of the parasite, κ is the constant rate for the elimination of the pathogen by the host's immune response, α and γ are the replacement and death rate of the immune cells, respectively, ρ is the maximum per capita replication rate of the immune response and φ represents the parasite density at which the rate of growth of immunity is half maximal.

2.2.2 Incorporating nutrition

Nutrient requirements are defined as the requirement for the most limiting nutrient in the food, which is assumed to be of the same kind for growth and immunity. We assume that nutrient requirements for growth (NG^*) and immune response (NI^*) are proportional to the resource demanding components of these processes. This is consistent with the principles of allocation of scarce resources adopted in many nutritional models [10], [35], [40] According to equations (1) and (2a) NG^* and NI^* are thus given by

$$N_G^* = e_G GP \tag{3a}$$

$$N_I^* = e_I(\alpha + \rho I_p \frac{P}{P+\phi}) \tag{3b}$$

where the values of the efficiency parameters e_I and e_G represent the resource cost (in units of the most limiting nutrient) per unit increase in *I* and *W* respectively.

Infection induced anorexia is included in the model through a reduction in the desired nutrient intake - which is defined as the total amounts of nutrients required to satisfy the genetic growth and resistance potentials $N^*=NG^*+NI^*$ - by a factor *r*. For simplicity and using empirical evidence, it is assumed that the degree of anorexia is proportional to the pathogen load (with proportionality constant ν) for small to moderate levels of infection [41], [42], and does not fall below a minimum $r_{\min}$ for a wide range of P values [21], [43]. Hence

We assume that the host distributes available nutrients (*N*) between immunity (*NI*) and growth (*NG*) according to

$$N_G = p_G N \quad \text{(4a)}$$

$$N_I = (1 - p_G) N \quad \text{(4b)}$$

where

$$p_G = \frac{N_G^*}{N_I^* + N_G^*} \quad \text{(5)}$$

This partitioning rule entails that the nutrient supplies for growth and immunity are reduced by the same proportion relative to their respective requirements if insufficient nutrients are available.

2.2.3 Actual growth and within-host parasite interaction

Combining equations 1, 3a & 4a, and equations 2, 3b & 4b, respectively, leads to the following expressions for the predicted actual growth and change in immune response in terms of allocated nutrients:

$$\frac{dW}{dt} = \frac{1}{e_G} \min(N_G, N_G^*) \quad \text{(6a)}$$

$$\frac{dI}{dt} = \frac{1}{e_I} \min(N_I, N_I^*) - \gamma I \quad \text{(6b)}$$

The change in pathogen load (*dP*/*dt*) is calculated according to equation (2b) after replacing I_p with I:

$$\frac{dP}{dt} = \lambda P - \kappa P I \quad \text{(6c)}$$

Equations 6a, 6b imply that actual growth and immune response equal their corresponding genetic potentials if sufficient nutrients are available, but are reduced by equal proportion if nutrients are scarce. Also note that the model for the within-host parasite dynamics (equations 6b, 6c) coincides with that of Antia et al. [39] when nutrient availability matches or exceeds nutrient requirements.

2.3 Simulations

Genetic differences in the potentials for growth and resistance were incorporated into the model by assigning different values to the parameters GP and ρ in the equations for growth (1) and immune response (2a), respectively. Thus, a genetically more resistant host, represented by a larger value of ρ, would be able to mount a more effective immune response, due to a faster immune cell replication rate (equation 2a) and due to allocating a larger proportion of nutrients towards immunity (equations 3b and 4b).

Simulations were performed for different amounts of available nutrients N and different values of the genetic parameters ρ and GP. For the immunological parameters α, Φ, γ, λ and κ (equations 2a & b) the scaled estimates from the original model of Antia et al. [39] were adopted, whilst for parameters related to nutrient intake and utilisation (e_G, e_I, r_{min}) the empirical estimates from Sandberg et al. [43], [44] were used. Sensitivity analysis was performed by modifying one parameter at a time or a combination of parameters within biological reasonable limits, and inspecting the impact of these changes on the model results. Nutrient availability (N) was varied from severe limitation (N=6), where the host was unable to control the pathogen (despite meeting maintenance requirements), to the point at which nutrient availability was sufficient to meet all requirements for growth and immunity in the absence of anorexia (N=20). In the case of severe limitations of available nutrients, within host pathogen load could not be controlled by the host's immune response and it was assumed that the host would eventually die (i.e. if pathogen load exceeded 1050). Since the focus of this study were interactive effects of nutrition and host genetics on growth and immune response, only results referring to parameter ranges for ρ and GP corresponding to host survival for both nutritional extremes are presented.

Infection starts with the host being infected with a single micro-parasite, which then replicates within the host. It was assumed that clearance of the pathogen occurs when P=1 is achieved and that re-infection does not take place over the course of the simulations which cover the time period required for clearing the pathogen or reaching the steady state. The duration of the infection was hence defined as the length of time until pathogen load has decreased to unity or reached the steady state.

3. RESULTS

3.1 Dynamic trends for pathogen load, immune response and growth

Provided that an infection can be established, the modelled pathogen load and host immune response exhibit one of the following three dynamic patterns: (i) the host immune response clears the infection (Figure 1A), (ii) pathogen load and immune response exhibit damped oscillations towards a persistent steady state with low pathogen load (Figure 1B), or (iii) the host immune response is too weak to control the pathogens and pathogen load increases until death of the host. As mentioned above, only results referring to the first two scenarios are of interest here. The immune response generally lags behind the pathogen load and is characterised by a slower decrease compared to that observed for pathogen load. Even in cases resulting in the eventual pathogen clearance, the immune response persists for an extended period of time (Figure 1A). This behaviour is observed for a wide range of model parameters. Only low values for the resistance parameters ρ or extremely low levels of nutrient intake, caused for example by high levels of anorexia (i.e. low values for r_{min}) or low values of N will result in host death. Compared to pathogen clearance, damped oscillations occur if the pathogen elimination rate by the host immune response κ is relatively large compared to the pathogen replication rate λ. Changes in the model parameter values of N, ρ and GP within the admissible range (i.e. leading to the eventual clearance of pathogens by the host or low persistent pathogen load) alters the time scales of the response curves without affecting the overall shape characteristics of the pathogen load and immune response curves.

Predicted growth rates are piecewise linear (or close to linear), with reduced growth rates during the periods of positive pathogen load.

3.2 Effect of host genetic potentials × nutrition on infection severity and duration

The combined effect of host genetic potentials and nutrient availability (N) on infection characteristics and on the predicted pathogen load (log scale) at different time points during the course of infection are shown in Figures 2A & 2B and Figure 3, respectively.

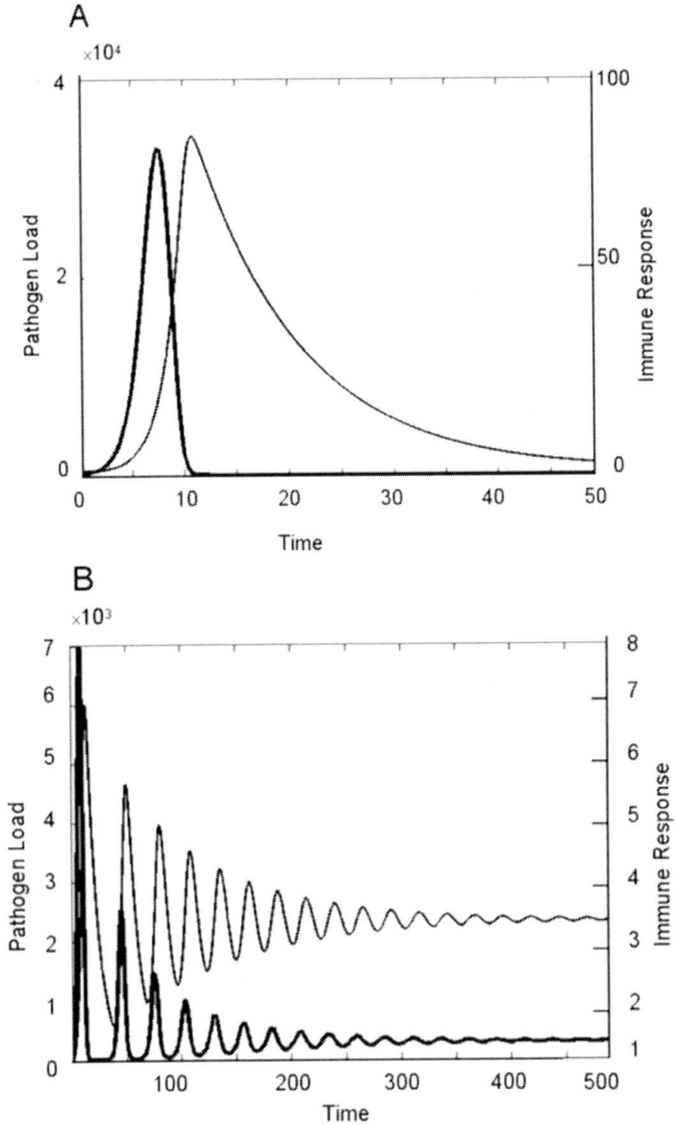

Figure 1. Predicted pathogen load (thick line) and immune response (thinner line) over time. Parameter values used for graph A were e_I and e_G=0.25, ν=0.01, r_{min}=0.3, λ=1, κ=0.05, α=γ=0.1, Φ=1000, ρ=1, GP=10, N=8; for graph B the same parameter values were used except for κ, which was set to 0.5 The initial size of the immune response (I_0) and Pathogen Load (P_0) were standardized to unity. For explanation of the parameters see text. All units are arbitrary.

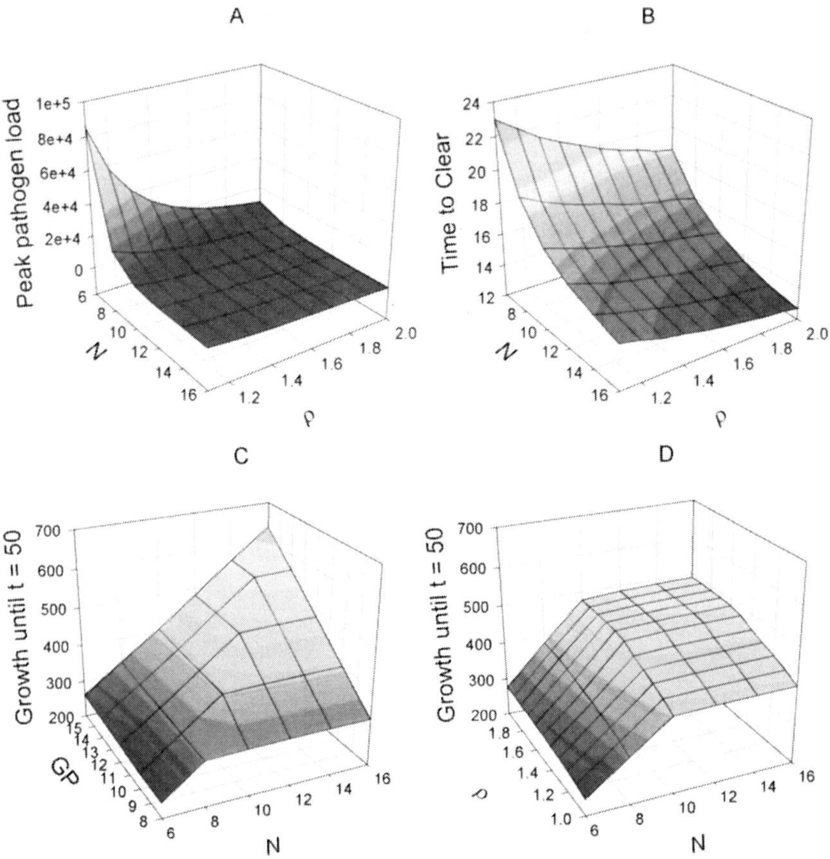

Figure 2. Effects of nutrient availability (*N*) and genetic potential for disease resistance (*ρ*) and growth (GP) on infection characteristics. Graph A: peak pathogen load over time course of infection as function of N and ρ; graph B: time to clear pathogens as function of *N* and *ρ*. Graphs C and D: and growth until t=50 as function of *N* and *GP* and *N* and *ρ*, resepectively. The growth potential in graphs A, B & D was *GP*=10, the resistance potential in graph C was *ρ*=1.5. For all parameter combinations depicted the infection was cleared at t=50. Values of other parameters are as in Figure 1. All units are arbitrary. Notice that the rotation angles differ between the individual graphs as they have been chosen for illustration clarity.

The model predicts that higher amounts of available nutrients generally enable a more effective immune response, resulting in lower pathogen loads at any stage of the infection and in a shorter duration of the infection (Figures 2 & 3). Simulations with gradual variations in nutrient availability and fixed combinations of *GP* or *ρ* revealed that the responses in severity (i.e. peak or cumulative pathogen load) and duration of infection show a diminishing trend: the beneficial influence of increasing nutrient availability on infection characteristics is strongest when nutrient availability is low.

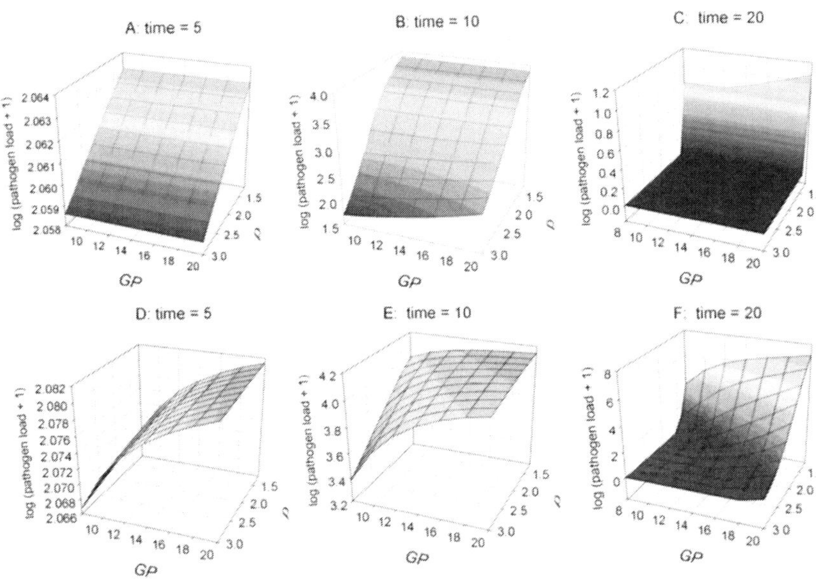

Figure 3. Effect of growth potential (*GP*) and resistance (*ρ*) on (log10 transformed) pathogen load at various times during infection. Graphs A–C: High nutrient availability (*N*=20); graphs D–F: Low nutrient availability (N=6). Values of other parameters are as in Figure 1. All units are arbitrary. Note that the value ranges for log (pathogen load+1) differ amongst individual graphs, as the value ranges have been chosen for illustration clarity.

The effect of altering nutrient availability also depends strongly on the host genetic potential. For example, as illustrated in Figures 2A & 2B, increasing nutrient availability is more beneficial for hosts with low genetic resistance potential.

Nutrient availability is predicted to have a strong influence on how growth and resistance potentials interact during the time course of infection. When nutrient availability is high, the within host pathogen load is independent of the growth potential at the early stages of infection, and also hardly depends on it at the later stages (Figures 3A–C). Infection severity and duration are primarily controlled by the resistance potential (*ρ*). The response surfaces of predicted pathogen load for hosts with different growth and resistance potentials change drastically when nutrient supply is limited (Figures 3D–F). In this case, the predicted pathogen load depends more on the genetic potential for growth than for resistance at the early and medium stages post infection (Figures 3D & E). Hosts with lower genetic growth potential are predicted to allocate a higher proportion of available nutrients towards the immune response, thus restricting the replication of pathogens within the host already during the early stages of the infection. The resistance potential has a significant, beneficial influence on pathogen load only when combined with a low growth potential (Figure 3E). However, the influence of the genetic resistance potential increases as time progresses: whereas a resistant host will have managed to clear all pathogens after a certain amount of time (e.g. t=20 in Figure 3F), pathogen load in susceptible hosts may continue to increase towards very high levels. Low resistance potentials are particularly detrimental for the late stage severity and duration of infection for hosts with simultaneously high growth potentials (Figure 3F).

3.3 Effect of host genetic potential × nutrition on growth under pathogen challenge

The model predicts that higher nutrient availability is beneficial for growth under pathogen challenge for any parameter combinations of ρ and GP (Figures 2C & D, Figure 4). When considered over a time period that spans the entire infection period, the response to an increase in nutrient availability is similar in infected and non-infected hosts: growth rate increases linearly with increasing nutrients (N) up to a plateau when the amount of available nutrients exceeds the levels needed for satisfying the growth potential (Figure 2C & D). Hosts with higher genetic growth potential reach this plateau at higher levels of N than hosts with lower growth potential and hence benefit from an increase of nutrient availability for a wider range of N values (Figure 2C). Similarly, genetically more resistant hosts benefit from increases in nutrient availability for a wider range of N, although the differences are less pronounced than for hosts with different growth potentials (Figure 2D). The impact of nutrient availability and host genetic potential on achieved growth generally increases over time (Figure 4).

The simulations further reveal that nutrient availability has a strong influence how the genetic potentials for growth and resistance interact towards the actual growth achieved by a challenged host. The model predicts that genetic growth potential is the main driver for achieved growth of infected hosts when nutrient availability (N) is high (Figures 4A–C). Only resistance levels below a certain threshold, which depends on N, produce substantial reductions in growth. These are the consequence of prolonged time periods of 'severe' anorexia produced by high pathogen loads over extended time periods.

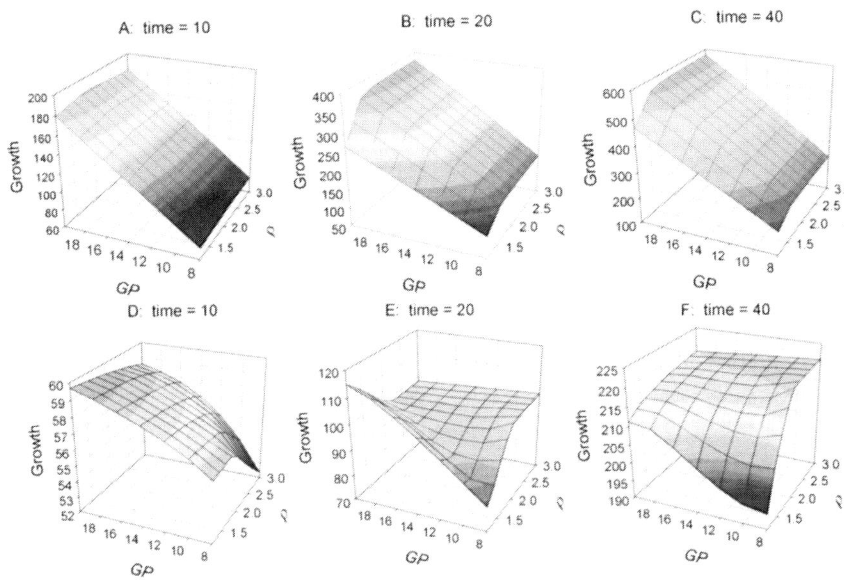

Figure 4. Effect of growth potential (*GP*) and resistance (ρ) on body weight growth at various times during infection. Graphs A–C: High nutrient availability (N=20); graphs D–F: Low nutrient availability (N=6). Values of other parameters are as in Figure 1. All units are arbitrary. Note that the value ranges for growth differ amongst individual graphs and that the rotation angle differs from that of Figure 3, as the value ranges and rotation angles have been chosen for illustration clarity.

In contrast, when nutrient availability is low the genetic growth potential is no longer the main driver of achieved growth. Instead, achieved growth depends on the combination of both genetic potentials (Figures 4 D–F). Also, in contrast to the relatively time stable sensitivity surfaces of achieved growth associated with high N (Figures 4A–C), the degree and direction of influence of the growth and resistance potentials change over time when nutrient availability is low (Figures 4D–F). Whereas the model predicts that high growth combined with low resistance potentials are preferable for growth at the earlier stages of infection (Figure 4D & E), low growth combined with high resistance potential appears to be the optimal strategy for achieved growth over the long-term (Figure 4F).

4. DISCUSSION

The thesis put forward in this paper was that the conflicting outcomes reported in the literature on the consequence of genetic selection for either host growth or resistance to pathogens may arise from the modification of the interrelationship between these genetic traits by the nutritional environment of the host. Our mathematical model aimed to shed some light on this conflict. The predictions of our model suggest that at higher planes of nutrition improvement in either of these two traits (e.g. by genetic selection) leads to a more effective immune response (seen as resulting low pathogen loads) and higher growth rates respectively, when hosts are exposed to micro-parasites. Thus infection characteristics are primarily determined by the genetic resistance potential (Figures 3A–C) and growth performance is mainly determined by the genetic potential for growth (Figures 4A–C), as would normally be expected [45], [46].

However, the outcome of improvement of exactly the same traits would be very different when the nutritional environment of the host is scarce. In these instances, growth performance and infection characteristics are influenced by both genetic potentials simultaneously, and the influence of either potential depends on the value of the other and varies over time (Figures 3D–F, 4D–F). In particular, the model predicts that an increase of the genetic growth potential would be detrimental to both actual growth and the ability to cope with pathogens when nutrient supply is scarce. This is consistent with the suggestions made by Rauw [27] that selection for production traits, such as growth may negatively influence the ability of animals to cope with pathogens. Our model emphasises that this would be the case, however, only under nutrient scarce environments.

Similarly, according to the model, a high resistance potential in nutrient scarce environments would be detrimental to growth and, if combined with a high growth potential, have no impact on the within-host pathogen load in the short term. Its beneficial effects on both growth and the animal's infectious state depend on the growth potential and may only become apparent over the long term. These predictions can account for previous contradictions in the literature regarding the outcome of selections for disease resistance on actual growth. In farm animal populations, where nutrient availability may be high, selection for resistance would be expected to lead to positive consequences on growth as previously observed [29]. The opposite suggestions about the trade offs between growth and resistance to pathogens made in the ecological literature may arise from the fact that wild animal populations are usually subjected to low or fluctuating nutrient availability.

The model results provide some early insights about the consequences of selection based on phenotypic information upon underlying genetic parameters. The results presented in Figures 3 & 4 imply that the same selection criterion applied in different (nutritional) environments or at different time points may lead to different genetic improvements. For example, in environments with high nutrient availability, selection for low severity and short duration of infection would be equivalent to selecting for high genetic resistance potential, ρ, with no selection pressure acting on the genetic growth potential (Figure 3A–C). Selection for high tolerance to infection in terms of growth in a nutrient rich environment would be equivalent to selecting for high *GP* with little selection pressure on ρ (Figure 4A–C). In particular, selection for both observed resistance and tolerance to micro-parasitic infections would be feasible if nutrient availability was sufficiently high (select hosts with simultaneously high ρ and GP).

In contrast, if nutrients are scarce, selection for low severity and short duration of infection would still favour animals with high genetic resistance potential ρ, but would simultaneously imply selection for low *GP* as only animals with low *GP* allocate the required amount of resources towards immunity (Figure 3D–F). Simultaneous selection for low severity and short duration of infection and tolerance (little impact of infection on growth) would only be possible if tolerance were evaluated over a sufficiently long time period, since in the shorter term, high tolerance implies high *GP* (Figure 4D–F).

The model results also demonstrate the importance of evaluating phenotypic traits over appropriate time periods. For example, although an increase in the genetic resistance potential may produce temporary reductions in growth due to increased investment in the immune response at the early infection stage, growth will benefit in the long-term due to infection having a shorter duration of, and less being less severe (Figures 4D–F). These observations thus match a previous hypothesis that 'increased investment in immune response in the earlier stages of infection may serve to limit the total resource cost of infection' [12].

The key drivers determining how host genetic growth and resistance potential and nutrition influence growth and infection characteristics in our model are the pathogen induced anorexia which triggers a conflict between growth and immunity and the allocation of scarce resources towards growth and immune functions.

In our model the reduction of food intake was assumed to depend linearly only on pathogen load up to a minimum level of intake. As a consequence of this assumption, growth and immunity compete for scarce resources only during the time period in which the host is infected with pathogens. The assumption was based on various challenge studies, which showed that food intake reduces with increasing pathogen load until a minimum level is reached that is similar for different levels of infectious dose of microparasites [42], [43] and that food intake recovers almost instantaneously when pathogens are removed artificially (eg through drug administration) [47]. Alternatively, given that anorexia is frequently considered as part of the host immune response [20], [22], anorexia could also have been represented as being a function of the host immune response. Our sensitivity analysis indicated that the influence of different anorexia rules (e.g. partial to full dependence on the immune response) and of different values for the parameters ν and $r_{\min}$ is generally low compared to the relative influence of the growth and resistance potentials under given nutrient availability. In particular, more severe or more prolonged anorexia (produced e.g. by assuming partial dependence on the immune response) would cause a trade-off between growth and immune response for higher levels of nutrient availability and thus increase the relative influence of the genetic potentials on observed growth and infection characteristics.

The relative impact of the genetic potentials for growth and resistance on predicted growth and infection characteristics strongly depends on the way that available nutrients are allocated, which we assume to be relative to the nutrient requirements for either process (equations 4a, b). Substantial differences exist between the nutrient allocation rules adopted in different modelling studies [3], [19], [34]. The rule chosen here implies that resource allocation partly depends on the host genetic potential for growth and disease resistance and varies over time according to the physical state of the animal [19], [48]. Also, the immune response is not given absolute priority for nutrients. Evidence in support of the latter is provided by experiments which have observed simultaneous increases in growth and expression of an immune response, in response to increased 'resource intake' [49]–[51]. Medley [3] proposed a model to assess the optimum allocation of resources for different resource availabilities assuming constant exposure to parasites throughout a host's lifetime. He assumed that the immune response has total priority over other processes (growth and reproduction) for resources and that the proportion of resources allocated to the immune response depends on the pathogen load and does not exceed a fixed proportion of the total amount of resources available to the host. This led to the conclusion that individuals on low nutritional planes should put increasingly less resources into immunity [3]. Our model also predicts that the influence of the genetic potential for resistance on infection characteristics may be reduced as nutrient availability decreases, but this influence also depends on the animal's genetic growth potential.

Other studies have assumed that nutrient allocation is largely controlled by the host genotype [23], [25]. The allocation rule adopted here indirectly implies host genetic influence on the distribution of resources, as a host with higher growth potentials but similar resistance potential relative to another host will allocate a higher proportion of resources towards growth than the immune response under the same pathogen challenge. The role of direct genetic effects on nutrient allocation (i.e. hosts with similar genetic potentials for growth and resistance have different genetically determined preferences for allocating nutrients) has been investigated in a previous study [25]. It was found there that compared to the impact of host genetic potentials for growth and resistance, genetically determined nutrient allocation had a relatively small impact on observed growth and immune response.

A simple approach was adopted here to obtain insight into the interacting effects of host genetic potentials and nutrition on growth and infection characteristics without adding complexity caused by other contributing factors (e.g. other environmental stressors or other biological processes such as reproduction). This approach allowed new interpretations of the apparently ambiguous outcomes reported in selection studies. It also revealed the possible side effects of selecting for one or several genetic traits that compete for nutritional resources. The simplistic representation of pathogen challenge, host genetic potentials, nutrition and immunity by single entities makes direct comparison of our model results with real data from challenge or selection studies difficult. The proposed framework could however easily be adapted to include more complex representations of the influencing factors and of the biological processes affected by them (e.g. different arms of immune response, reproduction). The hope is that such models can be then expanded to predict performance of animals under exposure to specific pathogens.

ACKNOWLEDGMENTS

The authors would like to thank Dr. Beatriz Villanueva for helpful comments to an earlier version of this manuscript.

REFERENCES

1. Knap PW, Roehe R, Kolstad K, Pomar C, Luiting P (2003) Characterizations of pig genotype for growth modelling. J Anim Sci 81: E187–195E.PW KnapR. RoeheK. KolstadC. PomarP. Luiting2003Characterizations of pig genotype for growth modelling.J Anim Sci81E187195E
2. Sheldon BC, Verhulst S (1996) Ecological immunology: costly parasite defences and trade-offs in evolutionary ecology. Trends in Ecology & Evolution 11: 317–321.BC SheldonS. Verhulst1996Ecological immunology: costly parasite defences and trade-offs in evolutionary ecology.Trends in Ecology & Evolution11317321
3. Medley GF (2002) The epidemiological consequences of optimisation of the individual host immune response. Parasitol 125: S61–S70.GF Medley2002The epidemiological

consequences of optimisation of the individual host immune response. Parasitol125S61S70

4. Black JL, Davies GT, Bray HJ, Giles LR, Chapple RP (1986) Simulation of energy and amino acid utilisation in the pig. Research and Development in Agriculture 3: 121–145.JL BlackGT DaviesHJ BrayLR GilesRP Chapple1986Simulation of energy and amino acid utilisation in the pig.Research and Development in Agriculture3121145
5. Green DM, Whittememore CT (2003) Architecture of a harmonized model of the growing pig for the determination of dietary net energy and protein requirements and of excretions into the environment (IMS Pig). Anim Sci 77: 113–130.DM GreenCT Whittememore2003Architecture of a harmonized model of the growing pig for the determination of dietary net energy and protein requirements and of excretions into the environment (IMS Pig).Anim Sci77113130
6. Wellock IJ, Emmans GC, Kyriazakis I (2003) Modelling the effects of thermal environment and dietary composition of pig performance: model logic and concepts. Anim Sci 77: 255–266.IJ WellockGC EmmansI. Kyriazakis2003Modelling the effects of thermal environment and dietary composition of pig performance: model logic and concepts.Anim Sci77255266
7. McNamara JP, Baldwin RL (2000) Estimation of parameters describing lipid metabolism in lactation: challenge of existing knowledge described in a model of metabolism. J Dairy Sci 83: 128–143.JP McNamaraRL Baldwin2000Estimation of parameters describing lipid metabolism in lactation: challenge of existing knowledge described in a model of metabolism.J Dairy Sci83128143
8. Bryant JR, Lopez-Villalobos N, Holmes CW, Pryce JE, Pitman GD, Davis SR (2007) The effect of level of feeding, genetic merit, body condition score and age on biological parameters of a mammary gland model. Animal 1: 175–183.JR BryantN. Lopez-VillalobosCW HolmesJE PryceGD PitmanSR Davis2007The effect of level of feeding, genetic merit, body condition score and age on biological parameters of a mammary gland model.Animal1175183
9. Scrimshaw NS, Taylor CE, Gordon JE (1968) Interactions of Nutrition and Infection. Geneva: World Health Organization (WHO Monograph Series No. 57). NS ScrimshawCE TaylorJE Gordon1968Interactions of Nutrition and InfectionGenevaWorld Health Organization (WHO Monograph Series No. 57)Ref Type: Serial (Book,Monograph). Ref Type: Serial (Book,Monograph).
10. Coop RL, Kyriazakis I (2001) Influence of host nutritions on the development and consequences of nematode parasitism in ruminants. Trends Parasitol 17: 325–330.RL CoopI. Kyriazakis2001Influence of host nutritions on the development and consequences of nematode parasitism in ruminants.Trends Parasitol17325330
11. Ritz BW, Gardner EM (2006) Malnutrition and Energy Restriction Differentially Affect Viral Immunity. J Nutr 136: 1141–1144.BW RitzEM Gardner2006Malnutrition and Energy Restriction Differentially Affect Viral Immunity.J Nutr13611411144
12. Lochmiller RL, Deerenberg C (2000) Trade-offs in evolutionary immunology: just what is the cost of immunity? Oikos 88: 87–98.RL LochmillerC. Deerenberg2000Trade-offs in evolutionary immunology: just what is the cost of immunity?Oikos888798
13. Stearns SC (1992) The evolution of life histories. New York: Oxford Univ. Press. SC Stearns1992The evolution of life historiesNew YorkOxford Univ. Press
14. Roff DA (1992) The evolution of life histories: theory and analysis. New York: Chapman and Hall. DA Roff1992The evolution of life histories: theory and analysisNew YorkChapman and Hall
15. Reznick D (1985) Costs of reproduction: an evaluation of the empirical evidence. Oikos 44: 257–267.D. Reznick1985Costs of reproduction: an evaluation of the empirical evidence.Oikos44257267
16. Mangel M, Stamps J (1001) Trade-offs between growth and mortality and the maintenance of individual variation in growth. Evol Ecol Res 3: 583–593.M. MangelJ. Stamps1001Trade-offs between growth and mortality and the maintenance of individual variation in growth.Evol Ecol Res3583593

17. Soler JJ, de Neve L, Perez-Contreras T, Soler M, Sorci G (2003) Trade-off between immunocompetence and growth in magpies: an experimental study. Proc R Soc Lond B 270: 241–248.JJ SolerL. de NeveT. Perez-ContrerasM. SolerG. Sorci2003Trade-off between immunocompetence and growth in magpies: an experimental study.Proc R Soc Lond B270241248
18. Shudo E, Iwasa Y (2001) Inducible defense against pathogens and parasites: optimal choice among multiple options. J Theor Biol 209: 233–247.E. ShudoY. Iwasa2001Inducible defense against pathogens and parasites: optimal choice among multiple options.J Theor Biol209233247
19. Houston AI, McNamara JM, Barta Z, Klasing KC (2007) The effect of energy reserves and food availability on optimal immune defence. Proc R Soc B 274: 2835–2842.AI HoustonJM McNamaraZ. BartaKC Klasing2007The effect of energy reserves and food availability on optimal immune defence.Proc R Soc B27428352842
20. Exton MS (1997) Infection-induced anorexia: active host defence strategy. Appetite 29: 369–383.MS Exton1997Infection-induced anorexia: active host defence strategy. Appetite29369383
21. Kyriazakis I, Tolkamp BJ, Hutchings MR (1998) Towards a functional explanation for the occurrence of anorexia during parasitic infection. Anim Behav 56: 265–274.I. KyriazakisBJ TolkampMR Hutchings1998Towards a functional explanation for the occurrence of anorexia during parasitic infection.Anim Behav56265274
22. Kyriazakis I, Doeschl-Wilson AB (2009) Anorexia during infection in mammals: variation and its sources. In: Torrallardona D, Roura E, editors. Voluntary feed intake in pigs. Wageningen Academic Publishers. pp. 307–318.I. KyriazakisAB Doeschl-Wilson2009Anorexia during infection in mammals: variation and its sources.D. TorrallardonaE. RouraVoluntary feed intake in pigsWageningen Academic Publishers 307318
23. De Jong G, Van Norrdwijk J (1992) Acquisition and allocation of resources: genetic (co)variances, selection and life histories. Am Nat 139: 749–770.G. De JongJ. Van Norrdwijk1992Acquisition and allocation of resources: genetic (co)variances, selection and life histories.Am Nat139749770
24. Friggens NC, Newbold JR (2007) Towards a biological basis for predicting nutrient partitioning: the dairy cow as an example. Animal 1: 87–97.NC FriggensJR Newbold2007Towards a biological basis for predicting nutrient partitioning: the dairy cow as an example.Animal18797
25. Doeschl-Wilson AB, Vagenas D, Kyriazakis I, Bishop SC (2008) Challenging the assumptions underlying genetic variation in host nematode resistance. Genet Sel Evol 40: 241–264.AB Doeschl-WilsonD. VagenasI. KyriazakisSC Bishop2008Challenging the assumptions underlying genetic variation in host nematode resistance.Genet Sel Evol40241264
26. Beilharz RG, Luxford BG, Wilkinson JL (1993) Quantitative genetics and evolution: Is our understanding of genetics sufficient to explain evolution? J Anim Breed Genet 110: 161–170.RG BeilharzBG LuxfordJL Wilkinson1993Quantitative genetics and evolution: Is our understanding of genetics sufficient to explain evolution?J Anim Breed Genet110161170
27. Rauw WM, Kanis E, Noordhuizen-Stassen EN, Grommers FJ (1998) Undesirable side effects of selection for high production efficiency in farm animals: a review. Livest Prod Sci 56: 15–33WM RauwE. KanisEN Noordhuizen-StassenFJ Grommers1998Undesirable side effects of selection for high production efficiency in farm animals: a review.Livest Prod Sci561533
28. Houdijk JG, Bunger L (2006) Selection for growth increases the penalty of parasitism on growth performance in mice. Proc Nut Soc 65: 68A.JG HoudijkL. Bunger2006Selection for growth increases the penalty of parasitism on growth performance in mice.Proc Nut Soc6568A
29. Mallard BA, Wilkie BN, Kennedy BW, Gibson J, Quinton M (1998) Immune responsiveness in swine: eight generations of selection for high and low immune response in Yorkshire pigs. BA MallardBN WilkieBW KennedyJ. GibsonM. Quinton1998Immune responsiveness in swine: eight generations of selection for high and low immune response in Yorkshire

pigs.Proceedings of the 6th world congress on genetics applied to livestock production 257–264. Proceedings of the 6th world congress on genetics applied to livestock production 257–264.

30. Imsland AK, Jonassen TM, Langston A, Hoare R, Wergeland H, et al. (2002) The interrelation of growth and disease resistance of different populations of juvenile Atlantic halibut (Hippoglossus hippoglossus L.) Aquaculture 204(1–2): 167–177.AK ImslandTM JonassenA. LangstonR. HoareH. Wergeland2002The interrelation of growth and disease resistance of different populations of juvenile Atlantic halibut (Hippoglossus hippoglossus L.)Aquaculture204(1–2)167177
31. Gross WB, Siegel PB, Pierson FW (2002) Effects of genetic selection for high or low antibody response on resistance to a variety of disease challenges and the relationship of resource allocation. Avian Diseases 46: 1007–1010.WB GrossPB SiegelFW Pierson2002Effects of genetic selection for high or low antibody response on resistance to a variety of disease challenges and the relationship of resource allocation.Avian Diseases4610071010
32. Morris CA, Vlassoff A, Bisset SA, Baker AL, Watson TG, et al. (2000) Continued selection of Romney sheep for resistance or susceptibility to nematode infection: estimates of direct and correlated responses. Anim Sci 70: 17–27.CA MorrisA. VlassoffSA BissetAL BakerTG Watson2000Continued selection of Romney sheep for resistance or susceptibility to nematode infection: estimates of direct and correlated responses.Anim Sci701727
33. Van der Waaij EH, Bijma P, Bishop SC, van Arendonk JAM (2000) Modeling selection for production traits under constant infection pressure. J Anim Sci 78: 2809–2820.EH Van der WaaijP. BijmaSC BishopJAM van Arendonk2000Modeling selection for production traits under constant infection pressure.J Anim Sci7828092820
34. Van der Waaij EH (2004) A resource allocation model describing the consequences of artificial selection under metabolic stress. J Anim Sci 82: 973–981.EH Van der Waaij2004A resource allocation model describing the consequences of artificial selection under metabolic stress.J Anim Sci82973981
35. Vagenas D, Bishop SC, Kyriazakis I (2007a) A model to account for the consequences of host nutrition on the outcome of gastrointestinal parasitism in sheep: logic and concepts. Parasitol 134: 1263–1277.D. VagenasSC BishopI. Kyriazakis2007aA model to account for the consequences of host nutrition on the outcome of gastrointestinal parasitism in sheep: logic and concepts.Parasitol13412631277
36. Vagenas D, Bishop SC, Kyriazakis I (2007b) A model to account for the consequences of host nutrition on the outcome of gastrointestinal parasitism in sheep: model evaluation. Parasitol 134: 1279–1289.D. VagenasSC BishopI. Kyriazakis2007bA model to account for the consequences of host nutrition on the outcome of gastrointestinal parasitism in sheep: model evaluation.Parasitol13412791289
37. Anderson RM, May RM (1991) Infectious Diseases of Humans: Dynamics and Control. Oxford, UK: Oxford University Press. RM AndersonRM May1991Infectious Diseases of Humans: Dynamics and ControlOxford, UKOxford University Press
38. Knap PW (1999) Simulation of growth in pigs: evaluation of a model to relate thermoregulation to body protein and lipid content and deposition. Anim Sci 70: 39–49.PW Knap1999Simulation of growth in pigs: evaluation of a model to relate thermoregulation to body protein and lipid content and deposition.Anim Sci703949
39. Antia R, Koella JC, Perrot V (1996) Models of the within-host dynamics of persistent mycobacterial infections. Proc Biol Sci 263: 257–263.R. AntiaJC KoellaV. Perrot1996Models of the within-host dynamics of persistent mycobacterial infections. Proc Biol Sci263257263
40. Ferguson NS (2007) Basic concepts describing animal growth and feed intake. In: Gous R, Morris T, Fisher C, editors. Mechanistic modeling in pig and poultry production. UK: CABI. pp. 22–53.NS Ferguson2007Basic concepts describing animal growth and feed intake.R. GousT. MorrisC. FisherMechanistic modeling in pig and poultry productionUKCABI2253
41. Frank JW, Mellencamp MA, Carroll JA, Boyd RD, Allee GL (2005) Acute feed intake and acute-phase protein responses following a lipopolysacharide challenge in pigs from two dam lines. Vet Immunol Immunopath 107: 179–187.JW FrankMA MellencampJA

CarrollRD BoydGL Allee2005Acute feed intake and acute-phase protein responses following a lipopolysacharide challenge in pigs from two dam lines.Vet Immunol Immunopath107179187

42. Greiner LL, Stahly S, Stabel TJ (2000) Quantitative relationship of systemic virus concentration on growth and immune response in pigs. J Anim Sci 78: 2690–2695.LL GreinerS. StahlyTJ Stabel2000Quantitative relationship of systemic virus concentration on growth and immune response in pigs.J Anim Sci7826902695
43. Sandberg FB, Emmans GC, Kyriazakis I (2006) A model for predicting food intake of growing animals during exposure to pathogens. J Anim Sci 84: 1552–1566.FB SandbergGC EmmansI. Kyriazakis2006A model for predicting food intake of growing animals during exposure to pathogens.J Anim Sci8415521566
44. Sandberg FB, Emmans GC, Kyriazakis I (2005) Partitioning of limiting protein and energy in the growing pig: testing quantitative rules against experimental data. Br J Nut 93: 213–224.FB SandbergGC EmmansI. Kyriazakis2005Partitioning of limiting protein and energy in the growing pig: testing quantitative rules against experimental data.Br J Nut93213224
45. Bishop SC, Stear MJ (1999) Genetic and epidemiological relationships between productivity and disease resistance: gastro-intestinal parasite infection in growing lambs. Anim Sci 69: 515–524.SC BishopMJ Stear1999Genetic and epidemiological relationships between productivity and disease resistance: gastro-intestinal parasite infection in growing lambs.Anim Sci69515524
46. Bishop SC, Morris CA (2007) Genetics of disease resistance in sheep and goats. Small Rum Res 70(1): 48–59.SC BishopCA Morris2007Genetics of disease resistance in sheep and goats.Small Rum Res70(1)4859
47. Kyriazakis I, Anderson DH, Oldham JD, Coop RL, Jackson F (1996) Long-term subclinical infection with Trichostrongylus colubriformis: effects on food intake, diet selection and performance of growing lambs. Vet Parasitol 61: 297–313.I. KyriazakisDH AndersonJD OldhamRL CoopF. Jackson1996Long-term subclinical infection with Trichostrongylus colubriformis: effects on food intake, diet selection and performance of growing lambs.Vet Parasitol61297313
48. Hutchings MR, Kyriazakis I, Gordon IJ (2001) Herbivore physiological state affects foragind trade-off decisions between nutrient intake and parasite avoidance. Ecology 82: 1138–1150.MR HutchingsI. KyriazakisIJ Gordon2001Herbivore physiological state affects foragind trade-off decisions between nutrient intake and parasite avoidance. Ecology 8211381150
49. Bhargava KK, Hanson RP, Sunde ML (1970) Effects of methionine and valine on antibody production in chicks infected with Newcastle disease virus. J Nutr 100: 241–248.KK BhargavaRP HansonML Sunde1970Effects of methionine and valine on antibody production in chicks infected with Newcastle disease virus.J Nutr100241248
50. Datta FU, Nolan JV, Rowe JB, Gray GD (1998) Protein supplementation improves the performance of parasitised sheep fed a straw-based diet. Int J Paras 28: 1269–1278.FU DattaJV NolanJB RoweGD Gray1998Protein supplementation improves the performance of parasitised sheep fed a straw-based diet.Int J Paras2812691278
51. Tsiagbe VK, Cook ME, Harper AE, Sunde ML (1987) Enhanced immune responses in broiler chicks fed methionine-supplemented diets. Poult Sci 66: 1147–1154.VK TsiagbeME CookAE HarperML Sunde1987Enhanced immune responses in broiler chicks fed methionine-supplemented diets.Poult Sci6611471154

CHAPTER 6

Immunohistochemical investigation of Foxp3 expression in the intestine in healthy and diseased dogs

Johannes Junginger[1], Ulrike Schwittlick[1], Frederik Lemensieck[2], Ingo Nolte[3] and Marion Hewicker-Trautwein[1*]

[1]*Institute of Pathology, University of Veterinary Medicine Hannover, Bünteweg 17, D-30559 Hannover, German*

[2]*Small Animal Clinic Duisburg-Asterlagen, Dr.-Detlev-Karsten-ohwedder-Str. 11, D-47228 Duisburg-Rheinhausen, Germany*

[3]*Small Animal Clinic, University of Veterinary Medicine Hannover, Bünteweg 9, D-30559 Hannover, Germany.*

ABSTRACT

Intestinal immune regulation including development of oral tolerance is of great importance for the maintenance of intestinal homeostasis. Concerning this, regulatory T cells (Tregs) occupy a pivotal role in cell-mediated immunosuppression. Dysregulation of mucosal immunology leading to an abnormal interaction with commensal bacteria is suggested to play a key role in the pathogenesis of Inflammatory Bowel Disease (IBD) in men and dogs. The aim of this study was to characterise the expression of Foxp3 in the normal canine gut of 18 dogs (mean age: 6.03 years), in 16 dogs suffering from IBD (mean age: 5.05 years), and of 6 dogs with intestinal nematode infection (mean age: 0.87 years) using immunohistochemistry. In the duodenum, Tregs in healthy dogs declined from villi (median: 10.67/62 500 μm²) to crypts (median: 1.89/62 500 μm²). Tregs were further increased in the villi of middle-aged dogs (median: 18.92/62 500 μm²) in contrast to juvenile (median: 3.50/62 500 μm²) and old (median: 9.56/62 500 μm²) individuals. Compared to healthy controls, animals suffering from IBD revealed reduced numbers of Tregs in duodenal villi (median: 4.13/62 500 μm²). Dogs with intestinal nematode infection displayed increased numbers of Tregs (median: 21.06/62 500 μm²) compared to healthy animals.

Age-related changes indicate a progressive establishment of oral tolerance and immunosenescence in the canine elderly. The results further suggest that a defect in Treg homeostasis may be involved in the pathogenesis of canine IBD. In contrast, increased numbers of Tregs in the duodenum may be due to nematode infection.

1. INTRODUCTION

The vertebrate's gut is perpetually exposed to an enormous amount of different microorganisms and food proteins with antigenic properties that are essential for the individual's life. Concerning this high antigenic load, a complex immunological network has been developed representing an effective protection composed of more structured areas (lymphoid aggregates, Peyer's patches, lymphoglandular complexes, and mesenteric lymph nodes) as well as diffuse parts (lamina propria leucocytes and intraepithelial lymphocytes) [1, 2, 3, 4]. Due to the intestinal colocalisation of commensal bacteria and pathogens, mechanisms preventing specific immune responses against orally administrated antigens (e.g. food proteins and commensal bacteria) termed as oral tolerance are as much as essential as the induction of specific immune responses against pathogenic organisms.

In general, immune regulation can be either mediated by regulatory cytokines (e.g. tissue growth factor (TGF)-β, interleukin (IL)-10) or cellular interactions. Today, it is well-known that specialised T helper cells with regulatory properties (Tregs) play an important part in immune regulation [5, 6]. Depletion of these cells or mutations of related transcription factors provokes severe cases of autoimmune diseases and chronic inflammatory disorders as described in rats, mice, and men [7, 8, 9, 10]. Furthermore, oral tolerance is supposed to be mainly mediated by the induction and expansion of Tregs in the context of specialised tolerogenic dendritic cells in the gut [11, 12, 13]. Alongside their regulatory capacity, Tregs are characterised by the expression of CD4, CD25 [7], and the highly conserved transcription factor Forkhead box P3 (Foxp3) [14] serving a pivotal role in stabilising their regulatory properties [15].

Canine idiopathic Inflammatory Bowel Disease (IBD) is a term used for a variety of disorders characterised by chronic persistent or recurrent gastrointestinal signs (e.g. vomiting, diarrhoea, and weight loss) with no underlying cause (e.g. parasitosis, bacterial infection, food allergy, lymphoma, or lymphangiectasia) and histological evidence of mucosal inflammatory infiltration [16, 17, 18].

In humans, idiopathic IBD is known as Crohn's disease (CD) or ulcerative colitis (UC) [19, 20]. CD is characterised by transmural neutrophilic inflammation followed by granuloma formation commonly present in the terminal jejunum and ileum and is often complicated by intestinal or anal fistulae, strictures, obstructions, or intestinal perforations [21]. In UC, neutrophilic infiltration is restricted to the colonic and rectal mucosa and is associated with ulceration and crypt abscesses. In addition to intestinal lesions, extraintestinal manifestations (e.g. arthritis, renal disease, pyrexia, mucocutaneous lesions, hepatobiliary complications, or osteopenia) are often reported in human IBD [22].

In dogs, mucosal inflammation is the main histopathological feature of idiopathic IBD and variants are classically distinguished regarding their dominant cellular infiltrates [17, 23]. Lymphoplasmacytic enteritis (LPE) represents the most common form of idiopathic IBD in dogs [24]. Additionally, eosinophilic gastroenteritis (EGE) characterised by a mixed infiltration of inflammatory cells with dominance of eosinophilic granulocytes is the second most frequently diagnosed form of canine IBD. Furthermore, several variants of LPE and EGE exist since inflammation can be restricted to one or more anatomical sites (e.g. stomach, small intestine, or colon). Additionally, histiocytic ulcerative colitis (HUC) is a rare colonic disorder reported in boxers that is characterised by the histological presence of PAS-positive macrophages within the colonic mucosa [25, 26] that is supposed to be associated with *Escherichia coli*[27]. Furthermore, idiopathic granulomatous intestinal inflammation characterised by infiltration of macrophages with granuloma formation similar to CD is rarely reported in dogs [28, 29]. In contrast to human IBD, complications are rare in canine patients and only thrombocytopenia is discussed as an extraintestinal manifestation of IBD in dogs [30]. Although canine idiopathic IBD differs from CD and UC in humans in several aspects, a common pathogenesis is suggested. Idiopathic IBD is supposed to develop in a multilayer model including genetic factors [31, 32, 33, 34, 35], dysregulation of the immune system, and disturbances of the intestinal microflora leading to a breakdown of immunological tolerance to luminal antigens [36, 37].

Parasitic infections constitute common intestinal diseases with zoonotic potential in companion animals [38, 39]. Possible clinical signs in dogs include vomiting, diarrhoea, abdominal pain, and weight loss, but clinical symptoms may be absent. Histopathologically, affected tissues reveal varying degrees of cellular infiltrations with prominence of eosinophilic granulocytes [40] and therefore share some histological features with canine idiopathic EGE. Immunologically, intestinal helminths trigger downregulation of the immune system ensuring longer survival in the host [41, 42]. In murine models of nematode infection, Foxp3+ Tregs expand after application of parasites and correlate with the period of worm survival, but decrease after application of anti-CD25 antibodies that is associated with increased intestinal pathology [43, 44, 45]. Furthermore, excretory-secretory products of several parasites (*Heligmosomoides polygyrus, Haemonchus contortus*, and *Teladorsagia circumcincta*) are known to drive TGF-β-like signalling resulting in expansion of murine Foxp3+ Tregs in vitro and in the gut-associated lymphoid tissue (GALT) in vivo [46]. Interestingly, the administration of parasitic antigens is discussed as a possible treatment of human IBD due to their regulatory properties [47, 48].

Since the availability of specific antibodies, Foxp3 has been established as a specific marker for Tregs in different species [49, 50] and the literature concerning regulatory T cells in domestic animals has been increased [51]. In cats, CD4+ CD25+ Tregs have been characterised in detail in the context of the pathogenesis of feline immunodeficiency virus [52, 53, 54, 55]. Tregs are also described and characterised in pigs [56, 57, 58], especially in linkage to foetal tolerance [59], renal and cardiac allotransplantation [60, 61, 62], and infection with porcine reproductive and respiratory syndrome virus [63, 64]. Expression of Foxp3 is also reported in cows. Most of bovine Foxp3-positive cells coexpress CD4 and CD25 [65], although minor populations of CD8β+ and γδ+ Foxp3+ T cells with Foxp3 expression have been described [66]. However, the regulatory properties of CD4+ CD25+ Foxp3+ bovine T cells are controversially discussed, since γδ+ T cells and CD14+ monocytes but not CD4+ CD25+ Foxp3+ cells exhibited regulatory function in one study [67]. Additionally, ovine Foxp3 has recently been described and Foxp3 expressing cells are significantly increased in the skin of sheep infected with *Psoroptes ovis*[68]. Other species in which Foxp3 expression is described include the horse [69], baboon [70, 71], macaque [72], chimpanzee [73], harbour seals and walrus [74], and zebrafish [75].

In dogs, the principle of peripheral tolerance was first suggested in 1976 as a specific population of T cells seeming to prevent graft-versus-host reaction [76]. In 2007, a subset of CD4+ T cells in blood and/or lymph nodes of dogs was detected using anti-mouse/rat Foxp3 monoclonal antibodies (clone FJK-16s) and authors suggested crossreactivity of these antibodies with canine Foxp3 [49]. Since then, the crossreactivity of this anti-mouse/rat Foxp3 antibody with canine Foxp3 has been confirmed by investigation of cell lines that overexpress the canine Foxp3 gene [77]. Additionally to Foxp3, specific antibodies against canine CD25 have been available for a few years [77, 78, 79] and therefore CD4+ CD25+ Foxp3+ T cells have been described in the canine peripheral blood [77, 80]. Additionally, canine Foxp3+ Tregs were investigated in the context of neoplasia, atopic dermatitis, and adverse food reactions [49, 81, 82, 83, 84, 85, 86, 87]. Recently, the regulatory function of canine CD4+ CD25+ Foxp3+ T cells was described in vitro [79, 80]. Furthermore, there is evidence for heterogeneity of canine Tregs since those with high expression of Foxp3 are possibly activated Tregs in contrast to Tregs with intermediate expression of Foxp3, which are suggested to be a more heterogeneous population of predominantly activated conventional T cells [80]. Regarding the canine gastrointestinal tract, there is only one report about the expression of cytokines and transcription factors of different T cell subsets including Tregs in the duodenum of dogs. In this study, no alterations in gene expression were present in dogs with cutaneous food hypersensitivity compared to healthy controls [86]. Until today, no investigations of Foxp3 expression in canine idiopathic IBD, nor in canine nematode infections have been available.

The aim of the study was to investigate the expression of Foxp3 in the canine GALT in healthy animals and in dogs with chronic idiopathic IBD. Additionally, we investigated cases of canine intestinal nematode infections that share some histopathological features with canine idiopathic EGE.

2. MATERIALS AND METHODS

2.1 Animals

For the evaluation of Foxp3-positive cells in the normal canine gut, 18 dogs without gastrointestinal diseases routinely submitted for necropsy to the Institute of Pathology (University of Veterinary Medicine, Hannover) were used (Table 1). These dogs were presented to the Small Animal Clinic of the University of Veterinary Medicine (Hannover) because of other problems than gastrointestinal symptoms. A detailed clinical workup (e.g. case history, clinical investigation, haematology, serum biochemistry, ultrasonics, and taking X-rays and MRI if necessary) was performed by the clinicians on these dogs and no gastrointestinal symptoms were present during this procedure. Additionally, only dogs without any patho-morphological signs of gastrointestinal disorders during necropsy were used as healthy controls.

Table 1 Dogs without gastrointestinal diseases.

No.	Breed	Age (years)	Age group	WSAVA total	WSAVA part	Reason for euthanasia
1	Leonberger	0.08	juvenile	1	1	Oesophageal dilatation
2	Labrador	0.20	juvenile	4	2	Oesophageal dilatation
3	Alsatian	0.22	juvenile	0	0	Seizures
4	Pinscher	0.52	juvenile	3	1	Ataxia
5	Briard	0.99	juvenile	0	0	Ataxia
6	Jack Russel Terrier	1.94	middle-aged	3	3	Right hemiparesis
7	Siberian Husky	4.00	middle-aged	6	3	Seizures
8	Kuvasz	4.28	middle-aged	0	0	Vertebral sarcoma
9	Australian Shepherd	4.43	middle-aged	3	3	GME
10	Pointer	5.85	middle-aged	3	0	Hypophyseal adenoma
11	Pyrenean Shepherd	6.09	middle-aged	2	2	Spinal disk herniation
12	Yorkshire Terrier	6.25	middle-aged	4	4	NLE
13	Beagle	8.98	old	4	4	Tonsillar SCC
14	Crossbreed	10.16	old	2	2	GME
15	Crossbreed	12.00	old	4	3	Haemangiosarcoma
16	Alsatian	12.85	old	6	4	Spinal disk herniation
17	Crossbreed	14.77	old	3	2	HCM
18	Crossbreed	15.22	old	1	1	Pancreatic adenocarcinoma

WSAVA = histopathological score regarding the guidelines for evaluation and scoring of canine intestinal inflammation of the World Small Animal Veterinary Association (WSAVA) Gastrointestinal Standardization Group. WSAVA total = WSAVA score of the total gastrointestinal tract (including stomach, duodenum, jejunum, ileum, and colon). WSAVA part = WSAVA score of the stomach, duodenum, and colon (for comparison with WSAVA scores of dogs with gastrointestinal diseases). *GME* granulomatous meningoencephalitis. *NLE* necrotizing leucoencephalitis. *SCC* squamous cell carcinoma. *HCM* hypertrophic cardiomyopathy.

Tissue samples collected from different locations (gastric fundus, descending duodenum, middle of jejunum, ileum, descending colon, and mesenteric lymph node) were taken as soon as possible after euthanasia (approximately between 15 and 30 minutes post mortem), being immediately fixed in 10% neutral buffered formalin, and embedded in paraffin. Animals were grouped as juvenile (0.08 to 0.99 years), middle-aged (1.94 to 6.25 years), and old individuals (8.98 to 15.22 years).

For the investigation of Tregs in canine IBD, gastrointestinal biopsy specimen were obtained from 16 dogs suffering from chronic gastrointestinal disorders of unknown aetiology (ages ranging from 2.00 to 7.22 years) that were admitted to the Small Animal Clinic of the University of Veterinary Medicine (Hannover) as well as to another Small Animal Clinic (Duisburg Asterlagen, Table 2). Clinically, dogs were presented with chronic (at least 3 weeks) or recurrent vomiting (no. 1- 4, 6- 8, 13, 15, and 16) and/or diarrhoea (no. 1, 2, 4-12, 14, and 15) partially in conjunction with emaciation (no. 7, 9, 12, 14, and 16). Detailed clinical workup was performed for every patient (including clinical examination, taking X-rays, ultrasonics, haematology, serum biochemistry comprising determination of cobalamin, urinanalysis, microbiology and parasitology of faecal samples, and antimicrobial therapy including application of metronidazole). Several endoscopical biopsies (at least four per localisation) taken from the gastric fundus (n = 16), duodenum (n = 16), and colon (n = 8; no. 1, 2, 4, 9, 12, 13, 15, and 16) during routine diagnostics were fixed in 10% neutral buffered formalin and embedded in paraffin.

Furthermore, 6 dogs with spontaneous intestinal nematode infection detected by macroscopical and/or microscopical findings (age ranges from 0.18 to 1.25 years) were evaluated (Table 2). Clinically, no gastrointestinal symptoms were present in these dogs. Fox Terrier dogs (no. 17 and 18) were euthanised due to progressive paraparesis and routinely submitted for necropsy to the Institute of Pathology (University of Veterinary Medicine, Hannover). Beagle dogs (no. 19-22) were humanely destroyed for reasons unrelated to the present study followed by macroscopical and histopathological examination. Tissue specimens were taken as soon as possible after euthanasia (approximately between 15 and 30 min post mortem) from gastric fundus, descending duodenum, and descending colon and each were immediately processed as described above.

2.2 Histopathological examination

For histopathological examination, 2-4 μm thin sections were prepared and stained with haematoxylin and eosin (H&E) by standard histological procedures.

Table 2 Dogs with inflammatory gastrointestinal diseases.

No.	Breed	Age (years)	Age group	WSAVA score	Disease
1	Crossbreed	2.00	middle-aged	6	EGE
2	Hungarian Vizslar	2.75	middle-aged	14.5	EGEC
3	Crossbreed	3.00	middle-aged	7	EE
4	Crossbreed	4.60	middle-aged	12	EGEC
5	Border Collie	4.76	middle-aged	5	EE
6	Irish Terrier	7.00	middle-aged	7	EGE
7	Crossbreed	7.00	middle-aged	5.5	EGE
8	Hovawart	2.33	middle-aged	8	LPGEC
9	Malinois	4.00	middle-aged	8.5	LPGE
10	Belgian Shepherd	4.50	middle-aged	7.5	LPGE
11	Jack Russel Terrier	5.24	middle-aged	8	LPGEC
12	Boxer	6.00	middle-aged	9	LPEC
13	Crossbreed	6.50	middle-aged	7.5	LPGE
14	WHWT	6.82	middle-aged	15.5	LPGEC
15	Poodle	7.00	middle-aged	4.5	LPGEC
16	Crossbreed	7.22	middle-aged	4.5	LPGE
17	Fox Terrier	0.18	juvenile	11	NI
18	Fox Terrier	0.18	juvenile	8	NI
19	Beagle	1.17	juvenile	3	NI
20	Beagle	1.17	juvenile	6	NI
21	Beagle	1.25	juvenile	10	NI
22	Beagle	1.25	juvenile	11	NI

WHWT = West Highland White Terrier. EE = IBD, eosinophilic enteritis. EGE = IBD, eosinophilic gastroenteritis. EGEC = IBD, eosinophilic gastroenterocolitis. LPE = IBD, lymphoplasmacytic enteritis. LPGE = IBD, lymphoplasmacytic gastroenteritis. LPEC = IBD, lymphoplasmacytic enterocolitis. LPGEC = IBD, lymphoplasmacytic gastroenterocolitis. NI = intestinal nematode infection.

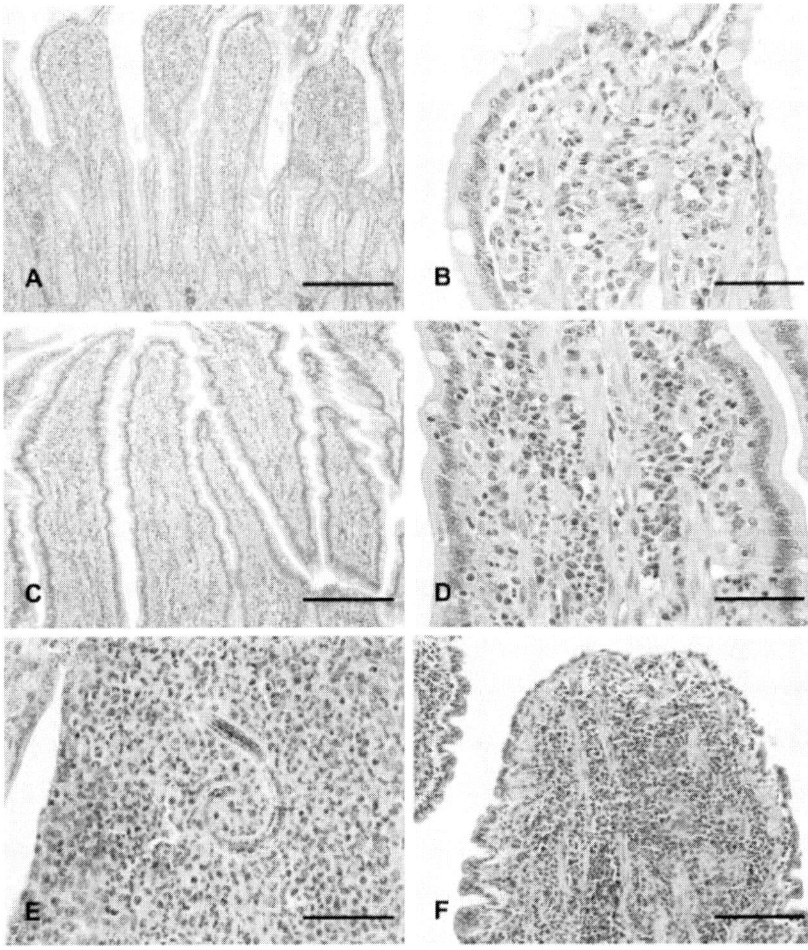

Figure 1 Histology of dogs without gastrointestinal symptoms and of dogs suffering from gastrointestinal diseases. Intestinal samples of dogs used as healthy controls (A: dog no. 8, bar: 200 μm; B: dog no. 8, bar: 50 μm) and of dogs with intestinal nematode infection (C: dog no. 20, bar: 200 μm; D: dog no. 20, bar: 50 μm) revealed comparable freshness to intestinal biopsy specimens. In dogs with intestinal nematode infection, mucosal inflammation was dominated by eosinophilic granulocytes in addition to increased amounts of lymphocytes, plasma cells, and neutrophils (D: dog no. 20, bar: 50 μm). Furthermore, parasitic structures resembling nematode larvae were occasionally present within inflammatory foci (E: dog no. 20, bar: 50 μm). In dogs with LPE, intestinal mucosa was infiltrated by lymphocytes and plasma cells (F: dog no. 9, bar: 100 μm).

Slices were independently evaluated by 3 pathologists (JJ, US, MHT) using light microscopes (Carl Zeiss) according to the histopathological guidelines for the evaluation and scoring of canine intestinal inflammation recently published by the World Small Animal Veterinary Association (WSAVA) Gastrointestinal Standardisation Group [88]. Concerning dogs used as healthy controls and dogs with intestinal nematode infection, only samples with no or minimal signs of autolysis were included in this study ensuring that they were comparable to endoscopical biopsy specimens (Figure 1a-d).

2.3 Immunohistochemistry

For immunohistochemistry, 2-4 μm thin sections were dewaxed in Rotihistol (Roth C. GmbH & Co. KG, Karlsruhe, Germany) and rehydrated in descending ethanol series. Blocking of endogenous peroxidase was achieved by incubation of the sections in H_2O_2 0.5% in ethanol for 30 min. Antigen retrieval was performed with citrate buffer (pH 6.0, 20 min, 95°C) followed by the application of normal rabbit serum for 30 min. Sections were then incubated with anti-mouse/rat Foxp3 antibodies (1 in 400 dilution; clone FJK-16s, eBioscience, San Diego, CA) overnight at 4°C that have shown to detect canine Foxp3 using flow cytometry [49, 77] and immunohistochemistry [89], followed by secondary antibody application (biotinylated rabbit anti-rat IgG [H + L]; Vector Laboratories, Burlingame, California) for 30 min. Subsequently, avidin-biotin complex (Vector Laboratories) was utilised for 30 min. Accordingly, biotinylated tyramine was applied to intensify the reaction followed by a second incubation with an avidin-biotin complex. Between each step, sections were rinsed three times in phosphate buffered saline (PBS; pH 7.2, 0.15 M). 3,3'-diaminobenzidine and H_2O_2 was applied for 5 min to generate a brown colour reaction. Sections were counterstained with Mayer haematoxylin. Sections of normal canine thymus and mandibular lymph node were used as positive controls. The primary antibody solution was substituted with normal rat serum as a negative control.

2.4 Examination of sections

Cells revealing clear lymphocyte morphology with distinct nuclear staining for Foxp3 but unstained cytoplasm (interpreted as Tregs) were quantified in different compartments (gastric fundus: lamina propria; small intestine: villus, basal crypt area, villus-crypt junction; large intestine: apical crypt area, basal crypt area) using a light microscope (Carl Zeiss), a × 40 objective, a × 10 eyepiece, and a square eyepiece graticule (10 × 10 squares, with a total area of 62 500 μm^2). Ten appropriate sites were chosen for each compartment and arithmetic means were calculated for each one. The results are expressed as positive cells per 62 500 μm^2.

2.5 Statistical analysis

Statistical analyses were performed using R version 2.13.1 (The R Foundation for Statistical Computing, Vienna, Austria). Primarily, each group was assessed for normal distribution using a histogram, Q-Q plot, and Shapiro-Wilk test. Kruskal-Wallis test and/or Wilcoxon rank-sum test were used to evaluate differences in WSAVA scores between different age groups in healthy dogs as well as between dogs with gastrointestinal diseases and controls. Furthermore, statistical significant differences in the numbers of Foxp3-positive cells between groups concerning different ages, localisations, and compartments as well as between diseased dogs and healthy controls were assessed using the Kruskal-Wallis test and Wilcoxon rank-sum test for independent samples or the Friedman test and Wilcoxon signed-rank test for related ones. Furthermore, the Spearman's rank correlation coefficient was used to investigate the relationship between intestinal Foxp3-positive cells in dogs with IBD and the WSAVA score for the appropriate intestinal section. P values ≤ 0.05 were defined as statistically significant.

3. RESULTS

3.1 Histopathological findings

In dogs lacking clinical and pathomorphological signs of gastrointestinal diseases, mild histopathological findings were present in the gut (e.g. mild increases in lamina propria lymphocytes and plasma cells, crypt abscesses, crypt dilation, and mucosal fibrosis). These findings are summarised in Table 1 given as WSAVA scores. In these animals, no significant differences between the WSAVA scores of different groups of age were present.

Endoscopical biopsy specimens of dogs with chronic or recurrent gastrointestinal disorders revealed histopathological evidence of mucosal inflammation (Figure 1f). Therefore idiopathic IBD was diagnosed since other possible aetiologies were excluded (e.g. parasitic disease, bacterial infection, food allergy, antibiotic-responsive diarrhoea, lymphangiectasia, or neoplasia). Cellular infiltrates were either composed of lymphocytes and plasma cells (n = 9) or mixed inflammatory cells with prominence of eosinophilic granulocytes (n = 7).

In dogs with intestinal nematode infection, mucosal inflammation was dominated by eosinophilic granulocytes in addition to increased amounts of lymphocytes, plasma cells, and neutrophils (Figure 1d). Furthermore, parasitic structures resembling nematode larvae were occasionally present within inflammatory foci (Figure 1e).

Compared to healthy controls, WSAVA (summarised in Table 2) were significantly increased in dogs suffering from IBD ($p < 0.001$) and in animals with intestinal nematode infections ($p < 0.01$).

3.2 Foxp3 expression in the canine gut of healthy dogs

In healthy dogs, Foxp3 was expressed in both structured and diffuse parts of the canine GALT. Concerning Peyer's patches, Foxp3-positive cells were numerously present in interfollicular areas and lower amounts were visible in dome areas (Figure 2a). Only a few cells were located at the margin of B cell follicles. In gastric lymphoid aggregates as well as in colonic lymphoglandular complexes, Foxp3-positive cells were randomly distributed around the follicles. Mesenteric lymph nodes revealed high amounts of Foxp3-positive Tregs that were largely present in the paracortex as well as in the medullary sinus. In diffuse parts of the canine GALT, Foxp3 was expressed by both lamina propria lymphocytes as well as intraepithelial lymphocytes.

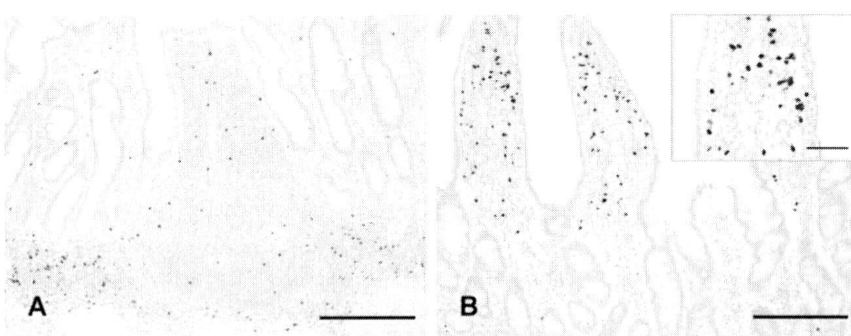

Figure 2 Foxp3 expression in the canine healthy GALT. In canine Peyer's patches, Foxp3-positive cells were numerously present in interfollicular areas and lower amounts were visible in dome areas. Only few cells were located at the margin of B cell follicles (A: jejunum, dog no. 10, bar: 200 µm). Regarding the small intestinal lamina propria, Foxp3-expressing lymphocytes were more numerous in the villus compared to crypts (B: ileum, dog no. 8, bar: 200 µm, inset bar: 50 µm).

Numbers of Foxp3 expressing lymphocytes in the gut of healthy dogs are summarised in Table 3 given as medians per group. Without consideration of different age groups, the numbers of lamina propria Foxp3-positive lymphocytes were significantly increased in villi compared to basal crypt areas and the villus-crypt junction ($p < 0.001$, Figure 2b and 3a-c) in healthy animals. Furthermore, the density of Foxp3-positive lymphocytes was significantly higher in the basal crypt area than in the villus-crypt junction ($p < 0.05$). In the colon, Foxp3-expressing cells were increased in the basal crypt areas compared to the apical areas ($p < 0.05$, Figure 3d).

Table 3 Foxp3 expressing lymphocytes in dogs without gastrointestinal diseases.

Localisation	Compartment	Total	Juvenile	Middle-aged	Old
stomach	la pro	1.13	0.60	1.31	0.98
duodenum	vi	10.67	3.50	18.92	9.56
duodenum	cry-b	2.49	0.50	2.33	3.14
duodenum	jun	1.31	1.20	2.06	1.00
duodenum	cry-total	1.89	0.85	2.85	1.95
jejunum	vi	6.45	2.18	13.13	3.53
jejunum	cry-b	1.78	0.54	2.29	2.05
jejunum	jun	1.26	1.27	1.46	1.00
ileum	vi	4.61	4.00	8.54	3.53
ileum	cry-b	2.62	0.93	2.88	2.89
ileum	jun	1.96	1.36	2.50	1.96
colon	cry-a	0.80	0.25	1.10	0.80
colon	cry-b	1.95	0.66	2.00	1.98

Medians of Foxp3+ lymphocytes per group are given as cells/62 500 μm^2. Total = all dogs disregarding their age. la pro = lamina propria. vi = villus. cry-b = basal crypt area. jun = villus-crypt junction. cry-total = total crypt area. cry-a = apical crypt area.

In healthy dogs disregarding their age, villus lamina propria Foxp3-positive cells were significantly decreased in the jejunum and the ileum compared to the duodenum ($p < 0.001$, Figure 3e). In contrast to this, differences between the localisations of the small intestine were neither detected in the basal crypt area nor the villus-crypt junction.

Furthermore, canine intestinal Tregs revealed differences in their staining intensity for Foxp3 independent of their localisation, i.e. some lymphocytes revealed weak nuclear staining and some lymphocytes had strongly labelled nuclei.

3.3 Age-related changes in canine intestinal Foxp3 expression

In healthy dogs, the number of villus Foxp3-positive lymphocytes were significantly increased in middle-aged dogs compared to juvenile as well as old animals in the duodenum and the jejunum ($p < 0.05$, Figure 3f-h). In the basal crypt area, an increase in Foxp3-expressing lymphocytes was detected in the jejunum between middle-aged and juvenile dogs ($p < 0.05$). In the animals examined, no significant differences in Foxp3 expression between different age groups were present in the villus-crypt junction, ileum, gastric fundus, and colonic lamina propria.

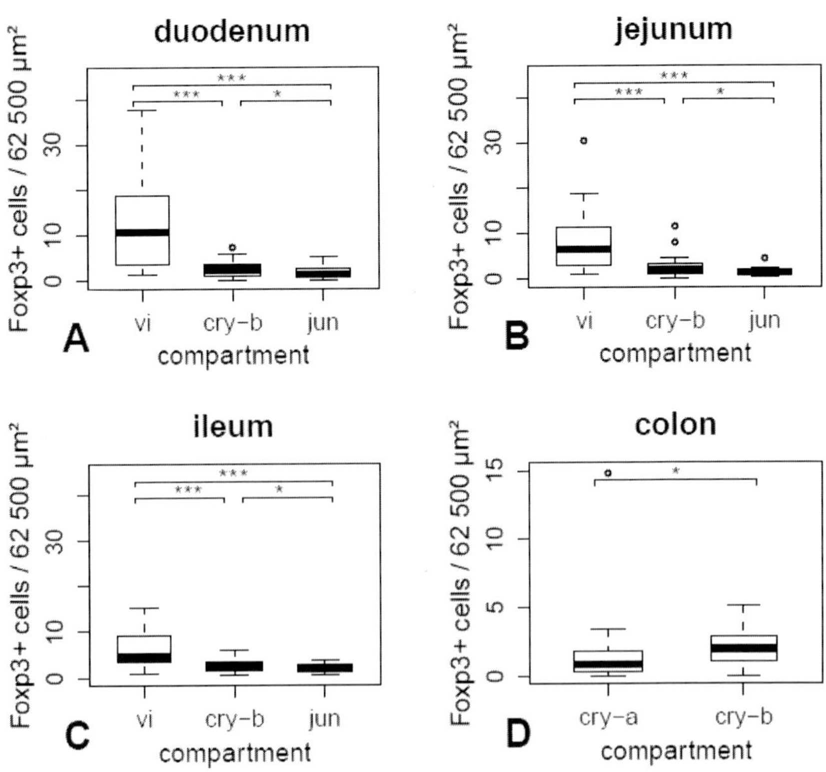

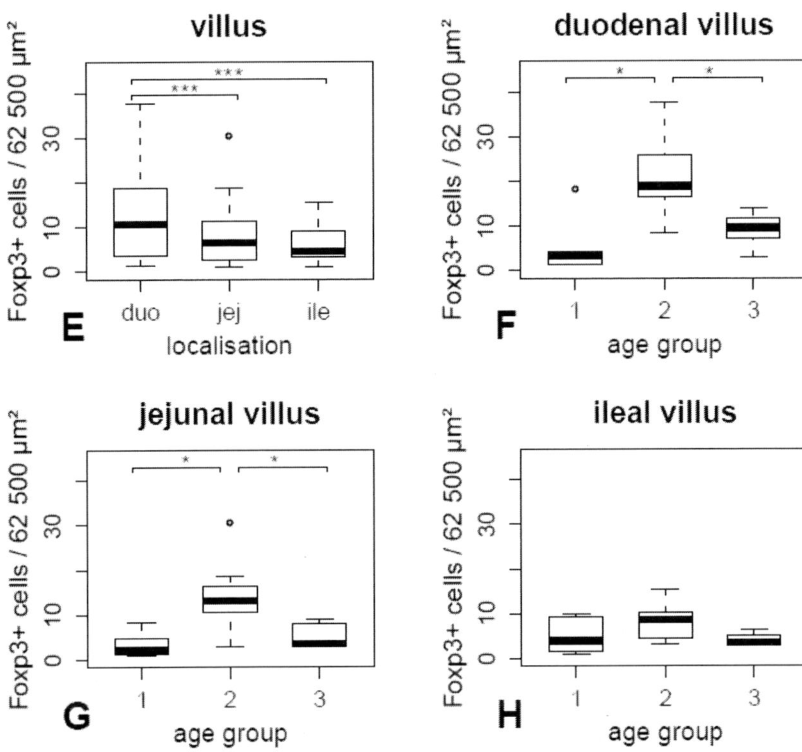

Figure 3 Intestinal Foxp3 expression in dogs without gastrointestinal diseases. Without consideration of their age, numbers of lamina propria Foxp3-positive lymphocytes were significantly increased in the villus compared to the basal crypt area as well as the villus-crypt junction and the density of Foxp3-positive lymphocytes was significantly higher in the basal crypt area than in the villus-crypt junction (A-C). In the colon, Foxp3-positive cells were increased in the basal crypt area compared to the apical part (D). Regarding the villus, lamina propria Foxp3-positive cells were significantly decreased in the jejunum and the ileum compared to the duodenum in healthy dogs disregarding their age (E). Concerning age-related changes, numbers of villus Foxp3-positive lymphocytes were significantly increased in middle-aged dogs compared to juvenile as well as old animals in the duodenum and the jejunum (F-G). In contrast, age-related changes were absent in the ileum (H). vi = villus. cry-b = basal crypt area. jun = villus-crypt junction. cry-a = apical crypt area. duo = duodenum. jej = jejunum. ile = Ileum. 1 = juvenile. 2 = middle-aged. 3 = old. * = $p < 0.05$. *** = $p < 0.001$.

3.4 Foxp3 expression in canine gastrointestinal diseases

Numbers of Foxp3 positive intestinal lymphocytes of dogs with gastrointestinal diseases are summarised in Table 4 given as medians per group. In dogs suffering from IBD, Foxp3-expressing Tregs of duodenal villi were significantly decreased in cases with lymphoplasmacytic infiltration ($p < 0.001$) and in those predominated by eosinophilic inflammation ($p < 0.01$) compared to controls (Figure 4a).

No differences in Foxp3 expression were detected between cases with lymphoplasmacytic and eosinophilic infiltrations. Indeed, differences in Foxp3 expression between dogs with IBD and control animals were absent in duodenal crypt area, gastric fundus, and colonic lamina propria. Regarding the Spearman's rank correlation coefficient, there was no significant correlation between the number of Foxp3-expressing lamina propria lymphocytes and the WSAVA score in the stomach, duodenum, and colon of dogs with IBD.

Table 4 Foxp3 expressing lymphocytes in dogs suffering from gastrointestinal diseases.

Localisation	Compartment	IBD total	IBD-E	IBD-LP	Nematode
stomach	la pro	0.85	1.00	0.70	1.90
duodenum	vi	4.13	4.37	1.85	21.06
duodenum	cry-total	1.41	1.81	0.57	3.16
colon	cry-a	1.63	2.38	1.50	2.02
colon	cry-b	2.81	3.00	1.60	1.86

Medians of Foxp3+ lymphocytes per group are given as cells/62 500 μm^2. la pro = lamina propria. vi = villus. cry-total = total crypt area. cry-a = apical crypt area. cry-b = basal crypt area. IBD-E = IBD dominated by eosinophilic inflammation. IBD-LP = IBD characterised by lymphoplasmacytic infiltration.

In dogs with intestinal nematode infection, Foxp3 expression was significantly increased in the duodenal villus compared to controls ($p < 0.05$, Figure 4b). However, no significant differences in the number of Tregs were present in the gastric fundus, the duodenal crypt area, and the colonic mucosa in animals with parasite infections compared to healthy individuals.

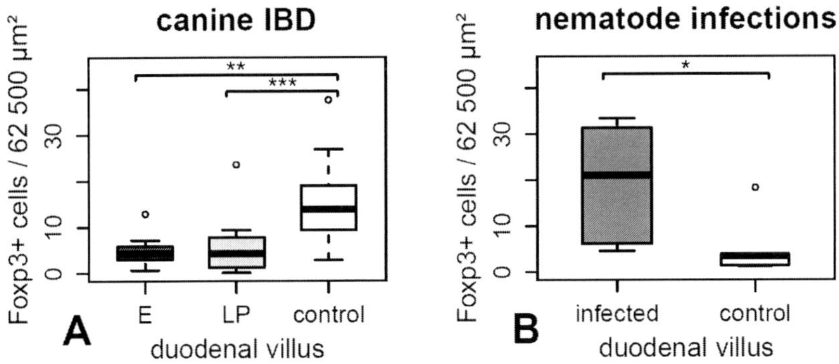

Figure 4 Intestinal Foxp3 expression in dogs suffering from gastrointestinal diseases. In canine IBD (age ranged from 2.00 to 7.22 years), Foxp3-positive cells of the duodenal villus were significantly decreased in cases with lymphoplasmacytic infiltration and in those predominated by eosinophilic inflammation compared to middle-aged controls (A). In dogs with spontaneous intestinal nematode infection (age ranged from 0.18 to 1.25 years), Foxp3 expression was significantly increased in the duodenal villus compared to juvenile healthy animals (B). E = IBD dominated by eosinophilic inflammation. LP = IBD characterised by lymphoplasmacytic infiltration. * = $p < 0.05$. ** = $p < 0.01$. *** = $p < 0.001$.

4. DISCUSSION

Intestinal immune regulation including development of oral tolerance is of great importance for the maintenance of gut homeostasis [90]. Concerning this, Tregs occupy a pivotal role in cell-mediated immunosuppression [5, 6]. Therefore, the aim of the study was to investigate the expression of Foxp3 in canine GALT of healthy and gastrointestinal diseased dogs.

Animals lacking clinical and pathomorphological signs of gastrointestinal diseases were used as healthy controls. Interestingly, minimal histopathological findings were present in the gastrointestinal tract of these dogs. Due to the absence of clinical and morphological correlates, these findings were interpreted as background variances of the canine gut. Although several reports on age-related changes and immunosenescence in dogs [91] would suggest an increasing inflammatory background in the canine elderly, no differences between the WSAVA scores of different age groups were observed. In contrast, WSAVA scores were significantly increased in dogs with gastrointestinal diseases compared to controls.

Immunohistochemically, application of anti-Foxp3 antibodies produced clear nuclear staining of cells revealing distinct lymphocyte morphology.

Although Foxp3 has been established as a specific marker for classical Tregs further characterised by the expression of CD4 and CD25, heterogeneity of this cell population has been described [92]. In dogs, recent analyses on canine Tregs illustrate CD4-positive cells with intermediate Foxp3 expression and authors postulate that they are activated heterogeneous conventional T cells instead of Tregs [80]. In our study, differences in staining intensity for Foxp3 were noticed immunohistochemically. This suggests differences in Foxp3 expression in situ supporting the concept of canine Treg heterogeneity. However, this result could also be due to the localisation of Tregs in different tissue layers and needs to be further investigated.

Structured lymphoid tissue in the intestine plays a crucial role in mucosal immunology especially in induction and gut homing of intestinal Tregs and therefore for the establishment of oral tolerance [13, 93]. The high density of Foxp3-expressing lymphocytes in those areas suggests similar roles of these anatomical compartments in dogs.

Additionally, Foxp3 was expressed by intraepithelial lymphocytes (IEL). It is well established that canine IEL are mainly characterised by the expression of CD8. However, less than 15 percent of these cells are CD4-positive [94]. Although most canine Tregs are known to express CD4, CD25, and Foxp3, a minor subpopulation of CD8-positive lymphocytes with strong Foxp3 expression is described [80]. In spite of their correct phenotype, our results confirm the immunoregulatory phenotype of canine IEL, which has recently been observed in vitro [94].

Furthermore, Foxp3-positive lymphocytes were present in the lamina propria. Concerning different compartments (vertical distribution), Tregs were significantly increased in the villus lamina propria compared to basal crypt area and villus-crypt junction. Previous studies about intestinal T cells in dogs report similar differences in their distribution [95, 96, 97]. Referring to this, a stronger exposition of the intestinal apical surface to luminal antigens is discussed. Analogously, increasing numbers of Tregs in the villus area may be related to the great importance of immunoregulation in areas characterised by high antigenic load, antigen sampling by dendritic cells, and priming of immune responses.

Additionally, the numbers of Foxp3-positive cells were significantly higher in the basal crypt area than in the villus-crypt junction. This finding was in contrast to previous observations of a constant increase in CD3-positive cells towards the villus tip [95, 96, 97]. Possibly, the increase in Tregs in the basal crypt area compared to the villus-crypt junction is due to increased numbers of IgA-secreting plasma cells towards the crypt [96] requiring a more intense local tuning of immunological tolerance in basal areas.

Interestingly, a vertical decline in Tregs was found in the colonic mucosa since the numbers of Foxp3-positive lymphocytes were significantly lower in the apical crypt area than in the basal part. Potentially, this is related to general differences of colonic Tregs in comparison to those located in the small intestine highlighting the heterogeneity of regulatory T cells. Otherwise, these differences may be due to the individual anatomy and physiology of the small and large intestines. However, further studies are needed to confirm possible differences in the vertical distribution of lymphocytes in the colonic mucosa and to evaluate its relevance.

Concerning the horizontal distribution of Tregs along the small intestinal villus, numbers of Foxp3-positive lymphocytes were significantly decreased from duodenal to jejunal as well as to ileal mucosa. In contrast to this, no horizontal differences in the expression of CD3, CD4, or CD8 have been reported [95, 97]. The horizontal decline in expression of Foxp3 may be explained by decreasing amounts of food antigens progressively destroyed along the small intestine, although the numbers of bacteria progressively increase from the duodenum to the large bowel [98]. Regarding the lower frequencies of Tregs at the luminal site of the colonic mucosa compared to those in the small intestine, it is possible that food proteins are more potent generators of intestinal Tregs than commensal bacteria. In mice, it has recently been shown that T cell receptors of colonic Tregs differ from those used by Tregs in other locations [99]. Therefore, canine colonic Tregs may generally differ from those in other areas including the small intestine.

In healthy dogs, age-related changes in Foxp3 expression were observed since Tregs increased in middle-aged animals compared to juvenile as well as older individuals. Age-dependent alterations of the immune system are well-known in companion animals [91, 100] and several reports on alterations of the canine gastrointestinal tract concerning different age groups exist [97, 101, 102]. For instance, intestinal lamina propria CD3-positive T cells and macrophages decline in older animals, whereas IgA-containing plasma cells increase with age [97]. Although age-related differences in Tregs are controversially discussed [103] and only age groups with small numbers of animals (5 to 7 dogs per age group) were examined in this study, our data suggest an increase of intestinal Tregs in middle-age dogs that may be an indication for an increase of oral tolerance in these dogs compared to juvenile individuals. Nevertheless, the numbers of Tregs may change in the canine gut during the first year of life since the canine immune system including the GALT undergoes maturation during this period. Additionally, the decline in Foxp3 expression in old animals could be a sign of immunosenescence and could possibly mean a loss of oral tolerance in the canine elderly.

Although age-related changes in Foxp3 expression were of statistical significance in this study, further investigations are needed to confirm these results using higher numbers of animals and to check the course of Tregs in juvenile dogs.

Dysregulation of mucosal immunology including a failure of oral tolerance against commensal bacteria and food proteins seems to play a crucial role in the pathogenesis of both human and canine IBD [36, 104]. As it is already known, depletion of regulatory T cells leads to chronic colitis in immunodeficient mice [11]. Furthermore, intravenous application of Tregs can cure established intestinal inflammation in animal models for experimental colitis [105, 106]. In Long-Evans Cinnamon rats (deficient in thymocyte development) showing spontaneous development of IBD-like colitis, Tregs are significantly reduced in the colonic lamina propria [9]. Interestingly, Foxp3-positive Tregs are present at higher densities in the lamina propria of human patients suffering from IBD than in healthy controls [107, 108, 109]. This indicates an impaired function of Tregs or a resistance of T effector cells to Treg-mediated suppression instead of a numerical aberrance in the Treg population. However, investigated dogs suffering form IBD revealed a significant reduction of Foxp3 expression in the duodenal villi in cases of both eosinophilic and lymphoplasmacytic inflammation compared to controls. Further studies using full-thickness intestinal samples are needed to check if similar differences in Tregs are also present in the jejunum and ileum in canine IBD. Interestingly, no differences in Treg numbers were present in the stomach and colon of dogs with IBD in comparison to controls. This may be an indication of different roles of small intestinal Tregs compared to those located in the stomach and the colon for canine IBD. Under physiological conditions, enrichment of Tregs in areas of inflammation is mediated by a negative feedback mechanism via IL-2 produced by T effector cells ensuring a balanced cooperation between protective immune responses and immunosuppression [110]. Therefore, our results may be a sign of a primary defect in Treg homeostasis leading to chronic inflammation due to a lack of immune regulation as a pathogenetic factor in canine IBD. Further studies on Treg function are needed to confirm this possibility.

Unlike dogs with IBD, dogs with intestinal nematode infection revealed increased numbers of Tregs in the duodenal villus compared to controls. Interestingly, no differences in Treg numbers were present in the stomach and colon of dogs with nematode infection in comparison to controls. Therefore, further studies are needed to determine whether this is related to the nematode species and its localisation in the gut or due to general differences in canine Tregs in the small intestine compared to those located in the stomach and colon. Intestinal parasites are known to trigger Th2 dominated immune responses and cause down regulation of immune responses ensuring a better survival in their hosts [42, 111]. Particularly, suspension of Treg activity leads to exacerbated intestinal pathology and enhances parasite clearance [43, 44]. Furthermore, helminth secretions are able to induce Foxp3 expression and enhance the regulatory function of T cells [46]. Therefore, our results suggest a parasite-related increase in intestinal immune regulation, although IL-2 mediated enrichment of Tregs may also be involved. Interestingly, parasites are discussed as potential strategies for the treatment of human IBD due to their ability to prevent intestinal pathology and their immunoregulatory properties [47, 48].

This study elucidates that regulatory T cells are expressed in the canine GALT with a decline of Tregs from the villus to the crypt and age-related changes suggesting a progressive establishment of oral tolerance and immunosenescence including decreased oral tolerance in the canine elderly. A reduced number of Tregs in canine IBD suggests a defect in Treg homeostasis leading to a loss of oral tolerance. In addition, Tregs were enriched in the duodenal villi of dogs with intestinal nematode infection indicating a parasite-induced enhancement of intestinal immune regulation.

ACKNOWLEDGEMENTS

We thank Dr P. Wefstaedt (Small Animal Clinic of the University of Veterinary Medicine, Hannover) for providing endoscopical biopsies from diseased dogs and B. Buck and K.-P. Kuhlmann for excellent technical assistance. Additionally, we thank Dr M. Beyerbach for advising on statistical analyses. Johannes Junginger received a scholarship from the Studienstiftung des deutschen Volkes.

REFERENCES

1. Atkins AM, Schofield GC: Lymphoglandular complexes in the large intestine of the dog. J Anat. 1972, 113: 169-178.
2. HogenEsch H, Felsburg PJ: Isolation and phenotypic and functional characterization of cells from Peyer's patches in the dog. Vet Immunol Immunopathol. 1992, 31: 1-10.
3. Stokes C, Waly N: Mucosal defence along the gastrointestinal tract of cats and dogs. Vet Res. 2006, 37: 281-293.
4. Ramiro-Puig E, Perez-Cano FJ, Castellote C, Franch A, Castell M: The bowel: a key component of the immune system. Rev Esp Enferm Dig. 2008, 100: 29-34. in Spanish
5. Lan RY, Mackay IR, Gershwin ME: Regulatory T cells in the prevention of mucosal inflammatory diseases: patrolling the border. J Autoimmun. 2007, 29: 272-280.
6. Sakaguchi S, Yamaguchi T, Nomura T, Ono M: Regulatory T cells and immune tolerance. Cell. 2008, 133: 775-787.
7. Sakaguchi S, Sakaguchi N, Asano M, Itoh M, Toda M: Immunologic self-tolerance maintained by activated T cells expressing IL-2 receptor alpha-chains (CD25). Breakdown of a single mechanism of self-tolerance causes various autoimmune diseases. J Immunol. 1995, 155: 1151-1164.
8. Bennett CL, Christie J, Ramsdell F, Brunkow ME, Ferguson PJ, Whitesell L, Kelly TE, Saulsbury FT, Chance PF, Ochs HD: The immune dysregulation, polyendocrinopathy,

enteropathy, X-linked syndrome (IPEX) is caused by mutations of FOXP3. Nat Genet. 2001, 27: 20-21.
9. Ishimaru N, Yamada A, Kohashi M, Arakaki R, Takahashi T, Izumi K, Hayashi Y: Development of inflammatory bowel disease in Long-Evans Cinnamon rats based on CD4+CD25+Foxp3+ regulatory T cell dysfunction. J Immunol. 2008, 180: 6997-7008.
10. Moes N, Rieux-Laucat F, Begue B, Verdier J, Neven B, Patey N, Torgerson TT, Picard C, Stolzenberg MC, Ruemmele C: Reduced expression of FOXP3 and regulatory T-cell function in severe forms of early-onset autoimmune enteropathy. Gastroenterology. 2010, 139: 770-778.
11. Veltkamp C, Ruhwald R, Giesem T, Autschbach F, Kaden I, Veltkamp R, Sartor RB, Stremmel W: CD4+CD25+ cell depletion from the normal CD4+ T cell pool prevents tolerance toward the intestinal flora and leads to chronic colitis in immunodeficient mice. Inflamm Bowel Dis. 2006, 12: 437-446.
12. Coombes JL, Maloy KJ: Control of intestinal homeostasis by regulatory T cells and dendritic cells. Semin Immunol. 2007, 19: 116-126.
13. Hadis U, Wahl B, Schulz O, Hardtke-Wolenski M, Schippers A, Wagner N, Muller W, Sparwasser T, Forster R, Pabst O: Intestinal tolerance requires gut homing and expansion of FoxP3+ regulatory T cells in the lamina propria. Immunity. 2011, 34: 237-246.
14. Brunkow ME, Jeffery EW, Hjerrild KA, Paeper B, Clark LB, Yasayko SA, Wilkinson JE, Galas D, Ziegler SF, Ramsdell F: Disruption of a new forkhead/winged-helix protein, scurfin, results in the fatal lymphoproliferative disorder of the scurfy mouse. Nat Genet. 2001, 27: 68-73.
15. Fontenot JD, Gavin MA, Rudensky AY: Foxp3 programs the development and function of CD4+CD25+ regulatory T cells. Nat Immunol. 2003, 4: 330-336.
16. Jergens AE, Moore FM, Haynes JS, Miles KG: Idiopathic inflammatory bowel disease in dogs and cats: 84 cases (1987-1990). J Am Vet Med Assoc. 1992, 201: 1603-1608.
17. Hall EJ, German AJ: Inflammatory Bowel Disease. Small Animal Gastroenterology. Edited by: Steiner JM. 2008, Hannover: Schluetersche Verlagsgesellschaft, 312-329.
18. Washabau RJ, Day MJ, Willard MD, Hall EJ, Jergens AE, Mansell J, Minami T, Bilzer TW: Endoscopic, biopsy, and histopathologic guidelines for the evaluation of gastrointestinal inflammation in companion animals. J Vet Intern Med. 2010, 24: 10-26.
19. Podolsky DK: Inflammatory bowel disease. N Engl J Med. 2002, 347: 417-429.
20. Xavier RJ, Podolsky DK: Unravelling the pathogenesis of inflammatory bowel disease. Nature. 2007, 448: 427-434.
21. Galandiuk S, Kimberling J, Al-Mishlab TG, Stromberg AJ: Perianal Crohn disease: predictors of need for permanent diversion. Ann Surg. 2005, 241: 796-801.
22. Williams H, Walker D, Orchard TR: Extraintestinal manifestations of inflammatory bowel disease. Curr Gastroenterol Rep. 2008, 10: 597-605. 10.
23. German AJ, Hall EJ, Day MJ: Chronic intestinal inflammation and intestinal disease in dogs. J Vet Intern Med. 2003, 17: 8-20.
24. Craven M, Simpson JW, Ridyard AE, Chandler ML: Canine inflammatory bowel disease: retrospective analysis of diagnosis and outcome in 80 cases (1995-2002). J Small Anim Pract. 2004, 45: 336-342.
25. Van Kruiningen HJ, Montali RJ, Strandberg JD, Kirk RW: A granulomatous colitis of dogs with histologic resemblance to Whipple's disease. Pathol Vet. 1965, 2: 521-544.
26. Craven M, Mansfield CS, Simpson KW: Granulomatous colitis of boxer dogs. Vet Clin North Am Small Anim Pract. 2011, 41: 433-445.
27. Mansfield CS, James FE, Craven M, Davies DR, O'Hara AJ, Nicholls PK, Dogan B, MacDonough SP, Simpson KW: Remission of histiocytic ulcerative colitis in Boxer dogs correlates with eradication of invasive intramucosal Escherichia coli. J Vet Intern Med. 2009, 23: 964-969.
28. Bright RM, Jenkins C, DeNovo RM, McCrackin MA, Toal R: Chronic diarrhoea in a dog with regional granulomatous enteritis. J Small Anim Pract. 1994, 35: 423-426.
29. Lewis DC: Successful treatment of regional enteritis in a dog. J Am Anim Hosp Assoc. 1995, 31: 170-173.

30. Ridgway J, Jergens AE, Niyo Y: Possible causal association of idiopathic inflammatory bowel disease with thrombocytopenia in the dog. J Am Anim Hosp Assoc. 2001, 37: 65-74.
31. German AJ, Hall EJ, Day MJ: Relative deficiency in IgA production by duodenal explants from German shepherd dogs with small intestinal disease. Vet Immunol Immunopathol. 2000, 76: 25-43.
32. Hostutler RA, Luria BJ, Johnson SE, Weisbrode SE, Sherding RG, Jaeger JQ, Guilford WG: Antibiotic-responsive histiocytic ulcerative colitis in 9 dogs. J Vet Intern Med. 2004, 18: 499-504.
33. Littler RM, Batt RM, Lloyd DH: Total and relative deficiency of gut mucosal IgA in German shepherd dogs demonstrated by faecal analysis. Vet Rec. 2006, 158: 334-341.
34. Kathrani A, House A, Catchpole B, Murphy A, German A, Werling D, Allenspach K: Polymorphisms in the TLR4 and TLR5 gene are significantly associated with inflammatory bowel disease in German shepherd dogs. PLoS One. 2010, 5: e15740.
35. Kathrani A, Werling D, Allenspach K: Canine breeds at high risk of developing inflammatory bowel disease in the south-eastern UK. Vet Rec. 2011, 169: 635
36. Maloy KJ, Powrie F: Intestinal homeostasis and its breakdown in inflammatory bowel disease. Nature. 2011, 474: 298-306.
37. Allenspach K: Clinical immunology and immunopathology of the canine and feline intestine. Vet Clin North Am Small Anim Pract. 2011, 41: 345-360.
38. Robertson ID, Thompson RC: Enteric parasitic zoonoses of domesticated dogs and cats. Microbes Infect. 2002, 4: 867-873.
39. Lee AC, Schantz PM, Kazacos KR, Montgomery SP, Bowman DD: Epidemiologic and zoonotic aspects of ascarid infections in dogs and cats. Trends Parasitol. 2010, 26: 155-161.
40. Hayden DW, Kruiningen HJ: Experimentally induced canine toxocariasis: laboratory examinations and pathologic changes, with emphasis on the gastrointestinal tract. Am J Vet Res. 1975, 36: 1605-1614.
41. Anthony RM, Rutitzky LI, Urban JF, Stadecker MJ, Gause WC: Protective immune mechanisms in helminth infection. Nat Rev Immunol. 2007, 7: 975-987.
42. Allen JE, Maizels RM: Diversity and dialogue in immunity to helminths. Nat Rev Immunol. 2011, 11: 375-388.
43. Taylor MD, LeGoff L, Harris A, Malone E, Allen JE, Maizels RM: Removal of regulatory T cell activity reverses hyporesponsiveness and leads to filarial parasite clearance in vivo. J Immunol. 2005, 174: 4924-4933.
44. D'Elia R, Behnke JM, Bradley JE, Else KJ: Regulatory T cells: a role in the control of helminth-driven intestinal pathology and worm survival. J Immunol. 2009, 182: 2340-2348.
45. Othman AA, El-Shourbagy SH, Soliman RH: Kinetics of Foxp3-expressing regulatory cells in experimental Toxocara canis infection. Exp Parasitol. 2011, 127: 454-459.
46. Grainger JR, Smith KA, Hewitson JP, McSorley HJ, Harcus Y, Filbey KJ, Finney CA, Greenwood EJ, Knox DP, Wilson MS, Belkaid Y, Rudensky AY, Maizels RM: Helminth secretions induce de novo T cell Foxp3 expression and regulatory function through the TGF-beta pathway. J Exp Med. 2010, 207: 2331-2341.
47. Moreels TG, Pelckmans PA: Gastrointestinal parasites: potential therapy for refractory inflammatory bowel diseases. Inflamm Bowel Dis. 2005, 11: 178-184.
48. Ruyssers NE, De Winter BY, De Man JG, Loukas A, Herman AG, Pelckmans PA, Moreels TG: Worms and the treatment of inflammatory bowel disease: are molecules the answer?. Clin Dev Immunol. 2008, 2008: 567314.
49. Biller BJ, Elmslie RE, Burnett RC, Avery AC, Dow SW: Use of FoxP3 expression to identify regulatory T cells in healthy dogs and dogs with cancer. Vet Immunol Immunopathol. 2007, 116: 69-78.
50. Banham AH, Lyne L, Scase TJ, Blacklaws BA: Monoclonal antibodies raised to the human FOXP3 protein can be used effectively for detecting Foxp3(+) T cells in other mammalian species. Vet Immunol Immunopathol. 2009, 127: 376-381.

51. Garden OA, Pinheiro D, Cunningham F: All creatures great and small: regulatory T cells in mice, humans, dogs and other domestic animal species. Int Immunopharmacol. 2011, 11: 576-588.
52. Vahlenkamp TW, Tompkins MB, Tompkins WA: Feline immunodeficiency virus infection phenotypically and functionally activates immunosuppressive CD4+CD25+ T regulatory cells. J Immunol. 2004, 172: 4752-4761.
53. Joshi A, Vahlenkamp TW, Garg H, Tompkins WA, Tompkins MB: Preferential replication of FIV in activated CD4(+)CD25(+)T cells independent of cellular proliferation. Virology. 2004, 321: 307-322.
54. Lankford S, Petty C, LaVoy A, Reckling S, Tompkins W, Dean GA: Cloning of feline FOXP3 and detection of expression in CD4+CD25+ regulatory T cells. Vet Immunol Immunopathol. 2008, 122: 159-166.
55. Mexas AM, Fogle JE, Tompkins WA, Tompkins MB: CD4+CD25+ regulatory T cells are infected and activated during acute FIV infection. Vet Immunol Immunopathol. 2008, 126: 263-272.
56. Kaser T, Gerner W, Hammer SE, Patzl M, Saalmuller A: Phenotypic and functional characterisation of porcine CD4(+)CD25(high) regulatory T cells. Vet Immunol Immunopathol. 2008, 122: 153-158.
57. Kaser T, Gerner W, Hammer SE, Patzl M, Saalmuller A: Detection of Foxp3 protein expression in porcine T lymphocytes. Vet Immunol Immunopathol. 2008, 125: 92.
58. Bolzer K, Kaser T, Saalmuller A, Hammer SE: Molecular characterisation of porcine Forkhead-box p3 (Foxp3). Vet Immunol Immunopathol. 2009, 132: 275-281.
59. Georgieva R: Dynamics of T-suppressor and T-helper lymphocytes and haemolytic plaque-forming cells during normal pregnancy in the sow. J Reprod Immunol. 1984, 6: 151-156.
60. Mezrich JD, Kesselheim JA, Johnston DR, Yamada K, Sachs DH, Madsen JC: The role of regulatory cells in miniature swine rendered tolerant to cardiac allografts by donor kidney cotransplantation. Am J Transplant. 2003, 3: 1107-1115.
61. Wu A, Yamada K, Ierino FL, Vagefi PA, Sachs DH: Regulatory mechanism of peripheral tolerance: in vitro evidence for dominant suppression of host responses during the maintenance phase of tolerance to renal allografts in miniature swine. Transpl Immunol. 2003, 11: 367-374.
62. Wu A, Yamada K, Baron C, Mathes DW, Monajati LM, Vagefi PA, Sachs DH: Detection of regulatory cells as an assay for allograft tolerance in miniature swine. J Heart Lung Transplant. 2004, 23: 210-217.
63. Wongyanin P, Buranapraditkun S, Chokeshai-Usaha K, Thanawonguwech R, Suradhat S: Induction of inducible CD4+CD25+Foxp3+ regulatory T lymphocytes by porcine reproductive and respiratory syndrome virus (PRRSV). Vet Immunol Immunopathol. 2010, 133: 170-182.
64. Silva-Campa E, Flores-Mendoza L, Resendiz M, Pinelli-Saavedra A, Mata-Haro V, Mwangi W, Hernandez J: Induction of T helper 3 regulatory cells by dendritic cells infected with porcine reproductive and respiratory syndrome virus. Virology. 2009, 387: 373-379.
65. Seo KS, Davis WC, Hamilton MJ, Park YH, Bohach GA: Development of monoclonal antibodies to detect bovine FOXP3 in PBMCs exposed to a staphylococcal superantigen. Vet Immunol Immunopathol. 2009, 128: 30-36.
66. Gerner W, Stadler M, Hammer SE, Klein D, Saalmuller A: Sensitive detection of Foxp3 expression in bovine lymphocytes by flow cytometry. Vet Immunol Immunopathol. 2010, 138: 154-158.
67. Hoek A, Rutten VP, Kool J, Arkesteijn GJ, Bouwstra RJ, Van Rhijn I, Koets AP: Subpopulations of bovine WC1(+) gammadelta T cells rather than CD4(+)CD25(high) Foxp3(+) T cells act as immune regulatory cells ex vivo. Vet Res. 2009, 40: 6.
68. McNeilly TN, McIntyre J, Frew D, Griffiths DJ, Wattegedera SR, van den Broek A, Huntley JF: Infestation of sheep with Psoroptes ovis, the sheep scab mite, results in recruitment of Foxp3(+) T cells into the dermis. Parasite Immunol. 2010, 32: 361-369.

69. Hamza E, Steinbach F, Marti E: CD4(+)CD25(+) T cells expressing FoxP3 in Icelandic horses affected with insect bite hypersensitivity. Vet Immunol Immunopathol. doi:10.1016/j.vetimm.2011.05.033.
70. Porter CM, Horvath-Arcidiacono JA, Singh AK, Horvath KA, Bloom ET, Mohiuddin MM: Characterization and expansion of baboon CD4+CD25+ Treg cells for potential use in a non-human primate xenotransplantation model. Xenotransplantation. 2007, 14: 298-308.
71. Singh AK, Horvath KA, Mohiuddin MM: Rapamycin promotes the enrichment of CD4(+)CD25(hi)FoxP3(+) T regulatory cells from naive CD4(+) T cells of baboon that suppress antiporcine xenogenic response in vitro. Transplant Proc. 2009, 41: 418-421.
72. Allers K, Loddenkemper C, Hofmann J, Unbehaun A, Kunkel D, Moos V, Kaup FJ, Stahl-Hennig C, Sauermann U, Epple HJ, Schneider T: Gut mucosal FOXP3+ regulatory CD4+ T cells and Nonregulatory CD4+ T cells are differentially affected by simian immunodeficiency virus infection in rhesus macaques. J Virol. 2010, 84: 3259-3269.
73. Manigold T, Shin EC, Mizukoshi E, Mihalik K, Murthy KK, Rice CM, Piccirillo CA, Rehermann B: Foxp3+CD4+CD25+ T cells control virus-specific memory T cells in chimpanzees that recovered from hepatitis C. Blood. 2006, 107: 4424-4432.
74. Seibel H, Stimmer L, Siebert U, Beineke A: Immunohistochemical characterization of selected cell markers for the detection of hematopoietic cells in formalin-fixed, paraffin wax-embedded lymphoid tissues of harbor seals (Phoca vitulina) and walruses (Odobenus rosmarus rosmarus). Vet Immunol Immunopathol. 2010, 137: 305-309.
75. Mitra S, Alnabulsi A, Secombes CJ, Bird S: Identification and characterization of the transcription factors involved in T-cell development, t-bet, stat6 and foxp3, within the zebrafish, Danio rerio. FEBS J. 2010, 277: 128-147.
76. Weiden PL, Storb R, Tsoi MS, Graham TC, Lerner KG, Thomas ED: Infusion of donor lymphocytes into stable canine radiation chimeras: implications for mechanism of transplantation tolerance. J Immunol. 1976, 116: 1212-1219.
77. Mizuno T, Suzuki R, Umeki S, Okuda M: Crossreactivity of antibodies to canine CD25 and Foxp3 and identification of canine CD4+CD25 +Foxp3+ cells in canine peripheral blood. J Vet Med Sci. 2009, 71: 1561-1568.
78. Rissetto KC, Rindt H, Selting KA, Villamil JA, Henry CJ, Reinero CR: Cloning and expression of canine CD25 for validation of an anti-human CD25 antibody to compare T regulatory lymphocytes in healthy dogs and dogs with osteosarcoma. Vet Immunol Immunopathol. 2010, 135: 137-145.
79. Abrams VK, Hwang B, Lesnikova M, Gass MJ, Wayner E, Castilla-Llorente C, Georges GE, Nash RA: A novel monoclonal antibody specific for canine CD25 (P4A10): Selection and evaluation of canine Tregs. Vet Immunol Immunopathol. 2009, 135: 257-265.
80. Pinheiro D, Singh Y, Grant CR, Appleton RC, Sacchini F, Walker KR, Chadbourne AH, Palmer CA, Armitage-Chan E, Thompson I, Williamson L, Cunningham F, Garden OA: Phenotypic and functional characterization of a CD4(+) CD25(high) FOXP3(high) regulatory T-cell population in the dog. Immunology. 2011, 132: 111-122.
81. O'Neill K, Guth A, Biller B, Elmslie R, Dow S: Changes in regulatory T cells in dogs with cancer and associations with tumor type. J Vet Intern Med. 2009, 23: 875-881.
82. Horiuchi Y, Tominaga M, Ichikawa M, Yamashita M, Okano K, Jikumaru Y, Nariai Y, Nakajima Y, Kuwabara M, Yukawa M: Relationship between regulatory and type 1 T cells in dogs with oral malignant melanoma. Microbiol Immunol. 2010, 54: 152-159.
83. Tominaga M, Horiuchi Y, Ichikawa M, Yamashita M, Okano K, Jikumaru Y, Nariai Y, Kadosawa T: Flow cytometric analysis of peripheral blood and tumor-infiltrating regulatory T cells in dogs with oral malignant melanoma. J Vet Diagn Invest. 2010, 22: 438-441.
84. Horiuchi Y, Tominaga M, Ichikawa M, Yamashita M, Jikumaru Y, Nariai Y, Nakajima Y, Kuwabara M, Yukawa M: Increase of regulatory T cells in the peripheral blood of dogs with metastatic tumors. Microbiol Immunol. 2009, 53: 468-474.
85. Keppel KE, Campbell KL, Zuckermann FA, Greeley EA, Schaeffer DJ, Husmann RJ: Quantitation of canine regulatory T cell populations, serum interleukin-10 and allergen-specific IgE concentrations in healthy control dogs and canine atopic dermatitis patients

receiving allergen-specific immunotherapy. Vet Immunol Immunopathol. 2008, 123: 337-344.
86. Veenhof EZ, Rutten VP, van Noort R, Knol EF, Willemse T: Evaluation of T-cell activation in the duodenum of dogs with cutaneous food hypersensitivity. Am J Vet Res. 2010, 71: 441-446.
87. Veenhof EZ, Knol EF, Schlotter YM, Vernooij JC, Rutten VP, Willemse T: Characterisation of T cell phenotypes, cytokines and transcription factors in the skin of dogs with cutaneous adverse food reactions. Vet J. 2011, 187: 320-324.
88. Day MJ, Bilzer T, Mansell J, Wilcock B, Hall EJ, Jergens A, Minami T, Willard M, Washabau R: Histopathological standards for the diagnosis of gastrointestinal inflammation in endoscopic biopsy samples from the dog and cat: a report from the World Small Animal Veterinary Association Gastrointestinal Standardization Group. J Comp Pathol. 2008, 138 (Suppl 1): 1-43.
89. Boozer LB, Davis TW, Borst LB, Zseltvay KM, Olby NJ, Mariani CL: Characterization of Immune Cell Infiltration Into Canine Intracranial Meningiomas. Vet Pathol. doi: 10.1177/0300985811417249.
90. Garrett WS, Gordon JI, Glimcher LH: Homeostasis and inflammation in the intestine. Cell. 2010, 140: 859-870.
91. Day MJ: Ageing, immunosenescence and inflammageing in the dog and cat. J Comp Pathol. 2010, 142 (Suppl 1): 60-69.
92. Murai M, Krause P, Cheroutre H, Kronenberg M: Regulatory T-cell stability and plasticity in mucosal and systemic immune systems. Mucosal Immunol. 2010, 3: 443-449.
93. Kraal G, Samsom JN, Mebius RE: The importance of regional lymph nodes for mucosal tolerance. Immunol Rev. 2006, 213: 119-130.
94. Luckschander N, Pfammatter NS, Sidler D, Jakob S, Burgener IA, Moore PF, Zurbriggen A, Corazza N, Brunner T: Phenotyping, functional characterization, and developmental changes in canine intestinal intraepithelial lymphocytes. Vet Res. 2009, 40: 58.
95. Elwood CM, Hamblin AS, Batt RM: Quantitative and qualitative immunohistochemistry of T cell subsets and MHC class II expression in the canine small intestine. Vet Immunol Immunopathol. 1997, 58: 195-207.
96. German AJ, Hall EJ, Day MJ: Analysis of leucocyte subsets in the canine intestine. J Comp Pathol. 1999, 120: 129-145.
97. Kleinschmidt S, Meneses F, Nolte I, Hewicker-Trautwein M: Distribution of mast cell subtypes and immune cell populations in canine intestines: evidence for age-related decline in T cells and macrophages and increase of IgA-positive plasma cells. Res Vet Sci. 2008, 84: 41-48.
98. Smith HW: Observations on the flora of the alimentary tract of animals and factors affecting its composition. J Pathol Bacteriol. 1965, 89: 95-122.
99. Lathrop SK, Bloom SM, Rao SM, Nutsch K, Lio CW, Santacruz N, Peterson DA, Stappenbeck TS, Hsieh CS: Peripheral education of the immune system by colonic commensal microbiota. Nature. 2011, 478: 250-254.
100. Day MJ: Immune system development in the dog and cat. J Comp Pathol. 2007, 137 (Suppl 1): 10-15.
101. Baum B, Meneses F, Kleinschmidt S, Nolte I, Hewicker-Trautwein M: Age-related histomorphologic changes in the canine gastrointestinal tract: a histologic and immunohistologic study. World J Gastroenterol. 2007, 13: 152-157.
102. Kleinschmidt S: Investigations about age-related changes in the canine gastrointestinal mucosal immune system and about the role of mast cell subtypes and immune cells in dogs with inflammatory bowel disease (IBD). PhD thesis. 2007, University of Veterinary Medicine, Hannover, Pathology Department
103. Gruver AL, Hudson LL, Sempowski GD: Immunosenescence of ageing. J Pathol. 2007, 211: 144-156.
104. Cerquetella M, Spaterna A, Laus F, Tesei B, Rossi G, Antonelli E, Villanacci V, Bassotti G: Inflammatory bowel disease in the dog: differences and similarities with humans. World J Gastroenterol. 2010, 16: 1050-1056.

105. Liu H, Hu B, Xu D, Liew FY: CD4+CD25+ regulatory T cells cure murine colitis: the role of IL-10, TGF-beta, and CTLA4. J Immunol. 2003, 171: 5012-5017.
106. Mottet C, Uhlig HH, Powrie F: Cutting edge: cure of colitis by CD4+CD25+ regulatory T cells. J Immunol. 2003, 170: 3939-3943.
107. Maul J, Loddenkemper C, Mundt P, Berg E, Giese T, Stallmach A, Zeitz M, Duchmann R: Peripheral and intestinal regulatory CD4+ CD25(high) T cells in inflammatory bowel disease. Gastroenterology. 2005, 128: 1868-1878.
108. Uhlig HH, Coombes J, Mottet C, Izcue A, Thompson C, Fanger A, Tannapfel A, Fontenot JD, Ramsdell F, Powrie F: Characterization of Foxp3+CD4+CD25+ and IL-10-secreting CD4+CD25+ T cells during cure of colitis. J Immunol. 2006, 177: 5852-5860.
109. Yu QT, Saruta M, Avanesyan A, Fleshner PR, Banham AH, Papadakis KA: Expression and functional characterization of FOXP3+ CD4+ regulatory T cells in ulcerative colitis. Inflamm Bowel Dis. 2007, 13: 191-199.
110. Almeida AR, Zaragoza B, Freitas AA: Indexation as a novel mechanism of lymphocyte homeostasis: the number of CD4+CD25+ regulatory T cells is indexed to the number of IL-2-producing cells. J Immunol. 2006, 177: 192-200.
111. Maizels RM, Yazdanbakhsh M: Immune regulation by helminth parasites: cellular and molecular mechanisms. Nat Rev Immunol. 2003, 3: 733-744.

CHAPTER 7

Differential macrophage function in Brown Swiss and Holstein Friesian cattle

Amanda Jane Gibson[a], Sally Woodman[a], Christopher Pennelegion[a], Robert Patterson[a], Emma Stuart[a], Naomi Hosker[a], Peter Siviter[a], Chloe Douglas[a], Jessica Whitehouse[a], Will Wilkinson[a], Sherri-Anne Pegg[a], Bernardo Villarreal-Ramos[b], Dirk Werling[a,*]

[a]*Molecular Immunology Group, Department of Pathology and Pathogen Biology, Royal Veterinary College, Hawkshead Lane, North Mymms, AL9 7TA, UK*

[b]*Animal and Plant Health Agency, Bovine TB, Weybridge, New Haw, Addlestone, Surrey KT15 3NB, UK*

ABSTRACT

There is strong evidence that high yielding dairy cows are extremely susceptible to infectious diseases, and that this has severe economic consequences for the dairy industry and welfare implications. Here we present preliminary functional evidence showing that the innate immune response differs between cow breeds. The ability of macrophages (MØ) to kill pathogens depends in part on oxygen-dependent and independent mechanisms. The oxygen-dependent mechanisms rely on the generation of reactive oxygen and nitrogen species (ROS/RNS, respectively). ROS production has been shown to activate the inflammasome complex in MØ leading to increased production of the pro-inflammatory cytokine Interleukin-1β (IL-1β). Conversely RNS inhibits inflammasome mediated IL-1β activation, indicating a division between inflammasome activation and RNS production. In the present study MØ from Brown Swiss (BS) cattle produce significantly more RNS and less IL-1β when compared to cells from Holstein Friesian (HF) cattle in response to bacterial or fungal stimuli. Furthermore, BS MØ killed ingested *Salmonella typhimurium* more efficiently, supporting anecdotal evidence of increased disease resistance of the breed. Inhibition of autophagy by 3-methyladenine (3-MA) stimulated IL-1β secretion in cells from both breeds, but was more pronounced in HF MØ. Blocking RNS production by L-arginase completely abolished RNS production but increased IL-1β secretion in BS MØ. Collectively these preliminary data suggest that the dichotomy of inflammasome activation and RNS production exists in cattle and differs between these two breeds. As pattern recognition receptors and signaling pathways are involved in the assessed functional differences presented herein, our data potentially aid the identification of *in vitro* predictors of appropriate innate immune response. Finally, these predictors may assist in the discovery of candidate genes conferring increased disease resistance for future use in combination with known production traits.

Keywords: Innate immunity, Cattle breeds, ROS Inflammasome, Macrophages

1. INTRODUCTION

Macrophages (MØ) are the mature form of circulating monocytes (Mo) derived from a common myeloid progenitor. Inflammatory cytokines trigger adherence and rolling of circulating Mo along the endothelium of adjacent blood vessels before extravasation, maturation to MØ and migration to sites of inflammation, where they have both pro-inflammatory and inflammation resolution activity (Aguilar-Ruiz et al., 2011; Cros et al., 2010; Hussen et al., 2013; Shi and Pamer, 2011; Wong et al., 2011, 2012). In humans Mo can be typified into three subsets based on the expression of CD14 and CD16: CD14++CD16−, CD14++CD16+ and CD14+CD16++. These are referred to as classical (cM), intermediate (intM) and non-classical (ncM), respectively (Ziegler-Heitbrock et al., 2010). It has been indicated that the three subsets have different roles (Grage-Griebenow et al., 2001) and there may be a developmental pathway from cM through intM to ncM (Wong et al., 2011; Zawada et al., 2011). Exposure to the cytokine Macrophage-Colony Stimulating Factor (M-CSF) may trigger this development (Korkosz et al., 2012).

In cattle, approximately 89% of monocytes are cM (Hussen et al., 2013). These express greater phagocytic activity than intM or ncM, whilst also producing anti-inflammatory cytokines such as IL-10 (Grage-Griebenow et al., 2001; Ziegler-Heitbrock, 2007; Ziegler-Heitbrock et al., 2010) and being involved in repair of damage tissue (Wong et al., 2011, 2012). Both subsets of CD16+ Mo increase with infection and disease, extravasating more efficiently and producing cytokines including TNFα and IL-1β, thus being described as inflammatory Mo (Aguilar-Ruiz et al., 2011; Ancuta et al., 2003; Belge et al., 2002; Grage-Griebenow et al., 2001; Tacke and Randolph, 2006; Thieblemont et al., 1995; Wong et al., 2012; Zawada et al., 2011; Ziegler-Heitbrock et al., 2010), although some findings indicate anti-inflammatory actions for ncM (Cros et al., 2010; Hussen et al., 2013). A range of pro-inflammatory actions are exhibited by murine and human intM, from inflammatory cytokine production to antigen presentation, requiring prior phagocytosis (Aguilar-Ruiz et al., 2011; Belge et al., 2002; Shi and Pamer, 2011; Thomas and Lipsky, 1994; Ziegler-Heitbrock, 2007).

Correlating with their role in antigen presentation, human intM display the highest MHC II expression levels of the three monocyte subsets (Wong et al., 2011), indicating a strong involvement in antigen presentation. In addition to the production of cytokines and their function as antigen presenting cells, MØ constantly sample the local environment and have antigen presenting capacity (Chesnut and Grey, 1985; Gordon, 1998; Gordon and Taylor, 2005).

Responses to pathogens by MØ are mediated by pattern recognition receptors (PRR) such as Toll-like Receptors (TLR) which recognise and interact with pathogen associated molecular patterns (PAMP) (Akira et al., 2001; Taylor and Gordon, 2003). PAMPs include, but are not restricted to pathogen surface structures, genetic material and secreted or released products (Taylor and Gordon, 2003). MØ are characteristically bactericidal in nature and utilise several approaches to mediate bacterial killing but mainly employing two mechanisms which are closely interlinked: production of reactive oxygen species and reactive nitrogen species (ROS and RNS, respectively). Their release results in cell- and DNA damage and thus shows antimicrobial effects (Bogdan, 2001; Forman and Torres, 2001; Nathan, 1992, 1987). Interaction of a PAMP with the corresponding PRR stimulates a variety of pathways, resulting in the activation with the subsequent cleavage of pro-IL-1β into its active, secreted form by caspase-1 (Lamkanfi et al., 2007), assembly of the NADPH oxidase complex, electron transfer from NADPH to oxygen and a series of enzymatic reactions to produce ROS (Bokoch and Knaus, 2003; Lambeth, 2004; Quinn and Gauss, 2004). Additionally, PAMP association with PRR results in transcription of inducible Nitric Oxide Synthase (iNOS) in MØ which interacts with NADPH and ROS intermediates to create RNS in bovine and murine MØ (Jungi et al., 1996a; Nathan, 1992; Green et al., 1990).

However, there are some infectious diseases where – due to the anatomical barrier present – the anti-bacterial activity of MØ is not sufficient to eliminate the pathogen. One such organ is the bovine udder, where mechanical, chemical and pathological irritation can lead to an increased somatic cell count (SCC) (Schukken et al., 2003). However, it has to be emphasised that somatic cells in the milk consist of a pleiotropha of cells, including MØ, neutrophils, lymphocytes and epithelial cells (Boulanger et al., 2001). In humans, the majority of cells in milk have been identified as unique breast milk MØ (Yagi et al., 2010). These have been shown to be more metabolically active compared with those in peripheral blood (Johnson et al., 1980). However, the killing mechanism was shown to involve the oxidation of glucose via the hexose-monophosphate (HMP) shunt, showing that cytotoxic effects are using the same mechanism as blood derived cells by the production of hydrogen peroxide and superoxide (Johnson et al., 1980). Indeed, bovine mammary MØ challenged with phagocytic stimuli produced ROS similarly to those produced by other MØ (Harmon and Adams, 1987). Thus milk-derived MØ are capable of ROS/RNS production, has been described in milk-derived MØ (Denis et al., 2006), and seems to be an approach of the innate immune cells to combat invading bacteria in the udder.

However, despite the fact that 1/4 of cows are suffering from mastitis at any given time, not all breeds show the same mastitis incidence rate (Begley et al., 2009). Indeed, *c*attle breed susceptibility or resistance to infection has long been studied to improve breeding strategies (Warner et al., 1987), and differences in the genetic resistance to infection have been identified for *Mycobacterium tuberculosis* (O'Reilly and Daborn, 1995; Vordermeier et al., 2012) and *Brucellosis* (Paixao et al., 2006).

However, it has to be emphasised that research results, especially analysis of immune responses, may not translate between breeds of the same species. Indeed, observations by farmers and veterinarians have noted that Brown Swiss (BS) cattle tend to have a lower somatic cell counts (SCC) than Holstein-Friesian (HF) cattle, and are thought to be less susceptible to mastitis. This anecdotal evidence has been reinforced by studies suggesting differences in SCC could be the result of improved microbial killing mechanisms in the innate cells of BS cattle (Busato et al., 2000; Kizilkaya, 2009; Rupp and Boichard, 2003). Work by Norimatsu et al. (2004), showing low RNS production from *Salmonella typhimurium* stimulated HF MØ compared to data published by Werling et al. (2004) demonstrating high RNS production from BS MØ with the same bacterium highlights that difference in disease resistance may be conferred by the micro-bicidal capabilities of diverse breeds. Furthermore, work by Jann et al. (2008, 2009) suggests that polymorphisms in TLR genes may be involved in disease resistance or susceptibility traits in domestic animals (Jann et al., 2009, 2008). Despite estimations in heritability of production traits varying (Rupp and Boichard, 2003), Lund et al. (1994) have shown the estimated heritability of SCC to be moderately high (Lund et al., 1994). This suggests the possibility of integrating low SCC into breeding strategies, particularly if lower SCCs in milk could be linked to an increased MØ killing ability.

2. MATERIALS AND METHODS

2.1. Sample collection

Blood for PBMC isolation and subsequent MØ generation was collected by puncture of the jugular vein from clinically healthy HF and BS cows housed at either the RVC Boltons Park Farm (Hertfordshire, UK) or at Cancourt Farm (Wiltshire, UK). Animals used were age- and lactation-matched (2nd or 3rd lactation, respectively), unless otherwise stated. All procedures were carried under Home Office Project licence which was approved by the College's Ethics and Welfare Committee. For biological assays, blood was drawn into sterile glass vacuum bottles containing 10% Acid Citrate Dextrose (ACD) as anticoagulant. For whole blood flow cytometric analysis, blood was collected from three cows of each breed into 10 ml EDTA Vacutainers (Beckton Dickinson, Oxford, UK) by puncture of the jugular vein. Whole milk was collected into sterile glass bottles from individual cows housed at either farm. All assays were performed in at least four breed matched pairs apart from bacterial killing assays, blood MØ NO induction and preliminary IL-1β stimulations, milk derived responses which were performed in 3, 2 and 1 breed matched pairs, respectively.

2.2. Isolation peripheral blood mononuclear cells and generation of macrophages

Peripheral blood mononuclear cells (PBMC) were isolated from whole blood as previously described (Jungi et al., 1996b). Briefly, blood was centrifuged at 700 × *g* for 20 min before buffy coats were removed and washed in citrate buffer. RBC lysis was performed using ammonium-chloride lysis buffer. Resultant cells were suspended in RPMI media (Life Technologies, UK) before being underlayed with Histopaque ($d = 1.083$ g l^{-1}, Sigma–Aldrich, UK) to isolate PBMCs by density centrifugation. PBMC were washed in PBS and counted by Trypan blue exclusion then incubated in Teflon bags for 7 days at 37 °C and 5% CO_2 at a concentration of 5×10^6 ml^{-1} in MØ medium containing 10% FCS. To harvest MØ, Teflon bags were placed in the fridge at 4 °C for 30 min and agitated to loosen cells. Cells were washed three times in PBS at 300 × *g* for 10 min, before counting as described above.

2.3. Determination of circulating monocyte populations

Mo populations were determined as previously described with modifications (Hubl et al., 2000). Briefly Briefly 50 µl blood samples were labelled with 5 µl mouse anti-human (mαh) CD14 APC (AbD Serotec, UK), 5 µl mαh CD16 FITC (AbD Serotec, UK) and 15 µl mαb BoLA-DR (clone CC108) labelled with PE-680 via Zenon labelling as per manufacturer's instructions. Isotype controls were prepared by direct addition of 10 µl Mouse IgG_{2a} APC, 10 µL Mouse IgG_{2a}FITC and 20 µl Mouse IgG_1 labelled with PE-680 as above to 50 µl of whole blood. Antibodies were allowed to incubate at room temperature for 20 min in the dark. Red blood cells were lysed by addition of ammonium chloride lysis buffer for 10 min as above before placing on ice. Remaining cells were washed in PBS (Gibco, UK) and suspended in FACSFlow (BD Biosciences, UK) for analysis. Samples were analysed by FACS Calibur (BD Biosciences, UK). Mo populations were gated for the presence of BoLA-DR and CD14 to exclude platelets, lymphocytes and granulocytes. BoLA-DR^{hi} or CD14+ populations were used to identify CD14+ and CD16+ population proportions of cM, intM and ncM. 50,000 events were recorded per sample using Cell Quest (BD Biosciences, UK). Data were subsequently analysed using FCS Express 4 (DeNovo Software, USA).

2.4. Isolation of bovine macrophages from milk

Whole milk samples were kept on ice for 10 min prior to centrifugation at 3000 × *g* for 5 min at 4 °C. Residual fat layer was removed using a spatula and supernatant discarded. Cell pellets were re-suspended in PBS/0.125 M EDTA/TE buffer (10 mM—Tris–HCl–1 mM EDTA pH 7.6) and allowed to stand at room temperature for 10 min. Cells were washed twice in TE buffer and finally suspended in PBS (PAA Laboratories, UK) for counting by Trypan blue exclusion as above.

2.5. Measurement of ROS production

ROS was measured as described previously (Conejeros et al., 2011). Briefly, production of ROS by MØ was measured by oxidation of 2,7-dichlorofluorescein diacetate (DCFH-DA) to fluorescent 2,7-dichlorofluorescein (DCFH). MØ were harvested as above and adjusted to a concentration of 1×10^7 per condition in HBSS (Gibco) before the addition of stimuli (1 mg ml^{-1} zymosan (Sigma); 2–5 µg ml^{-1} PMA (Sigma), 20 µg ml^{-1}*Listeria monocytogenes* (*L.m.*); 1×10^8bacteria ml^{-1}, *Staphylococcus aureus* (*S.a.*) *ST398* or HBSS as a negative control). DCFH-DA probe (1 mg ml^{-1}) was added at final concentration of 100 µg ml^{-1} before being incubated in the dark at 37 °C for 15 min. 150 μl aliquots were plated in triplicate on a black 96-well plate (Corning, UK). RFU were measured using a spectrophotometer (SpectraMax M2, Molecular Devices, UK) at 485 nm excitation and 530 nm emission. Cells were incubated at 37 °C, 5% CO_2 between readings in 15 min intervals for a two hour period.

2.6. Measurement of NO production using griess reagents

Determination of NO concentration in cell supernatants was carried out using Griess reagents as previously described (Werling et al., 2004). MØ were harvested as above and adjusted to 4×10^5 cells ml^{-1} in 2% FCS MØ medium and 8×10^4 cells added a flat-bottomed 96-well plate for 24 h at 37° C, 5% CO_2. Adherent cells were washed with PBS after 24 h to remove contaminating lymphocytes. Stimulants were prepared in 200 μl of 2% FCS MØ medium to give a final concentration of 250 μg ml^{-1} zymosan (Sigma, UK), 100 μg ml^{-1} *L.m.*, 5 μg ml^{-1} LPS or media controls. MØ were cultured for a further 24 h 37° C, 5% CO_2 and cell free supernatants were harvested and stored at −20° C until required for measurement of NO or cytokine analysis (IL-1β). Briefly, a two-fold standard curve of 128 μM sodium nitrite in 2% FCS MØ media and sample supernatants were placed in flat 96-well clear plate and mixed with equal volumes of Griess reagent (1:1 solutions A and B). After 10 min incubation absorbance was analysed at 550 nm using a spectrophotometer (Spectramax M2, Molecular Devices, UK).

2.7. Inhibition of RNS production by L-arginase

MØ were generated and harvested as described above, adjusted to 1×10^6 ml^{-1} in 2% FCS MØ medium and 2×10^5 MØ were added per well of 96-well plate and cultured as above. Adherent cells were washed after 24 h twice with PBS before addition of 2% MØ media, RNS inhibitor L-arginase (0.125 μg ml^{-1}) and Zymosan (1 mg ml^{-1}). Supernatants were collected 24 h later and stored until analysis of RNS and IL-1β production as described.

2.8. Inhibition of autophagy in macrophages with 3-methyladenine

MØ were generated and harvested as described above, adjusted to 1×10^6 ml^{-1} in 2% FCS MØ medium and 2×10^5 MØ were added per well of 96-well plate and cultured as above. After removal of contaminating lymphocytes and washing adherent cells twice in PBS, MØ were exposed to LPS (5 μg ml^{-1}) (Sigma, UK) and two-fold dilution of 3-methyladenine (3-MA) (Sigma, UK) from 10 mM to 1.25 nM, including medium, LPS and 3-MA alone controls. MØ were cultured for a further 24 h at 37° C, 5% CO_2 and cell free supernatants harvested and stored at −20 °C until assayed for IL-1β by ELISA.

2.9. IL-1β detection by ELISA

Cell-free supernatants were stored at −20 °C until sandwich ELISAs could be used to analyse cytokine production (IL-1β) in response to the ligand sets described (Zymosan, *L. monocytogenes* and LPS). ELISA kit specific for boIL-1β (Thermo-Scientific, IL, USA) was prepared using a Maxi-sorb 96-well ELISA plate (Nunc, Sigma, UK) and performed according the manufacturer's instructions. Samples were tested in duplicate and standard curves were generated using a serial 1:2 dilution from 2 ng ml^{-1} IL-1β. Samples diluted 1:10 with reagent diluent were assayed alongside standards as per manufacturers instructions without modifications. After final addition of stop solution, absorbance was measured at 450 nm with 550 nm reference using a spectrophotometer (SpectraMax M2, Molecular Devices, UK).

2.10. Macrophage phagocytosis of *S. aureus* bio-particles

MØ were harvested as described above and adjusted to 1×10^6 ml^{-1} in RPMI 1640 medium without phenol red (Gibco, UK) before placing 2×10^5 MØ per well of a 96 well U-bottomed low bind plate (Costar, USA). MØ were exposed to *S.a.* BioParticles® Alexa-594 (Life Technologies, UK) and were incubated at 37 °C and 5% CO_2. At two different timepoints (0 and 60 min) 100 µl of 4% paraformaldehyde was added to fix the cells and halt phagocytosis. Cells were then washed three times in PBS (Gibco, UK) and suspended in FACS flow for analysis. Samples were analysed by FACS Calibur (BD Biosciences UK), data collected with Cell Quest Pro (BD Biosciences, UK) and subsequently analysed using FlowJo v10 (Treestar, USA).

2.11. Preparation of *S. typhimurium* cultures

S. typhimurium SL1344 was inoculated onto LB agar and incubated at 37 °C for 16–18 h before storing at 4 °C as a source for future liquid cultures. On the day prior to each MØ infection, a new inoculum was created from a single colony in 10 ml sterile LB broth (Oxoid, UK) and cultured overnight at 37 °C for 16–18 h with agitation at 180 rpm. Cultures were centrifuged for 10 min at 3219 × *g*, 21 °C and suspended in RPMI to generate an OD_{600} reading between 0.8–1.0 (the equivalent of 10^8–10^9 CFU). Confirmation of CFU was determined by 10-fold serial dilutions of overnight culture plated onto LB agar, cultured overnight at 37 °C and counted.

2.12. Infection of macrophages with *S. typhimurium*

MØ were generated and harvested as described above, adjusted to 1 × 10^5/1 × 10^6 MØ ml^{-1} in antibiotic free 2% FCS MØ medium and seeded at 1 × 10^5/1 × 10^6 per well of 24-well plate for 24 h 37 °C, 5% CO_2. Contaminating lymphocytes were removed and cells were washed in PBS prior to inoculation with an MOI of 10 for 1 h at 37 °C, 5% CO_2. After 1 h incubation RPMI supplemented with 10% FCS and 50 mg ml^{-1} Gentamicin was added to eliminate extracellular bacteria and cells were incubated overnight at 37 °C, and 5% CO_2. Supernatants were collected after 1 h and MØ were washed with PBS before addition of culture medium for a further 24 h of culture at 37 °C, and 5% CO_2. Supernatants were harvested and MØ were washed with PBS prior to lysis using PBS +1% Triton X for 5 min. Supernatants from 1 h, 24 h and MØ lysis were serially diluted 10-fold before plating on LB agar for 24 h at 37 °C to determine bacterial uptake and killing.

2.13. Statistical analyses

Statistical analyses were performed using GraphPad Prism 6 using one-way ANOVA or two-way ANOVA (repeated measures for ROS assays) with Bonferroni post-*hoc* testing. Unpaired *t*-tests were used for analysis of bacterial killing with Bonferroni post-*hoc* testing. Significance was defined as $p \leq 0.05$ (*), $p \leq 0.01$ (**), $p \leq 0.001$ (***) and $p \leq 0.00001$ (****).

3. RESULTS

3.1. SCC, but not milk production or monocyte subsets differ between cells isolated from both breeds

Initially, it was assessed whether the above mentioned "resistance" to mastitis in BS compared to HF was purely based on differences in milk production, SCC or the number/composition of Mo subsets in the peripheral blood. Whereas SCC of BS cows was significantly lower over a one year period (Fig. 1A), neither milk production (Fig. 1B) nor number/composition of Mo subsets (Fig. 1C), as analysed by flow cytometry differed between HF and BS cows.

SCC was assessed over a five year period. Data were provided by the National Milk Record, and an annual average of BS and HF animals on the farms is shown (Fig. 1A). In contrast, average milk production (Fig. 1B) as well as composition of Mo subsets (Fig. 1C) was not statistically different.

Whole blood from BS and HF cattle were labelled with antibodies directed to human CD14:APC, human CD16:FITC and bovine HLA-DR: PE-680 alongside isotype controls as described. Cells were visualised post RBC lysis and washing using BD FACS Calibur, recording 50,000 events per sample with Cell Quest. Mo subsets were gated from HLA-DR^{hi} and CD14^{+} cells to identify cM (CD14^{++}/CD16^{-}), intM (CD14^{++}/CD16^{+}) and ncM (CD14^{+}/CD16^{++}) population proportions. Data were analysed by FCS Express 4 and population proportions exported to GraphPad Prism for statistical analysis. Differences between breeds were tested by one-way ANOVA, followed by Bonferroni *t*-test (*** $p < 0.001$)

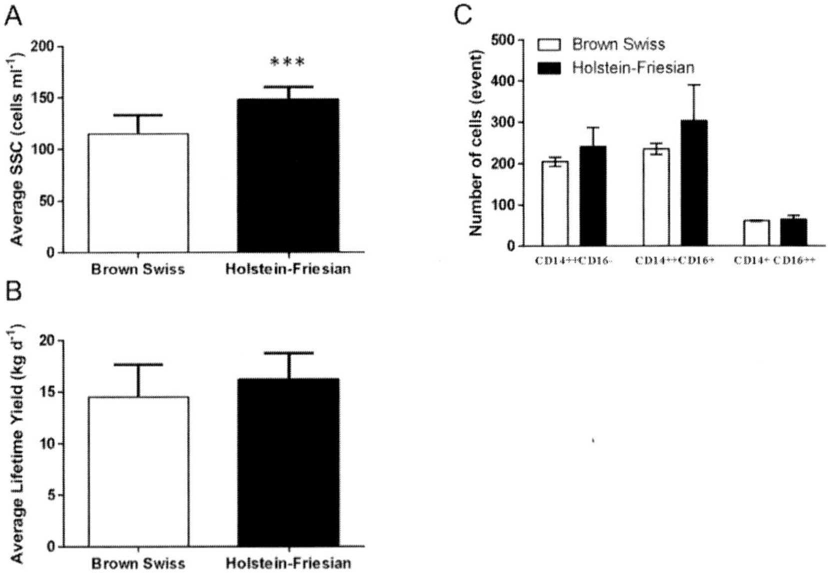

Figure. 1. SCC, but not milk production or monocyte subset distribution differ between Brown Swiss and Holstein Friesian cattle.

3.2. Ex-vivo blood derived macrophages from Brown Swiss cattle differentially produce ROS and NO compared to Holstein Friesian macrophages

Having established that the potential differences between BS and HF cows are not due to differences in milk production (and thus potentially less cells available) or differences in Mo subsets between both breeds, we next assessed direct anti-bacterial effector functions of matured MØ. Oxidative burst was found to be elevated in BS MØ exposed to *L.m.* and zymosan when compared to HF (Fig. 2A and B). ROS were produced in significantly greater amounts from 60 min post-stimulation for the duration of the assay (to 120 min). Reactivity appears to be stimulation specific and not an overall 'hyper' or constitutive activity as MØ from both breeds respond similarly to media or PMA controls (data not shown). Interestingly differences in NO production are less apparent towards the same ligands from the animals tested here, however BS MØ responses show reduced variation (Fig. 2C). Having established differences in ROS/NOS production to the same ligand in MØ derived from both breeds, we next wanted to assess whether these differences also exist in MØ derived from other tissues.

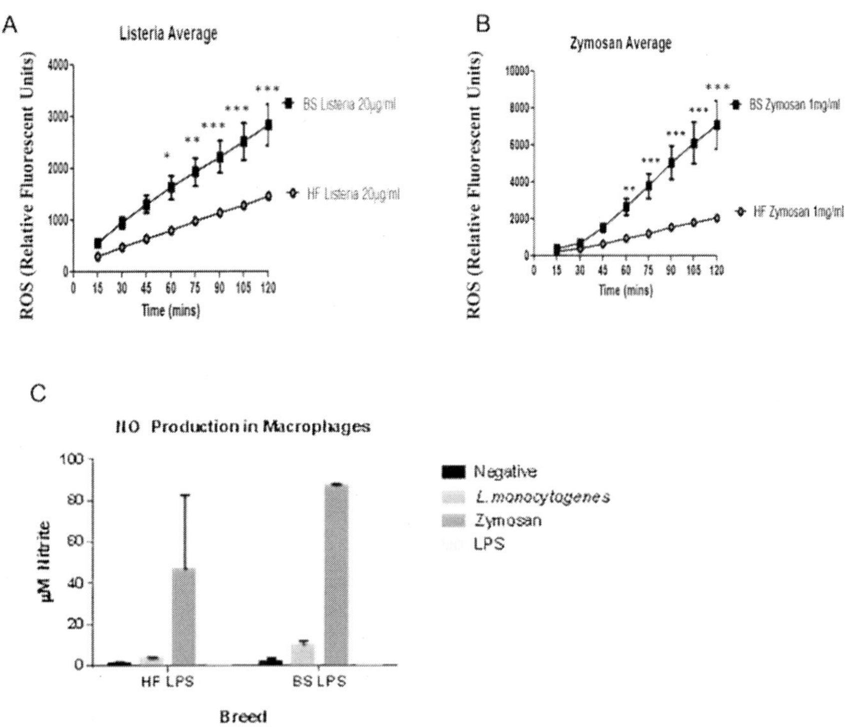

Figure. 2. Ex-vivo blood derived macrophages from Brown Swiss cattle differentially produce ROS and RNS compared to Holstein Friesian macrophages.

MØ were generated as described from BS and HF PBMC isolated from whole blood. ROS production was measured from MØ exposed to (A) *L. monocytogenes* (1×10^8 bacteria ml^{-1}) and (B) zymosan (1 mg ml^{-1}), by addition of DCFH-DA at 15 min intervals over 120 min period as relative fluorescent units. Cells were maintained at 37 °C, 5% CO_2throughout. Fluorescence was detected using SpectraMax M2 plate reader in black, glass bottomed micro-titre plates. RNS production was measured using Griess assay on overnight macrophage culture supernatant from both cattle breeds. Adhered MØ were exposed to (C) 2% medium, *L. monocytogenes* (100 μg ml^{-1}), zymosan (250 μg ml^{-1}) and LPS (5 μg ml^{-1}) for 24 h before addition of equal volumes of Griess reagent A and B and compared to 2-fold Nitrite standard curve. Absorbance was measured using a SpectraMax M2 plate reader. All data were post analysed with GraphPad Prism using two-way repeated measured ANOVA (ROS) with Bonferroni post-*hoc* correction and two-way non-repeated measures ANOVA (NO) with Bonferronic post-*hoc* testing.

3.3. Milk resident cells mirror ex-vivo derived macrophage responses: Brown Swiss cattle produce more ROS Holstein Friesian macrophages

A comparison of ROS production from milk derived cells revealed enhanced responses from cells derived from BS cattle when stimulated with *S.a.* and zymosan as compared to cells derived from, HF cattle (Fig. 3A and B). Increased levels of ROS observed for BS cells is not attributed to increased SCC as over the study period HF milk contained significantly greater somatic cells (Fig. 1A) and equal cell concentrations where stimulated per breed.

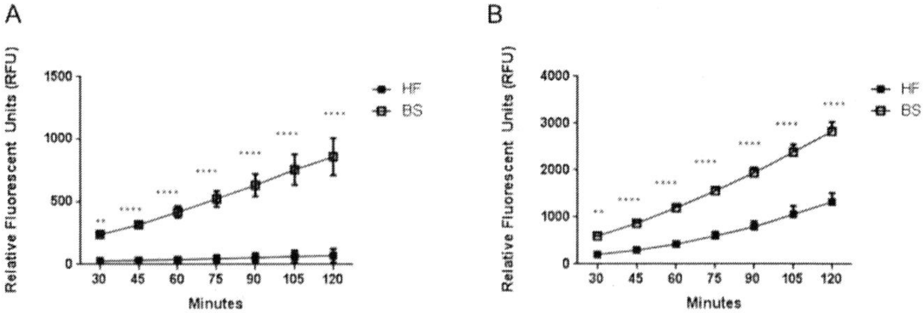

Figure. 3. Milk resident cells mirror ex-vivo blood derived macrophage responses: Brown Swiss cattle produce more ROS Holstein Friesian macrophages.

MØ from milk were isolated as described from BS and HF cattle, ROS was measured from MØ exposed to (A) *S. aureus* (1 × 10^8 ml^{-1}) and (B) Zymosan

(1 mg ml^{-1}) by addition addition of DCFH-DA at 15 min intervals over 120 min period. Cells were maintained at 37 °C, 5% CO_2 throughout. Fluorescence was detected using SpectraMax M2 plate reader in black, glass bottomed micro-titre plates. All data were post analysed with GraphPad Prism using two-way repeated measures ANOVA testing with Bonferroni post-*hoc* correction.

3.4. RNS blockade partially restores IL-1β Brown Swiss levels to those similar to Holstein Friesian macrophages

Having established that MØ and milk-derived cells from BS cows seem to produce more ROS/RNS compared to cells derived from HF cows, we next wanted to establish whether this increased reactivity is a general sign for an increased innate immune response in BS derived cells, and extends to other mediators of the innate immune response, such as production of IL-1β. Interestingly, stimulation of MØ derived from HF, but not BS cows with LPS resulted in increased amounts of IL-1β in the supernatant (Fig. 4A). Recently, a link between NOS and activation of the inflammasome was described (Mao et al., 2013; Mishra et al., 2013), indicating that production of RNS inhibits inflammasome activation. To assess whether such mechanism may also account for the reduced ability of BS MØ to produce IL-1β in response to LPS, we next inhibited RNS production using L-arginase (Garvey et al., 1997), which converts arginine into ornithine and urea, thus removing the substrate of iNOS. Indeed, inhibition of RNS production by addition of L-arginase partially restored the ability of BS MØ to produce IL-1β (Fig. 4B) indicating that activation of the inflammasome is potentially impaired due to high ROS/RNS production. In contrast, no effect was seen when using HF MØ (Fig. 4C), which was potentially due to the low production of RNS in the first instance by these cells.

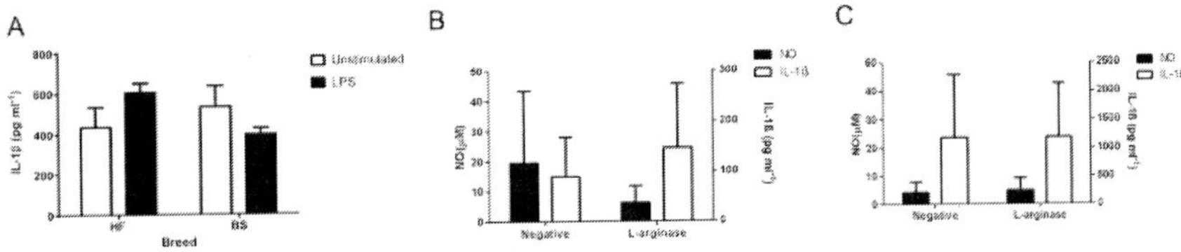

Figure. 4. NO blockade partially restores IL-1β Brown Swiss levels to those similar to Holstein Friesian macrophages.

MØ derived from PBMC isolated from BS and HF cattle were adhered overnight and stimulated with 2% medium or (A) LPS (5 μg ml^{-1}) for 24 h. Supernatants were collected and analysed for bovine IL-1β by ELISA.

Absorbance was measured using a SpectraMax M2 plate reader. RNS was inhibited in MØ cultures from (B) BS and (C) HF with L-arginase in the presence of Zymosan (1 mg ml^{-1}) for 24 h after adherence of MØ. Bovine IL-1β levels were determined by ELISA and RNS by addition of equal volumes of Griess reagent A and B and compared to 2-fold Nitrite standard curve. Absorbance was measured for both assays using a SpectraMax M2 plate reader. All data were post analysed with GraphPad Prism using two-way non-repeated measures ANOVA testing with Bonferroni post-*hoc* correction.

3.5. Autophagy inhibition increases LPS induced IL-1β secretion from both Brown Swiss and Holstein Friesian macrophages

Inflammasome activation and autophagy are thought to be two mutual exclusive processes within cells (Harris, 2011), and inhibition of autophagy has been described to increase inflammasome activation, resulting in increased IL-1β release (Harris, 2011). To further assess whether BS MØ have a generally impaired IL-1β production, autophagy was blocked in MØ of both breeds before cells were stimulated. As seen before, stimulation of MØ of either breed with LPS alone induced very little increase in IL-1β secretion compared to negative controls. Inhibition of autophagy before LPS stimulation increased IL-1β secretion by MØ from both breeds in a dose-dependent manner (Fig. 5A and B). However, the effect was more pronounced in HF MØ (Fig. 5C).

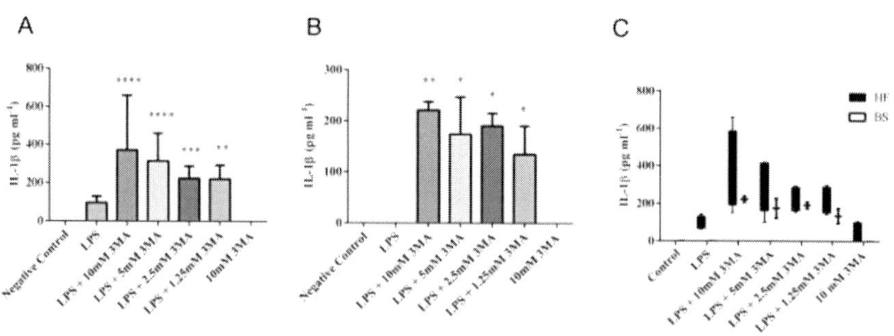

Figure. 5. Autophagy inhibition increases LPS induced IL-1β secretion from both Brown Swiss and Holstein Friesian macrophages.

3MA inhibits autophagy by blocking autophagosome formation via the inhibition of type III Phosphatidylinositol 3-kinases (PI-3 K), leading to a down-regulation of inflammaosome activation, and thus IL-1β production.

To assess whether MØ of both breeds have the capability to respond with IL-1 production to inflammasome activation, MØ were derived from PBMC isolated from (A) HF and (B) BS and adhered overnight prior to inhibition of autophagy with 3MA (from 10 mM to 1.25 mM) in the presence of LPS (5 $\mu g\ ml^{-1}$) for 24 h. Supernatants were collected and analysed for bovine IL-1β by ELISA. Absorbance was measured using a SpectraMax M2 plate reader. All data were post analysed using GraphPad Prism using two-way non-repeated measures ANOVA with Bonferroni post-*hoc* testing.

3.6. Phagocytosis and bacterial killing is enhanced in Brown Swiss macrophages compared to Holstein Friesian macrophages

As the increased production of ROS/RNS by BS MØ potentially indicated a greater anti-bacterial response, we finally assessed whether these differences could be attributed to phagocytic activity and bacterial killing. To do so, MØ of both breeds were assessed their ability to phagocytose *S.a.* bio-particles. After 60 min BS MØ engulfed more bio-particles compared to HF MØ, after normalisation to background phagocytosis (Fig. 6A). Lastly, we aimed to investigate whether all the parameters analysed increased the capability of BS MØ to kill bacteria. Therefore, MØ from both breeds were co-cultured with *S. typhimurium* and bacterial killing determined by recovered bacterial growth as described. Significantly fewer bacteria survived in BS MØ compared to HF MØ cows after 24 h of culture post-bacterial incubation (Fig. 6B).

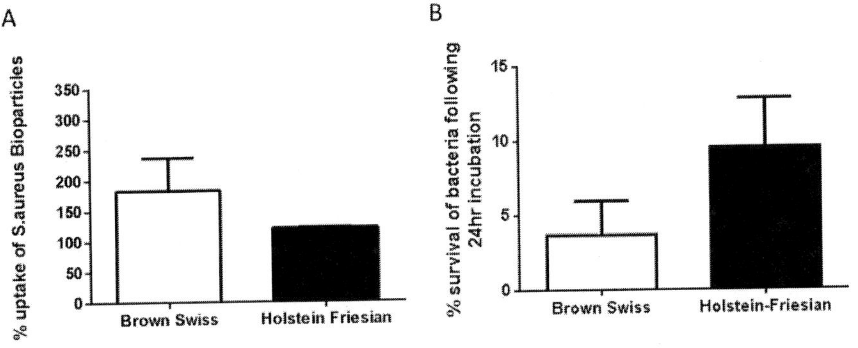

Figure. 6. Phagocytosis is enhanced in Brown Swiss macrophages compared to Holstein Friesian macrophages.

(A) MØ derived from PBMC isolated from BS and HF cattle were incubated with Alexa-594 labelled *S.a* bioparticles for up to 60 min before fixing with 4% PFA to suspend phagocytosis. Cells were analysed with BD FACS Calibur recording 10,000 events using Cell Quest Pro. Data are presented corrected against MFI at 0 min and post analysed with FlowJo v10 and GraphPad Prism using unpaired students *t*-test with Bonferroni post-*hoc* correction applied. (B) MØ were derived from PBMC isolated from BS and HF and adhered overnight prior to exposure to *S. typhimurium* (10CFU) for 1 h. Supernatants were collected, cells were washed and culture was continued in the presence of Gentamycin for 24 h. Supernatants (1 h and +24 h) and lysed MØ preparations where serially diluted 10-fold and plated for counting. Data were collated and post analysed using GraphPad Prism using unpaired students *t*-test with Bonferroni post-*hoc* testing.

4. DISCUSSION

In the present study, we examined potential functional differences to investigate anecdotal evidence suggesting BS cows are more resistant to bacterial infection compared to HF cows. We initially concentrated on the production and release of direct anti-bactericidal effectors indicative of oxidative burst and production of nitrogen species: ROS and RNS. Differences which we observed in blood derived cells were also mirrored in milk derived cells, leading us to investigate other immune parameters; production of pro-inflammatory cytokine and indicator of inflammasome activation, IL-1β, effects of RNS on IL-1β production, the ability of MØ from both breeds to phagocytose and finally the capacity of MØ to exert bactericidal effects on live bacteria.

An intriguing picture has emerged from this work, indicating that there is some functional evidence to support anecdotal differences between BS and HF cattle. Finding that BS MØ or milk derived MØ produce more ROS in response to the same stimulation compared to HF MØ indicates that differences between breeds may be reflected at the genetic level. Cattle breed resistance to infectious disease is not a new concept, there are descriptions detailing specific breed or traits for protection against infection with *Mycobacterium bovis* and *Brucella* spp. (O'Reilly and Daborn, 1995; Paixao et al., 2006; Vordermeier et al., 2012). Indirectly there are reports that BS and HF cattle respond differently to *S. typhimurium* in the production of free radicals; a study by Norimatsu et al. (2004) found HF MØ to be non-responsive however an independent study by Werling et al. (2004) showed BS MØ to produce high levels of RNS. We also observed greater ROS and RNS production from BS MØ with cells derived from age and lactation matched HF cows to help negate location, herd, and husbandry and pregnancy status differences not accounted for by the earlier studies.

We also examined the production of the pro-inflammatory cytokine, IL-1β, as a marker of inflammasome activation. We observed that HF MØ produce higher IL-1β levels compared to those derived from BS cows.

Making use of recent links between RNS, autophagy and IL-1β secretion, blocking both RNS and autophagy modulated IL-1β. Inhibition of autophagy by means of 3MA treatment exerted a dose-dependent increase in IL-1β production in both cattle breeds; however the blockade of RNS synthesis by depletion of L-arginine allowed a recovery or restoration of IL-1β responses in BS MØ only. These data suggest that RNS synthesis and associated negative feedback on IL-1β production may be more efficient in BS cattle. Additionally it appears that autophagy plays a similar role in the inhibition of inflammasome mediated IL-1β cleavage and release in both breeds, based on that MØ derived from BS and HF cows responded equally to 3MA blockade.

On determining these differences in secreted or released molecules upon stimulation of MØ derived from both cattle breeds, we sought to identify if these disparities followed through to bacterial killing. Mo subsets were found to be similar between breeds, yet *S.a.* bio-particle uptake was greater for BS MØ compared to HF MØ. Furthermore, BS MØ were better able to destroy live bacteria as demonstrated by decreased *S. typhimurium* survival rates after exposure to MØ from both breeds.

Our study examines several aspects of innate immune activation involved in bactericidal effects where BS MØ consistently out-performed HF counterparts. We postulate that these differences may be explained at the genetic level and governed by genes which directly affect recognition and uptake of bacteria; PRR such as C-type Lectins or TLRs, those involved in the production of ROS and RNS and finally genes pertaining to the activation of the inflammsome and IL-1β cleavage. Indeed, if such genetic differences would be identified, our assays would provide means to link genotypes with phenotypes by relatively easy to perform and cost-effective assays. Previous studies have suggested that polymorphsims on TLR genes may confer disease susceptibilities or resistance (Jann et al., 2009, 2008), thus future studies might be best placed to investigate both functional and genetic differences. Our finding that BS cattle were consistently lower over the study period for SCC detected in milk and were better able to kill live bacteria warrants further investigation as a trait indicative of increased immune function, irrespective of breed, and would potentially help to explain inflammation responses in different breeds. Indeed, some of these responses may be perpetuated or made more severe via a prolonged, "false" innate immune response, leading to increased tissue damage. Thus, it would perhaps be pertinent to include such immune function traits in combination with known production parameter traits to improve existing dairy and beef herd stocks in their genetic resistance to (bacterial) infection. Most importantly however, our data clearly show that analysis of immune responses may not be translated between breeds of the same species, and this should be kept in mind when assassing responses to pathogens as well as to vaccines.

ACKNOWLEDGMENTS

We thank Mr. M. Hussey (Cancourt Farm) for the support of our work, as well as the C. Verity and P. Christian (RVC Bolton Park Farm for their technical help. We also thank J. Harris (Monash University, Australia) for helpful discussion regarding the autophagy experiment. This study was supported by funds from the EMIDA-ANIHWA KOliMAstir project, grant BB/H0161104/1 from the BBSRC as well as several student projects grants from the Wellcome Trust to D.W. This manuscript represent publication number PPB_01069 of the RVC.

REFERENCES

1. Aguilar-Ruiz, S.R., Torres-Aguilar, H., Gonzalez-Dominguez, E., Narvaez, J.,Gonzalez-Perez, G., Vargas-Ayala, G., Meraz-Rios, M.A., Garcia-Zepeda, E.A.,Sanchez-Torres, C., 2011. Human CD16+ and CD16- monocyte subsets displayunique effector properties in inflammatory conditions in vivo. J. Leukoc. Biol.90, 1119–1131.
2. Akira, S., Takeda, K., Kaisho, T., 2001. Toll-like receptors: critical proteins linkinginnate and acquired immunity. Nat. Immunol. 2, 675–680.
3. Ancuta, P., Rao, R., Moses, A., Mehle, A., Shaw, S.K., Luscinskas, F.W., Gabuzda, D.,2003. Fractalkine preferentially mediates arrest and migration of CD16+monocytes. J. Exp. Med. 197, 1701–1707.
4. Begley, N., Buckley, F., Pierce, K.M., Fahey, A.G., Mallard, B.A., 2009. Differences inudder health and immune response traits of Holstein-Friesians, NorwegianReds, and their crosses in second lactation. J. Dairy Sci. 92, 749–757.
5. Belge, K.U., Dayyani, F., Horelt, A., Siedlar, M., Frankenberger, M., Frankenberger, B.,Espevik, T., Ziegler-Heitbrock, L., 2002. The proinflammatoryCD14+CD16+DR++ monocytes are a major source of TNF. J. Immunol. 168,3536–3542.
6. Bogdan, C., 2001. Nitric oxide and the immune response. Nat. Immunol. 2, 907–916.
7. Bokoch, G.M., Knaus, U.G., 2003. NADPH oxidases: not just for leukocytesanymore! Trends Biochem. Sci. 28, 502–508.
8. Boulanger, V., Bouchard, L., Zhao, X., Lacasse, P., 2001. Induction of nitric oxideproduction by bovine mammary epithelial cells and blood leukocytes. J. DairySci. 84, 1430–1437.
9. Busato, A., Trachsel, P., Schallibaum, M., Blum, J.W., 2000. Udder health and riskfactors for subclinical mastitis in organic dairy farms in Switzerland. Prev. Vet.Med. 44, 205–220.
10. Chesnut, R.W., Grey, H.M., 1985. Antigen presenting cells and mechanisms ofantigen presentation. Crit. Rev. Immunol. 5, 263–316.

11. Conejeros, I., Patterson, R., Burgos, R.A., Hermosilla, C., Werling, D., 2011. Inductionof reactive oxygen species in bovine neutrophils is CD11b, but notdectin-1-dependent. Vet. Immunol. Immunopathol. 139, 308–312.
12. Cros, J., Cagnard, N., Woollard, K., Patey, N., Zhang, S.Y., Senechal, B., Puel, A.,Biswas, S.K., Moshous, D., Picard, C., Jais, J.P., D'Cruz, D., Casanova, J.L., Trouillet,C., Geissmann, F., 2010. Human CD14dim monocytes patrol and sense nucleicacids and viruses via TLR7 and TLR8 receptors. Immunity 33, 375–386.
13. Denis, M., Parlane, N.A., Lacy-Hulbert, S.J., Summers, E.L., Buddle, B.M., Wedlock,D.N., 2006. Bactericidal activity of macrophages against Streptococcus uberis isdifferent in mammary gland secretions of lactating and drying off cows. Vet.Immunol. Immunopathol. 114, 111–120.
14. Forman, H.J., Torres, M., 2001. Redox signaling in macrophages. Mol. Asp. Med. 22,189–216.
15. Garvey, E.P., Oplinger, J.A., Furfine, E.S., Kiff, R.J., Laszlo, F., Whittle, B.J., Knowles,R.G., 1997. 1400W is a slow, tight binding, and highly selective inhibitor ofinducible nitric-oxide synthase in vitro and in vivo. J. Biol. Chem. 272,4959–4963.
16. Gordon, S., Taylor, P.R., 2005. Monocyte and macrophage heterogeneity. Nat. Rev. Immunol. 5, 953–964.
17. Gordon, S., 1998. The role of the macrophage in immune regulation. Res. Immunol.149, 685–688.
18. Grage-Griebenow, E., Flad, H.D., Ernst, M., 2001. Heterogeneity of humanperipheral blood monocyte subsets. J. Leukoc. Biol. 69, 11–20.
19. Green, S.J., Meltzer, M.S., Hibbs Jr., J.B., Nacy, C.A., 1990. Activated macrophagesdestroy intracellular Leishmania major amastigotes by an -arginine-dependentkilling mechanism. J. Immunol. 144, 278–283.
20. Harmon, B.G., Adams, L.G., 1987. Assessment of bovine mammary glandmacrophage oxidative burst activity in a chemiluminescence assay. Am. J. Vet.Res. 48, 119–125.
21. Harris, J., 2011. Autophagy and cytokines. Cytokine 56, 140–144.
22. Hubl, W., Ziegler-Heitbrock, L.H., Bayer, P.M., 2000. A simple method formeasurement of CD14weak CD16strong monocytes in peripheral blood.Immunobiology 202, 2–10.
23. Hussen, J., Duvel, A., Sandra, O., Smith, D., Sheldon, I.M., Zieger, P., Schuberth, H.J.,2013. Phenotypic and functional heterogeneity of bovine blood monocytes.PLoS One 8, e71502.
24. Jann, O.C., Werling, D., Chang, J.S., Haig, D., Glass, E.J., 2008. Molecular evolution ofbovine Toll-like receptor 2 suggests substitutions of functional relevance. BMCEvol. Biol. 8, 288.
25. Jann, O.C., King, A., Corrales, N.L., Anderson, S.I., Jensen, K., Ait-Ali, T., Tang, H., Wu,C., Cockett, N.E., Archibald, A.L., Glass, E.J., 2009. Comparative genomics ofToll-like receptor signalling in five species. BMC Genom. 10, 216.
26. Johnson, D.F., France, G.L., Marmer, D.J., Steele, R.W., 1980. Bactericidalmechanisms of human breast milk leukocytes. Infect. Immun. 28, 314–318.
27. Jungi, T.W., Adler, H., Adler, B., Thony, M., Krampe, M., Peterhans, E., 1996a.Inducible nitric oxide synthase of macrophages. Present knowledge andevidence for species-specific regulation. Vet. Immunol. Immunopathol. 54,323–330.
28. Jungi, T.W., Thony, M., Brcic, M., Adler, B., Pauli, U., Peterhans, E., 1996b. Inductionof nitric oxide synthase in bovine mononuclear phagocytes is differentiationstage-dependent. Immunobiology 195, 385–400.
29. Kizilkaya, A.K.a.K., 2009. Some factors influencing milk somatic cell count ofHolstein Friesian and Brown Swiss cows under the Mediterranean climaticconditions. Arch. Tierz. 52, 124–133.
30. Korkosz, M., Bukowska-Strakova, K., Sadis, S., Grodzicki, T., Siedlar, M., 2012.Monoclonal antibodies against macrophage colony-stimulating factor diminish the number of circulating intermediate and nonclassical(CD14(++)CD16(+)/CD14(+)CD16(++)) monocytes in rheumatoid arthritispatient. Blood 119, 5329–5330.
31. Lambeth, J.D., 2004. NOX enzymes and the biology of reactive oxygen. Nat. Rev.Immunol. 4, 181–189.
32. Lamkanfi, M., Kanneganti, T.D., Franchi, L., Nunez, G., 2007. Caspase-1inflammasomes in infection and inflammation. J. Leukoc. Biol. 82, 220–225.

33. Lund, T., Miglior, F., Dekkers, J.C.M., Burnside, E.B., 1994. Genetic relationshipsbetween clinical mastitis, somatic cell count, and udder conformation inDanish Holsteins. Livest. Prod. Sci. 39, 243–251.
34. Mao, K., Chen, S., Chen, M., Ma, Y., Wang, Y., Huang, B., He, Z., Zeng, Y., Hu, Y., Sun,S., Li, J., Wu, X., Wang, X., Strober, W., Chen, C., Meng, G., Sun, B., 2013. Nitricoxide suppresses NLRP3 inflammasome activation and protects againstLPS-induced septic shock. Cell Res. 23, 201–212.
35. Mishra, B.B., Rathinam, V.A., Martens, G.W., Martinot, A.J., Kornfeld, H., Fitzgerald,K.A., Sassetti, C.M., 2013. Nitric oxide controls the immunopathology oftuberculosis by inhibiting NLRP3 inflammasome-dependent processing ofIL-1beta. Nat. Immunol. 14, 52–60.
36. Nathan, C.F., 1987. Secretory products of macrophages. J. Clin. Invest. 79, 319–326.
37. Nathan, C., 1992. Nitric oxide as a secretory product of mammalian cells. FASEB J. 6,3051–3064.
38. Norimatsu, M., Chance, V., Dougan, G., Howard, C.J., Villarreal-Ramos, B., 2004. LiveSalmonella enterica serovar Typhimurium (S. typhimurium) elicit dendritic cellresponses that differ from those induced by killed S. typhimurium. Vet.Immunol. Immunopathol. 98, 193–201.
39. O'Reilly, L.M., Daborn, C.J., 1995. The epidemiology of Mycobacterium bovisinfections in animals and man: a review. Tuber. Lung Dis. 76 (Suppl. 1), 1–46.
40. Paixao, T.A., Ferreira, C., Borges, A.M., Oliveira, D.A., Lage, A.P., Santos, R.L., 2006.Frequency of bovine Nramp1 (Slc11a1) alleles in Holstein and Zebu breeds.Vet. Immunol. Immunopathol. 109, 37–42.
41. Quinn, M.T., Gauss, K.A., 2004. Structure and regulation of the neutrophilrespiratory burst oxidase: comparison with nonphagocyte oxidases. J. Leukoc.Biol. 76, 760–781.
42. Rupp, R., Boichard, D., 2003. Genetics of resistance to mastitis in dairy cattle. Vet.Res. 34, 671–688.
43. Schukken, Y.H., Wilson, D.J., Welcome, F., Garrison-Tikofsky, L., Gonzalez, R.N.,2003. Monitoring udder health and milk quality using somatic cell counts. Vet.Res. 34, 579–596.
44. Shi, C., Pamer, E.G., 2011. Monocyte recruitment during infection andinflammation. Nat. Rev. Immunol. 11, 762–774.
45. Tacke, F., Randolph, G.J., 2006. Migratory fate and differentiation of bloodmonocyte subsets. Immunobiology 211, 609–618.
46. Taylor, P.R., Gordon, S., 2003. Monocyte heterogeneity and innate immunity.Immunity 19, 2–4.
47. Thieblemont, N., Weiss, L., Sadeghi, H.M., Estcourt, C., Haeffner-Cavaillon, N., 1995.CD14lowCD16high: a cytokine-producing monocyte subset which expandsduring human immunodeficiency virus infection. Eur. J. Immunol. 25,3418–3424.
48. Thomas, R., Lipsky, P.E., 1994. Human peripheral blood dendritic cell subsets.Isolation and characterization of precursor and mature antigen-presentingcells. J. Immunol. 153, 4016–4028.
49. Vordermeier, M., Ameni, G., Berg, S., Bishop, R., Robertson, B.D., Aseffa, A.,Hewinson, R.G., Young, D.B., 2012. The influence of cattle breed onsusceptibility to bovine tuberculosis in Ethiopia. Comp. Immunol. Microbiol.Infect. Dis. 35, 227–232.
50. Warner, C.M., Meeker, D.L., Rothschild, M.F., 1987. Genetic control of immuneresponsiveness: a review of its use as a tool for selection for disease resistance.J. Anim. Sci. 64, 394–406.
51. Werling, D., Hope, J.C., Howard, C.J., Jungi, T.W., 2004. Differential production ofcytokines, reactive oxygen and nitrogen by bovine macrophages and dendriticcells stimulated with Toll-like receptor agonists. Immunology 111, 41–52.
52. Wong, K.L., Tai, J.J., Wong, W.C., Han, H., Sem, X., Yeap, W.H., Kourilsky, P., Wong,S.C., 2011. Gene expression profiling reveals the defining features of theclassical, intermediate, and nonclassical human monocyte subsets. Blood 118,e16–31.
53. Wong, K.L., Yeap, W.H., Tai, J.J., Ong, S.M., Dang, T.M., Wong, S.C., 2012. The threehuman monocyte subsets: implications for health and disease. Immunol. Res.53, 41–57.

54. Yagi, Y., Watanabe, E., Watari, E., Shinya, E., Satomi, M., Takeshita, T., Takahashi, H.,2010. Inhibition of DC-SIGN-mediated transmission of humanimmunodeficiency virus type 1 by Toll-like receptor 3 signalling in breast milkmacrophages. Immunology 130, 597–607.
55. Zawada, A.M., Rogacev, K.S., Rotter, B., Winter, P., Marell, R.R., Fliser, D., Heine, G.H.,2011. SuperSAGE evidence for CD14++CD16+ monocytes as a third monocytesubset. Blood 118, e50–61.
56. Ziegler-Heitbrock, L., Ancuta, P., Crowe, S., Dalod, M., Grau, V., Hart, D.N., Leenen,P.J., Liu, Y.J., MacPherson, G., Randolph, G.J., Scherberich, J., Schmitz, J.,Shortman, K., Sozzani, S., Strobl, H., Zembala, M., Austyn, J.M., Lutz, M.B., 2010.Nomenclature of monocytes and dendritic cells in blood. Blood 116, e74–80.
57. Ziegler-Heitbrock, L., 2007. The CD14+ CD16+ blood monocytes: their role ininfection and inflammation. J. Leukoc. Biol. 81, 584–592.

CHAPTER 8

Natural autoantibodies in *Bos taurus* calves during the first twelve weeks of life

N. Mayasari[a,b,*], A.T.M. Van Knegsel[a], G. de Vries Reilingh[a], B. Kemp[a], H.K. Parmentier[a]

[a]*Adaptation Physiology Group, Department of Animal Science, Wageningen University, P.O. Box 338, 6700 AH Wageningen, the Netherlands*

[b]*Faculty of Animal Husbandry, Universitas Padjadjaran, 45363 Bandung, Indonesia*

ABSTRACT

Natural autoantibodies (NAAb) have a role in maintaining physiological homeostasis and prevention of infections, and have been found in mammalian species tested so far. Albeit NAAb levels rise with age, little is known about the origin, function, regulation and initiation of NAAb in young animals. The present study addressed the presence of IgM and IgG NAAb binding glutamate dehydrogenase (GD), carbonic anhydrase (CA), myosin (MYO) and transferrin (TRANS) from before drinking colostrum until the first 12 weeks of life in plasma of female calves. In addition, NAAb to these four self-antigens were also measured in colostrum and in plasma of their mothers during three weeks before calving. Titers of NAAb binding GD, CA, MYO and TRANS were detected in plasma of cows before calving, in colostrum, and in plasma of calves before and after drinking of colostrum. Levels of NAAb in colostrum were positively related with levels of NAAb in plasma of cows. Before colostrum intake, levels of NAAb in plasma of calves were not related with levels of NAAb in plasma of their mother but were influenced by parity of their mother. After colostrum intake, levels of NAAb in plasma of calves in the first week of life were positively related with levels of NAAb in colostrum. Low NAAb levels in colostrum were related with low NAAb in plasma of calves in the first week of life, but after two weeks of life the relation between colostrum and plasma of calves was absent. In conclusion, NAAb are already present in the unborn calf, and levels of neonatal NAAb during the early weeks of life are affected by levels of maternal NAAb obtained via colostrum.

Keywords: Autoantibodies, Calf, Colostrum, Maternal antibody transfer, Dry period

1. INTRODUCTION

Natural antibodies (NAb) are a humoral part of innate immunity (Matter and Ochsenbein, 2008; Vollmers and Brändlein, 2009), and have been proposed to be involved in preventing infection and maintenance of physiological homeostasis (Ehrenstein and Notley, 2010; Lutz, 2012). Natural antibodies are polyreactive with low affinity binding with various antigens (Casali and Notkins, 1989), and are present in healthy animals in the absence of antigen stimulation (Avrameas, 1991; Baumgarth et al., 2005). Natural antibodies have been divided into two classes: overt and cryptic NAb. Overt NAb bind antigens that the individual has never encountered before, such as keyhole limpet hemocyanin (KLH). Cryptic NAb or so-called natural autoantibodies (NAAb) are antibodies that bind to self-antigens or slightly changed self-antigens (neo-epitopes). Natural autoantibodies are thought to be involved in inactivation of cytokines, prevention of inflammation, clearance of metabolic waste, perform various homeostatic roles within the immune response (Cojocura et al., 2009), and have a role in the prevention of autoimmunity (Nguyen et al., 2015). Antibodies binding a variety of self-antigens such as myosin, thyroglobulin (Lutz et al., 2009), heat shock proteins (Cohen, 2013), solubilized extracts of histologically normal organs were found in humans (Stahl et al., 2000) and in healthy calves (Khobondo et al., 2015). In dairy cows, NAb binding exo-antigens (Ploegaert et al., 2011; Thompson-Crispi et al., 2013; Mayasari et al., 2015) as well as self-antigens (Van Knegsel et al., 2012) were described earlier. Little is known of the routes that lead to the production of NAAb in healthy young individuals. It was proposed that NAAb may arise by cross reactivity with the intestinal microflora (Kamada et al., 2013), are initiated by dietary compounds, directed to neo-epitopes (Lutz et al., 2009; Lutz, 2012), rest on random VDJ recombination in B-cells maintained by self-antigens (Quintana and Cohen, 2004), or reflect levels of maternal antibodies after consumption of colostrum or milk. In dairy cows, maternal antibodies are not actively transferred to the fetus before birth, but calves obtain maternal antibodies passively from the dam through colostrum during the first 24 h of life. Thus, after first colostrum intake antibody titers in calves likely reflect maternal antibody titers in colostrum during the first weeks of life until neonatal antibody production starts. Until start of neonatal antibody production, adequate and sufficient passive immune transfer of maternal antibodies via colostrum intake provides immune competence of calves, which is a prerequisite to lower the risk of diseases and infections during the pre-weaning period (Oliveira et al., 2010).

To our knowledge, little information is present about the presence of NAAb in colostrum and in plasma of calves before colostrum intake and the relation between NAAb levels in cows and their calves during the early weeks of life. Earlier, we studied health and energy metabolism in cows with different dry period lengths. Omission of the dry period in dairy cows reduced NAb titers in colostrum compared with cows with a short (30-d) or conventional dry period (60-d) (Mayasari et al., 2015). Moreover, lower NAb titers in colostrum were reflected by lower titers of NAb binding keyhole limpet hemocyanin (KLH) and human serum albumin (HuSA) in plasma of calves in the first two weeks of life. No IgM and IgG NAb binding KLH or HuSA were detected in plasma of calves before colostrum intake. In the present study, we evaluated the level of NAAb to four self-antigens: glutamate dehydrogenase (GD), carbonic anhydrase (CA), myosin (MYO) and transferrin (TRANS) in plasma of calves before and after colostrum intake, and in plasma and in colostrum of their mother. Antibodies to those four self-antigens were found in plasma of dairy cows. Previous studies reported that self-antibodies to GD (Schulz et al., 2014) and CA (Puscas et al., 2001) are related with metabolic and health disorders in mammals. Cows with subclinical ketosis had higher non esterified fatty acid concentration and GD activation after calving (Schulz et al., 2014). Carbonic anhydrase also functions in regulation of pH and fluid balance (Badger and Price, 1994) as observed during inflammation (Mihaylova et al., 2008). Another study reported that NAAb binding MYO and TRANS in milk of cows were associated with sensitivity for mastitis (Van Knegsel et al., 2012). Myosin is a fibrous protein which has a role in muscle contraction (Schmitt, 1968). Transferrin is a serum globulin which works in the complexing and transport of iron (Crichton and Charloteauxwauters, 1987).

In the current study, it was hypothesized that NAAb binding GD, CA, MYO and TRANS are present in plasma of cows and in colostrum, but absent in plasma of calves before colostrum intake. Moreover, if omission of the dry period in cows reduced NAb titers in colostrum, NAAb titers in colostrum and in plasma of their calves may be reduced as well. Moreover, the NAAb titers in plasma of calves after colostrum intake are expected to reflect the NAAb titers of their mother (in plasma and colostrum). The present study aimed to evaluate NAAb to GD, CA, MYO and TRANS in mother and calves before and after colostrum intake until 12 weeks of age and to study relationships between these titers in mother, colostrum and the calves.

2. MATERIALS AND METHODS

2.1. Experimental design, animals and rations

The Institutional Animal Care and Use Committee of Wageningen University approved the experimental protocol. The registration number of the experimental protocol is 2010026. Blood and colostrum samples originated from an experiment that was designed to evaluate the effect of dry period length and ration composition on health of cows and calves. The experimental design, treatments of dry period lengths and ration composition in cows were described earlier by Van Knegsel et al. (2014). In short, Holstein-Friesian dairy cows (N = 167) were selected from the Dairy Campus research herd (WUR Livestock Research, Lelystad, the Netherlands) blocked according to parity, calving date, milk yield in previous lactation and body condition score (BCS), and randomly assigned to treatments. Treatment consisted of three dry period lengths: 0, 30 or 60 days; and two lactation rations (glucogenic or lipogenic ration). For the current study, female calves born to these cows (n = 63) were monitored.

2.1.1. Management of cows

Cows were housed in a freestall with slatted floor and cubicles and milked twice daily (at 0500 and 1630 h). Prepartum, dry cows received a dry cow ration and lactating cows received a lactating cow ration supporting 25 kg of milk per day. Forage was supplied ad libitum and consisted prepartum of grass silage, corn silage, wheat straw and a protein source (rapeseed meal or soybean meal) in a ratio of 39:25:25:11 (DM basis). Postpartum, forage consisted of grass silage, corn silage, straw, and a protein source (rapeseed meal or soybean meal) in a ratio 51:34:2:13 (DM basis). From d 10 before expected calving onwards, cows of all treatments were fed 1 kg/d glucogenic or lipogenic concentrate and increased postcalving stepwise with 0.5 kg/d until the concentration supply reached 8.5 kg/d. Main ingredient for glucogenic concentrate was corn and main ingredients for lipogenic concentrate were sugar beet pulp, palm kernel, and rumen protected palm oil. Concentrate and forage were supplied separately.

2.1.2. Colostrum sampling and management of calves

Colostrum sampling and management of the calves were described earlier (Mayasari et al., 2015). Distribution of cows, colostrum sampling, plasma samples of cows, plasma samples of female calves before and after colostrum intake across dry period length treatments group are shown in Table 1. Colostrum was collected and weighed immediately after calving from each quarter of mammary gland and pooled in one sample per cow. Calves received the colostrum from their mother via an artificial teat. Colostrum samples (n = 138; 10 ml) obtained after parturition were agitated and stored at −20 °C, until analysis.

Table 1. Distribution of cows, colostrum samples, plasma in cows and plasma of female calves (before and after colostrum intake) across the dry period length.

	n	Dry period length		
		0 days	**30 days**	**60 days**
Cows in the experiments	167	56	55	56
Colostrum	138	44	45	49
Plasma of cows (before calving)	92	30	32	30
Plasma of female calves before colostrum intake	52	19	16	17
Plasma of female calves after colostrum intake	63	23	19	21

Immediately after birth, calves were removed from the dam. When calves were born between 10.00 p.m. and 05.00 a.m. (n = 42), calves were removed from the dam at 05.00 a.m. Calves were weighed and within 24 h of life, they received four liter of colostrum in two portions from their mother. After 24 h after birth, female calves were fed with milk replacer (crude protein 22% and fat 17%) (two times a day two liter). After two days, calves were moved to the calf raising farm and were also fed milk replacer twice a day, until 60-d of life. After the first week of life, all calves were fed the same mixture of hay and grains (Agrifirm Feed, the Netherlands) based on requirements for preweaned Holstein calves. At 60-d of life, calves were weaned from milk replacer. From 60-d of life calves received a diet based on requirement for weaned Holstein calves. During first two weeks of life, calves were housed in individual hutches located approximately 60 cm apart. After two weeks, calves were kept in groups of six to eight calves with straw bedding. Body weight of female calves was recorded at birth and in week 2, 4, 6, 8, 10 and 12 after birth and reported earlier (Mayasari et al., 2015). Disease incidences and treatments of the female calves were recorded during the first 12 weeks of life.

In an earlier study, we evaluated the effect of maternal dry period on specific antibody responses (SpAb) to keyhole limpet hemocyanin (KLH) and human albumin serum (HuSA) in plasma of calves (Mayasari et al., 2015). Therefore, calves received primary and secondary immunizations with KLH and HuSA in week 6 and 10, respectively. Briefly, after 6 weeks of life, primary SpAb titers binding KLH and HuSA did not differ between calves from cows with different dry period lengths. However, secondary antibody responses were enhanced in calves from cows with a 0-d dry period (Mayasari et al., 2015).

2.2. Blood sampling

For cows, blood samples were taken from the tail vain for NAAb measurement (n = 92 cows), weekly from week −3 till −1 relative to calving. For calves, blood samples were taken from jugular vein for NAAb measurement at several time points: before colostrum intake (n = 52), after colostrum intake at day two (after 24 h) (n = 59) and every Wednesday in weeks 1, 2, 4, 6, 7, 8, 10, 11, and 12 of life (n = 63). Blood samples were taken in Vacutainer tubes (10 ml; Greiner Bio-One GmbH, Kremsmunster, Austria) containing heparin as anticoagulant. Blood samples were centrifuged at 3000*g* at 4 °C, for 15 min within two hours after collection. Plasma was transferred and stored at −20 °C until analysis.

2.3. Analysis of NAAb in plasma of cows, colostrum, and plasma of calves

Titers of NAAb binding the four self-antigens GD, CA, MYO and TRANS in plasma of calves and cows and in colostrum were measured by an indirect enzyme-linked immunosorbent assay (ELISA) technique as outlined by Van Knegsel et al. (2012). In brief, plates were coated with 4 μg/ml of GD (G7882, Sigma), CA (C3934, Sigma), MYO (M0531, Sigma) or TRANS (T1408, Sigma), respectively, in 100 μl/well. Natural autoantibodies of the IgG isotype were detected either using 1:20,000 (cow) or 1:40.000 (calf, colostrum) diluted sheep polyclonal anti-bovine IgG-heavy chain conjugated to horseradish peroxidase (PO) (Cat. No. E10-118P, Bethyl). Natural autoantibodies of the IgM isotype in plasma of cows and calves and in colostrum were detected using 1:20,000 diluted rabbit polyclonal anti-bovine IgM-whole molecule conjugated to PO (Cat. No. A10-100P, Bethyl). Four step serial dilutions for both IgG and IgM in colostrum and in plasma samples of cows and calves started at 1:40. After washing, a substrate containing tetra methyl benzidine (Sigma Aldrich Chemie, Steinheim, Germany) and 0.05 percent hydrogen peroxide was added, and incubated for 10 min at room temperature. The reaction was stopped by adding 1.25

M sulfuric acid. Extinctions were measured with a Multiskan reader (Lab Systems, Helsinki, Finland) at a wavelength of 450 nm. Titers were expressed as log 2 values of the dilutions that gave an extinction closest to 50% of Emax, where Emax represents the highest mean extinction of a standard positive (pooled) serum present on every microtiter plate (Ploegaert et al., 2007).

2.4. Statistical analyses

All data of NAAb titers in cows (plasma and colostrum) and in plasma of calves approximated normality of residuals by examining whether skewness and kurtosis were in a range of −2 until 2. Moreover, the data was analysed using PROC MIXED in SAS 9.2 (SAS Inst. Inc. Cary, NC). The NAAb titers for IgG and IgM binding GD, CA, MYO and TRANS in plasma of cows before calving were analysed with dry period length (0, 30 or 60 days), parity (1, 2 or >2), week (week −3 till −1 relative to calving), and their 2 way-interactions included as fixed effect. Cow was considered as the repeated subject. A first order auto-regressive [AR(1)] variance-covariance structure was the best fit according to Akaike's corrected information criterion (Model 1). The NAAb titers for IgG and IgM binding GD, CA, MYO and TRANS in colostrum and in plasma of calves before colostrum intake were analysed with dry period length (0, 30 or 60 days), parity of the cow (2, 3 or >3) and their 2-way interaction included as fixed effect (Model 2).

The NAAb titers for IgG and IgM binding GD, CA, MYO and TRANS after colostrum intake (week 0 until 6) and specific antibody titers after primary (week 6 until 10) and secondary immunization (week 10 until 12) in plasma were analysed with dry period length (0, 30 or 60 days), parity (2, 3 or >3), week (1, 2,... or 12) and their 2-way interactions included as fixed effects. Calf was considered as the repeated subject. A first order auto-regressive [AR(1)] variance-covariance structure was determined to be the best fit according to Akaike's corrected information criterion (Model 3). To compare dry period length effects, *P*-values are presented after a Bonferroni adjustment. Values are presented as least squares means (LSMs) with their pooled standard errors of the mean (SEMs), unless otherwise stated. Preliminary result showed primary and secondary immunization (in week 6 and 10, respectively) of KLH and HUSA did not affect the titers for IgG and IgM binding GD, CA, MYO and TRANS.

To assess the relationship between NAAb titers in plasma of cows with NAAb titers in colostrum or with NAAb titers in plasma of female calves (before and after colostrum intake), the average NAAb of the cows was included as continuous variable in model 2 with all 2-way interactions (Model 4). Model 4 was applied for four different time periods and the availability of plasma in cows and their female calves; (1) titers before colostrum intake (naïve calves) (n = 27), (2) titers after colostrum intake at day two (after 24 h) (n = 32), (3) the average titers in weeks 1 and 2 (n = 34), and (4) the average titers in weeks 4 till 12 of life (n = 34). Preliminary analysis showed there was no interaction between NAAb titers in cows and dry period length, therefore this interaction was excluded from the model. To assess the relationship between NAAb titers in colostrum with NAAb titers in plasma of calves after colostrum intake, the NAAb titers in colostrum was included as continuous variable in model 3 with all 2-way interactions (Model 5). The regression coefficients (β) from the statistical model and the *P*-value corresponding to the β are displayed.

3. RESULTS

Natural autoantibody (NAAb) titers for isotypes IgG and IgM binding glutamate dehydrogenase (GD), carbonic anhydrase (CA), myosin (MYO), and transferrin (TRANS) were detected in plasma of cows before calving (Table 2), in colostrum (Fig. 1) and in plasma of calves before and after colostrum intake (Fig. 2 and Table 3)

3.1. Natural autoantibodies in plasma of cows before calving (Model 1)

From week 3 till 1 before calving, cows with a 30-d dry period had higher plasma titers of IgG binding GD and IgM binding CA compared with cows with a 60-d dry period (Table 2). There was a dry period length x week interaction for titers of IgG binding GD and CA ($P < 0.05$). Cows with a 60-d dry period had lower IgG binding GD or CA one week before calving compared with week two or three before calving (GD; 8.2 vs. 8.0 vs. 7.3 ± 0.03, CA; 6.3 vs. 6.0 vs. 5.5 ± 0.2 for weeks −3 vs. −2 vs. −1). Cows with a 0-d dry period or cows with a 30-d dry period had no difference of IgG binding GD or CA titers between different weeks prepartum. Titers of IgM binding CA and IgM binding TRANS decreased with parity (CA; 6.5 vs. 6.3 vs. 5.8 ± 0.1, TRANS 6.9 vs. 6.5 vs. 5.4 ± 0.1 for parity 1 vs. 2 vs. >2, respectively).

3.2. Natural autoantibodies in colostrum (Model 2)

Cows with a 0-d dry period had lower titers for both IgM and IgG NAAb binding GD, CA, MYO and TRANS in colostrum compared with cows with a 30-d or 60-d dry period (Fig. 1). A dry period length x parity interaction was present for both titers of IgM binding GD and CA ($P < 0.05$).

Table 2. Plasma natural autoantibody titers (IgG and IgM) binding glutamate dehydrogenase (GD), carbonic anhydrase (CA), myosin (MYO), and transferrin (TRANS) in plasma of dairy cows with different dry period lengths (0, 30 or 60 days) during the precalving period* (LSM ± pooled SEM).

Item	Dry period length			SEM	*P*-values**				
	0	30	60		Dry period	Parity	Week*	D × P	D × W
Cows (n = 92)	30	32	30						
NAAb IgG-GD[1]	8.09[ab]	8.67[b]	7.81[a]	0.23	0.03	0.39	0.05	0.88	0.01
NAAb IgM-GD[1]	6.33	6.67	6.35	0.15	0.18	0.41	0.75	0.61	0.30
NAAb IgG-CA[1]	6.14	6.45	5.94	0.18	0.12	0.71	<0.01	0.93	0.03
NAAb IgM-CA[1]	6.02[a]	6.49[b]	6.01[a]	0.15	0.03	0.01	0.35	0.11	0.58
NAAb IgG-MYO[1]	6.20	5.88	5.71	0.24	0.34	0.96	0.90	0.91	0.08
NAAb IgM- MYO[1]	6.20	6.57	6.30	0.20	0.38	0.63	0.38	0.31	0.11
NAAb IgG-TRANS[1]	4.82	4.89	4.78	0.19	0.93	0.94	<0.01	0.08	0.30
NAAb IgM- TRANS[1]	6.51	6.72	6.64	0.15	0.64	0.05	<0.01	0.27	0.21

[a,b] Values within dry period length in the same row with different superscripts differ ($P<0.05$).

* Week −3, −2, and −1 relative to calving.

** D = dry period; P = parity (1, 2 and >2); W = week (−3 till −1 relative to calving); [1] Antibody names are combination of natural autoantibody (NAAb), antigen abbreviation (GD, CA, MYO or TRANS) and immunoglobulin isotype (IgG or IgM).

Older cows (parity ≥3) had higher titers of IgM binding GD and CA when cows had 0-d dry period compared with younger cows (GD; 4.4 vs. 5.4 vs. 5.8 ± 0.4 CA; 4.0 vs. 5.3 vs. 5.3 ± 0.5 for parity 2 vs. 3 vs. >3, respectively). Contrary, older cows had lower titers of IgM binding GD and CA when cows had a 60-d dry period compared with younger cows (GD; 7.9 vs. 7.5 vs. 7.3 ± 0.3 CA; 8.1 vs. 7.4 vs. 7.4 ± 0.3 for parity 2 vs. 3 vs. >3, respectively).

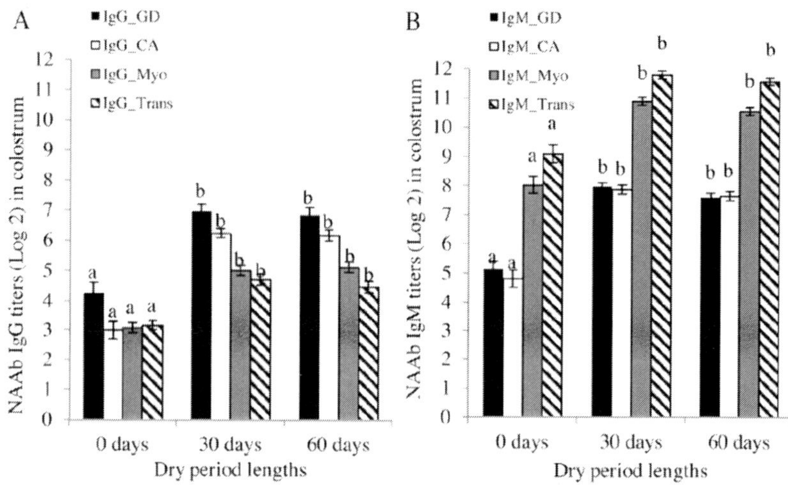

Figure. 1. Natural autoantibodies titers (Log 2) for isotype IgG (A) and IgM (B) titers binding glutamate dehydrogenase (GD), carbonic anhydrase (CA), myosin (MYO) and transferrin (TRANS) in colostrum from dairy cows with different dry period lengths (0, 30 or 60 days) (Means ± SEM). [a,b,c] Values between dry period length with different superscripts differ (P < 0.05).

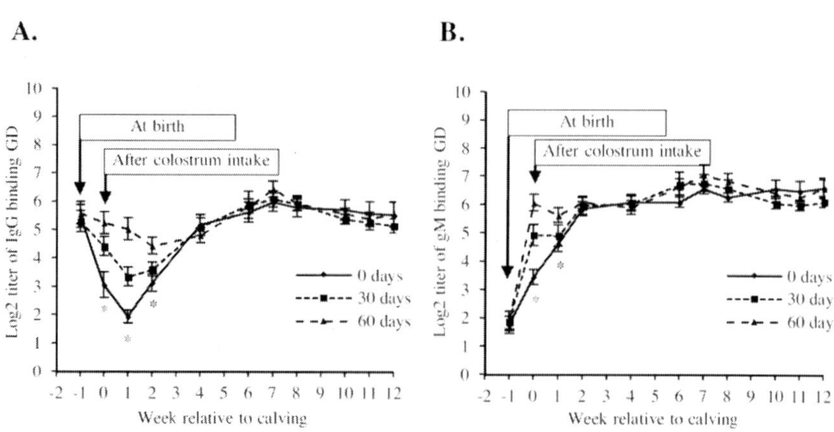

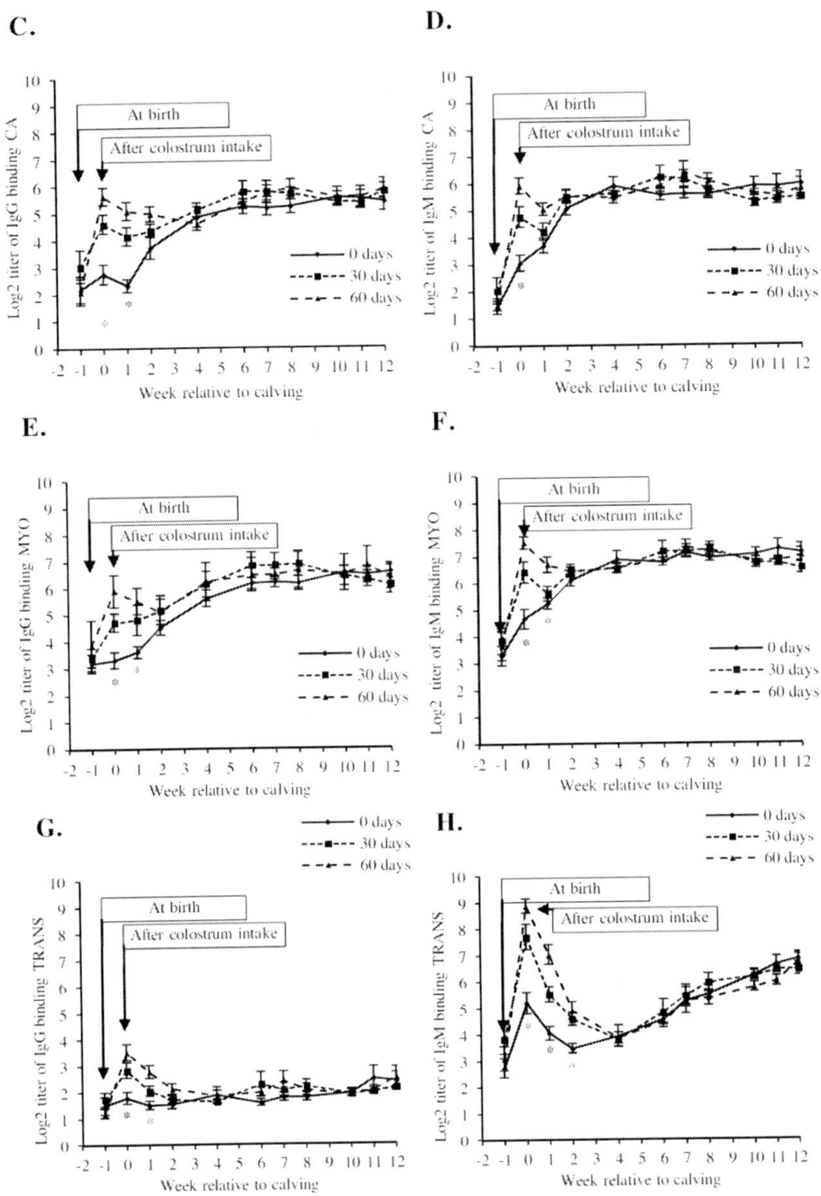

Figure. 2. Plasma natural autoantibodies (IgG and IgM) titers binding glutamate dehydrogenase (GD; A and B), carbonic anhydrase (CA; C and D), myosin (MYO; E and F) and transferrin (TRANS; G and H) of calves at birth (before colostrum intake), after drinking colostrum from cows with three different dry period lengths (0, 30, or 60 days). Values are means (±SEM) per dry period length (DPL) per week (W). * P-value < 0.05.

Table 3. Plasma natural autoantibody titers (IgG and IgM) binding glutamate dehydrogenase (GD), carbonic anhydrase (CA), myosin (MYO), and transferrin (TRANS) in plasma of female calves from cows with different dry period lengths (0, 30 or 60 days) before colostrum intake and after colostrum intake till week 6 of life (LSM ± pooled SEM).

Item	Dry period length			SEM	*P*-values[*]				
	0	30	60		Dry period	Parity	Week	D x P	D x W
Calves before colostrum intake (n = 52)	19	16	17						
NAAb IgG-GD[1]	6.09	5.66	5.85	0.39	0.73	0.08	NA	0.99	NA
NAAb IgM-GD[1]	1.33	1.43	1.71	0.29	0.62	0.04	NA	0.79	NA
NAAb IgG-CA[1]	2.44	2.15	2.21	0.56	0.93	0.38	NA	0.13	NA
NAAb IgM-CA[1]	1.02	1.52	1.41	0.34	0.54	0.37	NA	0.53	NA
NAAb IgG-MYO[1]	2.91	3.26	3.91	0.65	0.55	0.35	NA	0.22	NA
NAAb IgM- MYO[1]	2.89	3.16	3.30	0.43	0.78	0.01	NA	0.46	NA
NAAb IgG-TRANS[1]	1.56	1.59	1.17	0.34	0.64	0.57	NA	0.76	NA
NAAb IgM- TRANS[1]	2.30	3.32	2.93	0.46	0.28	0.19	NA	0.72	NA
Calves after colostrum intake (n = 63)	23	19	21						
NAAb IgG-GD[1]	3.86^{a}	4.47^{ab}	5.06^{b}	0.25	<0.01	0.16	<0.01	0.80	<0.01
NAAb IgM-GD[1]	5.24^{a}	5.74^{b}	6.06^{b}	0.17	<0.01	0.20	<0.01	0.25	<0.01
NAAb IgG-CA[1]	3.82^{a}	4.81^{b}	5.18^{b}	0.21	<0.01	0.71	<0.01	0.21	<0.01
NAAb IgM-CA[1]	4.68^{a}	5.29^{b}	5.36^{b}	0.17	0.01	0.10	<0.01	0.17	0.01
NAAb IgG-MYO[1]	4.67^{a}	5.53^{a}	5.93^{b}	0.39	0.06	0.54	<0.01	0.41	0.02
NAAb IgM- MYO[1]	5.96^{a}	6.46^{b}	6.84^{b}	0.16	<0.01	0.36	<0.01	0.45	<0.01
NAAb IgG-TRANS[1]	1.65^{a}	2.13^{b}	2.44^{b}	0.16	<0.01	0.80	<0.01	0.72	<0.01
NAAb IgM- TRANS[1]	4.25^{a}	5.30^{b}	5.86^{b}	0.21	<0.01	0.80	<0.01	0.44	<0.01

[a,b] Values within dry period length in the same row with different superscripts differ ($P<0.05$).

[*] D = dry period, P = parity (2, 3 and >3), W = week (1, 2,..6 weeks of life), NA = not applicable.

[1] Antibody names are combination of natural autoantibody (NAAb), antigen abbreviation (GD, CA, MYO or TRANS) and immunoglobulin isotype (IgG or IgM).

3.3. Natural autoantibodies in plasma of calves before colostrum intake (Model 2)

Natural autoantibodies for both isotypes (IgG and IgM) binding GD, CA, MYO and TRANS were detected in plasma of calves at birth before colostrum intake (Fig. 2 and Table 2). Dry period length of the mother did not affect NAAb titers in calves before colostrum intake. Before colostrum intake, calves from younger cows had higher titers of IgM binding GD and MYO in plasma compared with calves from older cows (GD; 2.1 vs. 1.1 vs. 1.3 ± 0.1, MYO; 4.2 vs. 2.7 vs. 2.4 ± 0.1 for parity 1 vs. 2 vs. >2, respectively, $P < 0.05$).

3.4. Natural autoantibodies in plasma of calves after colostrum intake (Model 3)

After colostrum intake (week 1) until immunizations of KLH and HuSA at week 6 of life, NAAb for isotypes IgG and IgM binding GD, CA, MYO and TRANS in plasma of calves were affected by dry period treatment of the mother especially in the first two weeks of life (Fig. 2). Primary and secondary immunizations of KLH and HuSA in week 6 and 10 of life did not affect NAAb titers for IgG and IgM binding GD, CA, MYO or TRANS in plasma of calves. After colostrum intake (week 1) until 6 weeks of life, calves from cows with a 0-d dry period had lower NAAb IgG and IgM binding GD, CA, MYO and TRANS in plasma compared with calves from cows with a 60-d (Table 2). Calves from cows with a 30-d dry period had no difference in IgM and IgG NAAb titers binding GD, CA, MYO or TRANS in plasma compared with calves from cows with a 60-d dry period, except for IgG binding MYO. From week 1 until 6 of life, NAAb for isotypes IgG and IgM binding GD, CA, MYO and TRANS in plasma of calves were affected by dry period and week interaction.

3.5. Relationship between NAAb titers in plasma of cows and NAAb titers in colostrum and in plasma of calves before colostrum intake (Model 4)

NAAb titers (IgG and IgM) binding GD, CA, MYO and TRANS in plasma of cows before calving were all positively related with NAAb titers (IgG and IgM) binding GD, CA, MYO and TRANS in colostrum (Table 4). There were no relationships between NAAb titers to four self-antigens in plasma of cows before calving with NAAb titers to four self-antigens in plasma of their calves before colostrum intake.

Table 4. Relationship between the average titers of natural auto antibodies (NAAb) binding glutamate dehydrogenase (GD), carbonic anhydrase (CA), myosin (MYO), and transferrin (TRANS) in plasma of cows before calving[a] with NAAb in colostrum, NAAb in plasma of calves before colostrum intake and after colostrum intake[b]. Values represent regression coefficients (β) and *P*-value.

	n[d]	Cows (week −3 until −1 relative to calving)[a,c]							
		IgG_GD	IgM_GD	IgG_CA	IgM_CA	IgG_MYO	IgM_MYO	IgG_TRANS	IgM_TRANS
Colostrum	74	0.454(0.01)	0.364(0.02)	0.388(0.02)	0.570(<0.01)	0.846(<0.01)	0.482(<0.01)	0.533(<0.01)	0.467(0.01)
Calves (before colostrum intake)	26	0.541(0.09)	−0.403(0.26)	−0.805(0.13)	−0.376(0.20)	−0.189(0.50)	0.131(0.71)	−0.466(0.07)	0.297(0.42)
Calves (after colostrum intake at day 2)	32	0.220(0.42)	0.336(0.31)	0.690(0.05)	0.531(0.09)	0.393(0.11)	0.325(0.13)	0.364(0.07)	0.702(0.09)
Calves (week 1-2 weeks of life)	34	0.360(0.05)	0.168(0.41)	0.482(0.04)	0.012(0.96)	0.193(0.24)	0.029(0.82)	0.243(0.02)	0.363(0.11)
Calves (4–12 weeks of life)	34	−0.134(0.46)	0.126(0.54)	0.192(0.35)	−0.077(0.69)	−0.090(0.51)	0.099(0.32)	0.128(0.30)	−0.074(0.68)

[a] NAAb titers in plasma of cows are the average titer in weeks −3 till −1 relative to calving.

[b] Calves were classified based on weeks to NAAb (titer before colostrum intake, titer after colostrum intake at day 2 (after 24 h), the average titer in the first 2 weeks (week 1 and 2) and the average titer in weeks 4 till 12 of life).

[c] Antibody names are combination of natural autoantibody (NAAb), antigen abbreviation (GD, CA, MYO, or TRANS) and immunoglobulin isotype (IgG or IgM).

[d] Samples available in plasma of cows and colostrum (n = 74, 0-d = 21; 30-d = 25 and 60-d = 28) and in calves (in several time points: plasma of calves before colostrum intake (n = 26, 0-d = 8; 30-d = 10 and 60-d = 8), plasma of calves after colostrum intake (at day 2) (n = 32, 0-d = 11; 30-d = 10 and 60-d = 11), plasma of calves after colostrum in week 1 and 2, plasma of calves after colostrum intake from week 4 till 12 (n = 34, 0-d = 12; 30-d = 11 and 60-d = 11).

3.6. Relationship between NAAb titers in plasma of cows and NAAb titers in plasma of calves after colostrum intake (Model 4)

Titer of IgG binding CA in plasma of calves after colostrum intake at day two was positively related with titer of IgG binding CA in plasma of cows (Table 4). The other titers were not correlated. In weeks one and two after birth, titers of IgG binding GD, CA and TRANS in plasma of calves were positively related with titers of IgG binding GD, CA and TRANS in plasma of cows. However, NAAb titers in plasma of calves from week 4 until 12 of life were not related with NAAb titers in plasma of cows.

3.7. Relationship between NAAb titers in colostrum and in plasma of calves after colostrum intake (Model 5)

Titers of NAAb (IgG and IgM) binding GD, CA, MYO and TRANS in colostrum correlated positively with titers of NAAb (IgG and IgM) binding GD, CA, MYO and TRANS in plasma of calves after colostrum intake (at day two) (Table 5). No relations were found between NAAb titers present in colostrum and NAAb titers in plasma of calves after two weeks of life.

4. DISCUSSION

In the present study IgM and IgG antibodies binding four self-antigens: GD, CA, MYO and TRANS, were detected in plasma of cows before calving, in colostrum, and in plasma of their calves both before and after colostrum intake. Previous studies showed that antibodies reacting to self-antigens can be found in plasma and milk from healthy non-immunized dairy cows (Van Knegsel et al., 2012) and in plasma from calves of 20–26 of weeks old (Khobondo et al., 2015). Together, these findings show that NAAb are present in young and old dairy cows.

The present study showed that NAAb of both antibody isotypes (IgG and IgM) binding GD, CA, MYO and TRANS were present in plasma of calves before colostrum intake. Earlier, NAb binding the exo-antigens KLH and HuSA were found in plasma from calves after colostrum intake but not before colostrum intake (Mayasari et al., 2015). The study of Mayasari et al. (2015) suggested that NAb binding exo-antigens during the first weeks of life depend on maternal transfer via colostrum and subsequent sensitization to environmental antigens, whereas the current study suggest that NAAb may be already initiated by self-antigens during the fetal phase. Knowledge of origin, induction and transfer of NAAb is scarce.

Table 5. Relationship between the average of natural auto antibodies (NAAb) binding glutamate dehydrogenase (GD), carbonic anhydrase (CA), myosin (MYO), and transferrin (TRANS) in colostrum and in plasma of calves after drinking colostrum. Values represent regression coefficient (β) and *P*-value.

	n	Colostrum							
		IgG_GD	IgM_GD	IgG_CA	IgM_CA	IgG_MYO	IgM_MYO	IgG_TRANS	IgM_TRANS
Day 2	50	0.658(<0.01)	0.600(<0.01)	0.570(<0.01)	0.519(<0.01)	0.270(<0.01)	0.557(0.02)	0.506(<0.01)	0.717(<0.01)
Week 1	49	0.568(<0.01)	0.082(0.51)	0.430(<0.01)	0.104 (0.32)	0.477(0.04)	0.176(0.13)	0.309(<0.01)	0.250(0.09)
Week 2	51	0.481(<0.01)	0.223(0.03)	0.224(0.06)	0.030(0.75)	0.281(0.21)	0.140(0.08)	0.261(<0.01)	0.312(0.02)
Week 4	50	0.205(0.14)	−0.009(0.94)	0.116(0.32)	0.032(0.75)	0.029(0.92)	−0.084 (0.40)	0.105(0.36)	−0.062(0.68)
Week 6	51	0.180(0.29)	0.286(0.05)	0.229(0.05)	0.209(0.08)	0.110(0.67)	0.142 (0.17)	0.146(0.27)	0.080(0.60)
Week 7	49	0.126(0.35)	0.245(0.07)	0.152(0.19)	0.178(0.14)	0.117(0.45)	0.098 (0.28)	0.218(0.10)	−0.017(0.92)
Week 8	51	0.130(0.30)	0.114(0.27)	0.138(0.17)	0.106(0.23)	0.133(0.62)	0.115(0.17)	0.086(0.39)	0.006(0.96)
Week 10	51	0.071(0.55)	−0.063(0.58)	0.037(0.71)	−0.038(0.66)	0.035(0.89)	0.009(0.91)	0.046(0.49)	−0.051(0.55)
Week 11	51	0.065(0.61)	−0.020(0.82)	0.142(0.12)	0.068(0.40)	−0.045(0.85)	0.020(0.79)	0.041(0.46)	−0.030(0.71)
Week 12	50	0.230(0.20)	0.034(0.76)	0.249(0.04)	0.089(0.48)	−0.002(0.99)	−0.110(0.27)	0.323(0.01)	−0.100(0.37)

[1] Antibody names are combination of natural autoantibody (NAAb), antigen abbreviation (GD, CA, MYO, or TRANS) and immunoglobulin isotype (IgG or IgM).

The presence of NAAb may reflect cross reactivity to microbiota (Kamada et al., 2013), or represent idiotypes either initiated or maintained by self-antigens, or alternatively initiated and maintained by maternal antigens transferred from the mother to the fetus. The presence of NAAb in antigen-free mice (Haury et al., 1997), in human fetal cord blood (Lacroix-Desmazes et al., 1998) and in plasma from calves before colostrum intake (this study) suggest that NAAb do reflect auto-responsiveness. In fetal calf's, lymph nodes IgM+ cells are present at 60 days in gestation, and from day 130 plasma immunoglobulins are found (Schultz et al., 1973). It is tempting to speculate that auto-antigens in the fetus are involved in the production of auto-antibodies. VDJ recombinations and somatic mutations provide a large degree of antigen receptor diversification that is required to combat infections, but as a consequence high levels of autoimmune B cells are found in healthy humans (Wardemann and Nussenzweig, 2007) indicating that yet to be identified mechanisms prevent auto-immune diseases. In addition, repeated immunizations with exo-antigens in this study did not affect NAAb levels in the calves. This study confirms previous studies in man that NAAb binding auto-antigens like DNA, hormones, histones, enzymes, receptors, major histocompatibility antigens, and intercellular matrix components are present in healthy individuals (Poletaev and Osipenko, 2003) even without any challenges and even without transfer of passive maternal antibodies via colostrum or intrauterine IgG transfer (Madi et al., 2009).

In the current study, titers of NAAb binding GD, CA, MYO and TRANS increased in plasma of calves after colostrum intake. A week after colostrum intake the levels of NAAb in plasma decreased in all calves, but especially in calves that received colostrum with lower levels of IgG binding GD and CA, and IgM binding TRANS, i.e. from cows with a 0-d dry period. Natural autoantibody titers for calves from cows with a 0-d dry period were specifically lower immediately after colostrum intake and in week one, but after four weeks the differences in titers between the three groups of calves were absent, suggesting an increased production of NAAb to GD, CA, MYO and TRANS in calves from cows with a 0-d dry period. Earlier we suggested that the lower NAb titers in plasma in the first two weeks of life of calves, which were born from cows with a 0-d dry period, were likely caused by lower immunoglobulin content in colostrum (Mayasari et al., 2015). Moreover, in this earlier study NAb titers in plasma of calves from cows with a 0-d dry period increased more between weeks two and six of age, compared with calves from cows with a 30-d or 60-d dry period. In the current study, there was no difference in the level of antibodies against self-antigens for calves from cows with a 30-d dry period, compared with calves born to cows a 60-d dry period. This is in line with the earlier study that calves born to cows with a 30-d dry period had no differences in antibodies against exo-antigens in plasma compared with calves born to cows with a

60-d dry period (Mayasari et al., 2015). Together, these findings suggest that the dynamics of NAAb binding GD, CA, MYO and TRANS in plasma of calves after colostrum intake were similar with the dynamics of NAb binding exo-antigens in calves after colostrum intake. Calves from cows with a 0-d dry period had lower NAb or NAAb in the first two weeks of life but eventually they had similar titers of NAb or NAAb from six weeks of life onwards.

In the current study, NAAb for isotypes IgG and IgM binding four autoantigens in plasma of calves increased after colostrum intake, except for IgG binding GD. We have no explanation for different behavior of IgG binding GD. Our observation suggested that calves already have NAAb before colostrum intake, which is markedly affected by colostral NAAb, but the level of IgG binding GD titers decreased very fast within a week after calves produced NAAb themselves. Since after a few weeks there were no differences anymore between the calves, it is questioned, whether NAAb have a specific function in colostrum and are selectively transfer to plasma. In addition, it is also questioned whether NAAb are waste products in the colostrum since calves seem to be able to make NAAb by themselves. A sudden increase in maternal antibodies may result in a response of the calf to get rid of IgG binding GD but this is all very speculative. A previous study showed that the mechanisms of low levels of antibodies in plasma after colostrum intake were either antigen specific or non-antigen specific (Banks, 1982). Another study showed that the concentration of colostral antibodies absorbed, the half-life of each class (isotypes) and the catabolism rate of each isotype determined the level of antibodies in plasma of calves (Murphy et al., 2014).

The low NAAb titers and subsequent higher increase in NAAb levels in calves from the 0-d dry period treated cows are in line with earlier studies which reported a negative relation between NAb and antibody responses following vaccination or immunization (Sinyakov and Avtalion, 2009). In man, colostrum of healthy mothers contains a broad spectrum of IgA (Pribylova et al., 2012; Quan et al.,1997) and IgM (Vassilev and Veleva, 1996) binding various autoantigens, which may modulate or supplement neonatal innate immunity (Pribylova et al., 2012). On the other hand, maternal antibodies, apart from protection of neonates to infection, also perform long term self-perpetuating imprinting of the neonatal B-cell repertoire by preventing development of activated self-reactive B cells by masking cross-reactive self during infections (Fink et al., 2008).

In our study, age (parity) of the mother affected the level of NAAb in plasma of calves before colostrum intake. Older cows with no dry period compared with young cows had lower IgM NAAb titers in plasma and they had higher IgM NAAb in colostrum. A previous study reported that old and young cows had no differences in NAAb in plasma or milk (Van Knegsel et al., 2012).

Other studies reported that older cows had higher NAb titers binding exo-antigens like KLH and HuSA in colostrum than young cows (Muller and Ellinger, 1981; Pritchett et al., 1991; Mayasari et al., 2015). The present study is not in line with Conneely et al. (2013) who suggested that older cows have higher antibody levels because they are likely to be exposed more to a greater number of antigens in their life (Larson et al., 1980). Low NAAb in plasma of cows may result from transfer of antibodies from blood to colostrum. Furthermore, IgM NAAb in plasma of calves before colostrum intake were lower in plasma of calves from older cows. In man, several genes (e.g. V_H genes) encoding some NAAb in the mother were shown to predominate in the newborn (Lacroix-Desmazes et al., 1998). In mice, the level of IgM binding self-antigens was high in plasma during fetal development and in newborns (Nobrega et al., 1996). Yet, we have no explanation how parity of the cows may affect the level of NAAb in plasma of their calves, because cows do not transfer antibodies to the fetus. Therefore, it is tempting to speculate that the level of NAAb in plasma of naïve calves may represent a health status biomarker of themselves or their mother.

To our best knowledge, this is the first study presenting the relationship between cows (plasma and colostrum) and their calves during the first weeks of life with respect to NAAb. In the current study, omitting the dry period resulted in lower levels of IgM binding CA in plasma of cows three weeks before calving and lower levels of IgG and IgM binding GD, CA, MYO or TRANS in colostrum compared with cows with a 30-d or a 60-d dry period. Earlier studies (Rastani et al., 2005; Verweij et al., 2014; Mayasari et al., 2015) reported that omission of the dry period reduced colostrum immunoglobulin contents and reduced NAb levels in colostrum at day of calving (Mayasari et al., 2015). Although titers of NAAb were detected in plasma of calves immediately after birth, the titers of NAAb in plasma of the calves before colostrum intake had no correlations with the averages of NAAb in plasma of their mother. The NAAb titers in plasma of calves increased directly after colostrum intake indicating that the high plasma NAAb titers in calves immediately after colostrum intake are likely derived from colostrum (Srinivasan et al., 1999). The levels of NAAb in plasma of calves during the first week of life were positively correlated with the level of NAAb in colostrum. This suggests that the level of antibodies in plasma of calves during the first weeks of life reflect the level of antibodies in colostrum and plasma of their mothers. It is suggested that Fc receptor mediated transfer for IgG and IgM (Lascelles, 1979; Hurley and Theil, 2011) has a role in transferring antibodies from plasma to colostrum and milk in dairy cows (Kacskovics et al., 2000). The level of NAAb in colostrum had no relation with the levels of NAAb in plasma of calves after two weeks of life. Low NAAb levels after colostrum intake coincided with a faster increase of NAAb levels within three till four weeks after birth. Thus, dry period treatment of cows had temporary transgenerational effects on calves. Similar with the increase of NAb to exo-antigens (Mayasari et al., 2015), calves may build up their own immune repertoire after the first two weeks of life regardless of their mother. In humans, high titers of NAAb binding self-antigens in plasma of the mother might reduce the incidence of diseases of their offspring (Tzioufas and Routsias, 2010). Our data indicate that the two weeks period in which calves received less colostrum from (0-d dry period) mothers did not result in enhanced disease sensitivity during the 12 weeks observation period.

5. CONCLUSIONS

In conclusion, IgG and IgM antibodies binding GD, CA, MYO and TRANS were detected in plasma of cows before calving, in colostrum and in plasma of calves before and after colostrum intake. The level of NAAb in plasma of calves before colostrum intake was not affected by the level of NAAb of their mother but were affected by parity of their mothers. After colostrum intake, calves from cows with no dry period (i.e. receiving the lower amount of maternal NAAb after drinking) had a rapid increase in NAAb to similar levels as the other calves. Our data indicate that; (1) low levels of maternal NAAb in colostrum affect production of neonatal NAAb positively and (2) independent of dry period length, healthy calves reach a specific level of NAAb from six weeks of age. Future studies should reveal the consequences of low or high levels of (other) NAAb from either neonatal or maternal of origin on disease sensitivity of calves later in life.

ACKNOWLEDGMENTS

The authors wish to thank the Dutch Dairy Board (PZ, Zoetermeer, the Netherlands), the Product Board Animal Feed (PDV, Zoetermeer, the Netherlands), and CRV (Arnhem, the Netherlands), for financing the experiment. The authors also wish to acknowledge their thanks for the scholarship of Novi Mayasari from the Directorate General of Higher Education, Ministry of National Education, Jakarta, Indonesia. The authors thank Gerrit Remmelink, Wilke Rijks, Xiaoyu Han, Mike Nieuwland, Joop Arts and staff of Dairy Campus (Lelystad, the Netherlands) for technical support during the experiment and the laboratory analysis.

REFERENCES

1. Avrameas, S., 1991. Natural autoantibodies: from 'horror autotoxicus' to 'gnothiseauton'. Immunol. Today 12, 154–159.
2. Badger, M.R., Price, G.D., 1994. The role of carbonic-anhydrase in photosynthesis.Annu. Rev. Plant Physiol. Plant Mol. Biol. 45, 369–392.
3. Banks, K.L., 1982. Host defense in the newborn animal. J. Am. Vet. Med. Assoc. 181,1053–1056.
4. Baumgarth, N., Tung, J.W., Herzenberg, L.A., 2005. Inherent specificities in naturalantibodies: a key to immune defense against pathogen invasion. SpringerSemin. Immunopathol. 26, 347–362.
5. Casali, P., Notkins, A.L., 1989. CD5+ B lymphocytes, polyreactive antibodies and thehuman B-cell repertoire. Immunol. Today 10, 364–368.
6. Cohen, I.R., 2013. Autoantibody repertoires, natural biomarkers, and systemcontrollers. Trend Immunol. 34, 620–625.
7. Cojocura, M., Cojocura, I.M., Silosi, I., 2009. The significance of naturalautoantibodies. M-edica 4, 22–25.
8. Conneely, M., Berry, D., Sayers, R., Murphy, J., Lorenz, I., Doherty, M., Kennedy, E.,2013. Factors associated with the concentration of immunoglobulin G in thecolostrum of dairy cows. Animal, 1–9.
9. Crichton, R.R., Charloteauxwauters, M., 1987. Iron transport and storage. Eur. J.Biochem. 164, 485–506.
10. Ehrenstein, M.R., Notley, C.A., 2010. The importance of natural IgM: scavenger,protector and regulator. Nat. Rev. Immunol. 10, 778–786.
11. Fink, K., Zellweger, R., Weber, J., Manjarrez-Orduno, N., Holdener, M., Senn, B.M.,Hengartner, H., Zinkernagel, R.M., Macpherson, A.J., 2008. Long-term maternalimprinting of the specific B cell repertoire by maternal antibodies. Eur. J.Immunol. 38, 90–101.
12. Haury, M., Sundblad, A., Grandien, A., Barreau, C., Coutinho, A., Nobrega, A., 1997.The repertoire of serum IgM in normal mice is largely independent of externalantigenic contact. Eur. J. Immunol. 27, 1557–1563.
13. Kacskovics, I., Wu, Z., Simister, N.E., Frenyó, L.V., Hammarström, L., 2000. Cloningand characterization of the bovine MHC class I-Like Fc receptor. J. Immunol.164, 1889–1897.
14. Kamada, N., Seo, S.G.Y., Chen Nunez, G., 2013. Role of the gut microbiota inimmunity and inflammatory disease. Nat. Rev. Immunol. 13, 321–335.
15. Khobondo, J.O., Nieuwland, M.G.B., Webb, L.E., Bokkers, E.A.M., Parmentier, H.K.,2015. Natural (auto) antibodies in calves are affected by age and diet. Vet. Q.35, 64–75.
16. Lacroix-Desmazes, S., Kaveri, S.V., Mouthon, L., Ayouba, A., Malanchère, E.,Coutinho, A., Kazatchkine, M.D., 1998. Self-reactive antibodies (naturalautoantibodies) in healthy individuals. J. Immnunol. Methods 216, 117–137.
17. Larson, B.L., Heary, H.L., Devery, J.E., 1980. Immunoglobulin production andtransport by the mammary gland. J. Dairy Sci. 63, 665–671.
18. Lascelles, A.K., 1979. The immune system of the ruminant mammary gland and itsrole in the control of mastitis. J. Dairy Sci. 62, 154–160.
19. Lutz, H.U., Binder, C.J., Kaveri, S., 2009. Naturally occurring auto-antibodies inhomeostasis and disease. Trend. Immunol. 30, 43–51.
20. Lutz, H.U., 2012. Naturally occurring antibodies. Adv. Exp. Med. Biol. 750, 7–10.
21. Madi, A., Hecht, I., Bransburg-Zabary, S., Merbl, Y., Pick, A., Zucker-Toledano, M.,Quintana, F.J., Tauber, A.I., Cohen, I.R., Ben-Jacob, E., 2009. Organization of theautoantibody repertoire in healthy newborns and adults revealed by systemlevel informatics of antigen microarray data. Proc. Natl. Acad. Sci. 106,14484–14489.
22. Matter, M.S., Ochsenbein, A.F., 2008. Natural antibodies target virus-antibodycomplexes to organized lymphoid tissue. Autoimmun. Rev. 7, 480–486.

23. Mayasari, N., de Vries Reilingh, G., Nieuwland, M.G.B., Remmelink, G.J., Parmentier,H.K., Kemp, B., van Knegsel, A.T.M., 2015. Effect of maternal dry period lengthon colostrum immunoglobulin content and on natural and specific antibodytiters in calves. J. Dairy Sci. 98, 3969–3979.
24. Mihaylova, N.M., Dimitrov, J.D., Djoumerska-Alexieva, I.K., Vassilev, T.L., 2008. Inflammation-induced enhancement of IgG immunoreactivity. Inflam. Res. 57,1–3.
25. Muller, L., Ellinger, D., 1981. Colostral immunoglobulin concentrations amongbreeds of dairy cattle. J. Dairy Sci. 64, 1727–1730.
26. Murphy, J.M., Hagey, J.V., Chigerwe, M., 2014. Comparison of serumimmunoglobulin G half-life in dairy calves fed colostrum, colostrum replaceror administered with intravenous bovine plasma. Vet. Immun. Immunopathol.158, 233–237.
27. Nguyen, T.T.T., Elsner, R.A., Baumgarth, N., 2015. Natural IgM preventsautoimmunity by enforcing B cell central tolerance induction. J. Immunol. 194,1489–1502.
28. Nobrega, A., Haury, M., Gueret, R., Coutinho, A., Weksler, M.E., 1996. Theage-associated increase in autoreactive immunoglobulins reflects aquantitative increase in specificities detectable at lower concentrations inyoung mice. Scandinavian J. Immunol. 44, 437–443.
29. Oliveira, R.A., Narciso, C.D., Bisinotto, R.S., Perdomo, M.C., Ballou, M.A., Dreher, M.,Santos, J.E.P., 2010. Effects of feeding polyphenols from pomegranate extracton health, growth, nutrient digestion, and immunocompetence of calves. J.Dairy Sci. 93, 4280–4291.
30. Ploegaert, T.C.W., Reilingh, G.D.V., Nieuwland, M.G.B., Lammers, A., Savelkoul,H.F.J., Parmentier, H.K., 2007. Intratracheally administeredpathogen-associated molecular patterns affect antibody responses of poultry.Poultry Sci. 86, 1667–1676.
31. Ploegaert, T.C.W., Tijhaar, E., Lam, T.J.G.M., Taverne-Thiele, A., van der Poel, J.J., vanArendonk, J.A.M., Savelkoul, H.F.J., Parmentier, H.K., 2011. Natural antibodies inbovine milk and blood plasma variability among cows, repeatability withincows, and relation between milk and plasma titers. Vet. Immunol.Immunopathol. 144, 88–94.
32. Poletaev, A., Osipenko, L., 2003. General network of natural autoantibodies asimmunological homunculus (Immunculus). Autoimmun. Rev. 2, 264–271.
33. Pribylova, J., Krausova, K., Kocourkova, I., Rossman, P., Klimesova, K., Kverka, M.,Tlaskalova-Hogenova, H., 2012. Colostrum of healthy mothers contains broadspectrum of secretory IgA autoantibodies. J. Clin. Immunol. 32, 1372–1380.
34. Pritchett, L.C., Gay, C.C., Besser, T.E., Hancock, D.D., 1991. Management andproduction factors influencing immunoglobulin G1 concentration in colostrumfrom holstein cows1. J. Dairy Sci. 74, 2336–2341.
35. Puscas, I., Coltau, M., Baican, M., Domuta, G., Hecht, A., 2001. Calcium, carbonicanhydrase and gastric acid secretion. Physiol. Res. 50, 359–364.
36. Quan, C.P., Berneman, A., Pires, R., Avrameas, S., Bouvet, J.P., 1997. Naturalpolyreactive secretory immunoglobulin A autoantibodies as a possible barrierto infection in humans. Infect. Immun. 65, 3997–4004.
37. Quintana, F.J., Cohen, I.R., 2004. The natural autoantibody repertoire andautoimmune disease. Biomed. Pharmacother. 58, 276–281.
38. Rastani, R.R., Grummer, R.R., Bertics, S.J., Gümen, A., Wiltbank, M.C., Mashek, D.G.,Schwab, M.C., 2005. Reducing dry period length to simplify feeding transitioncows: milk production, energy balance, and metabolic profiles. J. Dairy Sci. 88,1004–1014.
39. Schmitt, F.O., 1968. Fibrous proteins—neuronal organelles. P. Natl. Acad. Sci. U. S. A.60, 1092.Schultz, R.D., Dunne, H.W., Heist, C.E., 1973. Ontogeny of the bovine immuneresponse. Infect. Immun. 7, 981–991.
40. Schulz, K., Frahm, J., Meyer, U., Kersten, S., Reiche, D., Rehage, J., Dänicke, S., 2014.Effects of prepartal body condition score and peripartal energy supply of dairycows on postpartal lipolysis, energy balance and ketogenesis: an animal modelto investigate subclinical ketosis. J. Dairy Res. 81, 257–266.
41. Sinyakov, M.S., Avtalion, R.R., 2009. Vaccines and natural antibodies: a link to beconsidered. Vaccine 27, 1985–1986.
42. Srinivasan, A., Ni, Y., Tizard, I., 1999. Specificity and prevalence of natural bovineantimannan antibodies. Clin. Diagn. Lab. Immunol. 6, 946–952.

43. Stahl, D., Lacroix-Desmazes, S., Mouthon, L., Kaveri, S.V., Kazatchkine, M.D., 2000.Analysis of human self-reactive antibody repertoires by quantitativeimmunoblotting. J. Immunol. Methods 240, 1–14.
44. Thompson-Crispi, K.A., Miglior, F., Mallard, B.A., 2013. Genetic parameters fornatural antibodies and associations with specific antibody and mastitis inCanadian Holsteins. J. Dairy Sci. 96, 3965–3972.
45. Tzioufas, A.G., Routsias, J.G., 2010. Idiotype, anti-idiotype network ofautoantibodies: pathogenetic considerations and clinical application.Autoimmun. Rev. 9, 631–633.
46. Van Knegsel, A.T.M., Hostens, M., de Vries Reilingh, G., Lammers, A., Kemp, B.,Opsomer, G., Parmentier, H.K., 2012. Natural antibodies related to metabolicand mammary health in dairy cows. Prev. Vet. Med. 103, 287–297.
47. Van Knegsel, A.T.M., Remmelink, G.J., Jorjong, S., Fievez, V., Kemp, B., 2014. Effect ofdry period length and dietary energy source on energy balance, milk yield, andmilk composition of dairy cows. J. Dairy Sci. 97, 1499–1512.
48. Vassilev, T.L., Veleva, K.V., 1996. Natural polyreactive IgA and IgM autoantibodiesin human colostrum. Scand. J. Immunol. 44, 535–539.
49. Verweij, J.J., Koets, A.P., Eisenberg, S.W.F., 2014. Effect of continuous milking onimmunoglobulin concentrations in bovine colostrum. Vet. Immunol.Immunophatol. 160, 225–229.
50. Vollmers, H.P., Brändlein, S., 2009. Natural antibodies and cancer. New Biotechnol.25, 294–298.
51. Wardemann, H., Nussenzweig, M.C., 2007. B-cell self tolerance in humans. Adv.Immunol. 95, 83–109.

CHAPTER 9

Maternal and foetal immune responses of cattle following an experimental challenge with *Neospora caninum* at day 70 of gestation

Paul M BartleyEmail author, Stephen E Wright, Stephen W Maley, Colin N Macaldowie, Mintu Nath, Clare M Hamilton, Frank Katzer, David Buxton and Elisabeth A Innes

ABSTRACT

The immune responses of pregnant cattle and their foetuses were examined following inoculation on day 70 of gestation either intravenously (iv) (group 1) or subcutaneously (sc) (group 2) with live NC1 strain tachyzoites or with Vero cells (control) (group 3). Peripheral blood mononuclear cell (PBMC) responses to *Neospora* antigen and foetal viability were assessed throughout the experiment. Two animals from each group were sacrificed at 14, 28, 42 and 56 days post inoculation (pi). At post mortem, maternal lymph nodes, spleen and PBMC and when possible foetal spleen, thymus and PBMC samples were collected for analysis. Inoculation with NC1 (iv and sc) lead to foetal deaths in all group 1 dams (6/6) and in 3/6 group 2 dams from day 28pi; statistically significant ($p \leq 0.05$) increases in cell-mediated immune (CMI) responses including antigen-specific cell proliferation and IFN-γ production as well as increased levels of IL-4, IL-10 and IL-12 were observed in challenged dams compared to the group 3 animals. Lymph node samples from the group 2 animals carrying live foetuses showed greater levels of cellular proliferation as well as significantly ($p \leq 0.05$) higher levels of IFN-γ compared to the dams in group 2 carrying dead foetuses. Foetal spleen, thymus and PBMC samples demonstrated cellular proliferation as well as IFN-γ, IL-4, IL-10 and IL-12 production following mitogenic stimulation with Con A from day 14pi (day 84 gestation) onwards. This study shows that the generation of robust peripheral and local maternal CMI responses (lymphoproliferation, IFN-γ) may inhibit the vertical transmission of the parasite.

1. INTRODUCTION

The protozoan parasite *Neospora caninum* is a major cause of abortion and reproductive failure in cattle worldwide. The most common route of infection with *N. caninum* appears to be the transplacental (vertical) transmission of the parasite from mother to foetus; this may result in abortion or the birth of clinically normal but persistently infected offspring [1, 2]. Horizontal transmission of the parasite may occur in intermediate hosts through the ingestion of oocysts (shed by a definitive host i.e. dog) in contaminated feed and water [3], potentially leading to point source outbreaks (abortion storms) of neosporosis. Previous studies in cattle have shown that *N. caninum* infections can be maintained over several generations through vertical transmission of the parasite [1, 4], Moen *et al.* (1998) demonstrated that as a result of a primary infection, cattle were 3–7 times more likely to abort than uninfected animals [5]. However, animals that have aborted due to neosporosis are less likely to abort due to the parasite during subsequent pregnancies, compared to cows undergoing *N. caninum* infection during their first pregnancy [6], suggesting that a certain level of protective immunity builds following infection. Experimental data by Innes *et al.*, (2001) [7] demonstrated that exposure of cattle to *Neospora* prior to pregnancy protected against the vertical transmission of the parasite following an experimental challenge with *N. caninum* during pregnancy. Other factors influencing the outcome of *N. caninum* infections in pregnant cattle include; the quantity and duration of the parasitaemia [8], the parasite strain (as some have been shown to be more virulent than others, in cattle) [9], the immune status of the dam and the gestational age of the foetus at the time of infection [7, 8]. Experimental infections of pregnant cattle have shown that foetal death may occur when dams were challenged with *N. caninum* tachyzoites at day 70 of gestation [10, 11], while a challenge administered around mid gestation resulted in the vertical transmission of the parasite, but no foetal death [12, 13]. These observations would suggest that the timing of a parasitaemia during pregnancy is critical in the clinical outcome and will likely be influenced by both the maternal and foetal immune responses to the parasite. Work carried out by Williams *et al.* (2000); Collantes-Fernandez *et al.*, (2006) and Rosbottom *et al.*, (2007) [10, 14, 15] would support this conclusion.

The intracellular nature of *N. caninum* suggests that a cell-mediated immune (CMI) response is likely to be important to protect the host [12]. Increasing experimental data from pregnant cattle has confirmed this [7, 15, 16, 17, 18]. Work by Bartley *et al.*, (2004) [13] demonstrated a strong CMI response in dams and foetuses challenged with *N. caninum* on day 140 of gestation. Although, no foetal deaths were recorded, vertical transmission of the parasite occurred and the maternal and foetal immune responses appeared to contribute to the resolution of infection. Numerous other studies have illustrated the importance of a pro-inflammatory T-helper (Th)-1 type response, interferon-γ (IFN-γ) in particular has been shown to be crucial in controlling infection both *in vivo*[15, 16, 19] and *in vitro*[20, 21, 22]. The timing and location of these pro-inflammatory immune responses has also been shown to be critical to the clinical outcome of a *Neospora* infection in cattle. Work by Maley *et al.*, (2006) [23] demonstrated in cattle experimentally challenged with *Neospora* on day 70 of gestation; that the infiltration of large numbers of immune cells and increased levels of expression of IFN-γ mRNA in the placenta lead to foetal death and abortion.

In this study, we compared the maternal and foetal immune responses in cattle inoculated either intravenously (iv) or subcutaneous (sc), with live *N. caninum* (NC1 strain) tachyzoites at day 70 of gestation. A serial examination of the maternal and foetal immune responses was conducted looking at *Neospora* specific cell proliferation and cytokine production in PBMC and lymph node samples following experimental challenge.

2. MATERIALS AND METHODS

2.1 Animals, inoculum and experimental design

Twenty four pregnant Holstein-Friesian cattle aged 1.3 to 4 years and seronegative for *N. caninum, Toxoplasma gondii*, bovine viral diarrhoea virus, infectious bovine rhinotracheitis and *Leptospira hardjo* were assigned into three groups. Pregnancy and foetal viability was confirmed in all experimental animals by ultrasound scanning 36 days after insemination. On day 70 of gestation, group 1 dams (n = 8) received an intravenous (iv) inoculation in the right jugular vein of 5×10^8 live *N. caninum* (NC1 isolate) tachyzoites. Group 2 dams (n = 8) received a subcutaneous (sc) inoculation of 5×10^8 live *N. caninum* (NC1 isolate) tachyzoites over the left pre-femoral lymph node. Group 3 (n = 8), the control animals each received an iv inoculation of 5×10^{6}Vero cells. This dose of Vero cells was used, as it was the equivalent number of cells present in the parasite inocula. Blood was collected by weekly jugular venipuncture throughout the experiment for immunological analysis. Two animals from each group were sacrificed at days 14, 28, 42 and 56 post inoculation (pi). At post mortem samples of left pre-femoral lymph node (LPF), right pre-femoral lymph node (RPF), left uterine lymph node (LUL), right uterine lymph node (RUL), mesenteric lymph node (MLN), retropharyngeal lymph node (RLN), spleen and peripheral blood mononuclear cells (PBMC) were collected from each dam; When possible spleen, thymus and PBMC samples were collected from the foetuses.

2.2 Preparation of cells for immunological assays

Single cell suspensions of PBMC were prepared as previously described [7]. Samples from lymph nodes collected at post mortem were prepared as previously described [13]. In brief, excess fat was trimmed from the tissues, which were then cut into small pieces and placed in 10 ml wash buffer (Hank's balanced salt solution (HBSS) supplemented with 2% foetal bovine serum (FBS) (Labtech International, Ringmer, UK) 100 IU/ml penicillin and 50µg/ml streptomycin) (Northumbria Biologicals, Cramlington, UK), placed in a stomacher bag (Seward Medical, Northampton, UK) and homogenised for 10 seconds. The resultant cell suspension was decanted into a sterile universal through sterile lens tissue to remove clumps of cells, washed twice by repeated centrifugation at 260 × *g*, counted using a Neubauer haemocytometer and resuspended at a final concentration of 2×10^6 cells/ml in cell culture media (CCM) (Iscoves modified Dulbecco's media (IMDM) (Gibco, Paisley, UK) supplemented with 10% FBS, 100 IU/ml penicillin and 50µg/ml streptomycin).

2.3 Cell proliferation assays

Single cell suspensions of both PBMC and lymph node tissues were treated as previously described [13]. In brief, equal volumes (100 µl) of cells (2×10^6/ml) and antigen were added in quadruplicate to 96-well round bottom plates (Nunc, Roskilde, Denmark). Water-soluble *N. caninum* tachyzoite antigen (NCA) [7] was used at a final protein concentration of 1µg/ml, the T-cell mitogen concanavalin A (Con A) was used as a positive control at a final concentration of 5µg/ml, CCM alone was used as a negative control to determine the background level of proliferation. A Vero cell lysate antigen at 1 µg/ml was used as a control antigen. The cultures were incubated at 37°C in a humidified 5% CO_2 atmosphere for 5 days. The cultures were pulsed with 18.5 kBq ^{3}H Thymidine/well (Amersham

Biosciences, Little Chalfont, UK) for the final 18 hours, before being harvested onto glass-fibre filters (Canberra Packard, Meriden, CT, USA) and the cell-associated radioactivity was quantified using a MATRIX 96™ gas proportional counter.

2.4 Cytokine responses

Duplicate cell proliferation assays were prepared to those described above. Cell free supernatants were collected after 4 days incubation, to measure the levels of secreted cytokines (interferon-gamma (IFN-γ), interleukin-4 (IL-4), IL-10 and IL-12). The supernatant samples were stored at −20°C prior to analysis.

2.5 IFN-γ

Levels of IFN-γ production were quantified using a commercially available enzyme linked immunosorbent assay (ELISA) kit (CSL Veterinary, Parkville, Australia). A standard curve was generated using doubling dilutions of known quantities (ng/ml) of recombinant bovine IFN-γ (rBoIFN-γ) (Pfizer Animal Health, Parkville, Australia). Mean optical density (OD) values were plotted against ng/ml rBoIFN-γ and a standard regression curve was fitted to the data. Experimental samples were extrapolated against the standard regression curve to determine the levels of IFN-γ in the test samples.

2.6 IL-10

Levels of bovine IL-10 were quantified using an ELISA method as previously described [24]. In brief, 96-well ELISA plates (Greiner, Stonehouse, UK) were coated with 50µl (4µg/ml) per well with a primary anti-bovine IL-10 capture antibody and incubated at room temperature overnight. The plates were washed x 5 using phosphate buffered saline (PBS) supplemented with 0.05% Tween 20 (PBS-T) between each step, with the exception of the final TMB – H_2SO_4 stage. The plates were blocked at room temperature for 1 hour with PBS-T supplemented with 3% bovine serum albumin (BSA). Samples and standards (50µl each) were added and incubated at room temperature for 1 hour. Plates were then coated with (1µg/ml) secondary biotinylated anti IL-10 antibody (Diluted in PBS-T 1%BSA) (50µl per well) and incubated at room temperature for 1 hour. Streptavidin-horseradish peroxidase (HRP) (Dako Cytomation, Glostrup, Denmark) diluted 1:500 in PBS-T 1%BSA (50µl/well) was added and incubated at room temperature for 45 minutes. Colour was developed by the addition of TMB (3,3′,5,5′-tetramethylbenzidine) substrate (Insight Biotech. Ltd., Wembley, UK) (100µl/well) and incubated in the dark for 10–15 minutes. Reactions were stopped by adding 50μl/well 1 M H_2SO_4. The plates were read at 450/650 nm using a MRX II plate reader (Dynex, East Grinstead, UK). Doubling dilutions of known quantities of recombinant bovine IL-10 (rBoIL-10) were used to generate a standard regression curve against which the test sample concentrations were extrapolated.

2.7 IL-12

Interleukin-12 (IL-12) was quantified using the same method described above for the detection of IL-10, with the following changes. A primary anti IL-12 capture antibody (4μg/ml) was used along with a secondary biotinylated anti IL-12 antibody (8μg/ml). Known quantities of recombinant ovine IL-12 (rOvIL-12) were used as standards and standard regression curve was fitted to the data [25].

2.8 IL-4

Interleukin-4 (IL-4) was quantified using the same method described above for the detection of IL-10 and IL-12, with the following changes. A primary anti IL-4 capture antibody (6μg/ml) was used along with a secondary biotinylated anti IL-4 antibody (2μg/ml). Known quantities of recombinant bovine IL-4 (rBoIL-4) were used as standards and a standard regression curve was fitted to the data [26].

All primary and secondary antibodies used for the capture and detection of IL-4, IL-10 and IL-12 were purchased from AbD Serotec, (Oxford, UK). All rBoIL-4, rBoIL-10 and rOvIL-12 cytokines (Moredun Research Institute, Edinburgh, UK)

2.9 Foetal serology

At post mortem examination blood was drawn from the foetuses (when available) into non-heparinised evacuated tubes (Vacutainer, Becton Dickinson, Oxford,UK) and allowed to clot before centrifugation at 2000 *xg* for 15 minutes, the serum was removed and stored at −20°C prior to being tested for IgM and IgG to *N. caninum* by an indirect fluorescent antibody test (IFAT), as previously described [27]; an IFAT titre of ≥1:64 was considered positive.

2.10 Statistical analysis

The maternal cell proliferation data (PBMC, lymph nodes and spleen) and IFN-γ ELISA data were analysed with a linear mixed model, using a first-order autoregressive model to specify the temporal covariance structure. Both the proliferation and IFN-γ ELISA data were normalised by logarithmic transformation (base 10) prior to the analysis. The linear mixed model included the animal effect as a random effect and the treatment, day and the interaction effect of treatment and day as fixed effects. Parameters of the linear mixed models were estimated using the REML method and *p*-values were estimated using the modified *F*-statistic. If the *F*-statistic was statistically significant ($p \leq 0.05$), two-sided probabilities for each treatment comparison were obtained; these probabilities were then adjusted using a False Discovery Rate approach [28]. This value, denoted in this paper as p_f, therefore summarises the strength of evidence for there being a real difference in a way analogous to a standard *p*-value.

The foetal proliferation and IFN-γ data (PBMC and spleen) were analysed using non-parametric Kruskal-Wallis one way analysis of variance (ANOVA) using treatment as a grouping factor. No data was available for group 1 for the foetal thymus, (proliferation and IFN-γ ELISA) hence a two sample non-parametric Mann–Whitney test was conducted. All statistical analyses were carried out using GenStat 13th Edition software (VSN International, Hemel Hempstead, UK)

3. RESULTS

3.1 Clinical observations, pathology and foetal mortality

Foetal viability following experimental challenge is shown in Additional file 1. In group 1 (iv) viable foetuses were only seen on day 14 pi, both foetuses were found dead on day 28 pi and no foetuses were found on days 42 and 56 pi. In group 2 (sc) two viable foetuses were found on day 14 pi and one live and one dead foetus on days 28, 42 and 56 pi. In the group 3 (control) animals, two live foetuses were found at each time point. The maternal serology and histopathology data from this experiment is described in Macaldowie *et al.*, (2004) *Neospora caninum* parasites were only demonstrated in placental and foetal tissues from challenged animals carrying dead foetuses and they were associated with lesions [11].

3.2 Maternal PBMC

3.2.1 Cell proliferation

The results from the PBMC proliferation assays from group 1 (iv) demonstrated statistically significantly higher mean levels of antigen-specific proliferation on days 14 ($p_f = 0.020$), 21 ($p_f = 0.004$), 42 (p_f

= 0.043) and 56 pi (p_f < 0.001) compared to group 3 (control). The mean cell proliferation results from group 2 (sc) were found to be statistically significantly higher than group 3 on days 7 (p_f = 0.004), 14 (p_f = 0.011), 21 (p_f < 0.001), 42 (p_f < 0.001), 49 (p_f = 0.020) and 56 pi (p_f < 0.001). A comparison of the data from group 1 and group 2 showed that the mean antigen-specific proliferation in group 2 was higher than group 1 on day 7 pi (p_f= 0.003). At subsequent time points, the mean level of antigen specific PBMC proliferation of group 2 animals were higher in comparison to group 1 animals, though these differences were not statistically significant. (Additional file 2).

3.2.2 Cytokine responses

The levels of antigen-specific cytokine production was determined using the cell-free supernatants from maternal PBMC, lymph nodes and spleen samples following stimulation with NCA for 4 days

3.2.3 IFN-γ

The $\log_{10}$ transformed antigen specific-IFN-γ data from PBMC cultured with NCA (Additional file 3). Group 1 (iv) showed a rise in IFN-γ from days 7 to 21 pi and reached a peak on day 28 pi, then declined to baseline levels on day 56 pi. Though group 1 maintained an elevated level of IFN-γ compared to the group 3 (control) animals, there was no evidence of a difference between the mean IFN- γ levels of the two groups (p_f = 0.107). In group 2 (sc), levels of IFN-γ were consistently higher than other groups at all time points and reached a peak on day 21 pi. The mean level of IFN-γ from group 2 was statistically significantly different from that of group 3 (p_f = 0.005), though there was no evidence of a difference between the mean IFN-γ levels of group 1 and 2 (p_f = 0.118).

3.2.4 IL-4

In group 1 (iv) demonstrable antigen-specific IL-4 was observed in PBMC samples of one dam on day 14 pi (1.543U/ml), IL-4 production was seen in PBMC from both group 1 dams on day 28 pi (8.37U/ml and 2.54U/ml); while on days 42 and 56 pi PBMC samples from group 1 were below the detection level of the ELISA. Detectable levels of IL-4 were only seen in the PBMC from one of the group 2 (sc) animals on day 28 pi (carrying a dead foetus), these levels were much lower (0.518U/ml) than those seen in group 1, though no statistical differences were observed between groups 1 and 2. The levels of IL-4 in the group 3 (control) PBMC were either below the detection threshold of the ELISA at 0.114U/ml or only just detectable (data not shown).

3.2.5 IL-10

Demonstrable antigen-specific IL-10 was observed in PBMC from one group 2 (sc) animal carrying a dead foetus on 42 pi (0.286U/ml). All PBMC samples from groups 1 (iv) and 3 (control) were below the detection threshold (0.17U/ml) of the ELISA at all time points (data not shown).

3.2.6 IL-12

Levels of antigen-specific IL-12 were demonstrable in most PBMC samples collected from all three groups on days 14, 28 and 42 pi. Group 2 (sc) animals demonstrated higher levels of IL-12 than the other two groups. On day 28 pi the group 2 dam carrying the dead foetus demonstrated 39.1U/ml while, 94.3U/ml was recorded from the dam carrying the live foetus. By day 56 pi levels of IL-12 were either below the detection threshold (0.3U/ml) or only just detectable in all three groups (data not shown).

3.3 Maternal lymph nodes and spleen

3.3.1 Cell proliferation

The $\log_{10}$ transformed antigen specific proliferation in maternal lymph nodes and spleen samples following stimulation with NCA. On day 14 pi, samples of LPF from group 2 (sc) had significantly ($p_f = 0.010$) higher mean proliferation than group 3 (control). While mean proliferation from spleen samples from groups 1 (iv) and group 2 were significantly higher ($p = 0.05$ and $p = 0.047$ respectively) than group 3 (Additional file 4).

On day 28 pi, samples of LPF from group 1 and 2 had significantly higher mean proliferation than group 3 ($p_f = 0.006$ and $p_f = 0.025$ respectively), while samples of RLN and RPF from group 1 showed higher values than groups 2 and 3 ($p_f = 0.023$ and $p = 0.05$ respectively) and ($p = 0.033$ and $p = 0.015$ respectively) (Additional file 4).

On day 42 pi, samples of LPF from group 2 and 3 had significantly higher mean proliferation than the group 1 ($p_f = 0.005$ and $p_f = 0.031$ respectively) (Additional file 4).

On day 56 pi, group 2 demonstrated statistically significantly higher mean proliferation than group 3 from RPF ($p_f = 0.051$), LPF ($p_f < 0.001$), RLN ($p_f = 0.023$) and RUL ($p = 0.021$) and MLN ($p = 0.007$) samples. Group 2 also demonstrated significantly higher mean proliferation than group 1 for LPF samples ($p_f < 0.001$), while group 1 had significantly higher mean proliferation than group 3 for RLN ($p_f = 0.023$) and RUL ($p = 0.021$) samples (Additional file 4).

3.3.2 Cytokine responses

3.3.2.1 IFN-γ

Following the stimulation of cells from spleen and lymph nodes with NCA for 4 days; group 1 (iv) and group 2 (sc) both showed detectable levels of antigen specific-IFN-γ (ng/ml) in all samples tested from day 14 pi onwards, while group 3 (control) produced almost undetectable levels of IFN-γ at each of the time points tested (Additional file 5). The mean levels of IFN-γ from group 1 was statistically significantly higher than group 3 for RPF ($p_f = 0.001$), RUL ($p_f < 0.001$), LUL ($p_f = 0.013$), RLN ($p_f = 0.003$) and spleen ($p_f = 0.001$). Similarly, the mean levels of IFN-γ from group 2 were statistically significantly higher than group 3 for RPF ($p_f = 0.004$), LPF ($p_f = 0.005$), RUL ($p_f = 0.005$), RLN ($p_f = 0.009$) and spleen ($p_f = 0.004$). There was no evidence in the present data that the mean IFN-γ values of group 1 and 2 were different for any of the lymph node or spleen.

3.3.2.2 IL-4

Group 1 (iv) showed demonstrable antigen-specific IL-4 from both dams for all spleen and lymph node samples (except for LPF and RPF – only one dam responded) on day 28 pi and in LUL of one dam on 42 pi. All samples tested from group 2 (sc) and group 3 (control) were either below the detection threshold of the ELISA at 0.114U/ml or only just detectable and hence no statistical analysis was undertaken (data not shown).

3.3.2.3 IL-10

Group 1 (iv) was the only group with demonstrable antigen-specific IL-10, this was seen in LUL (2.61U/ml) and RLN (1.27U/ml) samples on day 42 pi, All samples tested from group 2 (sc) and group 3 (control) were below the detection threshold of the ELISA and hence no statistical analysis was undertaken (data not shown).

3.3.2.4 IL-12

Group 1 (iv) had demonstrable levels of antigen-specific IL-12 in RPF, LUL, RUL, MLN, RLN and spleen on days 14, 28 and 42 pi, with the highest levels being observed on day 42pi (126.75U/ml, 210.55U/ml, 98.00U/ml, 114.71U/ml, 307.9U/ml and 137.09U/ml respectively). By day 56 pi the levels of IL-12 had dropped below the detection threshold of the ELISA (0.3U/ml). Group 2 (sc) demonstrated levels of IL-12 below or barely above the threshold in most samples on days 14 and 28 pi. Levels of IL-12 peaked on day 42 pi, in the RLN and spleen, (23.64U/ml and 33.23U/ml respectively); by day 56 pi levels of IL-12 had dropped below threshold. Group 3 (control) showed low but detectable levels of antigen specific IL-12 from samples at all time points. Due to paucity of data, no statistical analysis was undertaken (data not shown).

3.3.3 Foetal PBMC, thymus and spleen

3.3.3.1 Cell proliferation

Samples collected from the foetuses for immunological analysis are shown in Table 1. No *Neospora* antigen-specific cell proliferation was demonstrated from any of the samples from any of the foetuses tested.

Table 1: Foetal samples collected at post mortem examination

	Number of foetal samples collected/Total number of foetuses								
Group 1 (iv)				Group 2 (sc)			Group 3 (Control)		
Day pi	PBMC	Thymus	Spleen	PBMC	Thymus	Spleen	PBMC	Thymus	Spleen
14	2/2	0/2	2/2	2/2	0/2	2/2	2/2	0/2	2/2
28	0/2‡	0/2‡	0/2‡	0/2	1/2‡	1/2‡	0/2	2/2	2/2
42	0/2†	0/2†	0/2†	1/2‡	1/2‡	1/2‡	2/2	2/2	2/2
56	0/2†	0/2†	0/2†	1/2‡	1/2‡	1/2‡	0/2	2/2	2/2

‡ – Foetus found dead *in utero* (No Samples able to be collected).† – No foetus found (due to resorption or abortion).

From day 14 pi (day 84 of gestation) PBMC samples showed proliferation following stimulation with Con A (Figure 1). On day 28 pi (day 98 of gestation) mitogenic proliferation was demonstrated in the spleen and thymus of foetuses from both group 2 (sc) and group 3 (control). On day 42 pi (day 112 of gestation) proliferative responses to Con A were continued to be seen in PBMC, thymus and spleen samples from the foetuses from both group 2 and group 3. On day 56 pi (day 126 of gestation) PBMC, spleen and thymus all showed proliferation following Con A stimulation (Figure 1). When the responses of the foetal tissues were compared, no statistical differences were observed between the groups in the levels of mitogenic proliferation.

3.3.4 Cytokine responses

Following stimulation with NCA, levels of antigen-specific IFN-γ IL-4, IL-10 and IL-12 were below detectable limits in all the foetal samples tested.

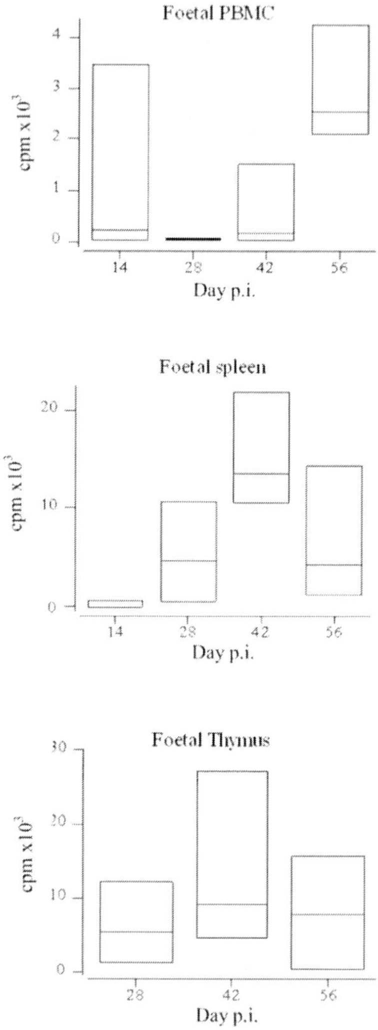

Figure 1 proliferative responses of foetal pbmc, spleen and thymus samples following stimulation with con a for 5 days.Samples of PBMC, spleen and thymus were collected at post mortem examination when available. Samples were stimulated with Con A for 5 days (37°C in a humidified 5% CO_2 atmosphere), with 18.5 kBq ^{3}H Thymidine/well being added for the final 18 hours, before being harvested onto glass-fibre filters. The box plots illustrate the combined counts per minute (cpm) ×10^3 values from the foetal samples from all three group s at each time point. The boxes represent the upper and lower values obtained for each sample, the horizontal line = the median value.

3.3.4.1 IFN-γ

The levels of IFN-γ following Con A stimulation on Day 14 pi (day 84 of gestation) were below detectable limits for all the samples tested. On days 28, 42 and 56 pi (days 98, 112 and 126 of gestation respectively) PBMC, spleen and thymus samples produced demonstrable quantities of IFN-γ, following stimulation with Con A. There were no significant differences observed in the levels of IFN-γ produced in Con A stimulated cells from any of the groups, at any of the time points tested (data not shown).

3.3.4.2 IL-4

Following Con A stimulation, on day 14 pi all samples tested from all three groups were below the detection threshold of the ELISA. On day 28 pi (day 98 of gestation), production of IL-4 was demonstrated in one group 3 (control) foetal thymus sample (0.117U/ml). On day 42 pi (day 112 of gestation) one group 3 foetal thymus (0.178U/ml) and both group 3 spleen (0.221U/ml, 0.214U/ml) samples produced demonstrable IL-4. On day 56 pi (day 126 of gestation), IL-4 was produced by the thymus samples of all three foetuses (group 2 and 3) and in the PBMC samples from both group 3 foetuses (data not shown).

3.3.4.3 IL-10

Following Con A stimulation, on day 14 pi all samples tested from all three groups were below the detection threshold of the ELISA. On day 28 pi (day 98 of gestation), IL-10 was produced by one group 3 (control) foetal spleen sample (0.587U/ml). On day 42 pi (day 112 of gestation) IL-10 was produced by group 2 (sc) and group 3 thymus samples (0.201U/ml and 0.194U/ml respectively); and in thymus, (0.359U/ml) spleen (0.947U/ml) and PBMC (1.681U/ml) samples from both group 2 and group 3 on day 56 pi (day 126 of gestation) (data not shown).

3.3.4.4 IL-12

Following Con A stimulation, IL-12 was expressed in foetal spleen and PBMC samples from all three groups from as early as day 14 pi (day 84 of gestation). Production of IL-12 continued to be demonstrated in all tissues tested from both group 2 (sc) and group 3 (control); peaking in the thymus (8.66U/ml) and spleen (12.17U/ml) on day 28 pi (day 98 of gestation) and in PBMC (12.92U/ml) on day 56 pi (day 126 of gestation) (data not shown).

3.3.5 Foetal serology

The results from the foetal anti-*Neospora* IgG IFAT show that on day 28 pi (day 98 of gestation) a serum sample collected from one of the foetuses from group 1 (iv) which was found dead *in utero*, gave a positive IFAT result with a titre of 1:128. On days 42 and 56 pi (days 112 and 126 of gestation respectively) no foetuses were present for sample collection. In groups 2 (sc) and 3 (control), serum samples were collected from live foetuses, these all tested negative for IgG IFAT (titre ≤1:64). All the samples tested from all the groups were negative (titre ≤1:64) for anti-*Neospora* IgM (data not shown).

3.3.6 Comparison of maternal immune responses in group 2 (sc) dams carrying live or dead foetuses

The immune responses of the dams in group 2 (sc) carrying live or dead foetuses were compared; The dam carrying a live foetus demonstrated considerably higher levels of cellular proliferation in LUL and RUL on days 28, 42 and 56 pi compared to the dam carrying the dead foetus; while on days 42 and 56 pi higher levels of proliferation were also seen in the PBMC of the dams carrying the live foetus compared to the dams carrying the dead foetuses. Though there were no significant differences seen in the levels of maternal proliferation, when the $\log_{10}$ data were compared using a linear mixed model, the dams carrying the live foetuses generally demonstrated higher levels of cell proliferation in PBMC and lymph nodes stimulated with NCA, compared to the dams carrying the dead foetuses (data not shown). Comparisons were also made in the levels of antigen specific-cytokine responses (IFN-γ, IL-4, IL-10 and IL-12) in PBMC and lymph nodes of dams carrying live foetuses compared to the dams carrying dead foetuses in group 2. On day 28 pi, greater levels of antigen specific-IFN-γ were seen in samples from the LPF, LUL, RUL and RPF from dams carrying live foetuses compared to the dams carrying dead foetuses (data not shown). On days 42 (Figure 2A) and 56 pi, (Figure 2B) dams carrying live foetuses demonstrated increased levels of IFN-γ in all samples except RLN (on day 56 pi) compared to the dams carrying dead foetuses. When IFN-γ data from days 28, 42 and 56 pi were combined and analysed using a linear mixed model, the IFN-γ responses in the RPF and spleen of dams carrying live foetuses were statistically significantly higher than in dams carrying dead foetuses ($p = 0.008$ and $p = 0.005$ respectively). Dams carrying live foetuses also showed higher levels of antigen specific-IL-12 on day 28 pi (RPF, LUL, RUL, spleen and PBMC). On day 42 pi, the dam with the live foetus had higher levels of antigen specific-IL-12 in RLN (23.64U/ml), spleen (33.27U/ml) and PBMC (94.30U/ml) compared to the dam carrying a dead foetus (1.23U/ml, 13.9U/ml, and 39.11U/ml respectively). None of the dams carrying live foetuses had demonstrable antigen specific-IL-4 at any time point, while one dam carrying a dead foetus on day 42 pi showed antigen-specific IL-4 production in LUL, RUL, MLN, spleen and PBMC. No differences were seen in levels of antigen-specific IL-10 production between the dams carrying live and dead foetuses at any of the time points tested.

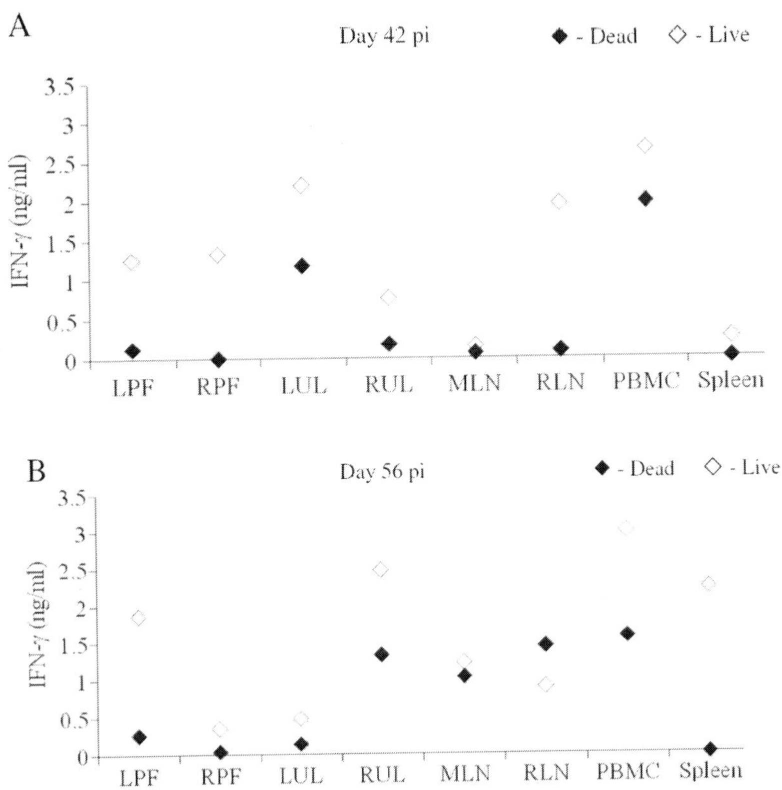

Figure 2 comparison of ifn-γ production by group 2 (sc) dams carrying live or dead foetuses. Maternal lymph node, spleen and PBMC samples were collected at post mortem examination. Following stimulation with NCA for 4 days (37°C in a humidified 5% CO_2 atmosphere), ELISA were performed to determine the concentration of IFN-γ produced. The data is expressed as ng/ml. -- Live, -- Dead.

4. DISCUSSION

This paper examines the maternal and foetal cell-mediated immune (CMI) responses of pregnant cattle experimentally inoculated with live *N. caninum* tachyzoites on day 70 of gestation (for a full description of the clinical observations and pathological findings see Macaldowie *et al.*, (2004) and Maley *et al.*, (2006) respectively) [11, 23]. The route of inoculation of *N. caninum* tachyzoites played an important role in the outcome of infection, with 100% of group 1 (iv) animals having dead, reabsorbed or aborted foetuses from day 28 pi, compared to only 50% of the group 2 (sc) animals. The dams in group 2 (sc) proved to be very interesting, allowing us to compare the CMI responses in dams carrying live compared to dead foetuses following an identical challenge. Although the number of animals in the experiment was limited these results will aid our understanding of a protective maternal immune response, which is critical for foetal survival, due to the immaturity of the foetal immune response at this stage of gestation. Mitogenic responses were detected in the foetuses from day 84 of gestation (day 14 pi) and anti-*Neospora* IgG was demonstrable from day 98 of gestation (day 28 pi).

4.1 Maternal immune response

The CMI responses observed in the dams in this study provides further evidence that cell-mediated immunity and in particular IFN-γ is important in protection against the rapidly multiplying intracellular tachyzoite stage of the parasite. Our data would suggest that a strong Th1 type cell-mediated immune response in both the peripheral and uterine lymph nodes as well as in PBMC, in particular in the group 2 (sc) dams carrying live foetuses, appears to have been sufficient to prevent the transplacental transmission of the parasite. When the immune responses of the dams in group 2 (sc) carrying live and dead foetuses were compared, it was seen that dams carrying live foetuses tended to exhibit stronger antigen-specific lymphocyte proliferation, significantly higher levels of IFN-γ as well as increased IL-12 production in samples of PBMC and lymph nodes, compared to dams in the same group (sc) carrying dead foetuses. Work carried out by Williams *et al*., (2000) [10] demonstrated that infection of pregnant cattle at 10 weeks gestation with *N. caninum* was accompanied by high levels of IFN-γ and lymphoproliferative responses indicating a profound Th1 helper T-cell like response in PBMC. Rosbottom *et al*., (2007) [15] observed proliferation of $CD4^+$ T-cells and the expression of IFN-γ and IL-4 in PBMC stimulated with *N. caninum* antigen in pregnant cattle challenged on day 70 of gestation.

However, when a strong immunological response occurs in the placenta during early gestation, this can be detrimental to the pregnancy. Davies *et al*., (2004) [29] demonstrated that the activation of the maternal mucosal immunity within the uterus during early placentome development, may lead to immune mediated abortion. During our study, the group 2 (sc) animals carrying dead foetuses (examined by Maley *et al*. (2006) [23]) demonstrated infiltration of large numbers of T lymphocytes, γδ T-cells, NK cells and IFN-γ mRNA, leading to extensive immune mediated damage (necrosis) of the placenta, compared to the dams carrying the live foetuses who showed mild or no infiltration of the immune cells, no demonstrable IFN-γ mRNA and consequently little or no immune mediated placental damage. Similar observations were made by Rosbottom *et al.*, (2008) [30] who found in pregnant cattle, following a challenge with *N. caninum* in early gestation, foetal death was associated with extensive placental necrosis, which corresponded to $CD4^+$ T-cell and macrophage infiltration and increased expression of IFN-γ.

4.2 Route of infection

Our data shows that the route of inoculation plays a critical role in the outcome of infection, with iv inoculation ($5x10^8$ live tachyzoites) leading to a greater incidence of foetal mortality than a sc inoculation with the same dose of parasites. This may be due in part to the haemotogenous nature of the iv inoculation, allowing the rapid transportation of the parasites to the placenta, before an effective maternal CMI response can be initiated to limit the spread of the tachyzoites. Comparing the immunological responses generated by the different routes of inoculation, group 2 animals (sc) demonstrated stronger proliferative responses in PBMC (by day 7 pi) and many of the lymph nodes (by day 14 pi) than the group 1 (iv) animals. Superior priming of the immune responses in the group 2 (sc) animals appears to have prevented the transplacental transmission of the parasite in 50% of cases. The animals in group 2 (sc) carrying live foetuses may have initiated an innate immune response more rapidly than the group 1 (iv) animals. This initiation is likely through the interaction of pathogen associated molecular patterns (PAMP) with toll like receptors (TLR) on the surface of antigen presenting cells (APC); this interaction induces a signalling pathway that will in turn lead to the activation of APC's and the production of pro-inflammatory cytokines, as well as leading to the maturation and migration of dendritic cells (DC) to local draining lymph nodes [31]. This APC migration may have been more rapid in the group 2 (sc) animals, as they were inoculated over a lymph node (left pre femoral lymph node). To date there is little experimental evidence demonstrating a role for the innate immune response in bovine neosporosis; work by Boysen *et al.*, (2006) [20] and Klevar *et al.*, (2007) [32] demonstrated a role for Natural killer (NK) cells during early infection in calves, in producing IFN-γ, and observed NK cytotoxicity in *N. caninum* infected fibroblast cells. Work in mice by Dion *et al.*, (2011) [33] demonstrated that both DC and macrophages secrete IL-12 following exposure to viable *N. caninum* parasites, and that interactions between T lymphocytes and DC were involved in inducing IFN-γ production by T lymphocytes. Evidence from the closely related parasite *Toxoplasma gondii* has demonstrated an important role for innate mechanisms in a protective immune response including TLR [34, 35, 36]. Our experiment demonstrates through serial analysis; differences in the PBMC as well as lymph node responses of animals experimentally challenged by two different routes of inoculation (iv and sc).

These differences appear to have been sufficient to protect against the vertical transmission of the parasite in some sc inoculated animals, compared to the iv challenged animals where vertical transmission and foetal death occurred in all animals from day 28 pi onwards.

4.3 Foetal immunology

As ruminants have a syndesmochorial placentation, which does not usually allow the transplacental passage of maternal immune factors including immunoglobulins and cytokines [37], any immune responses detected in the foetus are likely to be due to a response to an active infection *in utero*. Our study showed that foetuses at day 98 of gestation (day 28 pi) are capable of mounting a humoral (IgG) immune response against *Neospora*. This finding would agree with work by Senogles *et al.*, (1979) [38] who demonstrated that at 3 months (approx 90 days) gestation bovine foetal PBMC contained B cells that labelled positively for the presence of IgG. By day 98 gestation, components of a cell-mediated immune response are also starting to develop. Our study demonstrated the presence of lymphoproliferative responses as well as the production of IFN-γ, IL-4, IL-10, and IL-12 being observed in foetal spleen, thymus and PBMC samples following mitogenic (Con A) stimulation. Schultz *et al.*, (1973) [39] demonstrated in cattle the development of follicles of lymphoid thymus from as early as day 42 of gestation, while the spleen appears to develop from day 55 of gestation; while, Senogles *et al.*, (1979) [38] demonstrated the presence of T-cells in the bovine foetal thymus, spleen and PBMC from 3 months of gestation. Work carried out by Hein *et al.*, (1988) [40] showed that by day 120 of gestation bovine foetal lymphocytes were well developed and capable of mitogenic stimulation as well as the production of IL-2. Our work would suggest that this development occurs by around day 100 of gestation. The lack of any *Neospora*-specific CMI or humoral responses in the live foetuses from the group 2 (sc) dams was due to the fact that the foetuses were naïve to *Neospora*, due to no vertical transmission having occurred. The lack of histopathological lesions or demonstrable *Neospora* antigen in the placentomes or foetal tissues [11] indicates that vertical transmission had not occurred in the group 2 (sc) dams carrying live foetuses.

Other studies looking at the development of foetal immunology in cattle experimentally infected with *Neospora* have demonstrated that foetuses from infected dams at day 131 of gestation are capable of producing both Th1 and Th2 type cytokines [13, 16]. Bartley *et al.* (2004) [9] showed that the foetuses from dams inoculated with *N. caninum* at mid gestation (day 140) are capable of mounting a specific cell-mediated immune response from 14 days post inoculation (day 154 gestation). While Andrianarivo *et al.*

(2001) [17] demonstrated strong CMI responses 9 weeks post challenge in bovine foetuses (days 219–231 of gestation).

4.4 Timing of infection

We demonstrate that experimental inoculation of cattle at day 70 of gestation can lead to foetal mortality between 14–28 days post inoculation (pi) (days 84–98 of gestation). Studies by Barr *et al.*, (1994), Williams *et al.*, (2000), Collantes-Fernandez *et al.* (2006) and Rosbottom *et al.*, (2007, 2008) [14, 30, 41, 42] have demonstrated that experimental infections with *N. caninum* in the first trimester of pregnancy leads to increased foetal mortality and higher parasite burdens and greater dissemination of parasites in foetal tissues than infections at mid or late pregnancy in cattle. Though Rojo-Montejo *et al.*, 2009b observed no demonstrable foetopathy in cattle challenged on day 70 of gestation with the *Neospora* isolate Nc Spain 1 H [43]. While, Williams *et al.*, (2000), Maley *et al.* (2003), Rosbottom *et al.*, (2007) and Almeria *et al.*, (2011) [10, 12, 15, 44] demonstrated that a primary challenge in naïve pregnant heifers around mid gestation led to the transplacental transmission of the parasite but no foetal death. The results from these studies demonstrate that the timing of a primary infection during pregnancy is critical to the survival of the foetus. As gestation progresses the severity of a foetal infection decreases but the chances of congenital infection increases.

5. CONCLUSIONS

Our study demonstrates the development of maternal and foetal immune responses in lymph nodes and PBMC following challenge with live *N. caninum* tachyzoites. The innate immune response in the group 2 (sc) dams carrying live foetuses appears to have lead to superior priming of a cell mediated immune response, which inhibited the vertical transmission of the parasite, compared to the group 2 dams carrying dead foetuses. The route of inoculation of *N. caninum* tachyzoites has an impact on the clinical outcome of pregnancy in cattle, with an iv route of inoculation resulting in greater incidence of foetal mortality and less efficient immune priming compared to a sc inoculation of the same number of parasites. Foetuses from day 84 of gestation are capable of producing cellular responses to the mitogen Con A; these include lymphocyte proliferation as well as cytokine (IFN-γ, IL-4, IL-10, and IL-12) production. Our data has also shown that by day 98 of gestation foetuses are able to mount a humoral (IgG) response to *Neospora*. However should vertical transmission of the parasite occur, the foetuses are still too immunologically immature to mount a protective immune response, resulting in foetal mortality.

ACKNOWLEDGEMENTS

The authors would like to acknowledge the Scottish Government Rural and Environmental Research and Analysis Directorate (RERAD) for funding this study. The rBo IL-4, rBo IL-10 and rOvIL-12 were provided by Prof. Gary Entrican and Sean Wattegedera (Moredun Research Institute, Edinburgh, UK,(RERAD funded)). The recombinant bovine IFN-γ was kindly supplied by Dr. Steven Jones, (Pfizer Animal Health, Parkville, Australia).

REFERENCES

1. Anderson ML, Andrianarivo AG, Conrad PA: Neosporosis in cattle. Anim Reprod Sci. 2000, 60–61: 417-431.
2. Bjorkman C, Johansson O, Stenlund S, Holmdahl OJ, Uggla A: Neospora species infection in a herd of dairy cattle. J Am Vet Med Assoc. 1996, 208: 1441-1444.
3. McAllister MM, Dubey JP, Lindsay DS, Jolley WR, Wills RA, McGuire AM: Dogs are definitive hosts of Neospora caninum. Int J Parasitol. 1998, 28: 1473-1478. 10.1016/S0020-7519(98)00138-6.
4. Piergili FD, Pasquali P, Diaferia M, Mangili V, Rosignoli L: Neospora caninum infection and congenital transmission: serological and parasitological study of cows up to the fourth gestation. J Vet Med B Infect Dis Vet Public Health. 2003, 50: 399-404. 10.1046/j.1439-0450.2003.00686.x.
5. Moen AR, Wouda W, Mul MF, Graat EA, van Werven T: Increased risk of abortion following Neospora caninum abortion outbreaks: a retrospective and prospective cohort study in four dairy herds. Theriogenology. 1998, 49: 1301-1309. 10.1016/S0093-691X(98)00077-6.
6. Thurmond MC, Hietala SK: Effect of congenitally acquired Neospora caninum infection on risk of abortion and subsequent abortions in dairy cattle. Am J Vet Res. 1997, 58: 1381-1385.
7. Innes EA, Wright SE, Maley S, Rae A, Schock A, Kirvar E, Bartley P, Hamilton C, Carey IM, Buxton D: Protection against vertical transmission in bovine neosporosis. Int J Parasitol. 2001, 31: 1523-1534. 10.1016/S0020-7519(01)00284-3.
8. Innes EA, Buxton D, Maley S, Wright S, Marks J, Esteban I, Rae A, Schock A, Wastling J: Neosporosis. Aspects of epidemiology and host immune response. Ann N Y Acad Sci. 2000, 916: 93-101.
9. Rojo-Montejo S, Collantes-Fernandez E, Regidor-Cerrillo J, Alvarez-Garcia G, Marugan-Hernandez V, Pedraza-Diaz S, Blanco-Murcia J, Prenafeta A, Ortega-Mora LM: Isolation and characterization of a bovine isolate of Neospora caninum with low virulence. Vet Parasitol. 2009, 159: 7-16. 10.1016/j.vetpar.2008.10.009.
10. Williams DJ, Guy CS, McGarry JW, Guy F, Tasker L, Smith RF, MacEachern K, Cripps PJ, Kelly DF, Trees AJ: Neospora caninum-associated abortion in cattle: the time of experimentally-induced parasitaemia during gestation determines foetal survival. Parasitology. 2000, 121 (Pt 4): 347-358.
11. Macaldowie C, Maley SW, Wright S, Bartley P, Esteban-Redondo I, Buxton D, Innes EA: Placental pathology associated with fetal death in cattle inoculated with Neospora

caninum by two different routes in early pregnancy. J Comp Pathol. 2004, 131: 142-156. 10.1016/j.jcpa.2004.02.005.

12. Almeria S, Araujo RN, Darwich L, Dubey JP, Gasbarre LC: Cytokine gene expression at the materno-foetal interface after experimental Neospora caninum infection of heifers at 110 days of gestation. Parasite Immunol. 2011, 33: 517-523. 10.1111/j.1365-3024.2011.01307.x.
13. Bartley PM, Kirvar E, Wright S, Swales C, Esteban-Redondo I, Buxton D, Maley SW, Schock A, Rae AG, Hamilton C, Innes EA: Maternal and fetal immune responses of cattle inoculated with Neospora caninum at mid-gestation. J Comp Pathol. 2004, 130: 81-91. 10.1016/j.jcpa.2003.08.003.
14. Collantes-Fernandez E, Rodriguez-Bertos A, Arnaiz-Seco I, Moreno B, Aduriz G, Ortega-Mora LM: Influence of the stage of pregnancy on Neospora caninum distribution, parasite loads and lesions in aborted bovine foetuses. Theriogenology. 2006, 65: 629-641. 10.1016/j.theriogenology.2005.06.003.
15. Rosbottom A, Guy CS, Gibney EH, Smith RF, Valarcher JF, Taylor G, Williams DJ: Peripheral immune responses in pregnant cattle following Neospora caninum infection. Parasite Immunol. 2007, 29: 219-228. 10.1111/j.1365-3024.2007.00936.x.
16. Almeria S, De MT, Dawson H, Araujo R, Dubey JP, Gasbarre LC: Cytokine gene expression in dams and foetuses after experimental Neospora caninum infection of heifers at 110 days of gestation. Parasite Immunol. 2003, 25: 383-392. 10.1046/j.1365-3024.2003.00645.x.
17. Andrianarivo AG, Barr BC, Anderson ML, Rowe JD, Packham AE, Sverlow KW, Conrad PA: Immune responses in pregnant cattle and bovine fetuses following experimental infection with Neospora caninum. Parasitol Res. 2001, 87: 817-825. 10.1007/s004360100442.
18. Williams DJ, Guy CS, Smith RF, Guy F, McGarry JW, McKay JS, Trees AJ: First demonstration of protective immunity against foetopathy in cattle with latent Neospora caninum infection. Int J Parasitol. 2003, 33: 1059-1065. 10.1016/S0020-7519(03)00143-7.
19. Staska LM, McGuire TC, Davies CJ, Lewin HA, Baszler TV: Neospora caninum-infected cattle develop parasite-specific CD4+ cytotoxic T lymphocytes. Infect Immun. 2003, 71: 3272-3279. 10.1128/IAI.71.6.3272-3279.2003.
20. Boysen P, Klevar S, Olsen I, Storset AK: The protozoan Neospora caninum directly triggers bovine NK cells to produce gamma interferon and to kill infected fibroblasts. Infect Immun. 2006, 74: 953-960. 10.1128/IAI.74.2.953-960.2006.
21. Nishikawa Y, Mishima M, Nagasawa H, Igarashi I, Fujisaki K, Otsuka H, Mikami T: Interferon-gamma-induced apoptosis in host cells infected with Neospora caninum. Parasitology. 2001, 123: 25-31.
22. Nishikawa Y, Makala L, Otsuka H, Mikami T, Nagasawa H: Mechanisms of apoptosis in murine fibroblasts by two intracellular protozoan parasites, Toxoplasma gondii and Neospora caninum. Parasite Immunol. 2002, 24: 347-354. 10.1046/j.1365-3024.2002.00476.x.
23. Maley SW, Buxton D, Macaldowie CN, Anderson IE, Wright SE, Bartley PM, Esteban-Redondo I, Hamilton CM, Storset AK, Innes EA: Characterization of the immune response in the placenta of cattle experimentally infected with Neospora caninum in early gestation. J Comp Pathol. 2006, 135: 130-141. 10.1016/j.jcpa.2006.07.001.
24. Kwong LS, Hope JC, Thom ML, Sopp P, Duggan S, Bembridge GP, Howard CJ: Development of an ELISA for bovine IL-10. Vet Immunol Immunopathol. 2002, 85: 213-223. 10.1016/S0165-2427(02)00007-7.
25. Hope JC, Kwong LS, Entrican G, Wattegedera S, Vordermeier HM, Sopp P, Howard CJ: Development of detection methods for ruminant interleukin (IL)-12. J Immunol Methods. 2002, 266: 117-126. 10.1016/S0022-1759(02)00113-8.
26. Hope JC, Kwong LS, Thom M, Sopp P, Mwangi W, Brown WC, Palmer GH, Wattegedera S, Entrican G, Howard CJ: Development of detection methods for ruminant interleukin (IL)-4. J Immunol Methods. 2005, 301: 114-123. 10.1016/j.jim.2005.04.010.

27. Buxton D, Maley SW, Thomson KM, Trees AJ, Innes EA: Experimental infection of non-pregnant and pregnant sheep with Neospora caninum. J Comp Pathol. 1997, 117: 1-16. 10.1016/S0021-9975(97)80062-X.
28. Benjamini Y, Hochberg Y: Controlling the false discovery rate: a practical and powerful approach to multiple testing. J Roy Stat Soc Series B. 1995, 57: 289-300.
29. Davies CJ, Hill JR, Edwards JL, Schrick FN, Fisher PJ, Eldridge JA, Schlafer DH: Major histocompatibility antigen expression on the bovine placenta: its relationship to abnormal pregnancies and retained placenta. Anim Reprod Sci. 2004, 82–83: 267-280.
30. Rosbottom A, Gibney EH, Guy CS, Kipar A, Smith RF, Kaiser P, Trees AJ, Williams DJ: Upregulation of cytokines is detected in the placentas of cattle infected with Neospora caninum and is more marked early in gestation when fetal death is observed. Infect Immun. 2008, 76: 2352-2361. 10.1128/IAI.01780-06.
31. Werling D, Jungi TW: TOLL-like receptors linking innate and adaptive immune response. Vet Immunol Immunopathol. 2003, 91: 1-12. 10.1016/S0165-2427(02)00228-3.
32. Klevar S, Kulberg S, Boysen P, Storset AK, Moldal T, Bjorkman C, Olsen I: Natural killer cells act as early responders in an experimental infection with Neospora caninum in calves. Int J Parasitol. 2007, 37: 329-339. 10.1016/j.ijpara.2006.11.002.
33. Dion S, Germon S, Guiton R, Ducournau C, Dimier-Poisson I: Functional activation of T cells by dendritic cells and macrophages exposed to the intracellular parasite Neospora caninum. Int J Parasitol. 2011, 41: 685-695. 10.1016/j.ijpara.2011.01.008.
34. Debierre-Grockiego F, Campos MA, Azzouz N, Schmidt J, Bieker U, Resende MG, Mansur DS, Weingart R, Schmidt RR, Golenbock DT, Gazzinelli RT, Schwarz RT: Activation of TLR2 and TLR4 by glycosylphosphatidylinositols derived from Toxoplasma gondii. J Immunol. 2007, 179: 1129-1137.
35. Egan CE, Sukhumavasi W, Butcher BA, Denkers EY: Functional aspects of Toll-like receptor/MyD88 signalling during protozoan infection: focus on Toxoplasma gondii. Clin Exp Immunol. 2009, 156: 17-24. 10.1111/j.1365-2249.2009.03876.x.
36. Miller CM, Boulter NR, Ikin RJ, Smith NC: The immunobiology of the innate response to Toxoplasma gondii. Int J Parasitol. 2009, 39: 23-39. 10.1016/j.ijpara.2008.08.002.
37. Osburn BI, MacLachlan NJ, Terrell TG: Ontogeny of the immune system. J Am Vet Med Assoc. 1982, 181: 1049-1052.
38. Senogles DR, Paul PS, Johnson DW, Muscoplat CC: Ontogeny of T cells, B cells and monocytes in the bovine foetus. Clin Exp Immunol. 1979, 36: 299-303.
39. Schultz RD, Dunne HW, Heist CE: Ontogeny of the bovine immune response. Infect Immun. 1973, 7: 981-991.
40. Hein WR, Shelton JN, Simpson-Morgan MW, Seamark RF, Morris B: Flow and composition of lymph from the ovary and uterus of cows during pregnancy. J Reprod Fertil. 1988, 83: 309-323. 10.1530/jrf.0.0830309.
41. Barr BC, Rowe JD, Sverlow KW, BonDurant RH, Ardans AA, Oliver MN, Conrad PA: Experimental reproduction of bovine fetal Neospora infection and death with a bovine Neospora isolate. J Vet Diagn Invest. 1994, 6: 207-215. 10.1177/104063879400600212.
42. Rosbottom A, Guy CS, Gibney EH, Smith RF, Valarcher JF, Taylor G, Williams DJ: Peripheral immune responses in pregnant cattle following Neospora caninum infection. Parasite Immunol. 2007, 29: 219-228. 10.1111/j.1365-3024.2007.00936.x.
43. Rojo-Montejo S, Collantes-Fernandez E, Blanco-Murcia J, Rodriguez-Bertos A, Risco-Castillo V, Ortega-Mora LM: Experimental infection with a low virulence isolate of Neospora caninum at 70 days gestation in cattle did not result in foetopathy. Vet Res. 2009, 40: 49-10.1051/vetres/2009032.
44. Maley SW, Buxton D, Rae AG, Wright SE, Schock A, Bartley PM, Esteban-Redondo I, Swales C, Hamilton CM, Sales J, Innes EA: The pathogenesis of neosporosis in pregnant cattle: inoculation at mid-gestation. J Comp Pathol. 2003, 129: 186-195. 10.1016/S0021-9975(03)00032-X.

CHAPTER 10

A proteomics-based identification of putative biomarkers for disease in bovine milk

S.E.C. van Altena[a,*], B. de Klerk[b], K.A. Hettinga[c], R.J.J. van Neerven[a,d], S. Boeren[e], H.F.J. Savelkoul[a], E.J. Tijhaar[a]

[a]*Cell Biology and Immunology Group, Wageningen University, P.O. Box 338, Wageningen, The Netherlands*

[b]*Animal Breeding and Genomics Centre, Wageningen University, P.O. Box 338, Wageningen, The Netherlands*

[c]*Dairy Science and Technology Group, Wageningen University, 6700 EV Wageningen, The Netherlands*

[d]*FrieslandCampina, Stationsplein 4, 3818 LE Amersfoort, The Netherlands*

[e]*Laboratory of Biochemistry, Wageningen University, 6700 EV Wageningen, The Netherlands*

ABSTRACT

The objective of this study was to identify and characterize potential biomarkers for disease resistance in bovine milk that can be used to indicate dairy cows at risk to develop future health problems. We selected high- and low-resistant cows i.e. cows that were less or more prone to develop diseases according to farmers' experience and notifications in the disease registration data. The protein composition of milk serum samples of these high- and low-resistant cows were compared using NanoLC–MS/MS. In total 78 proteins were identified and quantified of which 13 were significantly more abundant in low-resistant cows than high-resistant cows. Quantification of one of these proteins, lactoferrin (LF), by ELISA in a new and much larger set of full fat milk samples confirmed higher LF levels in low- *versus* high-resistant cows. These high- and low-resistant cows were selected based on comprehensive disease registration and milk recording data, and absence of disease for at least 4 weeks. Relating the experienced diseases to LF levels in milk showed that lameness was associated with higher LF levels in milk. Analysis of the prognostic value of LF showed that low-resistant cows with higher LF levels in milk had a higher risk of being culled within one year after testing than high-resistant cows. In conclusion, LF in milk are higher in low-resistant cows, are associated with lameness and may be a prognostic marker for risk of premature culling.

Keywords: Dairy cattle, Biomarker Lactoferrin Milk

1. INTRODUCTION

The objective of this study was to identify biomarkers for disease resistance in bovine milk, thereby providing a prognostic tool to indicate dairy cows at risk to develop future health problems. The last decades dairy farming in the Netherlands has changed enormously and the number of cows per farm increased with 40% during the last 10 years (CRV, 2015). Clinical mastitis, one of the major health problems in dairy farming, has an incidence of about 33 cases per 100 cows annually (Santman-Berends et al., 2015) with associated annual costs of approximately €61 to €97 per cow based on worldwide estimations (Hogeveen et al., 2011). Also fertility problems and lameness are important issues in dairy farming (Huxley, 2013; Weaveret al., 2007). About 75% of the diseases in dairy cows occur in the first month after calving (LeBlanc et al., 2006). Around parturition, the immune system is compromised and the feed intake does not meet the energy requirements of the cow resulting in a negative energy balance (NEB), which makes the cow susceptible for diseases (Ingvartsen and Moyes, 2013; LeBlanc et al., 2006; van Knegsel et al., 2007).

To monitor the health status of cows, several studies were performed to obtain specific biomarkers. For example the energy balance, and thereby the risk of developing disease, can be measured by the levels of not-esterified fatty acids (NEFA) and beta-hydroxybutyrate (BHBA) in blood (Ospina et al., 2010). Pre-partum NEFA serum levels were shown to be positively correlated with the risk of mastitis after parturition (Holtenius et al., 2004; Moyes et al., 2009b). High post-partum NEFA levels are also a predictor for clinical ketosis, retained placenta, metritis and displaced abomasum (Ospina et al., 2010). Acute phase proteins (APP) in cows, like haptoglobin and serum amyloid, are common markers for infection and inflammation (Ceciliani et al., 2012; Eckersall and Bell, 2010; Eckersall et al., 2006). Haptoglobin and mammary-associated serum amyloid A (M-SAA3) were consistently increased in milk and subsequently in blood after a *Staphylococcus aureus*-induced subclinical mastitis (Eckersall et al., 2001; Eckersall et al., 2006). In milk, an increase in somatic cell counts (SCC) or lactate dehydrogenase (LDH) are markers for mastitis (Åkerstedt et al., 2011; Hiss et al., 2007) and are now routinely tested. Furthermore, ketosis can also be determined in milk by the rise in BHBA levels.

The risk for development of important diseases in dairy cattle can thus be monitored by the levels of some of these markers, which are used for regular screening in dairy farming already. A regularly used marker like SCC is specifically related to detection of mastitis, but does not indicate other diseases. Therefore, we are aiming for prognostic markers in bovine milk that are related to diseases different than mastitis. Markers in milk are preferred since milk samples are already collected regularly for routine screening, in contrast to blood samples. Nowadays, hundreds of unique proteins can be identified in different fractions of bovine milk by mass spectrometry (Hettinga et al., 2011; Nissen et al., 2013). This makes a proteomics approach a valuable tool for discovery of novel biomarkers (Boehmer et al., 2010; Ferreira et al., 2013). Here, we use shotgun proteomics (NanoLC–MS/MS) to compare milk samples of cows with a good health history (high-resistant cows) to milk samples of cows with a poorer health history (low-resistant cows). In this study, we consider high-resistant cows as having a low susceptibility to the development of disease. Likewise low-resistant cows have a high susceptibility for disease development. To exclude the detection of acute disease related markers, all samples were taken from cows that had not experienced health problems in the preceding 4 weeks. With this approach, we aimed to identify novel candidate biomarkers in milk for disease incidence in dairy cows, which were then evaluated in a larger number of milk samples from high- and low-resistant cows, selected on basis of comprehensive disease registration data collected during this study.

2. MATERIALS AND METHODS

2.1. Samples

Milk samples were obtained from the Resilient Cattle ("Weerbaar Vee") biobank established in the Netherlands from 2010 until 2015. Cows from 29 conventional Dutch dairy farms were sampled multiple times during this period with the highest sampling frequency in 2014. In 2014, all full fat milk samples tested in the general milk recording and monitoring program were also stored in Resilient Cattle biobank at −80° C (5–14 samples per cow). The average number of dairy cows per farm was 114 with a range of 63–266 cows. From 2010 until 2015 comprehensive disease registration data of these cows were collected. The disease registration data were carefully documented as instructed and supervised by one veterinarian and contained information about the diseases, applied treatments and medications the cows received including data about the duration of disease and treatment, vaccinations and hoof trimming. Diseases were categorised by the same veterinarian into: mastitis, other udder problems, lameness, retained placenta, metritis (uterus-related problems), respiratory diseases, metabolic diseases (e.g. ketosis) and "other" (diseases different than the previous categories for example trauma due to accidents).

First, milk serum (whey) samples used for proteomics analysis were selected based on the farmer's opinion on perceived disease resistance of the cows in combination with disease registration data. At that moment, the average number of dairy cows per farm was 108 with a range of 59–230 cows. In consultation with the veterinarian, farmers were asked to identify their five highest and five lowest performing cows in terms of health problems, which are henceforward called high- and low-resistant cows. These cows were checked for health problems using the recorded disease registration data and milk recording data. Cows with somatic cell counts above 250,000 cells/ml were excluded to reduce the chance on including cows with an ongoing mammary infection (Sampimon et al., 2010). In addition, cows were excluded with annotations in the disease registration data within one month before or after the moment of sampling. High-resistant cows had no or only minor health problems, while low-resistant cows had recurrent health problems. Four high-resistant and four low-resistant cows were selected for proteomics analysis. These two group of cows were matched for age, parity, milk production, somatic cell counts (SCC), fat percentage, protein percentage and days in milk (DIM). At the moment of milk sampling all cows in both groups were clinically healthy based on disease registration and milk recording data. The individual milk serum samples were compared to a pooled of milk serum sample derived from 26 cows. This randomly chosen "average group" is matched to both groups of low-resistant and high-resistant individual samples in terms of age, parity, milk production, SCC, fat percentage, protein percentage and DIM.

The second and larger group of 43 high- and 36 low-resistant cows were selected based on the disease registration data obtained from the beginning of 2010 until summer 2014. Cows in the high- and low-resistant groups were matched for farm (n = 9), age, parity, milk production, SCC, fat percentage, protein percentage and DIM. Other inclusion criteria for the cows were: raised on the selected farms, born between 2008 and 2011, more than 30 days in lactation, production above the average production per farm and somatic cell count at sampling below 250,000 cells/ml. High-resistant cows had no annotations in the comprehensive disease registration data except for vaccinations. Farmers were carefully instructed and coached by the same veterinarian in keeping the disease registration accurate and up to date. Low-resistant cows had at least two annotations in the disease registration data (excepting regular vaccinations).

2.2. NanoLC–MS/MS

Milk serum samples were prepared by centrifugation at 1500*g* for 10 min at 10° C. The supernatant was collected (without fat layer) and diluted 1:1 in 0.05 M ammonium bicarbonate buffer pH = 8.0 (ABC buffer, NH_4HCO_3 in water), then ultra-centrifuged at 100,000*g* for 90 min at 30° C. The clear supernatant (milk serum) was collected and prepared for proteomics analysis as described by (Zhang et al., 2015b). Milk serum samples were treated using the filter-aided sample preparation (FASP) method (Wisniewski et al., 2009) to clean the samples and perform trypsin digestion. After trypsin digestion, the resulting peptides were labelled by dimethyl labelling (Lu et al., 2011). The amine-group of each peptide reacts with formaldehyde (for light label) or deuterated formaldehyde (for heavy label) forming a so called Schiff base, which is subsequently reduced by cyanoborohydride resulting in a light or heavy label attached to each peptide (Boersema et al., 2009). The milk serum samples from high- and low-resistant cows were individually labelled with a light label and compared to a pool of milk serum from 26 cows containing a heavy label. Protein quantity is expressed as a $\log_2$ ratio of the individual milk serum samples to the pooled milk serum sample. All eight individual samples can be compared with each other due to this labelling approach.

NanoLC–MS/MS analysis was performed as described by (Zhang et al., 2015a). Full scan positive mode FTMS spectra were measured between m/z 380 and 1400 on a LTQ-Orbitrap XL (Thermo electron, San Jose, CA, USA) in the Orbitrap at high resolution (60,000). CID fragmented MSMS scans of the four most abundant 2+ and 3+ charged peaks in the FTMS scan were recorded in data dependent mode in the linear trap (MSMS threshold = 5.000, 45s exclusion duration for the selected m/z +/−25 ppm).

LC–MS runs with all MSMS spectra obtained were analysed with MaxQuant 1.3.0.5 (Cox and Mann, 2008) using default settings for the Andromeda search engine (Cox et al., 2011) except that extra variable modifications were set for de-amidation of N and Q.

A bovine database downloaded from Uniprot (http://www.uniprot.org) was used together with a contaminants database that contains sequences of common contaminants as for instance: BSA (P02769, bovine serum albumin precursor), Trypsin (P00760, bovine), Trypsin (P00761, porcine), Keratin K22E (P35908, human), Keratin K1C9 (P35527, human), Keratin K2C1 (P04264, human) and Keratin K1CI (P35527, human).

In the MaxQuant analysis, only peptides and proteins with a false discovery rate (FDR) of less than 1% were accepted. Reversed hits were deleted from the MaxQuant protein groups result table. The protein groups result was filtered further to keep only proteins with at least 2 identified peptides of which at least one should be unique and at least one should be unmodified.

2.3. Lactoferrin ELISA

The bovine lactoferrin ELISA was performed according to manufacturer instructions (Bovine Lactoferrin ELISA Quantitation Set, Bethyl Laboratories Inc., Montgomery USA). Milk samples were diluted 500, 1000 or 4000 times in Blocking Reagent (Roche Applied Science, Mannheim, Germany). All samples and standards were prepared and measured in duplicate at 450 nm using a Filtermax F5 Plate Reader (Molecular Devices, Sunnyvale, California). The within assay coefficient of variation is <5% (Soyeurt et al., 2012).

2.4. Gamma glutamyltransferase 1 ELISA

Levels of gGT1 were determined using the bovine gamma glutamyltransferase 1 ELISA kit according to the manufacturers' instructions (NovaTeinBio, Woburn, USA).

2.5. Statistical analysis

Groups of high- and low-resistant cows were matched for age, parity, milk production, SCC, fat percentage, protein percentage and DIM. The association of disease status or condition with LF levels were analysed with PROC MIXED in SAS 9.3 (SAS Inst. Inc. Cary, NC). To approximate normality, the natural logarithm (ln) of LF levels were calculated and analysed. Preliminary analyses indicated that the fixed continuous effect of a 305 days milk production in kg, age of the cows in years, DIM, the interaction between age and DIM, and the random class effect of farms (farm 1–9) significantly affected LF levels. These effects were, therefore, maintained as correction factors within the models (base model). Inspection of the distribution of the residuals and the Q-Q plot indicated 2 possible outliers which were LF levels below the detection limit. These observations were, therefore, removed from the dataset, resulting in 43 high-resistant and 34 low-resistant cows.

To assess whether disease status or condition status was associated with LF levels, disease and condition variables were categorized into a limited set of different classes as some categories only had 1 observation. For mastitis, lameness, metabolic diseases and retained placenta the data were converted to a binary trait consisting of either having a condition (1) or not having a condition (0). For metritis the data was categorized in three classes consisting of having metritis once (1), having metritis more than once (2) or not having metritis (0). The fixed class effect of disease status or condition status was then individually added to the base model in separate analyses. To assess whether significant differences exist between cows that are not ill and cows that are ill with or without a certain condition, cows were categorized for each individual condition as either not ill, ill with a condition, or ill without a condition. These categories were individually added to the base model in separate analyses. Results are displayed in vertical scatter plots as the original individual observation (uncorrected LF level in μg/ml) for each disease status and/or condition.

3. RESULTS

3.1. Comparison of milk serum samples of high- and low-resistant cows by proteomics

To characterize potential biomarkers for disease resistance in dairy cows, differences in milk serum proteins between 4 high- and 4 low-resistant cows were determined by NanoLC–MS/MS. In total, 78 proteins were identified and quantified in these samples (Supplementary Table 1). Each protein is represented in Fig. 1 as a $\log_2$ ratio of the protein abundance in high-resistant versus low-resistant cows. Log_2 protein ratios between −1 and 1 were considered as natural variation, which may be caused by, for example, differences in cow genetics and individual farms. The proteins outside these limits may indicate true variation between high- and low-resistant cows. The expression of 10 proteins was significantly ($p < 0.05$)

different between the high- and low-resistant cows (open squares in Fig. 1). All 10 proteins were 2.3–6 times more abundant in milk serum of low-resistant than high-resistant cows (Table 1). To reduce the chance of missing potential biomarkers, while testing only a small number of cows, initially also proteins with $0.05 < p < 0.07$ were included, which resulted in the inclusion of 3 additional proteins (open triangles in Fig. 1) and a total of 13 proteins. The cut-off at $p < 0.07$ was chosen to include potential biomarkers, but still outside the natural variation between cows ($\log_2$ protein ratios between −1 and 1). The proteins that were selected were detected in at least 3 out of 4 cows per group.

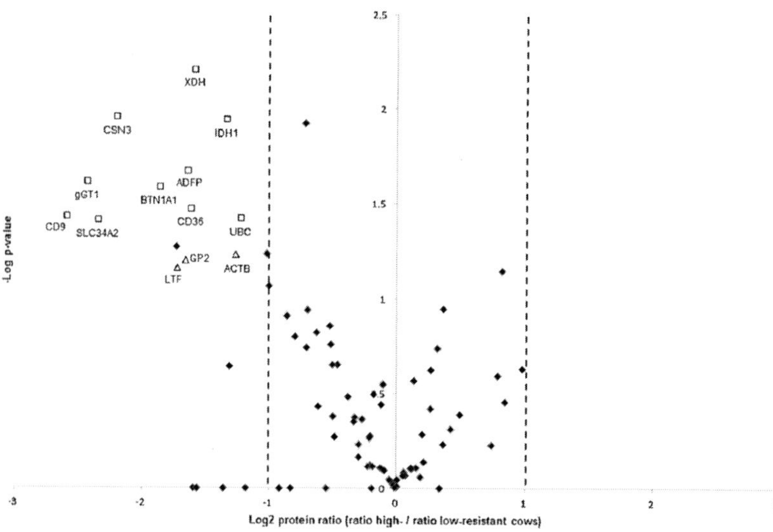

Figure. 1. Ratio of milk serum proteins of high- to low-resistant cows.

The diamonds, triangles and squares indicate the proteins found in milk serum samples from 4 high- and 4 low-resistant cows (n = 78). Proteins are represented as a $\log_2$ ratio of (ratio protein expression in high-resistant cows/protein expression in pooled milk sample) divided by (ratio protein expression in low-resistant cows/protein expression in pooled milk sample). Log_2 protein ratios between −1 and 1 are considered as natural variation. Open squares: proteins that differ between high-and low-resistant cows ($p < 0.05$). Open triangles: proteins that differ between high- and low-resistant cows ($0.05 < p < 0.07$). The abbreviations correspond to the proteins listed in Table 1.

Table 1. Milk serum proteins that differ between high- and low-resistant cows.

Protein	Full name	Uniprot ID	Subcellular location	Fold change	p-value
CD9	CD9 antigen	P30932	M	6.0	0.037
GGT1	Uncharacterised protein; Gamma-glutamyl transpeptidase 1, gamma-glutamyltransferase 1	G3N2D8	M, S	5.4	0.024
SLC34A2	Solute carrier family 34 (sodium-dependent phosphate transport protein 2B) member 2	F1N6D4	M	5.1	0.039
CSN3	Casein kappa	P02668	E	4.6	0.011
BTN1A1	Butyrophilin subfamily 1 member A1	P18892	M	3.6	0.026
LTF	Lactotransferrin	P24627	E, S	3.3	0.069
GP2	Uncharacterised protein; Glycoprotein 2, zymogen granule membrane, pancreatic secretory granule membrane major glycoprotein	F1N726	E, S	3.1	0.063
PLIN2/ADFP	Perilipin 2, adipophilin, adipose differentiation-related protein	F1N1N6	M	3.1	0.021
CD36	CD36 molecule, platelet glycoprotein IV, thrombospondin receptor	P26201	M	3.1	0.034
XDH	Xanthine dehydrogenase/oxidase	F1MUT3	C, E	3.0	0.006
IDH1	Isocitrate dehydrogenase 1 (NADP +)	Q9XSG3	C	2.5	0.011
ACTB	Actin, cytoplasmic 1	P60712	C	2.4	0.060
UBC	Polyubiquitin	P0CH28	C	2.3	0.037

C = cytosol, E = extra-cellular space, M = membrane, S = secreted.

3.2. Selection potential biomarkers

The 13 proteins that differed between high- and low-resistant cows ($p < 0.07$) were judged for their potential as a biomarker for bovine health. Most proteins were membrane-bound and were therefore less suitable as a biomarker since sample preparation and protein quantification with a high-throughput technique like ELISA is more difficult (Table 1). We therefore selected two soluble proteins for further investigation: gamma-glutamyl transpeptidase 1 (gGT1) and lactoferrin (LF). Gamma-GT1 is an enzyme involved in the Meister glutamyl cycle and is responsible for the transfer of glutathione across the cell membrane to maintain homeostasis (Meister and Anderson, 1983). Furthermore, gGT1 levels in blood are related to liver performance and intoxication (Pompella et al., 2007). Gamma GT1 levels measured in urine can be an indicator for renal damage (Ferguson et al., 2008). The association with disease and damage makes gGT1 in milk an interesting candidate marker for disease resistance. Lactoferrin is a glycoprotein produced by glandular epithelial cells and neutrophils and is present in all body fluids. It is part of the innate immune system and has antimicrobial, antiviral, antifungal, anti-inflammatory and anti-oxidative properties, among others by iron sequestering (García-Montoya et al., 2012; Kanwar et al., 2015) and can also act as an acute phase protein (Tothova et al., 2014).

3.3. LF levels in milk of high- and low-resistant cows, selected on disease registration data

New, full fat milk samples were selected from the "Resilient cattle" biobank to determine if LF and gGT1 were suitable markers for health-related problems in cows. Milk samples of 43 high-resistant cows and 34 low-resistant cows were collected based on comprehensive disease registration data and milk recording data from January 2010 until July 2015. These cows did not have annotations the disease registration data in the month before and after sampling. LF and gGT1 levels were determined using commercially available ELISAs. The gGT1 levels in full fat milk samples were below the detection limit (data not shown) and were therefore not further investigated. Lactoferrin levels were detectable using ELISA and were expressed in µg per ml (Fig. 2). For statistical analysis, the LF levels were ln transformed and corrected for milk production, days in milk (DIM), age, farm and the interaction between DIM and age as these variables significantly influenced LF levels in milk. No association was found for LF levels and somatic cell counts (data not shown). The least squares means (corrected mean) for LF levels in full fat milk were 244 µg/ml and 311 µg/ml for high- and low-resistant cows, respectively, and were significantly higher in low-resistant cows ($p = 0.047$) (Fig. 2). This is consistent with the previous results obtained by NanoLC–MS/MS for the 8 milk serum samples. This implies that being high- or low-resistant is significantly associated with LF levels in full fat milk, even after correction for milk production, DIM, age, farm and the interaction between DIM and age.

LF levels were measured in full fat milk samples from 43 high- and 34 low-resistant cows using ELISA. LF levels were expressed as the original LF concentration in µg/ml. The horizontal line indicates the mean. Statistics were performed on the ln transformed and corrected data. (Correction for farm, 305 days production, age, days in milk (DIM) and the interaction between age and DIM.) LF levels were significantly higher in low-resistant cows than high-resistant cows (p = 0.047).

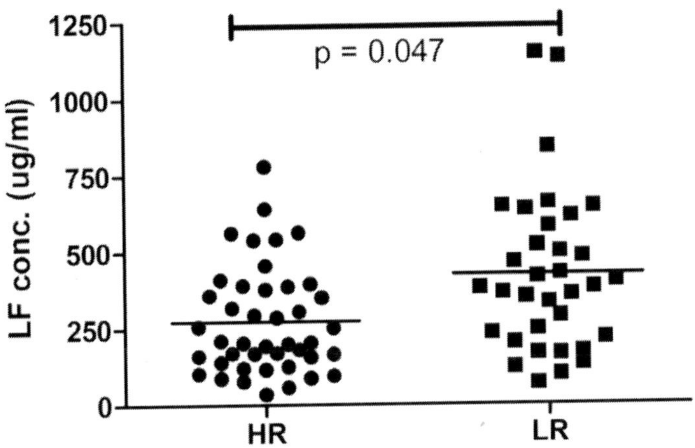

Figure. 2. LF levels in full fat milk samples of high- versus low-resistant cows.

3.4. Association of LF levels in milk and specific diseases

Next, we determined if the LF levels in full fat milk were associated with specific classes of diseases, independent of the resistance status of the cows. Cow performance was monitored by disease registration data and milk recording data. Diseases were categorised in: mastitis, other udder problems, lameness, retained placenta, metritis, respiratory diseases, metabolic diseases and "other" (diseases different than the previous categories or e.g. trauma). Cows with high disease resistance had none of the aforementioned health problems, while the low-resistant cows had at least two times an annotation in the disease registration data. Based on disease incidence, the association with LF levels in milk was determined for the following health-related problems in cows: mastitis, metabolic diseases, lameness, metritis and retained placenta (Supplementary Fig. 1).

We did not observe a significant difference in LF levels in milk between cows that had mastitis, metabolic diseases, metritis or a retained placenta in the past and cows without these particular diseases (Supplementary Fig. 1a–d). Nevertheless, cows that had suffered from lameness (n = 14) were significantly different from cows without lameness (p = 0.024) as cows with lameness had higher LF levels in their milk (Supplementary Fig. 1e).

3.5. Association of LF levels in milk with mastitis and lameness in high- and low-Resistant cows

Independent of being high- or low-resistant, LF levels were determined with respect to the different diseases. When the high- and low-resistance status was taken into account, the low-resistant cows that had suffered from mastitis showed similar LF levels in milk as the low-resistant cows without mastitis (Fig. 3a). Low-resistant cows with lameness had significantly higher LF levels in milk compared to high-resistant cows without lameness (p = 0.014) (Fig. 3b). LF levels appeared to be higher in low-resistant cows that had suffered from lameness compared to low-resistant cows without lameness, although, this difference was not significant (p = 0.13) (Fig. 3b).

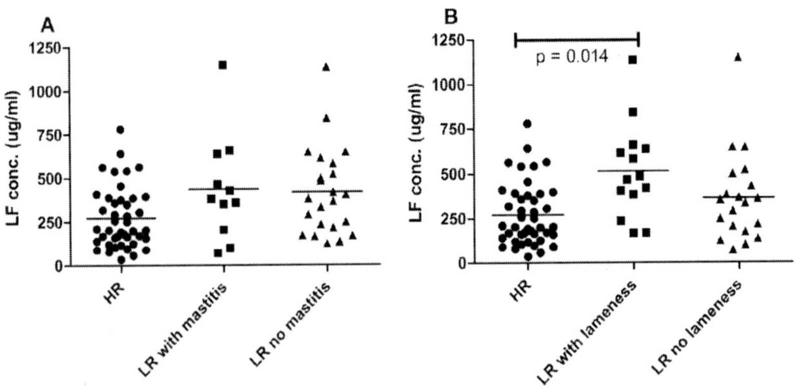

Figure. 3. LF levels in full fat milk related to mastitis and lameness.

LF levels were determined in full fat milk samples and related to disease thereby comparing high-resistant cows, low-resistant cows which had suffered from a specific disease and low-resistant cows without that particular disease. Individual LF levels were expressed in μg/ml. Statistics were performed on the ln transformed and corrected data. (3A) High resistant cows, low-resistant cows which had suffered from mastitis and low-resistant cows which had suffered from other diseases than mastitis had similar LF levels in milk. (3B) Low-resistant cows which had suffered from lameness had significantly higher LF levels in milk compared to high-resistant cows (p = 0.014).

3.6. Prognostic value lactoferrin

The LF levels were measured in milk samples collected during the summer of 2014. One year later, summer 2015, we investigated which of the high- and low-resistant cows were culled or not as an indicator for cow performance. Fig. 4 shows the LF levels in full fat milk of high- and low-resistant cows that were still alive and those that were culled within one year. The low-resistant cows that were culled within one year after sampling had on average higher LF levels (least squares means = 394.0 μg/ml) than high-resistant cows that were alive (265.2 μg/ml), high-resistant cows that were culled (166.1 μg/ml) and low-resistant cows that were still alive (288.6 μg/ml). This difference was significant between high-resistant cows that were still on the farm and low-resistant cows that were culled within one year ($p = 0.035$) and between high- and low-resistant cows that were culled ($p = 0.0017$) (Fig. 4). A similar trend was observed for low-resistant cows that were still alive and those that were culled ($p = 0.10$). In addition, high-resistant cows that were culled had lower LF levels in milk compared to high-resistant cows that were still alive ($p = 0.041$) and low-resistant cows that were still alive ($p = 0.017$). Taken together, these results indicate that low-resistant cows with higher LF levels had a higher chance of being culled.

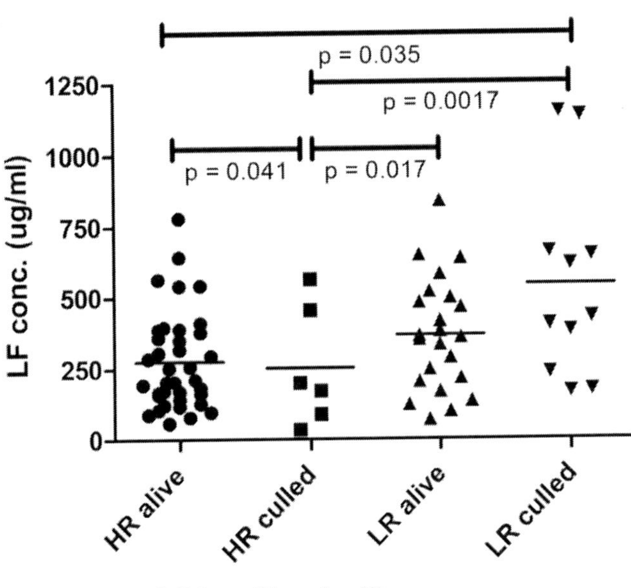

Figure. 4. LF levels in full fat milk and culling.

To determine the prognostic role of LF, LF levels in milk were related to culling. LF levels in milk were determined in high- and low-resistant cows that were culled or alive one year after sampling. Individual LF levels were expressed in μg/ml. Statistics were performed on the ln transformed and corrected data. LF levels were lower in high-resistant cows which were culled compared to high-resistant cows that were still alive ($p = 0.041$) or low-resistant cows that were still alive ($p = 0.017$). Low-resistant cows that were culled had higher LF levels in milk than high-resistant cows which were still alive ($p = 0.035$) or culled ($p = 0.0017$).

4. DISCUSSION

In this study we identified LF as a putative marker for lameness with prognostic value for early culling. Potential biomarkers were characterised in milk to indicate cows that are at risk to develop disease using a proteomics-based approach. The initial comparison of milk serum samples from 4 high- versus 4 low- disease-resistant cows revealed 13 proteins that were differentially expressed between both groups (ten $p < 0.05$ and three $p < 0.07$). These proteins were all more abundantly detected by proteomics in milk serum of low-resistant cows compared to high-resistant cows. Using ELISA we confirmed in a much larger set of full fat milk samples that the levels of one of these proteins, LF, were significantly increased in low-resistant cows. A positive association was observed between LF levels in milk and cows that had suffered from lameness. No significant associations were found for LF levels and metabolic disease, metritis, retained placenta or mastitis. The LF levels were also associated with culling and had a prognostic value for culling as low-resistant cows that were culled within one year had higher LF levels in milk.

The two potential biomarkers, gGT1 and LF, were selected from the proteins that were significantly different between high- and low-resistant cows, because these were soluble proteins which were related to disease in practice and literature. Elevated gGT1 levels in blood is an indicator for cholestasis and liver failure and is already used in practice. In literature, enhanced gGT1 activity in serum of *Rathi* cattle is a marker for stress and metabolic dysfunction (Kataria and Kataria, 2012). Increased gGT1 levels in urine were related with renal injury (Ferguson et al., 2008). Additionally, gGT1 levels were used as an indicator for colostrum uptake in young calves and lambs (Maden et al., 2003). Unfortunately, the gGT1 levels in milk were below the detection limit of the available capture ELISA, therefore the relevance of gGT1 levels in milk for bovine health could not be determined in this study. LF levels were shown to be significantly higher in low-resistant cows. Lactoferrin has different functions (García-Montoya et al.,

2012; Kanwar et al., 2015; Tothova et al., 2014) and plays e.g. an important role in the induction of innate immunity by sequestering iron and thereby limiting the availability of free iron, which is essential for bacterial growth. Therefore, it might appear counterintuitive that a protective agent like LF is increased in low-resistant cows, since higher LF levels imply a better protection against disease. However, a similar situation is seen for SCC. A higher SCC is usually caused by an influx of neutrophils in the udder to fight a bacterial infection (Paape et al., 2002). Similar to LF, the higher SCC level helps to control the infection, which is favourable for the cow. At the same time the higher SCC level is an indication of a recent or ongoing infection, which will occur more often in low-resistant cows.

Lactoferrin levels were significantly associated with lameness i.e. cows that had suffered from lameness in the past had higher LF levels in milk. Other studies showed relations between lameness and elevated levels of the acute phase proteins serum amyloid A, haptoglobin and fibrinogen in serum (Kujala et al., 2010; Tóthová et al., 2011). However, in our milk samples these proteins were not significantly different in high- and low-resistant cows or were not detected at all by NanoLC–MS/MS (Supplementary data 1), suggesting that LF was not produced due to an acute phase response. As we excluded cows with annotations in the months before and after sampling, this implies that LF could indicate lameness on a long term in contrast to e.g. serum amyloid A which indicates an ongoing infection.

Lactoferrin in milk is released by neutrophils and epithelial cells and its production is positively related with the influx of neutrophils and SCC (Lindmark-Månsson et al., 2006). Since SCC were on average equal between high- and low-resistant cows (data not shown), it is less likely that LF levels were increased due to neutrophil infiltration in the udder. Lactoferrin and haptoglobin levels in milk were also associated with the energy status of cows (Hiss et al., 2009). Cows with higher serum NEFAs levels around parturition, an indicator for increased fat mobilisation and a negative energy balance (NEB), had significantly higher LF and haptoglobin levels in milk for several weeks afterwards. Moreover, cows with more metabolic stress after calving had more health disorders (Gross et al., 2011), like lameness (Collard et al., 2000). In line with this, the low-resistant cows in our study had higher LF levels in milk, suffered at least two times from periparturient diseases, and showed a significant association between LF levels and lameness. Unfortunately, data on the cows‘ energy status was lacking in our study and a significant association between LF levels in milk and periparturient diseases was not observed (data not shown).

In contrast to other studies, no association was observed between LF levels in milk and SCC (Chaneton et al., 2013; Cheng et al., 2008; Liu et al., 2010). However, we selected for cows with SCC below 250.000 cells/ml and excluded cows with annotations in the disease registration data in the month before and after sampling, thereby excluding an association with mastitis. In addition, we found an increase in CD36, BTN1A1, IDH1 and kappa-casein in milk of low-resistant cows, while others showed these proteins to be decreased in milk from cows with ongoing mastitis (Huang et al., 2014; Moyes et al., 2009a). This supports the notion that the increased LF levels observed in this study were not caused by mastitis.

Initially, we found an association between LF levels and disease-resistance in cows, but a prognostic marker for disease-resistance would be more valuable. The prognostic value of LF was studied based on culling rates one year after our analysis. LF levels were significantly higher in low-resistant cows that were culled within one year compared to high-resistant cows. A similar trend was observed comparing low-resistant cows which were culled or still alive. This suggest that LF is an indicator for culling risk and might be a result of repeated health problems. High-resistant cows that were culled within one year surprisingly had significantly lower LF levels compared to the high-resistant cows that were still alive, suggesting that these cows were culled for other reasons than low-resistant cows. Although, it should be taken into account that the number of high-resistant that were culled were low (n = 6). Despite the variability in basal LF levels between cows (Stelwagen et al., 2009) and the short half-life of LF in milk (Kutila et al., 2002), we did show an association between LF levels in milk and lameness and culling. Based on our data, a clear cut-off value for LF to indicate "at risk" cows cannot yet be chosen, but they warrant a more extensive study to determine the full potential of LF as a biomarker to indicate cows with a higher risk to develop health problems.

5. CONCLUSION

In conclusion, using a proteomics approach, we have established a potential biomarker for cows with health problems. Low-resistant cows have higher lactoferrin levels in milk even though these cows were clinically healthy at the moment of sampling. The LF levels were positively associated with lameness and may be a long term indicator for this disease. Moreover, cows with high LF levels were more likely to have been culled one year after sampling, indicating that LF has potential as a prognostic biomarker for premature culling.

ACKNOWLEDGEMENTS

This study is part of a joint project Weerbaar Vee, which is funded by the Dutch Ministry of Economic Affairs (The Hague, the Netherlands), Dutch Dairy Product Board ("Productschap Zuivel", Zoetermeer, the Netherlands), Dutch Cooperative Cattle Improvement (CRV, Arnhem, the Netherlands), Dutch Federation of Agriculture and Horticulture ("LTO-Noord Fondsen", Zwolle, the Netherlands), GD Animal Health (Deventer, the Netherlands), and Wageningen University (Wageningen, the Netherlands). The authors also thank the herd owners for their collaboration and Danny de Koning for his excellent support in the statistical analysis of the data.

REFERENCES

1. Åkerstedt, M., Forsbäck, L., Larsen, T., Svennersten-Sjaunja, K., 2011. Naturalvariation in biomarkers indicating mastitis in healthy cows. J. Dairy Res. 78,88–96.
2. Boehmer, J.L., DeGrasse, J.A., McFarland, M.A., Tall, E.A., Shefcheck, K.J., Ward, J.L.,Bannerman, D.D., 2010. The proteomic advantage: label-free quantification ofproteins expressed in bovine milk during experimentally induced coliformmastitis. Vet. Immunol. Immunopathol. 138, 252–266.
3. Boersema, P.J., Raijmakers, R., Lemeer, S., Mohammed, S., Heck, A.J.R., 2009.Multiplex peptide stable isotope dimethyl labeling for quantitativeproteomics. Nat. Protoc. 4, 484–494.
4. Ceciliani, F., Ceron, J.J., Eckersall, P.D., Sauerwein, H., 2012. Acute phase proteins inruminants. J. Proteomics 75, 4207–4231.
5. Chaneton, L., Bontá, M., Pol, M., Tirante, L., Bussmann, L.E., 2013. Milk lactoferrin inheifers: influence of health status and stage of lactation. J. Dairy Sci. 96,4977–4982.
6. Cheng, J.B., Wang, J.Q., Bu, D.P., Liu, G.L., Zhang, C.G., Wei, H.Y., Zhou, L.Y., Wang,J.Z., 2008. Factors affecting the lactoferrin concentration in bovine milk. J. DairySci. 91, 970–976.
7. Collard, B.L., Boettcher, P.J., Dekkers, J.C., Petitclerc, D., Schaeffer, L.R., 2000.Relationships between energy balance and health traits of dairy cattle in earlylactation. J. Dairy Sci. 83, 2683–2690.
8. Cox, J., Mann, M., 2008. MaxQuant enables high peptide identification rates,individualized ppb-range mass accuracies and proteome-wide proteinquantification. Nat. Biotechnol. 26, 1367–1372.
9. Cox, J., Neuhauser, N., Michalski, A., Scheltema, R.A., Olsen, J.V., Mann, M., 2011.Andromeda: a peptide search engine integrated into the MaxQuantenvironment. J. Proteome Res. 10, 1794–1805.
10. CRV, 2015. Jaarstatistieken: MPR statistiek.
11. Eckersall, P.D., Bell, R., 2010. Acute phase proteins: biomarkers of infection andinflammation in veterinary medicine. Vet. J. 185, 23–27.
12. Eckersall, P.D., Young, F.J., McComb, C., Hogarth, C.J., Safi, S., Weber, A., McDonald,T., Nolan, A.M., Fitzpatrick, J.L., 2001. Acute phase proteins in serum and milkfrom dairy cows with clinical mastitis. Vet. Rec. 148, 35–41.
13. Eckersall, P.D., Young, F.J., Nolan, A.M., Knight, C.H., McComb, C., Waterston, M.M.,Hogarth, C.J., Scott, E.M., Fitzpatrick, J.L., 2006. Acute phase proteins in bovinemilk in an experimental model of staphylococcus aureus subclinical mastitis. J.Dairy Sci. 89, 1488–1501.

14. Ferguson, M.A., Vaidya, V.S., Bonventre, J.V., 2008. Biomarkers of nephrotoxic acutekidney injury. Toxicology 245, 182–193.
15. Ferreira, A.M., Bislev, S.L., Bendixen, E., Almeida, A.M., 2013. The mammary gland indomestic ruminants: a systems biology perspective. J. Proteomics 94, 110–123.
16. García-Montoya, I.A., Cendón, T.S., Arévalo-Gallegos, S., Rascón-Cruz, Q., 2012.Lactoferrin a multiple bioactive protein: an overview. Biochim. Biophys. Acta(BBA)—Gen. Subj. 1820, 226–236.
17. Gross, J., van Dorland, H.A., Bruckmaier, R.M., Schwarz, F.J., 2011. Performance andmetabolic profile of dairy cows during a lactational and deliberately inducednegative energy balance with subsequent realimentation. J. Dairy Sci. 94,1820–1830.
18. Hettinga, K., van Valenberg, H., de Vries, S., Boeren, S., van Hooijdonk, T., vanArendonk, J., Vervoort, J., 2011. The host defense proteome of human andbovine milk. PLoS One 6, e19433.
19. Hiss, S., Mueller, U., Neu-Zahren, A., Sauerwein, H., 2007. Haptoglobin and lactatedehydrogenase measurements in milk for the identification of subclinicallydiseased udder quarters. Vet. Med. (Praha) 52, 245–252.
20. Hiss, S., Weinkauf, C., Hachenberg, S., Sauerwein, H., 2009. Short communication: relationship between metabolic status and the milk concentrations ofhaptoglobin and lactoferrin in dairy cows during early lactation. J. Dairy Sci. 92,4439–4443.
21. Hogeveen, H., Huijps, K., Lam, T., 2011. Economic aspects of mastitis: newdevelopments. N. Z. Vet. J. 59, 16–23.
22. Holtenius, K., Persson Waller, K., Essén-Gustavsson, B., Holtenius, P., HallénSandgren, C., 2004. Metabolic parameters and blood leukocyte profiles in cowsfrom herds with high or low mastitis incidence. Vet. J. 168, 65–73.
23. Huang, J., Luo, G., Zhang, Z., Wang, X., Ju, Z., Qi, C., Zhang, Y., Wang, C., Li, R., Li, J.,2014. iTRAQ-proteomics and bioinformatics analyses of mammary tissue fromcows with clinical mastitis due to natural infection with Staphylococci aureus.BMC Genomics 15, 839.
24. Huxley, J.N., 2013. Impact of lameness and claw lesions in cows on health andproduction. Livest. Sci. 156, 64–70.Ingvartsen, K.L., Moyes, K., 2013. Nutrition, immune function and health of dairycattle. Animal 7, 112–122.
25. Kanwar, J.R., Roy, K., Patel, Y., Zhou, S.-F., Singh, M.R., Singh, D., Nasir, M., Sehgal, R.,Sehgal, A., Singh, R.S., 2015. Multifunctional iron bound lactoferrin andnanomedicinal approaches to enhance its bioactive functions. Molecules 20,9703–9731.
26. Kataria, N., Kataria, A., 2012. Use of serum gamma glutamyl transferase as abiomarker of stress and metabolic dysfunctions in Rathi cattle of arid tract inIndia. J. Stress Physiol. Biochem. 8.
27. Kujala, M., Orro, T., Soveri, T., 2010. Serum acute phase proteins as a marker ofinflammation in dairy cattle with hoof diseases. Vet. Rec. 166, 240–241.
28. Kutila, T., Pyorala, S., Kaartinen, L., Vahtola, K., Myllykoski, L., Saloniemi, H., 2002.Disposition kinetics of lactoferrin in milk after intramammary administration.J. Vet. Pharmacol. Ther. 25, 129–133.
29. LeBlanc, S.J., Lissemore, K.D., Kelton, D.F., Duffield, T.F., Leslie, K.E., 2006. Majoradvances in disease prevention in dairy cattle. J. Dairy Sci. 89, 1267–1279.
30. Lindmark-Månsson, H., Bränning, C., Aldén, G., Paulsson, M., 2006. Relationshipbetween somatic cell count, individual leukocyte populations and milkcomponents in bovine udder quarter milk. Int. Dairy J. 16, 717–727.
31. Liu, G., Zhang, C., Wang, J., Bu, D., Zhou, L., Zhao, S., Li, S., 2010. Canonicalcorrelation of milk immunoglobulins, lactoferrin concentration and dairy herdimprovement data of Chinese Holstein cows. Livest. Sci. 128, 197–200.
32. Lu, J., Boeren, S., de Vries, S.C., van Valenberg, H.J., Vervoort, J., Hettinga, K., 2011.Filter-aided sample preparation with dimethyl labeling to identify andquantify milk fat globule membrane proteins. J. Proteomics 75, 34–43.
33. Maden, M., Altunok, V., Birdane, F.M., Aslan, V., Nizamlioglu, M., 2003. Blood andcolostrum/milk serum gamma-glutamyltransferase activity as a predictor ofpassive transfer status in lambs. J. Vet. Med. B Infect. Dis. Vet. Public Health 50,128–131.

34. Meister, A., Anderson, M.E., 1983. Glutathione. Annu. Rev. Biochem 52, 711–760.
35. Moyes, K.M., Drackley, J.K., Morin, D.E., Bionaz, M., Rodriguez-Zas, S.L., Everts, R.E.,Lewin, H.A., Loor, J.J., 2009a. Gene network and pathway analysis of bovinemammary tissue challenged with Streptococcus uberis reveals induction of cellproliferation and inhibition of PPAR _ signaling as potential mechanism for thenegative relationships between immune response and lipid metabolism. BMCGenomics 10, 542.
36. Moyes, K.M., Larsen, T., Friggens, N.C., Drackley, J.K., Ingvartsen, K.L., 2009b.Identification of potential markers in blood for the development of subclinicaland clinical mastitis in dairy cattle at parturition and during early lactation. J.Dairy Sci. 92, 5419–5428.
37. Nissen, A., Bendixen, E., Ingvartsen, K.L., Rontved, C.M., 2013. Expanding the bovinemilk proteome through extensive fractionation. J. Dairy Sci. 96, 7854–7866.
38. Ospina, P.A., Nydam, D.V., Stokol, T., Overton, T.R., 2010. Evaluation ofnonesterified fatty acids and _-hydroxybutyrate in transition dairy cattle inthe northeastern United States: critical thresholds for prediction of clinicaldiseases. J. Dairy Sci. 93, 546–554.
39. Paape, M., Mehrzad, J., Zhao, X., Detilleux, J., Burvenich, C., 2002. Defense of thebovine mammary gland by polymorphonuclear neutrophil leukocytes. J.Mammary Gland Biol. Neoplasia 7, 109–121.
40. Pompella, A., Corti, A., Paolicchi, A., Giommarelli, C., Zunino, F., 2007. _-Glutamyltransferase, redox regulation and cancer drug resistance. Curr.Opin. Pharmacol. 7, 360–366.
41. Sampimon, O., van den Borne, B.H., Santman-Berends, I., Barkema, H.W., Lam, T.,2010. Effect of coagulase-negative staphylococci on somatic cell count inDutch dairy herds. J. Dairy Res. 77, 318–324.
42. Santman-Berends, I.M.G.A., Lam, T.J.G.M., Keurentjes, J., van Schaik, G., 2015. Anestimation of the clinical mastitis incidence per 100 cows per year based onroutinely collected herd data. J. Dairy Sci. 98, 6965–6977.
43. Soyeurt, H., Bastin, C., Colinet, F.G., Arnould, V.M., Berry, D.P., Wall, E., Dehareng, F.,Nguyen, H.N., Dardenne, P., Schefers, J., Vandenplas, J., Weigel, K., Coffey, M.,Theron, L., Detilleux, J., Reding, E., Gengler, N., McParland, S., 2012.Mid-infrared prediction of lactoferrin content in bovine milk: potentialindicator of mastitis. Animal 6, 1830–1838.
44. Stelwagen, K., Carpenter, E., Haigh, B., Hodgkinson, A., Wheeler, T.T., 2009. Immunecomponents of bovine colostrum and milk. J. Anim. Sci. 87, 3–9.
45. Tothova, C., Nagy, O., Seidel, H., Kovac, G., 2014. Acute phase proteins and their usein the diagnosis of diseases in ruminants: a review. Vet. Med.—Czech. 59,163–180.
46. Tóthová, C., Nagy, O., Seidel, H., Paulíková, I., Kováˇc, G., 2011. The influence of hoofdiseases on the concentrations of some acute phase proteins and othervariables of the protein profile in heifers. Acta Vet. (Beogr.) 61, 141–150.
47. van Knegsel, A.T., de Vries Reilingh, G., Meulenberg, S., van den Brand, H., Dijkstra,J., Kemp, B., Parmentier, H.K., 2007. Natural antibodies related to energybalance in early lactation dairy cows. J. Dairy Sci. 90, 5490–5498.
48. Weaver, A.D., Jean, G.S., Steiner, A., 2007. Bovine Surgery and Lameness. BlackwellPublishing Ltd., pp. 208–268.
49. Wisniewski, J.R., Zougman, A., Nagaraj, N., Mann, M., 2009. Universal samplepreparation method for proteome analysis. Nat. Methods 6, 359–362.
50. Zhang, L., Boeren, S., van Hooijdonk, A.C., Vervoort, J.M., Hettinga, K.A., 2015a. Aproteomic perspective on the changes in milk proteins due to high somatic cellcount. J. Dairy Sci. 98, 5339–5351.
51. Zhang, L., Boeren, S., Hageman, J.A., van Hooijdonk, T., Vervoort, J., Hettinga, K.,2015b. Bovine milk proteome in the first 9 days: protein interactions inmaturation of the immune and digestive system of the newborn. PLoS One 10,e0116710

CHAPTER 11

An immuno-epidemiological model for Johne's disease in cattle

Maia Martcheva[1], Suzanne Lenhart[2*], Shigetoshi Eda[3], Don Klinkenberg[4,7], Eiichi Momotani[5] and Judy Stabel[6]

[1]*Department of Mathematics, University of Florida, 358 Little Hall, Gainesville, FL 32611, USA*

[2]*Department of Mathematics, University of Tennessee, Knoxville, TN 37996, USA.*

[3]*Department of Forestry, Wildlife and Fisheries, University of Tennessee, Knoxville, TN 37996, USA*

[4]*Department of Farm Animal Health, Faculty of Veterinary Medicine, Utrecht University, Utrecht 3584CL, The Netherlands*

[5]*Department of Human-care, Tohto College of Health Sciences, Tokyo Medical and Dental University, Fukaya, Saitama 366-0052, Japan*

[6]*National Animal Disease Center, USDA, Ames, IA 50010, USA*

[7]*Centre for Infectious Disease Control, National Institute for Public Health and the Environment, Bilthoven, The Netherlands.*

ABSTRACT

To better understand the mechanisms involved in the dynamics of Johne's disease in dairy cattle, this paper illustrates a novel way to link a within-host model for *Mycobacterium avium* ssp. *paratuberculosis* with an epidemiological model. The underlying variable in the within-host model is the time since infection. Two compartments, infected macrophages and T cells, of the within-host model feed into the epidemiological model through the direct transmission rate, disease-induced mortality rate, the vertical transmission rate, and the shedding of MAP into the environment. The epidemiological reproduction number depends on the within-host bacteria load in a complex way, exhibiting multiple peaks. A possible mechanism to account for the switch in shedding patterns of the bacteria in this disease is included in the within-host model, and its effect can be seen in the epidemiological reproduction model.

1. INTRODUCTION

Johne's disease (JD) in dairy cattle is a chronic infectious disease in the intestines caused by the bacillus, *Mycobacterium avium* ssp. *paratuberculosis* (MAP). MAP in a contaminated environment infects cattle through oral route. Contaminated colostrum and milk from infected cows are important sources of infection for calves. Actual infection occurs when MAP bacilli are phagocytosed by M-cells covering the dome of the Peyer's patches [1] and transported to macrophages. In the early stages of the infection, some of the MAP will not be destroyed by macrophages and will grow in those cells until cellular immunity will be generated. To develop specific cellular immunity, macrophages differentiate into epithelioid cells and then intracellular growth of the MAP will be suppressed. Epithelioid cells form specific structures, granulomas, which act to restrict MAP growth inside and destroy them gradually. Some of the MAP in the granuloma will survive and enter a period of dormancy until reactivation. Reactivation of MAP will start slowly in the subclinical stage in which intermittent shedding of MAP will start. Granulomatous enteritis develops during the subclinical stage and accelerates in the clinical stage. Histological studies of infected areas reveal the presence of many infected macrophages, but very little extracellular bacteria have been observed [2].

This disease exhibits a variety of shedding patterns of MAP into feces, usually with a brief initial shedding period and then followed by a long latent period. In later stages, some infected cattle progress to weight loss, diarrhea, and reduced milk production. The mechanisms of the pathogen and the reaction of the immune system that cause this long latent period are not well understood. Also the connection of underlying immunology responses and the variable shedding patterns is difficult to explain. See the following papers in this issue related to these mechanisms and their relationship with data on shedding patterns, the growth of granulomas, and the length of the latent period [3,4].

To better understand the mechanisms of this disease and to later choose control strategies, we will illustrate through modeling how the immune system of infectious cows may have an effect at the epidemiological herd level. In particular, we are interested in understanding how the host immune responses influence the epidemiological reproduction number of

JD in a farm or a geographical region. We will study the complex immuno-epidemiology of JD through explicitly linking of epidemiological processes to the immune system dynamics.

Linking models at the two scales, immunological and epidemiological, has been done recently [5,6]. Different approaches have been used for such models, and some work has used decoupling assumptions to deal with the two scales separately [7-9]. In this paper, we are following the nested approach, introduced by Gilchrist, Coombs and Sasaki [10,11], in which the within-host model is independent of the between-host model but feeds into the between-host model. Our immuno-epidemiological model consists of two components: a time-since-infection dependent immunological model (within-host) and an epidemiological model (between hosts) whose transmission rates and virulence depend on the within-host variables. We will illustrate the basic ideas of linking a within-host model with a between-host model. For this illustration, we use a relatively simple model for the epidemiological dynamics. The representation of epidemiological reproduction number shows the dependence on specific within-host populations. Since there may be a type of switch in the within-host system that accounts for the variability of the shedding patterns, we give a way to illustrate a possible switching mechanism and show its effect on the epidemiological reproduction number. We use the stimulation rate of the immune response from infected macrophages as this switch.

In the next section, we introduce our within-host model, and then we discuss its estimated parameter values and stability analysis. We include a mechanism to switch from low bacterial load to high bacterial load. The third section gives the between-host model linked to the within-host model. We show how the epidemiological reproduction number can be represented in terms of the equilibrium points of the within-host model.

2. MATERIALS AND METHODS

2.1 A within-host immunological model of JD

We give some brief background of immunology of MAP leading to the within-host model.

2.2 Essential immunology of MAP

MAP enters the body of a cow from environmental sources, including fecal material, at birth or early in life through milk. After entry, MAP travels to the intestines of the affected animal and infects the macrophages located in the Peyer's patches [1]. In the early stages of the disease the bacterium prevents the macrophages from destroying it and the infected macrophages are kept dormant. The infection persists in the affected animal in the subclinical stage. Histological studies of infected areas reveal the presence of many infected macrophages but the amount of free bacteria has not been clearly documented. It appears that bacteria, that are released when an infected macrophage is destroyed, are immediately engulfed by new macrophages. To account for this observation, we include in the model uninfected macrophages and infected macrophages but no free bacteria. Immune responses to MAP are reviewed more thoroughly in [2] in this special issue. Early in the infection, a cellular immune response is activated through T cells and other cells. The cellular immune response is accompanied by proinflammatory cytokines, such as interferon-γ (IFN-γ) [12]. The cellular immune response is effective in controlling the infection, so that during the subclinical stage the shedding is often minimal or intermittent [13-16]. In the later stages of infection the cellular immune response wanes and a humoral response is activated in the form of B-cells and antibodies [17]. This response appears to be less effective in the controlling the infection and often increased shedding is observed at this stage of the infection. For this reason we include in the model the cellular immune response only.

2.3 The immunological model

Within-host mathematical models describe the dynamics of cells that participate in the infection process within a single individual. In our system of differential equations with τ as the underlying variable for time since infection, we denote by M_u the number of uninfected macrophages in the infected area, M_i the number of infected macrophages in the infected area, and C the number of T cells in the infected area. The rate of change of uninfected macrophages increases by the total recruitment rate r of new macrophages to the area. Furthermore, it decreases by the new infections and by the uninfected macrophages' per capita death rate d_1. The uninfected macrophages become infected by the bacteria that come out of infected macrophages that are bursting. We represent this as direct transmission from infected to uninfected macrophages. To account for possible saturation as the number of infected macrophages increases, we incorporate per capita infection rate that levels off with the increase in M_i. The proposed form of new infections per unit of time is a generalization of the mass action law, typically assumed in within-host models [18].

$$
\begin{cases}
\dfrac{dM_u}{d\tau} = r - \dfrac{\alpha M_u M_i}{1 + cM_i} - d_1 M_u, \\
\dfrac{dM_i}{d\tau} = \dfrac{\alpha M_u M_i}{1 + cM_i} - \dfrac{pCM_i}{1 + \varepsilon_1 M_i + \varepsilon_3 C} - \delta M_i, \\
\dfrac{dC}{d\tau} = \dfrac{kM_i}{1 + \varepsilon_2 M_i} - d_2 C,
\end{cases}
\tag{1}
$$

The rate of change of infected macrophages increases by the rate of newly infected macrophages and decreases through the killing rate of cellular immunity and the death rate of infected macrophages δ. In the M_i differential equation, this double-saturation form is based on the assumption that one killer T-cell can kill only finitely many infected macrophages, independently of how many are present [19]. In the special case when $\varepsilon_1=0$ and $\varepsilon_3=0$, the killing rate is a mass action law. The per capita killing rate of T-cells is saturating in infected macrophages. Saturating force of infection and killing rates have been previously used in modeling human tuberculosis [20]. The rate of change of C is increasing in saturating fashion with respect to infected macrophages and is decreasing at per capita death rate of T-cells d_2. We assume that in the presence of infected macrophages the immune response will be automatically activated, no additional conditions on the parameters are necessary. The dependent variables and the parameters are listed in Table 1.

2.4 Fitting the immune model to calf data and parameter estimation

Estimating parameters in within-host models can be achieved through fitting to time series data and estimates of lifespans. We use calf data and procedures reported in [21,22]. Neonatal Holstein dairy calves were obtained from status level 4 herds with no reportable incidence of JD in Minnesota at 1–2 days of age. Calves were housed in Biosafety Level-2 containment barns for the duration of the study and experimentally inoculated by feeding milk replacer containing 2.6×10^{12} live MAP obtained by scraping the ileal mucosa from a clinically infected cow as previously described in [21]. Calves were dosed on days 0, 7, and 14 of the study. All procedures performed on the calves were approved by the Institutional Animal Care and Use Committee (National Animal Disease Center (NADC), Ames, Iowa). The clinical isolate of MAP was obtained from the ileum of a clinical cow at necropsy that had shed high numbers of MAP in the feces.

Infection of calves was determined by measurement of shedding viable MAP in the feces and recovery of MAP from tissues as described in [21]. After 12 weeks of incubation at 39 °C, viable organisms were determined by counting the number of colonies on the agar slants. IFN-γ and interleukin-

10 (IL-10) were measured in cell-free supernatants to assess immune response to infection. Briefly, periperhal blood mononuclear cells (PBMCs) were isolated from the buffy coat fractions of whole blood and cultured at 2.0×10^6/mL in complete medium at 39 °C in 5% CO_2 in a humidified atmosphere. In vitro treatments consisted of no stimulation (medium only), concanavalin A (10 μg/mL), and a whole-cell sonicate of MAP (10 μg/mL) [22].

Table 1 Parameters and dependent variables and their interpretation

Variable or parameter	**Meaning**
$M_u(\tau)$	Number of uninfected macrophages at time τ
$M_i(\tau)$	Number of infected macrophages at time τ
$C(\tau)$	Number of immune cells at time τ
R	Total recruitment rate for uninfected cells
A	Infection rate of uninfected macrophages by infected macrophages
d_1	Natural death rate for uninfected macrophages
Δ	Death rate of infected macrophages
d_2	Clearance rate for immune cells
K	Stimulation rate of immune response from infected cells
ε_1	Reciprocal of half saturation constant for killing of infected macrophages
ε_2	Reciprocal of half saturation constant for immune response stimulation rate
ε_3	Reciprocal of half saturation constant for killing of infected macrophages
C	Reciprocal of half saturation constant for macrophages infection rate
P	Killing rate of infected macrophages by immune response.

Macrophages have a lifespan of several months. Wigginton and Kirschner [20] uses three months for lifespans of both uninfected and infected macrophages. We set d_1=0.25 months^{-1} and δ=0.3 months^{-1}. We take a longer lifespan of infected macrophages because MAP prevents the self-destruction of infected macrophages [23]. We also fix $M_u(0) = 5$and $M_i(0)=2$ so that they are consistent with the available data. We fit the remaining parameters to data. The data we use are colony forming unit (CFU) data and IFN-γ data for three calves (denoted by calf 9, calf 10 and calf 11). We complete one fit of the parameters for each data set for an individual calf. Since the model (1) does not have CFU or IFN-γ, we fit $c_1 M_i(\tau)$ to the CFU data and $c_2 C(\tau)$ to the IFN-γ data, where c_1 and c_2 are constants to be determined by the fitting. To reduce the number of parameters to be fitted we set ε_3=0 and ε_1=ε_2. The parameter values are given in Table 2 which were obtained by fitting with Mathematica. Since we were fitting more parameters than data points, proper identification of the parameters was not feasible and potentially more than one set of parameters may match the data with good fit. However, the parameters in Table 2 are consistent with the available data and similar between the calves. They identify sensible parameter ranges for simulations to illustrate the key points of linking with the epidemiological model.

Table 2 Estimated parameter values

Parameter	Calf 9	Calf 10	Calf 11	Units
R	114	600	1425	cells/month
A	0.47	0.4	0.6	Month^{-1}
d_2	0.5	0.4	0.3	month^{-1}
K	8.0	3.0	13.8	(cells × month)$^{-1}$
P	3.8	1.4	16.1	(cells × month)$^{-1}$
ε_1	0.105	0.111	0.65	cells^{-1}
ε_2	0.105	0.111	0.65	cells^{-1}
ε_3	0.0	0.0	0.0	cells^{-1}
C	0.0001659	0.0001	0.0001	cells^{-1}
$C(0)$	26.3	15.3	0.1	cells
c_1	0.005	0.01	0.01	unitless
c_2	0.048	0.057	0.05	unitless

2.5 Multiple equilibria and significance in terms of shedding

One of the key features of JD in cattle is that animals, when infected, may switch between shedding and non-shedding, thus identified because shedding data are collected as a positive (shedding) or negative (non-shedding) status. As an illustration, the shedding pattern for one of the calves in our data-set is depicted in Figure 1. This figure shows that animals may spend a significant amount of time in each status, during which the shedding is relatively constant. Such a pattern is characterized by the presence of long, stable periods of indefinite lengths, and possibly multiple stable periods with switching between. Mathematically such dynamical behavior is captured by the presence of multiple steady states in the dynamical model, at least two of which should be locally stable (solutions that start close to them, converge to them). In this section we show that the immunological model (1) exhibits such multiple steady states.

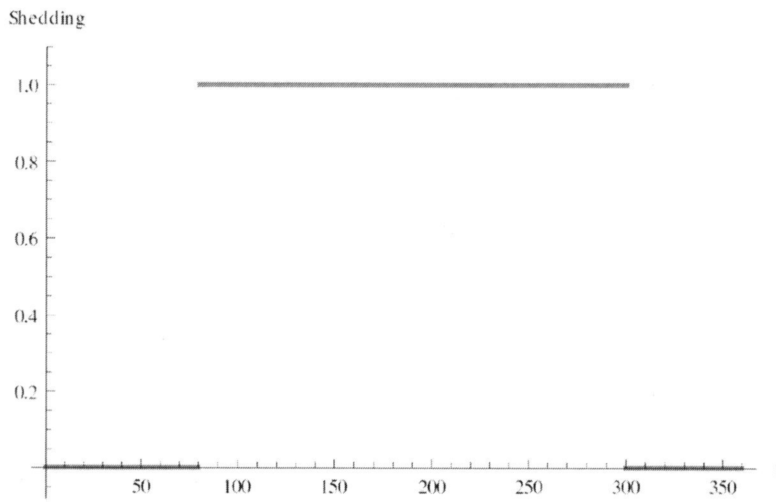

Figure 1 shedding pattern. The x-axis shows time-post-infection in days. Non-shedding is plotted as zero, shedding as one.

To understand the presence and absence of steady states (equilibria) and how their stabilities change we need to define the immunological reproduction number

$$\Re_0 = \frac{\alpha}{\delta} \cdot \frac{r}{d_1} \tag{2}$$

The reproduction number gives the numbers of secondarily infected macrophages that one infected macrophage will produce in an entirely uninfected macrophage population during its lifespan as infected. Since JD is a chronic infection, we expect that for realistic parameter values $\Re_0 > 1$. Model (1) always has an infection-free equilibrium $E_0 = \left(\frac{r}{d_1}, 0, 0\right)$, and if $\Re_0 > 1$, then there is at least one infected equilibrium $E_i = (M_u^*, M_i^*, \varepsilon_3 = 0, C^*)$. To find the equilibria we set the time derivatives in our system (1) equal to zero, resulting in the following equation in M_i^*:

$$\frac{\alpha r}{(d_1 + (d_1 c + \alpha)M_i^*)} = \frac{pkM_i^*}{\left(d_2(1+\varepsilon_1 M_i^*)(1+\varepsilon_2 M_i^*)\right)} + \delta.$$

The solution $M_i{}^*$ of this equation gives us the infected equilibria. If $f(x)$ denotes the left-hand side of this equation with x instead of $M_i{}^*$, and $g(x)$ denotes the right-hand side, the solutions of the above equation are given by the intersections of the curves $y=f(x)$ and $y=g(x)$. It can be seen in Figure 2 that if $\Re_0 > 1$ for some parameter values, there could be three intersections when $M_i{}^* > 0$. Each intersection gives one equilibrium value of $M_i{}^*$, and each equilibrium value of $M_i{}^*$ gives one equilibrium $Ei=(M^*_u, M^*_i, C^*)$ where

$$M_u^* = \frac{r(1+cM_i^*)}{(d_1 + (\alpha + cd_1)M_i^*)} \quad and$$
$$C^* = \frac{kM_i^*}{d_2(1+\varepsilon_2 M_i^*)} \tag{3}$$

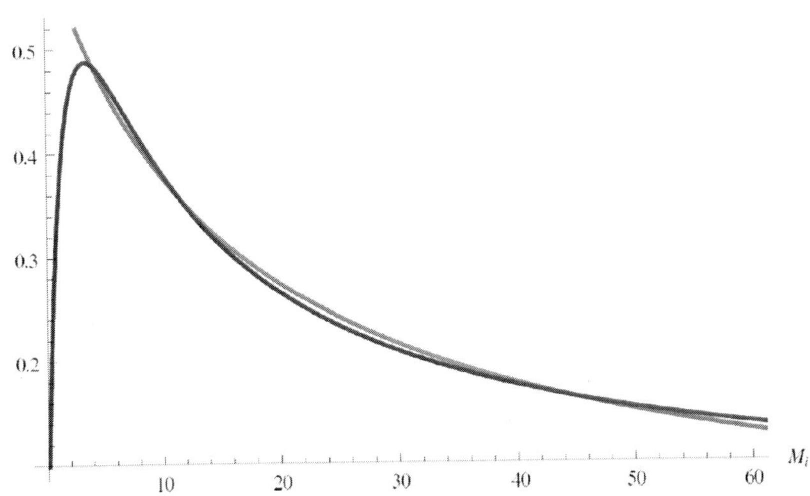

Figure 2 Multiple equilibria. Figure shows multiple intersections of the functions $f(x)$ (in red) and $g(x)$ (in blue). Each intersection gives one value of $M_i{}^*$ and therefore one equilibrium.

We label the equilibria in increasing value of M_i *. It often can be mathematically shown that the lowest one is locally stable, the middle one is unstable and the upper one is also locally stable. In biological terms the presence of multiple stable equilibria allows animals to switch between low bacterial load and no shedding (below the detection limit) to high bacterial load and shedding and vice versa. If we assume that stressful events may temporarily affect the immune system negatively, we can model this temporary disturbance by modifying the constant parameter k (stimulation rate for immune response from infected cells) into a time dependent function:

$$k(\tau) = \begin{cases} k_1 & \tau \notin [\tau_1, \tau_2] \\ k_2 & \tau \in [\tau_1, \tau_2] \end{cases} \quad (4)$$

where $k_1 > k_2$. This type of k (τ) may model for instance temporary reduction in the immune response strength due to stress. This minor disturbance may be very short-lived but may move the animal from a no-shedder to high-shedder, see Figure 3. Multiple equilibria do not occur for the values in Table 2 but for some modified values of the fitted parameters. However, in most cases for which d_1=0.25 and δ=0.3, the lower equilibrium appears to be unstable while the upper one is stable. In this case the bacterial load and the shedding still exhibit a switch, however, the switch occurs spontaneously, without external disturbance. The length of the non-shedding phase depends on various factors, including the status of the immune system at infection. This scenario is illustrated in Figure 4.

2.6 The immuno-epidemiological model of Johne's disease

We imbed the immunological model (1) from the previous section into an epidemiological model of JD. The epidemiological model and the immunological model of JD are linked through the time-post-infection variable as well as the dependence of some epidemiological parameters on the within-host pathogen load.

2.7 A brief description of JD epidemiology

JD is present in many countries with livestock industry and, in the United States, the causative agent of the disease, MAP, was found in about 70% of dairy farms [24,25]. After a long incubation time [26], dairy cattle infected with MAP start shedding the pathogen into feces, colostrum and milk [27-29]. MAP bacilli shed into feces can survive longer than a year in the environment [30] and the fecal-oral route was shown to be the major transmission pathway in dairy farms [31]. However, there is also evidence for transmission through contaminated milk, colostrum and placenta (vertical transmission) [29,31,32]. Young animals are more susceptible to MAP infection than adults [33], and therefore, good management practices to prevent MAP ingestion at young age (<1 year old) are suggested to be effective to control JD in dairy farms [34].

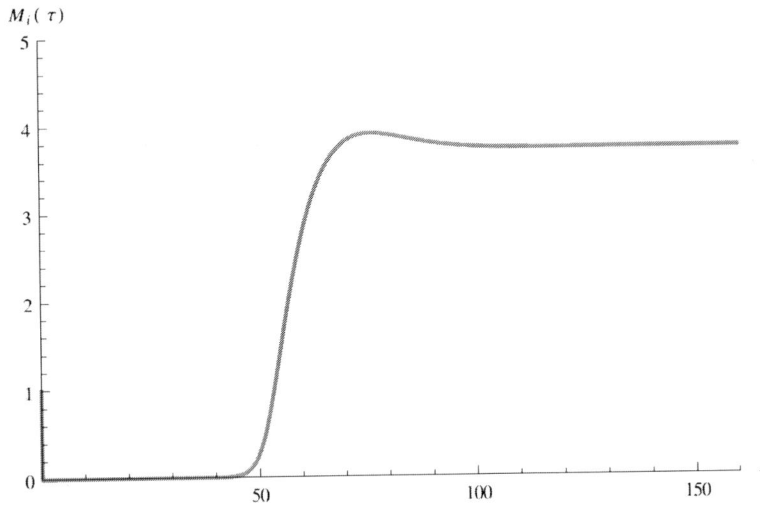

Figure 3 Switch between equilibria. Figure shows dynamic switch between the lower and upper equilibrium.

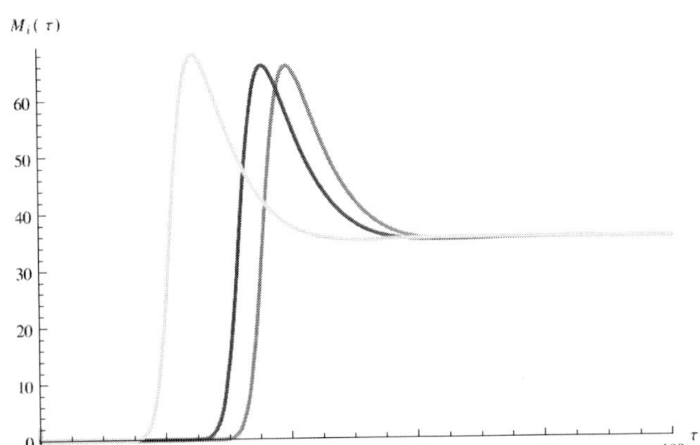

Figure 4 Spontaneous switch between equilibria. Figure shows dynamic spontaneous switch between the lower and upper equilibrium.

2.8 The epidemiological component of the JD model

To account for differential susceptibility between calves and cows, we separate the the number of healthy calves $S_C(t)$ from the number of healthy adults $S_A(t)$, where t is the time (in months) since some initial point called time zero. Because the infectivity of infected individuals depends more on their shedding and pathogen load, rather than chronological age, we merge infected calves and infected adults in one class whose density with respect to time-post-infection is given by $i(\tau,t)$. The quantity $i(\tau,t)$ is a density since

$$I(t) = \int_0^\infty i(\tau,t)d\tau$$

is the number of all infected individuals at time t. The units of $i(\tau,t)$ are number of individuals per unit of time. The newly infected individuals move to the class $i(0\ t)$. The amount of bacteria in the environment is given by $B(t)$. The model below has ordinary differential equations for S_C, S_A and B and a first order partial differential equation for $i(\tau,t)$.

$$\begin{cases} S_C'(t) = b_S(1-\gamma_C f_E(B)) + \int_0^\infty (1-q(\tau))b_I(\tau)i(\tau,t)d\tau \\ \qquad -\beta_C S_C \int_0^\infty \kappa(\tau)i(\tau,t)d\tau - \mu_C S_C - aS_C \\ S_A'(t) = aS_C - \beta_A S_A \int_0^\infty \kappa(\tau)i(\tau,t)d\tau - \gamma_A S_A f_E(B) - \mu_A S_A \\ \frac{\partial i}{\partial \tau} + \frac{\partial i}{\partial t} = -\nu(\tau)i - \mu i \\ i(0,t) = \int_0^\infty q(\tau)b_I(\tau)i(\tau,t)d\tau + \beta_C S_C \int_0^\infty \kappa(\tau)i(\tau,t)d\tau \\ \qquad + f_E(B)(\gamma_A S_A + \gamma_C b_S) + \beta_A S_A \int_0^\infty \kappa(\tau)i(\tau,t)d\tau \\ B' = \int_0^\infty \eta(\tau)i(\tau,t)d\tau - dB \end{cases}$$

We assume that b_S gives the number of susceptible calves born per unit of time. A proportion $\gamma_C f_E(B)$ are infected soon after birth through contact with the infected environment. These move to the newly infected individuals $i(0,t)$. Since newborn calves are separated after birth, we do not incorporate further environmental transmission of JD for calves. As a result of vertical transmission, from the number of calves born to infected individuals, a proportion $1-q(\tau)$ are susceptible and remain in the class S_C, while a proportion $q(\tau)$ are born infected and appear in the class $i(0, t)$. The dependence of q on τ will be explained below. Susceptible calves can also become infected through direct contact with infected individuals (with coefficient β_C) and also move to $i(0, t)$. Susceptible calves progress to adulthood at rate a. Healthy adult cows can become infected through direct contact with infected individuals as well as through contact with the contaminated environment. In both cases the newly infected individuals move to $i(0, t)$. We recall that calves are more susceptible to direct infection than adult cows. That fact implies that we need to have $\beta_C > \beta_A$. Infected individuals shed bacteria into the environment at rate $\eta(\tau)$, and bacteria are cleared from the environment at a rate d. See Table 3 for the parameters and the dependent variables for the epidemiological model.

Table 3 Parameters and dependent variables and their interpretation

Variable or parameter	Meaning
$S_C(t)$	Number of healthy calves at time t
$S_A(t)$	Number of healthy adults at time t
$i(\tau, t)$	Density of infected cattle with age-post-infection τ at time t
$B(t)$	Number of bacteria in the environment at time t
b_S	Birth rate for susceptible cattle
$q(\tau)$	Proportion of births of infected calves from infected cows
β_C	Susceptibility of calves to infection due to milk
μ_C	Natural death rate of healthy calves
μ_A	Natural death rate of healthy adults
M	Natural death rate of infected cattle
A	Rate of progression of calves to adulthood
β_A	Susceptibility of adults to infection
γ_C	Proportion of newborn calves getting infected from the environment
γ_A	Rate of infection of healthy adults from the environment
$b_I(\tau)$	Birth rate of infected adults τ units post infection
$\kappa(\tau)$	Infectivity of infectious cattle τ units post infection

For the general function of environmental bacteria infectiousness $f_E(B)$, we assume that $0 \leq f_E(B) \leq 1$. We consider the following specific form of the function

$$f_E(B) = \frac{1-e^{-\frac{B}{K_1}}}{1+K_2e^{-\frac{B}{K_1}}} \tag{5}$$

Where K_1 and K_2 are given constants. This function is chosen because it has a threshold effect: it is low for low values of B and at some threshold value of B increases fast to one.

2.9 Linking the immunological and the epidemiological models

Understanding the impact of within-host pathogen dynamics on the spread of JD in a population of cattle requires proper linking of the epidemiological parameters to the within-host dynamic variables $M_i(\tau)$ and $C(\tau)$. The epidemiological parameters linked to the within-host parameters are the τ-dependent birth rate of infected cattle $b_I(\tau)$, the transmission rate of infected cows $\kappa(\tau)$, the disease-induced death rate $\nu(\tau)$, the proportion of calves born infected $q(\tau)$ and the shedding into the environment $\eta(\tau)$. We assume that the cow's birth rate decreases slightly with the progression of the disease as the weight loss and milk reduction lead to suboptimal conditions [35,36]. We model the effect of the disease-progression through the following relationship between birth rate and infected macrophage load:

$$b_I(\tau) = \frac{b_S'}{1+bM_i(\tau)}$$

where b'_S is the per capita birth rate of healthy cows and b has to be specified. The constant b controls the change in birth rate due to bacterial load and illness. Furthermore, the proportion of calves born infected also depends on the mother's pathogen load and is represented by an increasing saturating function:

$$q(\tau) = \frac{qM_i(\tau)}{Q+M_i(\tau)}$$

where Q is the half-saturation constant. The half-saturation constant controls how fast the curve reaches half-saturation level. Direct transmission rate κ (τ) is assumed to be small in JD. We surmise that it occurs when shedded feces become immediately infectious to susceptible calves or cows without having to stay in the environment and be subject to degradation. In this case the direct transmission rate is directly correlated with the CFU shed in the environment. Prior research suggests that in the case of HIV [37] the dependence of the transmission rate on the pathogen is saturating at high pathogen loads, and we use the simplest saturation function:

$$\kappa(\tau) = \frac{M_i(\tau)}{\kappa_T + M_i(\tau)}$$

where κ_T is the half-saturation constant.

The disease-induced mortality is linked to the bacterial load as well as the immune response,

$$\nu(\tau) = mM_i(\tau) + vC(\tau)$$

where m and v are scaling constants. The immune response leads to inflammation and thickening of the intestinal wall which causes diarrhea and malnutrition. Finally, the shedding into the environment η (τ) is given by the CFU and is taken to be proportional to the bacterial load:

$$\eta(\tau) = c_1 M_i(\tau)$$

where c_1 is estimated from fitting.

The linking above describes how the epidemiological parameters change with the within-host bacterial load. The resulting immuno-epidemiological model assumes that the within-host progression of the disease is averaged among all individuals and gives the population-level spread which results from the averaged immune dynamics.

2.10 Reproduction number of the immuno-epidemiological model

Whether JD persists on epidemiological level depends on the epidemiological reproduction number R_0 which gives the number of secondary infections one infected individual will produce in an entirely susceptible cattle population. If $R_0 > 1$ then JD will persist in the population;

if however, $R_0 < 1$, JD may be eliminated from the population, even in the case when the infection persists in some individual infected animals.

To derive the epidemiological reproduction number, we first compute the disease-free equilibrium $E_0=(S^0_C,S^0_A,0,0)$ where

$$S_C^0 = \frac{b_s}{\mu_C + a} \qquad S_A^0 = \frac{ab_s}{\mu_A(\mu_C + a)}.$$

The epidemiological reproduction number is then defined as:

$$\begin{aligned} R_0 = {} & (\beta_C S_C^0 + \beta_A S_A^0)\int_0^\infty \kappa(\tau)\pi(\tau)d\tau \\ & + \frac{(\gamma_A S_A^0 + \gamma_C b_S)f_E'(0)}{d}\int_0^\infty \eta(\tau)\pi(\tau)d\tau \\ & + \int_0^\infty q(\tau)b_I(\tau)\pi(\tau)d\tau \end{aligned}$$

Where (6)

$$\pi(\tau) = e^{-\int_0^\tau v(\sigma)d\sigma} e^{-\mu\tau}$$

is the probability of survival in the infectious class. Each term of the reproduction number corresponds to a different mode of transmission of JD. The first two terms give the number of secondary infections of calves and adults who get infected through the direct transmission route. The third term gives the number of secondary infections, generated through environmental transmission. The integral in this term gives the amount of bacteria shedded by one infectious individual, $1/d$ gives the survival time of bacteria in the environment and f_E (0) gives the rate of infection when the bacteria in the environment are rare. The last term gives the number of xsecondary infections generated through vertical transmission.

The epidemiological reproduction number depends on the immune system parameters through its dependence on M_i and C. It is instructive to see this dependence explicitly at least in the case when the immune system (1) is at equilibrium. In this case the solution of the system is given by $E^*_i=(M^*_i, ,C^*)$ where $M_u{}^*$ and C^* are given by (3). In this special case the epidemiological number becomes:

$$R_0 = \left[(\beta_C S_C^0 + \beta_A S_A^0) \frac{M_i^*}{\kappa_T + M_i^*} + \frac{(\gamma_A S_A^0 + \gamma_C b_S) f_E'(0)}{d} c_1 M_i^* + \frac{q b_S' M_i^*}{(1 + b M_i^*)(Q + M_i^*)} \right] \frac{1}{m M_i^* + \nu C^* + \mu} \tag{7}$$

2.11 Estimation of epidemiological parameters

To illustrate the impact of within-host prevalence of MAP on the epidemiological reproduction number, we need estimates for some of the epidemiological parameters involved in the model. Reference [30] gives a detailed account of the survival properties of MAP in various environments. This allows us to estimate d which we take in the range 1/15–1/3 months^{-1}. We also derive the shedding rate from the fits of the immune model to CFU data. The constant that links the MAP amount within a host M_i to the CFU is given by c_1, and we set $c_1=0.01$.

Reference [38] states that up to 25% of the calves may be infected in utero if the mother shows clinical signs. The removal rates μ_C, μ_A and the age-progression rate a are well known for cattle [39]. The natural death rate μ is also known from [39]. The transmission parameters β_C, β_A, γ_C, γ_A are assumed based on the expectation for an epidemiological reproduction number of about 2. The rate of infection of healthy adults from the environment can be obtained from [40]. The clearance rate of MAP d is taken from [30]. The epidemiological parameters are given in Table 4.

Table 4 Parameters and dependent variables and their interpretation

Parameter	Meaning	Value	Range
b_S	Birth rate for susceptible cattle	100	0.1–1000
b_S	Per capita birth rate for susceptible cattle	100	0.01–100
B	Constant in $b_I(\tau)$	0.1	0.0001–1
Q	Proportion of infected calves born to infected cows	0.25	0–0.25
Q	Half saturation constant	100	0–100000
β_C	Susceptibility of calves to infection due to milk	0.0183333	0.001–1
μ_C	Natural death rate of healthy calves	1/(10 × 12)	1/(25 × 12)–1/(5 × 12)
μ_A	Natural death rate of healthy adults	1/(5 × 12)	1/(25 × 12)–1/(2 × 12)

M	Natural death rate of infected cattle	1/(3 × 12)	1/(5 × 12)–1/(2 × 12)
A	Rate of progression of calves to adulthood	1/12	1/(2 × 12)–1/(0.5 × 12)
β_A	Susceptibility of adults to infection	0.00733333	0.001–1
γ_C	Proportion of newborns infected from environment	0.0001	0–1
γ_A	Rate of infection of healthy adults from environment	0.0000183333	0.0–3
κ_T	Half-saturation constant for $\kappa(\tau)$	1000	10–10000
m	Proportionality constant for $\nu(\tau)$	0.3	0.0001–1
V	Proportionality constant for $\nu(\tau)$	0.01	0.0001–1
η_E	Shedding rate of infected individuals	0.01	0.001–0.1
D	Clearance rate of MAP from the environment	1/5	1/15–1/3
K_1	Constant in f_E	1	1–10000
K_2	Constant in f_E	1	1–10000

3. RESULTS

3.1 Implications of the within-host dynamics to the epidemiology of JD

Immuno-epidemiological models allow us to evaluate the impact of the within-host bacterial load on the epidemiological reproduction number. As we show, this dependence in the case of JD may be quite complicated. Intuitively, we may surmise that the epidemiological reproduction number is an increasing function of the bacterial load M_i^*. That is indeed the case when M_i^* is small. When internal bacterial load is larger, then "trade off" effects come into play, and host mortality due to illness takes precedence. As a result the epidemiological reproduction number starts decreasing. This creates the typical "hump"-shaped form of the reproduction number as a function of the pathogen load [41]. We observe that shape in the epidemiological reproduction number of JD in the case when the transmission occurs only directly (c_1=0, q=0) in Figure 5. Figure 5 also shows that the stronger the impact of the immune response on host mortality, that is the larger the ν, the smaller the reproduction number. Furthermore, for larger ν the peak in R_0 is less pronounced and occurs for larger values of the bacterial load. This rather surprising effect that increasing v requires higher bacterial load for the maximum R_0 to occur is produced by the saturation effect of the immune response in killing M_i^* given by ε_2. If ε_2 is small the maximum occurs at smaller values for M_i^* as v increases but as ε_2 becomes larger, the immune response plays a smaller role. When transmission occurs only environmentally, that is $\beta_C = \beta_A = 0$ and $q = 0$, the epidemiological reproduction number is an increasing function of M_i^* and there is no maximum of R_0 for intermediate values of M_i^*. We surmise that this is a result of the fact that shedding is directly proportional to the internal bacterial load and there is no saturation in the infection rate. When transmission occurs only vertically, then the epidemic reproduction number again rises for small internal bacterial load but reaches a peak quickly and then declines, see Figure 6. In this case the peak occurs for very small value of M_i^*. Furthermore, the weaker the impact of the immune response, the peak occurs for a larger value of M_i^*. When all three transmission mechanisms are combined, the reproduction number can experience several rises and declines, with several maximums for various values of the bacterial load M_i^*. For small values of M_i^*, vertical transmission dominates and leads to an early peak of R_0 at small values of M_i^*. The survival trade-offs compensate, but as M_i^* increases, the role of the vertical transmission declines and direct transmission takes on the dominant role, leading to a second peak in the reproduction number. Eventually the survival trade-off effect compensates for this increase also and R_0 declines. The behavior of R_0 with all three modes of transmission is given in Figure 7. This figure shows that as the effect of the immune system on the host v becomes smaller, the second peak becomes less pronounced and perhaps will disappear. To see the impact of the multiple within-host equilibria, we denote by $M_i^{(1)}$ the lower stable one and by $M_i^{(3)}$ the upper stable one. If the switch between the lower and the upper occurs at time T, we have

$$M_i(\tau) = \begin{cases} M_i^{(1)} & 0 < \tau < T \\ M_i^{(3)} & \tau > T \end{cases} \quad (8)$$

This step-wise dynamics, approximates the actual dynamics which is continuous with a fast transition. We can define $R^{(1)}_0$ and $R^{(3)}_0$ to be given by (8) with M_i replaced by $M_i^{(1)}$ or $M_i^{(3)}$ respectively. Substituting (9) in the reproduction number (7), we obtain:

$$\begin{aligned} R_0 = {} & R_0^{(1)}\left(1 - e^{-\left(mM_i^{(1)} + vC^{(1)} + \mu\right)T}\right) \\ & + R_0^{(3)} e^{-\left(mM_i^{(1)} + vC^{(1)} + \mu\right)T}. \end{aligned}$$

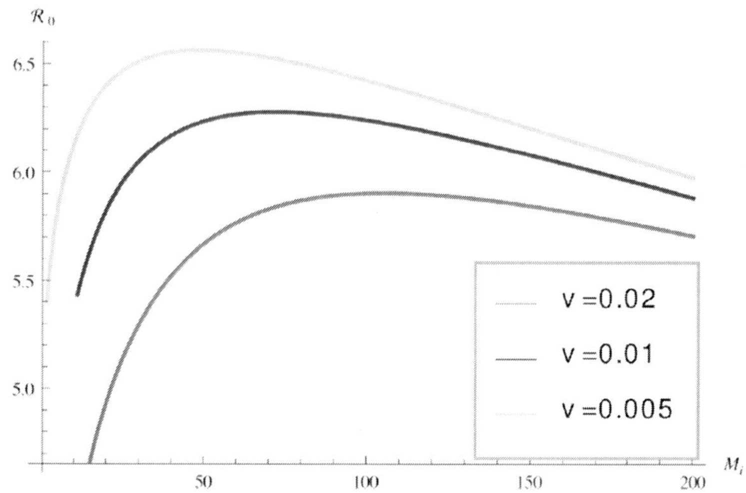

Figure 5 R_0 of direct transmission. Figure shows the epidemiological reproduction number R_0 as a function of the internal bacterial load M_i * when the only mechanism of transmission is direct transmission.

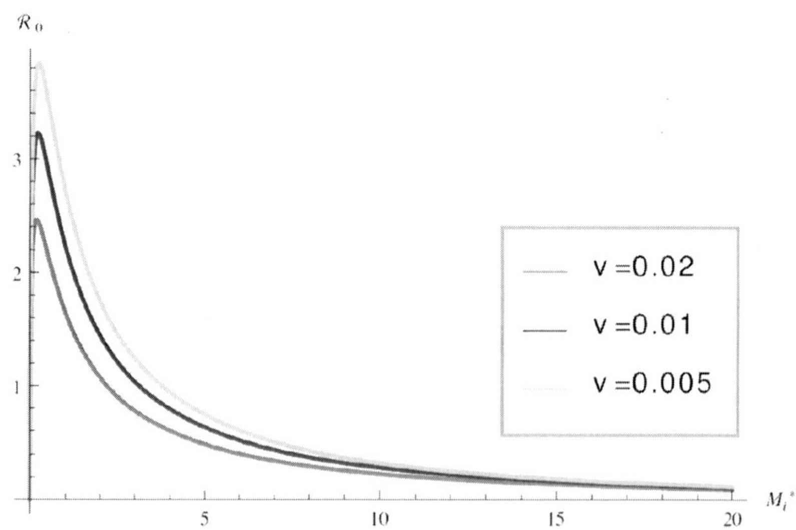

Figure 6 R_0 of vertical transmission. Figure shows the epidemiological reproduction number R_0 as a function of the internal bacterial load M_i * when the only mechanism of transmission is vertical transmission.

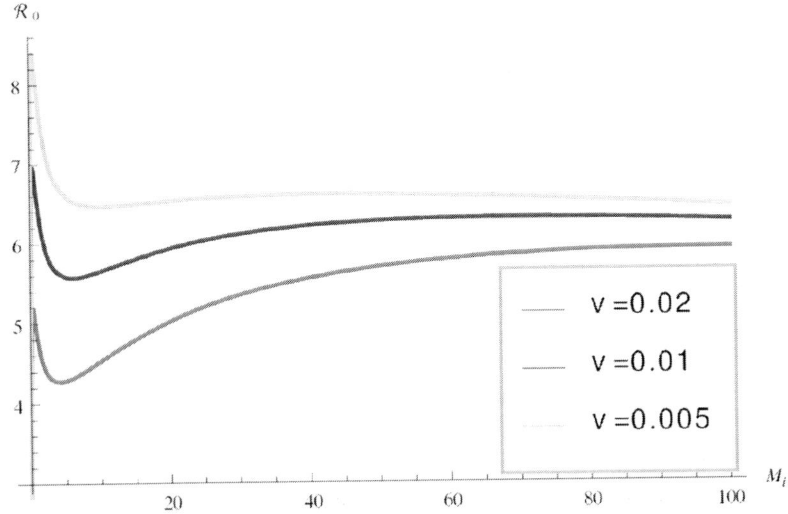

Figure 7 R_0 with all transmission modes. Figure shows the epidemiological reproduction number R_0 as a function of the internal bacterial load M_i^* when all transmission modes are included.

From this expression, we can tell that if $R^{(1)}_0 < R^{(3)}_0$, then $R_0$is decreasing as T increases. However, this relationship between $R^{(1)}_0$ and $R^{(3)}_0$ is not guaranteed since the reproduction number could be large for small values of the within-host bacterial load and smaller for larger values of the bacterial load. We plot the reproduction number as a function of $M_i^{(1)}$ or $M_i^{(3)}$ with $M_i^{(1)} < M_i^{(3)}$ and we see that it is changing dramatically (see Figure 8). Note that R_0 is large for small $M_i^{(1)}$ but is very small for very large $M_i^{(3)}$. For slightly higher values of $M_i^{(1)}$ it sharply dips but then increases and subsequently decreases. Computing the prevalence in the case including all three modes of transmission is too complicated to be useful but typically prevalence increases with the increase of the reproductive number. That suggests that the prevalence may be high when most cattle appears nearly healthy and it may be decreasing when all infected cattle seems sick. This last fact is most likely due to increased culling in that case and reduced fertility.

4. DISCUSSION

Mathematical modeling is an important tool in learning about infectious diseases. We believe that linking immunological and epidemiological models will give further important contributions to understanding of multi-scale biological processes and to leading the way in disease control.

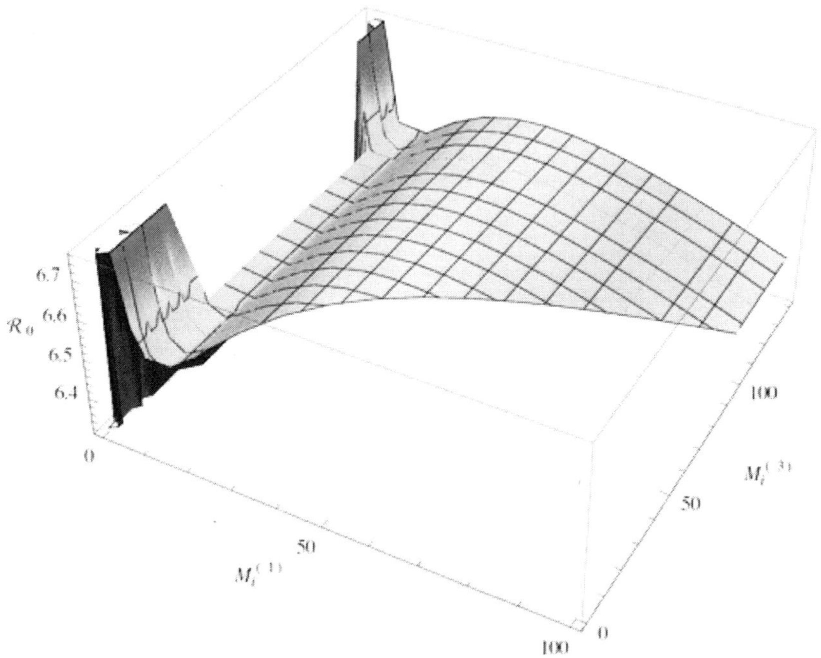

Figure 8 R_0 with switch in internal bacteria load**.** Figure shows the epidemiological reproduction number R_0 as a function of the internal bacterial load $M_i^{(1)}$ and $M_i^{(3)}$ when all transmission modes are included and switch time T=4.

4.1 Overview of insights of immunological model

The immunological model in the system describes the within-host dynamics of MAP. We fitted the model to CFU data of three calves. The fitting resulted in the estimation of a number of key quantities in the within-host dynamics of MAP. In particular we obtain values of the infection rate of uninfected macrophages, the stimulation rate of the immune response and others. These results are given in Table 2. The immunological model also reveals that saturating incidence in infected macrophages leads to multiple steady states of the bacterial load when $\Re 0 > 1$. In particular for some parameter regimes close to the estimated values, there are three nonzero equilibria, two of which are typically attracting. That allows for the MAP bacterial load to stabilize at two non-zero values, mimicking stable nonshedding and stable shedding. Switching between shedding and non-shedding can occur as a result of a short-term disturbance that a cow may experience. This suggests that prolonged no shedding, followed by prolonged shedding are two stable regimes where the switching occurs as a result of a stressor. Simulations show, however, that for some parameter regimes close to the estimated values, the lower equilibrium can be unstable and the switching between no shedding and shedding may occur spontaneously without any external disturbance. In this case the duration of the non phase depended on the initial infection and other factors.

4.2 Overview of insights of immuno- epidemiological model

The epidemiological reproduction number R_0, computed in terms of the within-host bacterial load, allows us to potentially infer when JD will persist or die out in a herd from the within-host bacterial load or shedding. At the very least, we can observe how the within-host bacterial load impacts the reproduction number, and hence, the prevalence of JD in a herd. JD is spread through three modes of transmission: environmental, vertical and direct. In general for directly transmitted diseases it is known that the epidemiological R_0 exhibits a humped shaped dependence with respect to the pathogen load. Surprisingly, we find that the stronger the virulence caused by the immune response, the larger the pathogen load needed for the maximum to occur. Furthermore, we find that R_0 also exhibits such a humped shape if the transmission is only vertical. In contrast with direct transmission, the peak in this case is more pronounced and occurs for lower pathogen load. The decline of R_0 in both cases is attributed to the virulence. As JD is spread via all three modes of transmission, JD's epidemiological R_0 exhibits complex dependence on the within-host bacterial load with two peaks – the first one caused by vertical transmission and the second one caused by direct transmission. The non-monotone dependence of R_0 on the within-host bacterial load does not allow us to infer in general that if the shedding of all cows in a herd is non-detectable, R_0 and prevalence of JD will be low. In fact, no shedding (low pathogen load) in all cows may very well lead to high prevalence of JD in the herd. Because of that, control measures that reduce shedding or extend the time of the cows in a herd being non-shedders may not reduce the prevalence of JD.

4.3 Significance of results

We introduce a new way to model long periods of shedding and no shedding as well as the switching between the two. This leads to understanding that these are due to stable stationary patterns of the within-host bacterial load. Furthermore, most likely the switching occurs due to temporary disturbance of the infected cow. Furthermore, the new model extends the modeling techniques to immuno-epidemiological models with multiple within-host equilibria and their impact on the epidemiology of disease. Finally, we uncover that in the case of multiple modes of transmission, as with JDs, the epidemiological reproduction number can depend on the pathogen load in a complex fashion, experiencing multiple peaks. Ultimately we conclude that no shedding (low pathogen load) does not necessarily imply low epidemiological reproduction number or low prevalence in the herd.

ACKNOWLEDGMENTS

The authors acknowledge the support of the Within-host modeling of MAP infections Working Group at the National Institute for Mathematical and Biological Synthesis, sponsored by the National Science Foundation, the U.S. Department of Homeland Security, and the U.S. Department of Agriculture through NSF Award DBI-1300426, with additional support from The University of Tennessee, Knoxville. Lenhart is partially supported by the University of Tennessee Center for Business and Economic Research.

REFERENCES

1. Momotani E, Whipple DL, Thiermann AB, Cheville NF (1988) Role of M cells and macrophages in the entrance of Mycobacterium paratuberculosis into domes of ileal Peyer's patches in calves. Vet Pathol 25:131–137
2. Koets A, Eda S, Sreevatsan S (2015) The within host dynamics of Mycobacterium avium spp. paratuberculosis infection in cattle: where time and place matter. Vet Res (in press)
3. Klinkenberg D, Koets AP (2015) The long latent subclinical phase of Mycobacterium avium ssp. paratuberculosis infections explained without adaptive immunity. Vet Res (in press)
4. Magombedze G, Eda S, Bakker D, Koets AP (2015) Can immune response mechanisms explain the fecal shedding patterns of cattle infected with Mycobacterium avium subspecies paratuberculosis. Vet Res (in press)
5. Martcheva M (2011) An immuno-epidemiological model of paratuberculosis. AIP Conference Proceedings, Vol. 1404. AIP Publishing, Melville, NY, pp 176–183
6. Numfor E, Bhattacharya S, Lenhart S, Martcheva M (2014) Optimal control applied in coupled within-host and between-host models. Math Models Nat Phenomena 9:171–203

7. Feng Z, Velasco-Hernandez J, Tapia-Santos B, Leite M (2012) A model for coupled within-host and between-host dynamics in an infectious disease. Nonlinear Dyn 68:401–411
8. Feng ZL, Velasco-Hernandez J, Tapia-Santos B (2013) A mathematical model for coupling within-host and between-host dynamics in an environmentally-driven infectious disease. Math Biosci 241:49–55
9. Ganusov VV, Bergstrom CT, Antia R (2002) Within-host population dynamics and the evolution of microparasites in a heterogeneous host population. Evolution 56:213–223
10. Gilchrist MA, Coombs D (2006) Evolution of virulence: Interdependence, constraints, and selection using nested models. Theor Popul Biol 69:145–153
11. Gilchrist MA, Sasaki A (2002) Modeling host-parasite coevolution: a nested approach based on mechanistic models. J Theor Biol 218:289–308
12. Begg DJ, de Silva K, Carter N, Plain KM, Purdie A, Whittington RJ (2011) Does a Th1 over Th2 dominancy really exist in the early stages of Mycobacterium avium subspecies paratuberculosis infections? Immunobiology 216:840–846
13. Mitchell RM, Medley GF, Collins MT, Schukken YH (2012) A meta-analysis of the effect of dose and age at exposure on shedding of Mycobacterium avium subspecies paratuberculosis (MAP) in experimentally infected calves and cows. Epidemiol Infect 140:231–246
14. Wu CW, Livesey M, Schmoller SK, Manning EJ, Steinberg H, Davis WC, Hamilton MJ, Talaat AM (2007) Invasion and persistence of Mycobacterium avium subsp. paratuberculosis during early stages of Johne's disease in calves. Infect Immun 75:2110–2119
15. de Silva K, Begg DJ, Plain KM, Purdie AC, Kawaji S, Dhand NK, Whittington RJ (2013) Can early host responses to mycobacterial infection predict eventual disease outcomes? Prev Vet Med 112:203–212
16. Sommer S, Pudrith CB, Colvin CJ, Coussens PM (2009) Mycobacterium avium subspecies paratuberculosis suppresses expression of IL-12p40 and iNOS genes induced by signalling through CD40 in bovine monocyte-derived macrophages. Vet Immunol Immunopathol 128:44–52
17. Nielsen SS, Toft N, Okura H (2013) Dynamics of specific anti-Mycobacterium avium subsp. paratuberculosis antibody response through age. PLoS One 8:e63009
18. Nowak MA, May RM (2000) Virus dynamics: mathematical principles of immunology and virology. Oxford University Press, Oxford; New York
19. Gadhamsetty S, Maree AF, Beltman JB, de Boer RJ (2014) A general functional response of cytotoxic T lymphocyte-mediated killing of target cells. Biophys J 106:1780–1791
20. Wigginton JE, Kirschner D (2001) A model to predict cell-mediated immune regulatory mechanisms during human infection with Mycobacterium tuberculosis. J Immunol 166:1951–1967
21. Stabel JR, Palmer MV, Harris B, Plattner B, Hostetter J, Robbe-Austerman S (2009) Pathogenesis of Mycobacterium avium subsp. paratuberculosis in neonatal calves after oral or intraperitoneal experimental infection. Vet Microbiol 136:306–313
22. Stabel JR, Robbe-Austerman S (2011) Early immune markers associated with Mycobacterium avium subsp. paratuberculosis infection in a neonatal calf model. Clin Vaccine Immunol 18:393–405
23. Kabara E, Coussens PM (2012) Infection of primary bovine macrophages with Mycobacterium avium subspecies paratuberculosis suppresses host cell apoptosis. Front Microbiol 3:215
24. USDA/NAHMS report, Johne's Disease on U.S. Dairies, 1991–2007. http://www.aphis.usda.gov/animal_health/nahms/dairy/downloads/dairy07/Dairy07_is_Johnes.pdf. Accessed 7 May 2015.
25. Salem M, Heydel C, El-Sayed A, Ahmed SA, Zschock M, Baljer G (2013) Mycobacterium avium subspecies paratuberculosis: an insidious problem for the ruminant industry. Trop Anim Health Prod 45:351–366
26. Espejo LA, Godden S, Hartmann WL, Wells SJ (2012) Reduction in incidence of Johne's disease associated with implementation of a disease control program in Minnesota demonstration herds. J Dairy Sci 95:4141–4152

27. Streeter RN, Hoffsis GF, Bech-Nielsen S, Shulaw WP, Rings DM (1995) Isolation of Mycobacterium paratuberculosis from colostrum and milk of subclinically infected cows. Am J Vet Res 56:1322–1324
28. Sweeney RW, Whitlock RH, Hamir AN, Rosenberger AE, Herr SA (1992) Isolation of Mycobacterium paratuberculosis after oral inoculation in uninfected cattle. Am J Vet Res 53:1312–1314
29. Sweeney RW (1996) Transmission of paratuberculosis. Vet Clin North Am Food Anim Pract 12:305–312
30. Elliott GN, Hough RL, Avery LM, Maltin CA, Campbell CD Environmental risk factors in the incidence of Johne's disease. Crit Rev Microbiol (in press)
31. Dore E, Pare J, Cote G, Buczinski S, Labrecque O, Roy JP, Fecteau G (2012) Risk factors associated with transmission of Mycobacterium avium subsp. paratuberculosis to calves within dairy herd: a systematic review. J Vet Intern Med 26:32–45
32. Whittington RJ, Windsor PA (2009) In utero infection of cattle with Mycobacterium avium subsp. paratuberculosis: a critical review and meta-analysis. Vet J 179:60–69
33. Windsor PA, Whittington RJ (2010) Evidence for age susceptibility of cattle to Johne's disease. Vet J 184:37–44
34. Pillars RB, Grooms DL, Gardiner JC, Kaneene JB (2011) Association between risk-assessment scores and individual-cow Johne's disease-test status over time on seven Michigan, USA dairy herds. Prev Vet Med 98:10–18
35. Smith RL, Strawderman RL, Schukken YH, Wells SJ, Pradhan AK, Espejo LA, Whitlock RH, Van Kessel JS, Smith JM, Wolfgang DR, Grohn YT (2010) Effect of Johne's disease status on reproduction and culling in dairy cattle. J Dairy Sci 93:3513–3524
36. Smith RL, Grohn YT, Pradhan AK, Whitlock RH, Van Kessel JS, Smith JM, Wolfgang DR, Schukken YH (2009) A longitudinal study on the impact of Johne's disease status on milk production in individual cows. J Dairy Sci 92:2653–2661
37. Lange A, Ferguson NM (2009) Antigenic diversity, transmission mechanisms, and the evolution of pathogens. PLoS Comput Biol 5:e1000536
38. Seitz SE, Heider LE, Heuston WD, Bech-Nielsen S, Rings DM, Spangler L (1989) Bovine fetal infection with Mycobacterium paratuberculosis. J Am Vet Med Assoc 194:1423–1426
39. Lu Z, Schukken YH, Smith RL, Mitchell RM, Grohn YT (2013) Impact of imperfect Mycobacterium avium subsp. paratuberculosis vaccines in dairy herds: a mathematical modeling approach. Prev Vet Med 108:148–158
40. Magombedze C, Ngonghala CN, Lanzas C (2013) Evaluation of the iceberg phenomenon in Johne's disease through mathematical modelling. PLoS One 8:e76636
41. Pugliese A (2011) The role of host population heterogeneity in the evolution of virulence. J Biol Dyn 5:104–119

CHAPTER 12

A Review: Interactions of Equine Herpesvirus-1 with Immune System and Equine Lymphocyte

Nor Dini Rusli, Khairiyah Binti Mat, Hasnita Che Harun

Faculty of Agro Based Industry, Universiti Malaysia Kelantan, Jeli Campus, Kelantan, Malaysia

ABSTRACT

Equine herpesvirus-1 (EHV-1) remains one of the most common viral pathogens affecting horses worldwide presenting as a persistent infection which can establish latency in nerve ganglia (trigeminal ganglion), lymphoid tissues of the respiratory tract and peripheral blood lymphocytes. EHV-1 infection induces both humoral and cellular immune responses in horses. Virus neutralising antibody, particularly in the nasopharynx, is to kill free virus shed from infected epithelial cells. Hence this antibody has important functions in reducing virus shedding and spreading infection to cohorts. Cellular immune responses, particularly those carried out by cytotoxic T lymphocyte (CTL), have been shown to be effective in killing virus-infected cells in vitro. This review underlines the state of knowledge regarding immunity to EHV-1 and also its interaction with equine lymphocyte. Finally, the review also includes the importance of the viral immediate early (IE) protein in the pathogenesis of EHV-1. This information can be used as the basis for future research.

Keywords: Equine Herpesvirus-1 (EHV-1), Lymphocyte, Cytotoxic T Lymphocyte (CTL), Immune Response, Peripheral Blood Mononuclear cell (PBMC)

1. INTRODUCTION

Virus infections are the most common human disease, particularly acute respiratory infections. In horses, viral respiratory infections are also common and the most notable viral pathogens are equine herpesvirus-1 (EHV-1), equine herpesvirus-4 (EHV-4), equine influenza virus (EIV) and equine arteritis virus (EAV) [1] [2] .

EHV-1 is a ubiquitous pathogen of all breeds of horses and other equids worldwide. It can cause respiratory and neurological disease as well as abortion in pregnant mares and it remains responsible for major economic and welfare problems. Studies of virus-specific immune responses have played essential roles in the development of diagnostic techniques and the evaluation of new strategies for vaccination. Both humoral and cellular immune responses are important in protection and recovery from EHV-1 infection and the relative importance of these responses has been the subject of much investigation. For example, although virus neutralising (VN) antibody specific for EHV-1 in serum can be associated with a reduction in the amount and duration of virus shed, there are no significant effects on cell-associated viraemia [3] , although a previous paper [4] indicated that VN antibody stimulated by vaccination did reduce the cell-associated viraemia but the mechanism is unknown. The perceived failure of VN antibody to affect the cell-associated viraemia is because most viruses become intracellular soon after landing on the respiratory mucosal surface and killing of cell associated virus is the major function of cytotoxic T lymphocytes (CTL) [5] - [8] .

Current commercial vaccines that contain inactivated virus give only partial clinical and virological protection against respiratory infections with EHV-1 because they do not stimulate CTLs and therefore do not control the cell-associated viraemia which disseminates virus from the respiratory tract [9] . CTLs are stimulated after both experimental and natural infection with EHV-1 [5] [9] and it has been shown that CTL precursor (CTLp) frequency is the major correlate of immunity to this virus [7] [8] . The product of the EHV-1 immediate early (IE) gene 64 (Figure 1) has been shown to be a target for EHV-1 specific CTL restricted by the B2 allele on the equine leukocyte antigen A3 (ELA-A3) haplotype [9] [10] .

2. PATHOGENESIS OF EHV-1

EHV-1 infections in horses may become established because of exposure to infectious virus by the respiratory route or by reactivation (recrudescence) of latent virus elsewhere in the body [11] . New infections are acquired by inhalation of infectious aerosols or contact with infectious secretions. After inhalation, viral replication occurs in the epithelia of the upper respiratory tract, including turbinates, pharynx, soft palate and tracheal epithelium [12] . The virus penetrates the epithelial barrier and by infecting a wide range of cell types, including immune cells, moves into deeper tissues of the respiratory tract and into the draining lymph nodes via lymphatic drainage (Figure 2). Subsequently, a second stage of viral replication occurs in which EHV-1 replication can be demonstrated in lymph nodes by the isolation of infectious virus and the presence of viral antigen in mononuclear cells of the retropharyngeal and submandibular lymph nodes from 12 hours post infection [13] [14] .

From 4 to 6 days post infection, the infected lymphocytes migrate from lymph nodes into the blood, resulting in a cell-associated viraemia and then EHV-1 is carried by infected peripheral blood mononuclear cells (PBMC) and spreads throughout the body. By this method, the cell associated virus may disseminate to internal organs such as the uterus and nervous system (Figure 2) [15] . In the uterus, infected lymphocytes adhere to the endothelial cells in small blood vessels supplying the placenta. This endothelial cell infection initiates the formation of protein clots, leading to thrombosis and vascular damage [16] [17] . Inflammation of blood vessels (vasculitis) develops, leading to damage of the placenta and thus the virus spreads to the fetus and consequently results in the abortion (Figure 2) [18] [19]. It is suggested that the uterine pathology and/or fetal distress initiates a physiological cascade, resulting in the separation of the placenta from the endometrium [9] [20] .

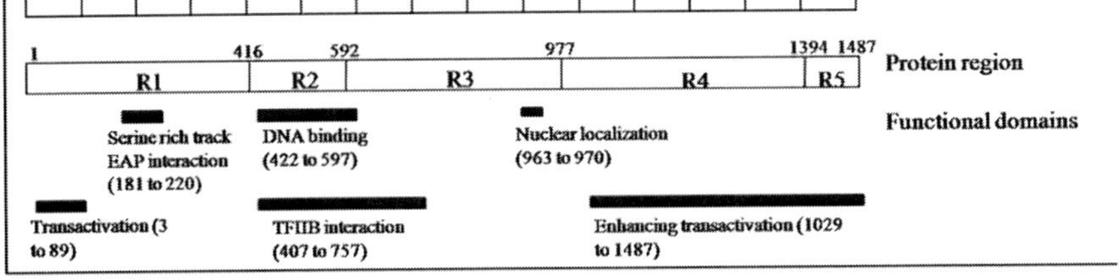

Figure 1. Schematic representation of the EHV-1 immediate early gene.

In the early 1990s, some studies reported that viral replication and vascular damage were more prevalent in the late pregnant uterus [17] than in the early pregnant uterus [21] . The causes of these differences in the early and late pregnancy are still unclear.

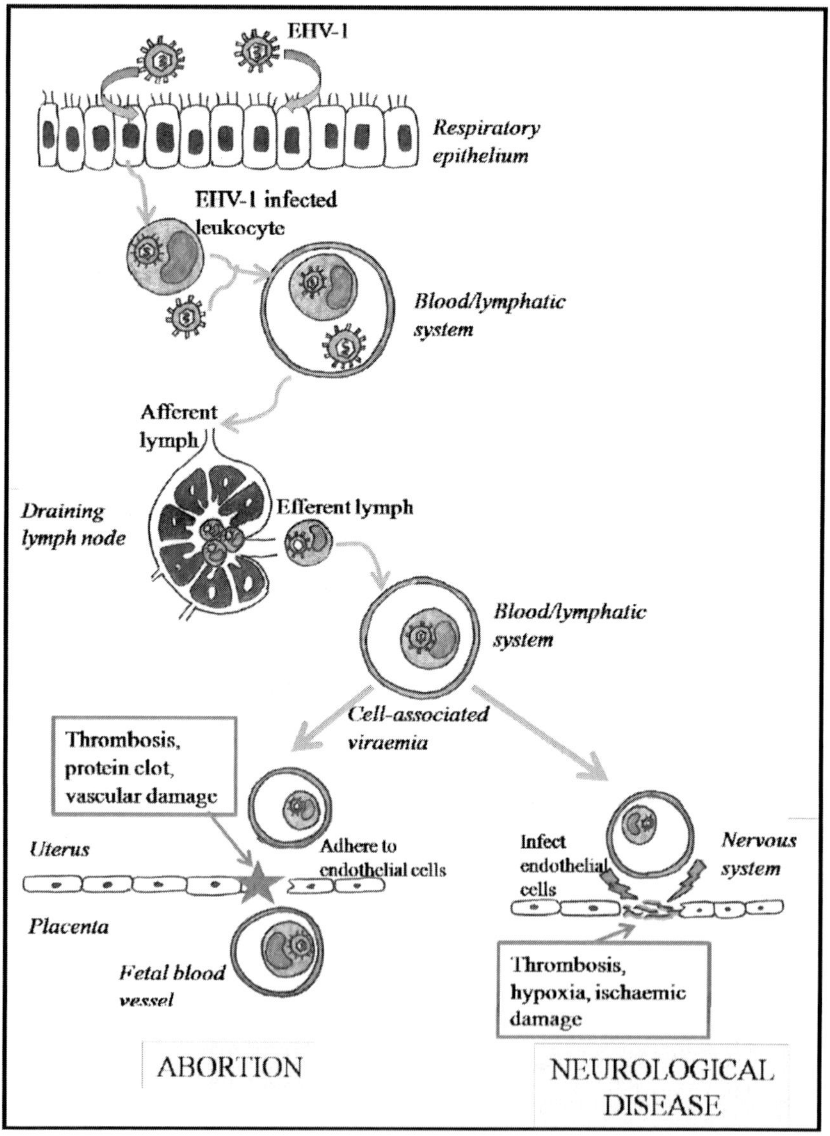

Figure 2. Schematic diagram of EHV-1 pathogenesis.

Neurological disease is a consequence of EHV-1 infected PBMC reaching the nervous system and causing infection of vascular endothelia of small arterioles and venules (Figure 2). Immune complexes may play a key role in this pathogenesis as some studies observed that recovery of virus from the central nervous system frequently failed [22] [23] . Edington et al. [24] provided evidence that EHV-1 can infect endothelial cells of the blood vessels of the nervous system and that this results in vasculitis, leading to thrombosis, hypoxia and secondary ischaemic damage, which compromises the blood supply to neurones and results subsequently in nervous system disorders (Figure 2) [15] .

As with other alpaherpesvirinae, EHV-1 establishes latent infection and persists in its host, following a primary infection. During latency, the expression of the EHV-1 genome is repressed and only a viral RNA transcribed from the IE gene occurs giving rise to a latency-associated transcript (LAT). However the precise function of LAT is not well understood and no infectious virus is produced [25] . The locations of EHV-1 latency were identified as the trigeminal ganglion [26] [27] , in lymphoid tissue draining the respiratory tract [26] [28] [29] and in peripheral blood leukocytes [28] -[30] . Smith et al. [30] reported that $CD5^{+}/CD8^{+}$ T lymphocytes were the predominant site of EHV-1 latency in peripheral blood leukocytes. In 2000, one study demonstrated that alveolar macrophages were also a site of latency by detecting the latent transforming growth factor β in alveolar macrophages [31] .

3. INTERACTION OF EHV-1 WITH THE IMMUNE SYSTEM

Attempts to identify immune responses which correlate with protective immunity against re-infection have been made, so that they can be stimulated by vaccination. However they have proved difficult to identify. Generally, herpesvirus infection, particularly EHV-1 infection involves a local immune response at the primary site of replication as well as a systemic immune response which consists of antibody, cytokine and cellular components.

3.1. Antibody Responses

In the past, serological assays were developed in order to measure serum antibodies with VN and complement fixing (CF) activity in vitro [32] [33] . These diagnostic techniques demonstrated VN, which is also called serum neutralising (SN) and CF antibodies appear in the sera of infected horses and are demonstrable from two weeks after experimental EHV-1 infection, up to 1 year or less than 3 months respectively [33] [34] . VN antibodies are largely type specific for a foal immunised with an inactivated EHV-1 vaccine, however there was cross reactivity after immunisation with inactivated EHV-4 [35] . VN antibodies are directed primarily against epitopes on glycoproteins gB and gC as recognised by the immune system of horses infected or vaccinated with EHV-1 [34] . In contrast to the VN antibodies, CF antibodies are usually undetectable by 3 months after infection. The epitopes recognised by CF antibodies are cross reactive with EHV-1 and EHV-4 proteins, resulting in an inaccurate diagnosis using CF antibodies alone [36] .

Previously, EHV-1 antibody in serum samples was tested by several methods; a serum neutralisation (SN) which is equivalent to a VN test, a CF test and the EHV-1 gG specific enzyme-linked immunosorbent assay (ELISA). The antibody titres increased in animals experimentally infected with EHV-1 on day 14 post-infection [4] [37] . Sera samples were previously assessed using EHV-1 gG specific ELISA, detecting antibodies against both EHV-1 and EHV-4 in Crabb et al. [38] . ELISA was also performed to detect EHV-1 specific IgG isotype and IgA antibodies in serum and nasal secretions [39] . The SN test can detect high SN titres following infection or vaccination within 3 months [40] and for up to 1 year after infection [34] . Goehring et al. [4] also found that the SN test could cross-react with the EHV-4 strain which is similar to the CF assay, therefore the results may vary. In the horse, there are five IgG subclasses; IgGa, IgGb, IgGc, IgG (T) and IgG (B), the latter is an aggregating immunoglobulin; and these have been investigated using mAbs specific to IgG subisotypes [41] [42] . In 2004, the horse Ig heavy chain constant region (IGHC) genes which encoded one IgD, one IgM, one IgA, one IgE and seven IgG isotypes were described. All seven genes of IgG heavy chain constant in horses were expressed in vivo [43] . In addition, it described the gene encoding equine IgD [43] . This molecular analysis resulted in new nomenclature of the equine immunoglobulins; equine IGHD and IGHG7 genes. The antibody isotypes produced as a result of infection with EHV-1 have been characterised. Many nasal secretions contain antibodies primarily of the EHV-1 specific IgA type, while IgGa and IgGb were the main isotype responses in serum [39] [44] [45] . Breathnach et al. [44] analysed EHV-1 specific antibody responses by monitoring equine IgGa, IgGb, IgG (T) IgA and IgM in nasal wash samples from weanling foals after primary and subsequent experimental infections with a virulent EHV-1 using an ELISA. IgA was the major antibody isotype detected throughout the sampling period for 13 weeks in nasal mucosa. The mucosal IgA with VN activity was stimulated after experimental primary EHV-1 infection. The IgA response was short lived after a single infection, but it persists longer with subsequent infections until at least 26 weeks. Other studies have also reported that IgA is the primary antibody isotype at the nasal mucosa for protection of horses against several upper respiratory tract pathogens such as Streptococcus equi [46] , EIV [47] [48] .

Comparing the SN titres with kinetic ELISA IgG subclass anti-EHV-1 antibody responses, the IgGa and IgGb responses showed similar results and IgGb isotype responses appeared to associate well with the pattern of SN responses [37] [39] . Although the role of equine IgGc isotype has not been described yet, this isotype response was also measured but it was negligible between infection groups in the kinetic ELISA. It was remarkable that there was a pre-existing IgG (T) response in all experimental horses although SN titers had gradually declined at this time point, indicating IgG (T) response is unlikely to be associated with humoral immunity to EHV-1 [37] . A study by Goodman et al. [49] demonstrated that the IgGb/IgG (T) ratio may be able to protect against EHV-1 infection as the IgGb responses was high in horses administered an inactivated virus vaccine and the horses were partially protected from neurological disease, viral shedding and clinical signs.

3.2. Cytokine Responses

Numerous studies have described cytokine responses to EHV-1 infection by screening for the presence of cytokines' nucleic acid or protein, for example interferon gamma (IFNγ), interferon alpha (IFNα), interleukin-4 (IL- 4), interleukin-10 (IL-10), tumor necrosis factor (TNF-α) [37] [50] - [53] . Interferons were detected in nasal secretions and serum during the first 10 days after experimental EHV-1 infection of ponies [54] . IFN type I which consists of IFNα and IFNβ was detected in the upper respiratory tract. In other species, both these IFNs are very important for protection against viral infection by increasing the synthesis of major histocompatibility complex (MHC) class I molecules in the infected cells, thus making them more susceptible to killing by $CD8^{+}$CTLs [55] -[58] . Apart from type I IFNs, a single type II IF N called IFNγ has an important role in the adaptive immune response to intracellular pathogens. A study demonstrated that a monoclonal antibody (mAb) cross reacts with equine IFNγ produced by PBMC [51] . This equine IFNγ marker, mouse anti bovine IFNγ mAb (clone CC302, IgG1 isotype) detects expression of this protein in PBMC following the flow cytometric analysis [51] [59] . Another study also used the same mAb to measure antigen-specific IFNγ synthesis by equine lymphocytes in order to compare post EHV-1 infection PBMC samples to pre-infection samples [60] . EHV-1 stimulated PBMC collected 10 days post-infection induced a higher percentage of $CD4^{+}$ and $CD8^{+}$ $IFN\gamma^{+}$ T lymphocyte subsets, suggesting that it may contribute to protective immunity to EHV-1 [60] . The percentage of $IFN\gamma^{+}$cells following in vitro stimulation with EHV-1 was undetectable in foals, but increased with age [51] or after vaccination with a vaccinia-virus construct coding for the EHV-1 gene 64 [61] . This increase in antigen specific T cells may indicate that EHV-1 specific memory cells can be boosted after subsequent challenge infection or vac- cination with appropriate antigens. Another study also reported that ponies repeatedly infected with EHV-1 showed IFNγ responses to EHV-1 [52] . The frequency of EHV-1 specific cells was also investigated in a large population of Thoroughbred horses and the results demonstrated higher numbers of IFNγ$^+$ cells in older horses; among 2 year olds and vaccinated horses [62] . After in vitro EHV-1 stimulation, the IFNγ$^+$ cells were phenotyped and predominantly associated with CD8$^+$ cells [50] [51] . These studies also compared the frequency of cytotoxic T lymphocyte precursors (CTLp) with the number of CD8$^+$/IFNγ$^+$ lymphocytes and showed a similar pattern. CTLp are the precursors of CTL found in peripheral blood which develop effector cytotoxic activity in vitro following incubation of PBMC with EHV-1 for approximately 7 days. Therefore, the IFNγ synthesis by CD8$^+$ cells may be used as an alternative indirect measure for CTL activity against EHV-1.

MAbs that specifically detect equine IL-4, IL-10 and IFNγ have been generated and tested in ELISA and intracellular staining to measure these important cytokines [53] [63] [64] . The first multiplex assay for equine cytokines has also been developed to allow simultaneous analysis of IL-4, IL-10 and IFN-α in horses [65] . The measurement of cytokine mRNA in conjunction with polymerase chain reaction (PCR) has also been widely performed to illustrate the stimulation of immune response after infection or vaccination against viruses such as EIV [66] - [68] , equine infectious anaemia virus (EIAV) [69] [70] or Rhodococcus equi [71] . In EHV-1, the upregulation of mRNA expression of IFNγ, IL-2, IL-4, IL-8, IL-10 has been shown in horses after experimental infection or vaccination against EHV-1 [37] [52] [72] [73] .

3.3. Cellular Immune Responses

The cellular immune response is effective in controlling intracellular EHV-1 and consequently to reduce or eli- minate the cell-associated viraemia [74] . Several techniques can be performed to assess the cellular immune responses including measurement of antigen-specific T cell proliferation, detection of CTL activity, cytokine mRNA expression and EHV-1 specific IFNγ synthesis.

Proliferation of lymphocytes is commonly used as an indicator of immune status because proliferation usually precedes the development of effector functions. Previous studies examined the immune response to primary EHV-1 infection by incubating lymphocytes from infected horses with EHV-1 antigens (virus-specific lymphocyte response) [75] - [77] or with phytohemagglutinin (PHA), a lectin which stimulates a non-specific lymphocyte response [75] [76] [78] and by subsequent determination of the uptake of methyl tritiated thymidine ([^{3}H]- thymidine). Generally, these studies showed the proliferation of lymphocytes from reputedly primary infected horses towards EHV-1 antigens, suggesting previous, undetected exposure. Following the PHA response, Dutta et al. [75] and Bumgardner et al. [78] reported an increase of non-specific lymphocyte proliferation, but several other studies described a suppression of the non-specific lymphocyte proliferation [76] [79] [80] . To date, there is no study reported on the response of CTL effectors following a primary EHV-1 infection.

The CTL response is an essential part of the immune defence against many virus infections including EHV-1. EHV-1 becomes intracellular in lymph nodes within hours of contact with host cells, suggesting the importance of the cell mediated immune response to EHV-1 in vivo [12] [14] . After infection, CD8$^+$ T lymphocytes increase in the blood and lung [81] [82] . However, studies of cellular immune response to EHV-1 in infection-immune horses have advanced by focussing on the role of CTL in the clearance of viraemic lymphocytes [83] . In Allen et al. [5] , PBMC were collected from infected immune horses, cultured and restimulated with live, cell-free EHV-1 in vitro for CTLp stimulation. This study demonstrated EHV-1 specific CTLp as early as 1 week post infection with the highest level after 2 to 3 weeks. CD8$^+$ T lymphocytes were also detected as the predominant lymphocyte population involving cytotoxic activity and restricted by MHC class I molecules. With regard to the local cell-mediated immune response, Breathnach et al. [44] analysed the response of CTLp and CTL effectors as well as the EHV-1-specific lymphoproliferative response indicative for Th cell activity from the nasal-asso- ciated lymphoid tissue and local draining lymph nodes nodes at 1 week post inoculation. They found the CTLp and CTL effectors were elicited ex vivo in all horses following EHV-1 inoculation, which suggested that EHV-1 infection can establish virus-specific memory CTL.

The frequency of CTLp was quantified by limiting dilution analysis (LDA) in PBMC of ponies before and after infection with EHV-1 [84] . The results showed that ponies with high numbers of CTLp were protected clinically and virologically against infection with EHV-1 whereas animals with low CTLp numbers were susceptible to infection [84] . When CTLp or memory CTLs are established by vaccination or prior to EHV-1 infection, they can differentiate and mature into CTL effectors which in turn will eliminate the infected target cells on re-infection. Instead of live, cell-free EHV-1 that was used in Allen et al. [5] , Siedek et al. [85] used EHV-1 infected dendritic cells to restimulate PBMC in vitro. Following this stimulation, cytotoxic activity of EHV-1 specific CTLp was restimulated by equine dendritic cells which act as antigen presenting cells in vivo, suggesting that EHV-1 can infect dendritic cells in vitro.

Some studies using lymphocytes from animals infected experimentally with virulent strains of EHV-1 demonstrated a long-term suppression of the proliferation of lymphocytes to mitogen and viral antigen for periods varying between 2 and 10 weeks [79] [86] . The suppressive factor was associated with the increased levels of circulating activated transforming growth factor-β and inhibited T cell response to PHA [79] . Hannant et al. [76] also described a decrease of the non-specific lymphocyte proliferation following the PHA response which was associated with cells, not serum. It was suggested that the immunosuppression may be due to the viral replication within the lymphocytes which interferes with the cells' ability to enter mitogenesis [87] .

Identification of EHV-1 proteins which contain ELA-A restricted CTL epitopes is important as they may be included in recombinant or DNA vaccines [10] . In this study, PBMC were collected from ponies which had been primed to EHV-1 by previous infection. PBMC were then incubated for 1 week with equine dendritic cells transfected with individual EHV-1 plasmids encoding individual EHV-1 proteins, in order to stimulate autologous CTLp populations. Cytotoxicity was measured against EHV-1-infected PWM lymphoblast targets. This demonstrated that the IE protein elicited CTL responses [10] and later studies demonstrated that presentation of the IE protein's epitopes was restricted by the B2 allele of the ELA-A3 haplotype [9] . Although cytotoxicity assays are very useful for the measurement of CTL activity and frequency during the recovery phase of EHV-1 infection, they are time consuming and demanding. As an alternative, IFNγ synthesis detected by intracellular IFNγ staining in PBMC or enzyme-linked immunosorbent spot (ELISPOT) has been used as a measure of lymphocyte activation [51] .

4. INTERACTION OF EHV-1 WITH THE PERIPHERAL BLOOD MONONUCLEAR CELLS (PBMC)

The majority of studies have focused on the role of PBMC in EHV-1 viraemia which transmits the infectious virus from the primary site of infection to the vasculature of susceptible tissues. Some studies have reported which PBMC subpopulations are infected. An in vivo study was first performed in 1983 to identify infected PBMC by using PBMC from EHV-1 inoculated pony foals. The sub-populations of T and B lymphocytes and monocytes were separated to detect and quantify the presence of virus within these cells. Infected PBMC were detected at day 2 to 14, whereas the number of infected cells was maximal on day 4 post inoculation. The T lymphocyte population was identified as the most predominant cell type infected with EHV-1 during viraemia [88] . Furthermore, this study also suggested that pokeweed mitogen (PWM) can promote virus replication in PBMC as the virus was detected in a 10-high fold higher number of PWM stimulated cells. A similar study used in vitro infection of PBMC [89] with similar results. Subsequent in vitro work by Van der Meulen et al. [90] discovered EHV-1 replication in the resting monocytes after in vitro infection of PBMC as well as in T lymphocyte from mitogen stimulated PBMC. In an in vitro study, Gryspeerdt et al. [91] used two EHV-1 strains; neurovirulent and non-neurovirulent to inoculate ponies intranasally. Interestingly, both strains infected cells of the monocyte lineage ($CD172\alpha^+$) rather than of T lymphocyte ($CD5^+$) in the mandibular lymph node and PBMC. In addition, Goodman et al. [92] also looked at the infected cells of all PBMC subpopulations in vitro using EHV-1 green fluorescent protein virus and reported that monocyte was the predominant infected cells. Another previous study demonstrated for the first time that EHV-1 is detectable in all PBMC subpopulations in vivo with $CD8^+$ lymphocytes and B lymphocytes the most frequently infected subpopulations [93] .

5. IMMUNOMODULATORY PROPERTIES OF EHV-1

During evolution, like other herpesviruses, EHV-1 has developed immune evasion mechanisms to avoid host immunity [94] - [98] . Numerous in vitro findings on EHV-1 immunomodulation have been reported [76] [79] [94] [99] - [101] . However, there are only few studies on the in vivo EHV-1 immune evasion mechanism [92] [97] [102] looking at the clinical outcomes and viraemia. Examples of the evasion methods are as follows. Immunosuppression involving a reduction of in vitro PBMC proliferation to mitogens and viral antigens following EHV-1 infection of ponies [76] [86] , avoiding antibody-dependent immune responses by inhibiting cell surface expression of viral antigens [103] , inducing down-regulation of MHC class I expression on infected cells [101] and inhibiting transporter associated with antigen processing (TAP) protein and blocking viral peptide transport to MHC class I molecules in the endoplasmic reticulum [94] .

6. EHV-1 PROSPECTS

Adaptive immune responses to EHV-1 have been widely studied [37] [44] [76] [83] [104] -[107] . EHV-1 promotes both humoral and cellular immune responses. VN antibodies appear approximately 2 weeks after EHV-1 infection [3] and function to reduce the amount and duration of nasopharyngeal virus shedding [3] [106] , however, they do not play a major role in reducing cell-associated viraemia. However, one study indicates that a commercial vaccine which contains inactivated virus and thus stimulates VN antibody was successful in reducing the viraemia [4] . Cellular immune responses, particularly MHC class I restricted $CD8^+$ CTLs are most crucial in protection against EHV-1 infection [7] -[9] [84] and it has been reported that the number of circulating CTLp offers an immunological correlate of protection from infection and recovery [7] [8] [84] . Consequently, rational design of new EHV-1 vaccines has focused on products with the best potential for stimulation of CTL responses and MLV which have been reported to perform the best [107] .

Previous studies have demonstrated that the EHV-1 viral gene 64 (IE gene) acts as a target for effector CTL within the equine ELA-A3 MHC class I haplotype B2 allele [9] [10] . This information has proved crucial to understanding the development of cell mediated immunity specific for EHV-1. Taken together, these published research findings emphasise the importance of defining the EHV-1 target peptides in the IE protein encoded by gene 64 that may function as stimulators and targets for CTL restricted by the B2 allele, expressed on the ELA- A3 genotype. The identification of target peptide is a logical step towards the development of new vaccines and offers great potential for construction of MHC class I virus peptide tetramers and the establishment of CTL lines and clones. These reagents will enable further understanding of the mechanisms which underlie the cellular immune response to EHV-1 in vivo and in vitro.

7. CONCLUSION

Nor Dini Rusli, Khairiyah Binti Mat, Hasnita Che Harun (s) of the IE protein which is recognised by the immune system.

REFERENCES

1. Ostlund, E.N. (1993) The Equine Herpesviruses. Veterinary Clinics of North America: Equine Practice, 9, 283-294.
2. Mumford, J.A. (1992) Progress in the Control of Equine Influenza. In: Plowright, W., Rossdale, P.D. and Wade, J.F., Eds., Equine Infectious Diseases, VI. Proceedings of the 6th International Conference, R & W Publications, Newmarket, 207-217.
3. Hannant, D., Jessett, D., O'Neill, T., Dolby, C.A., Cook, R.F. and Mumford, J.A. (1993) Responses of Ponies to Equid Herpesvirus-1 ISCOM Vaccination and Challenge with Virus to the Homologous Strain. Research in Veterinary Science, 54, 299-305. http://dx.doi.org/10.1016/0034-5288(93)90126-Z

4. Goehring, L.S., Wagner, B., Bigbie, R., Hussey, S.B., Morley, P.S. and Lunn, D.P. (2010) Control of EHV-1 Viremia and Nasal Shedding by Commercial Vaccines. Vaccine, 28, 5203-5211. http://dx.doi.org/10.1016/j.vaccine.2010.05.065
5. Allen, G.P., Yeargan, M., Costa, L.R.R. and Cross, R. (1995) Major Histocompatibility Complex Class I-Restricted Cytotoxic T-Lymphocyte Responses in Horses Infected with Equine Herpesvirus-1. Journal of Virology, 69, 606-612.
6. Allen, G.P., Kydd, J.H., Slater, J.D. and Smith, K.C. (2004) Equid Herpesvirus-1 (EHV-1) and -4 (EHV-4) Infections. In: Coetzer, J.A.W. and Tustin, R.C., Eds., Infectious Diseases of Livestock, 2nd Edition, Oxford Press, Cape Town, 829-859.
7. Kydd, J.H., Wattrang, E. and Hannant, D. (2003) Pre-Infection Frequencies of Equine Herpesvirus-1 Specific, Cytotoxic T Lymphocytes Correlate with Protection against Abortion Following Experimental Infection of Pregnant Mares. Veterinary Immunology and Immunopathology, 96, 207-217. http://dx.doi.org/10.1016/j.vetimm.2003.08.004
8. Allen, G.P. (2008) Risk Factors for Development of Neurologic Disease after Experimental Exposure to Equine Herpesvirus-1 in Horses. American Journal of Veterinary Research, 69, 1595-1600. http://dx.doi.org/10.2460/ajvr.69.12.1595
9. Kydd, J.H., Townsend, H.G.G. and Hannant, D. (2006) The Equine Immune Response to Equine Herpesvirus-1: The Virus and Its Vaccines. Veterinary Immunology and Immunopathology, 111, 15-30. http://dx.doi.org/10.1016/j.vetimm.2006.01.005
10. Soboll, G., Whalley, J.M., Koen, M.T., Allen, G.P., Fraser, D.G., Macklin, M.D., Swain, W.F. and Lunn, D.P. (2003) Identification of Equine Herpesvirus-1 Antigens Recognized by Cytotoxic T Lymphocytes. Journal of General Virology, 84, 2625-2634. http://dx.doi.org/10.1099/vir.0.19268-0
11. Allen, G.P., Bolin, D.C., Bryant, U., Carter, C.N., Giles, R.C., Harrison, L.R., Hong, C.B., Jackson, C.B., Poonacha, K., Wharton, R. and Williams, N.M. (2008) Prevalence of Latent, Neuropathogenic Equine Herpesvirus-1 in the Thoroughbred Broodmare Population of Central Kentucky. Equine Veterinary Journal, 40, 105-110.
12. Kydd, J.H., Smith, K.C., Hannant, D., Livesay, G.J. and Mumford, J.A. (1994) Distribution of Equid Herpesvirus-1 (EHV-1) in the Respiratory Tract of Ponies: Implication for Vaccination Strategies. Equine Veterinary Journal, 26, 466-469. http://dx.doi.org/10.1111/j.2042-3306.1994.tb04051.x
13. Patel, J.R., Edington, N. and Mumford, J.A. (1982) Variation in Cellular Tropism between Isolates of Equine Herpesvirus-1 in Foals. Archives of Virology, 74, 41-51. http://dx.doi.org/10.1007/BF01320781
14. Kydd, J.H., Smith, K.C., Hannant, D., Livesay, G.J. and Mumford, J.A. (1994) Distribution of Equid Herpesvirus-1 (EHV-1) in the Respiratory Tract-Associated Lymphoid Tissue: Implications for Cellular Immunity. Equine Veterinary Journal, 26, 470-473. http://dx.doi.org/10.1111/j.2042-3306.1994.tb04052.x
15. Allen, G.P. and Bryans, J.T. (1986) Molecular, Epizootiology, Pathogenesis and Prophylaxis of Equine Herpes Virus-1 Infections. Progress in Veterinary Microbiology and Immunology, 2, 78-144.
16. Edington, N., Smyth, B. and Griffiths, L. (1991) The Role of Endothelial Cell Infection in the Endometrium, Placenta and Foetus of Equid Herpesvirus 1 (EHV-1) Abortions. Journal of Comparative Pathology, 104, 379-387. http://dx.doi.org/10.1016/S0021-9975(08)80148-X
17. Smith, K.C., Whitwell, K.E., Binns, M.M., Dolby, C.A., Hannant, D. and Mumford, J.A. (1992) Abortion of Virologically Negative Foetuses Following Experimental Challenge of Pregnant Pony Mares with Equid Herpesvirus 1. Equine Veterinary Journal, 24, 256-259. http://dx.doi.org/10.1111/j.2042-3306.1992.tb02830.x
18. Smith, K.C. and Borchers, K. (2001) A Study of the Pathogenesis of Equid Herpesvirus-1 (EHV-1) Abortion by DNA In-Situ Hybridization. Journal of Comparative Pathology, 125, 304-310. http://dx.doi.org/10.1053/jcpa.2001.0513
19. Smith, K.L., Allen, G.P., Branscum, A.J., Frank Cook, R., Vickers, M.L., Timoney, P.J. and Balasuriya, U.B. (2010) The Increased Prevalence of Neuropathogenic Strains of EHV-1 in Equine Abortions. Veterinary Microbiology, 141, 5-11. http://dx.doi.org/10.1016/j.vetmic.2009.07.030

20. Smith, K.C., Whitwell, K.E., Blunden, A.S., Bestbier, M.E., Scase, T.J., Geraghty, R.J., Nugent, J., Davis-Poynter, N.J. and Cardwell, J.M. (2004) Equine Herpesvirus-1 Abortion: Atypical Cases with Lesions Largely or Wholly Restricted to the Placenta. Equine Veterinary Journal, 36, 79-82. http://dx.doi.org/10.2746/0425164044864732
21. Smith, K.C., Mumford, J.A. and Lakhani, K. (1996) A Comparison of Equid Herpesvirus-1 (EHV-1) Vascular Lesions in the Early versus Late Pregnant Equine Uterus. Journal of Comparative Pathology, 114, 231-247. http://dx.doi.org/10.1016/S0021-9975(96)80045-4
22. Jackson, J.C., Golden, G.T. and Benjamin, R.M. (1977) The Distribution of Olfactory Input in the Opossum Mediodorsal Nucleus. Brain Research, 138, 229-240. http://dx.doi.org/10.1016/0006-8993(77)90742-9
23. Platt, H., Singh, H. and Whitwellm, K.E. (1980) Pathological Observations on an Outbreak of Paralysis in Broodmares. Equine Veterinary Journal, 12, 118-126. http://dx.doi.org/10.1111/j.2042-3306.1980.tb03398.x
24. Edington, N., Bridges, C.G. and Patel, J.R. (1986) Endothelial Cell Infection and Thrombosis in Paralysis Caused by Equid Herpesvirus-1: Equine Stroke. Archives of Virology, 90, 111-124. http://dx.doi.org/10.1007/BF01314149
25. Roizman, B., Desrosiers, R.S., Fleckenstein, B., Lopez, C., Minson, A.C. and Studdert, M.J. (1992) The Family Herpesviridae: An Update. Archives of Virology, 123, 425-449. http://dx.doi.org/10.1007/BF01317276
26. Edington, N., Welch, H.M. and Griffiths, L. (1994) The Prevalence of Latent Equid Herpesviruses in the Tissues of 40 Abattoir Horses. Equine Veterinary Journal, 26, 140-142. http://dx.doi.org/10.1111/j.2042-3306.1994.tb04353.x
27. Slater, J.C., Borchers, K., Thackray, A.M. and Field, H. (1994) The Trigeminal Ganglion Is a Site of Equine Herpsvirus-1 (EHV-1) Latency and Reactivation in the Horse. Journal of General Virology, 75, 2007-2016. http://dx.doi.org/10.1099/0022-1317-75-8-2007
28. Welch, H.M., Bridges, C.G., Lyon, A.M., Griffiths, L. and Edington, N. (1992) Latent Equid Herpesviruses 1 and 4: Detection and Distinction Using the Polymerase Chain Reaction and Co-Cultivation from Lymphoid Tissues. Journal of General Virology, 73, 261-268. http://dx.doi.org/10.1099/0022-1317-73-2-261
29. Chester, P.M., Allsop, R., Purewal, A. and Edington, N. (1997) Detection of Latency-Associated Transcripts of Equid Herpesvirus 1 in Equine Leukocytes but Not in Trigeminal Ganglia. Journal of Virology, 71, 3437-3443.
30. Smith, P.M., Zhang, Y., Jennings, S.R. and O'Callaghan, D.J. (1998) Characterization of the Cytolytic T-Lymphocyte Response to a Candidate Vaccine Strain of Equine Herpesvirus 1 in CBA Mice. Journal of Virology, 72, 5366-5372.
31. Chesters, P.M., Hughes, A. and Edington, N. (2000) Equid Herpesvirus 1: Platelets and Alveolar Macrophages Are Potential Sources of Activated TGF-B1 in the Horse. Veterinary Immunology and Immunopathology, 75, 71-79. http://dx.doi.org/10.1016/S0165-2427(00)00187-2
32. Doll, E.R., Wallace, M.E., Bryans, J.T. and Richards, M.G. (1953) Complement-Fixation Antibody Response Following Administration of Equine Virus Abortion Vaccine. American Journal of Veterinary Research, 14, 46-48.
33. Thomson, G.R., Mumford, J.A., Campbell, J., Griffiths, L. and Clapham, P. (1976) Serological Detection of Equid Herpesvirus 1 Infections of the Respiratory Tract. Equine Veterinary Journal, 8, 58-65. http://dx.doi.org/10.1111/j.2042-3306.1976.tb03291.x
34. Allen, G., Coogle, L.D., Ostlund, E.N. and Yeargan, M.R. (1992) Molecular Dissection of Two Major Equine Herpesvirus-1 Glycoprotein Antigens (Gb and Gc) that Elicit Humoral Immune Responses in the Horse. In: Plowright, W., Rossdale, P.D. and Wade, J.F., Eds., Equine Infectious Diseases VI, R&W Publications, Newmarket, 181-193.
35. Fitzpatrick, D.R. and Studdert, M.J. (1984) Immunologic Relationships between Equine Herpesvirus Type 1 (Equine Abortion Virus) and Type 4 (Equine Rhinopneumonitis Virus). American Journal of Veterinary Research, 45, 1947- 1952.
36. Crabb, B.S. and Studdert, M.J. (1993) Epitopes of Glycoprotein G of Equine Herpesviruses 4 and 1 Located Near the C Termini Elicit Type Specific Antibody Responses in the Natural Host. Journal of Virology, 67, 6332-6338.

37. Soboll Hussey, G., Hussey, S.B., Wagner, B., Horohov, D.W., Van De Walle, G.R., Osterrieder, N., Goehring, L.S., Rao, S. and Lunn, D.P. (2011) Evaluation of Immune Responses Following Infection of Ponies with an EHV-1 ORF1/2 Deletion Mutant. Veterinary Research, 42, 23. http://dx.doi.org/10.1186/1297-9716-42-23
38. Crabb, B.S., Macpherson, C.M., Reubel, G.H., Browning, G.F., Studdert, M.J. and Drummer, H.E. (1995) A Type- Specific Serological Test to Distinguish Antibodies to Equine Herpesviruses 4 and 1. Archives of Virology, 140, 245- 258. http://dx.doi.org/10. 1007/BF01309860
39. Soboll, G., Hussey, S.B., Whalley, J.M., Allen, G.P., Koen, M.T., Santucci, N., Fraser, D.G., Macklin, M.D., Swain, W.F. and Lunn, D.P. (2006) Antibody and Cellular Immune Responses Following DNA Vaccination and EHV-1 Infection of Ponies. Veterinary Immunology and Immunopathology, 111, 81-95. http://dx.doi.org/10. 1016/j. vetimm.2006.01.011
40. Mumford, E.L., Traub-Dargatz, J.L., Carman, J., Callan, R.J., Collins, J.K., Goltz, K.L., Romm, S.R., Tarr, S.F. and Salman, M.D. (2003) Occurrence of Infectious Upper Respiratory Tract Disease and Response to Vaccination in Horses on Six Sentinel Premises in Northern Colorado. Equine Veterinary Journal, 35, 72-77. http://dx.doi.org/ 10.2746/ 042516403775467379
41. Sheoran, A.S., Lunn, D.P. and Holmes, M.A. (1998) Monoclonal Antibodies to Subclass-Specific Antigenic Determinants on Equine Immunoglobulin Gamma Chains and Their Characterization. Veterinary Immunology and Immunopathology, 62, 153-165. http:// dx.doi.org/10.1016/S0165-2427(97)00162-1
42. Lunn, D.P., Holmes, M.A., Schram, B. and Duffus, W.P. (1995) Monoclonal Antibodies Specific for Equine IgG Sub- Isotypes Including an Antibody Which Recognizes B Lymphocytes. Veterinary Immunology and Immunopathology, 47, 239-251. http:// dx.doi.org/10.1016/0165-2427(95)97067-J
43. Wagner, B., Miller, D.C., Lear, T.L. and Antczak, D.F. (2004) The Complete Map of the Ig Heavy Chain Constant Gene Region Reveals Evidence for Seven IgG Isotypes and for IgD in the Horse. Journal of Immunology, 173, 3230- 3242. http://dx.doi.org/10.4049/ jimmunol.173.5.3230
44. Breathnach, C.C., Yeargan, M.R., Sheoran, A.S. and Allen, G.P. (2001) The Mucosal Humoral Immune Response of the Horse to Infective Challenge and Vaccination with Equine Herpesvirus-1 Antigens. Equine Veterinary Journal, 33, 651-657. http://dx.doi.org/ 10.2746/042516401776249318
45. Ruitenberg, K.M., Love, D.N., Gilkerson, J.R., Wellington, J.E. and Whalley, J.M. (2000) Equine Herpesvirus 1 (EHV-1) Glycoprotein D DNA Inoculation in Horses with Pre-Existing EHV-1/EHV-4 Antibody. Veterinary Microbiology, 76, 117-127. http://dx. doi.org/10.1016/S0378-1135(00)00237-6
46. Sheoran, A.S., Sponseller, B.T., Holmes, M.A. and Timoney, J.F. (1997) Serum and Mucosal Antibody Isotype Responses to M-Like Protein (Sem) of Streptococcus equi in Convalescent and Vaccinated Horses. Veterinary Immunology and Immunopathology, 59, 239-251. http://dx.doi.org/10.1016/S0165-2427(97)00074-3
47. Nelson, K.M., Schram, B.R., Mcgregor, M.W., Sheoran, A.S., Olsen, C.W. and Lunn, D.P. (1998) Local and Systemic Isotype-Specific Antibody Responses to Equine Influenza Virus Infection versus Conventional Vaccination. Vaccine, 16, 1306-1313. http://dx.doi.org/ 10.1016/S0264-410X(98)00009-7
48. Hannant, D., Easeman, R. and Mumford, J.A. (1999) Equine Mucosal Immune System: Intranasal Vaccination with Inactivated Equine Influenza Virus Protects from Infection. In: Wernery, O., Wade, J.F., Mumford, J.A. and Kaaden, O.R., Eds., Proceedings of the 8th International Conference on Equine Infectious Diseases, R&W Publications Ltd., Newmarket, 50-56.
49. Goodman, L.B., Wagner, B., Flaminio, M.J., Sussman, K.H., Metzger, S.M., Holland, R. and Osterrieder, N. (2006) Comparison of the Efficacy of Inactivated Combination and Modified-Live Virus Vaccines against Challenge Infection with Neuropathogenic Equine Herpesvirus Type 1 (EHV-1). Vaccine, 24, 3636-3645. http://dx.doi.org/10. 1016/j. vaccine.2006.01.062

50. Paillot, R., Daly, J.M., Luce, R., Montesso, F., Davis-Poynter, N., Hannant, D. and Kydd, J.H. (2007) Frequency and Phenotype of EHV-1 Specific, IFN-Gamma Synthesizing Lymphocytes in Ponies: The Effects of Age, Pregnancy and Infection. Developmental and Comparative Immunology, 31, 202-214. http://dx.doi.org/10.1016/j.dci.2006.05.010
51. Paillot, R., Daly, J.M., Juillard, V., Minke, J.M., Hannant, D. and Kydd, J.H. (2005) Equine Interferon Gamma Synthesis in Lymphocytes after in Vivo Infection and in Vitro Stimulation with EHV-1. Vaccine, 23, 4541-4551. http://dx.doi.org/ 10.1016/ j.vaccine.2005.03.048
52. Coombs, D.K., Patton, T., Kohler, A.K., Soboll, G., Breathnach, C., Townsend, H.G. and Lunn, D.P. (2006) Cytokine Responses to EHV-1 Infection in Immune and Non-Immune Ponies. Veterinary Immunology and Immunopathology, 111, 109-116. http://dx.doi.org/ 10.1016/j.vetimm.2006.01.013
53. Wagner, B., Hillegas, J.M. and Antczak, D.F. (2006) A Monoclonal Antibody to Equine Interleukin 4. Veterinary Immunology and Immunopathology, 110, 363-367. http:// dx.doi.org/10.1016/j.vetimm.2006.01.001
54. Edington, N., Bridges, C.G. and Griffiths, L. (1989) Equine Interferons Following Exposure to Equid Herpesvirus-1 or -4. Journal of Interferon Research, 9, 389-392. http:// dx.doi.org/10.1089/jir.1989.9.389
55. Durbin, J.E., Fernandez-Sesma, A., Lee, C.K., Rao, T.D., Frey, A.B., Moran, T.M., Vukmanovic, S., García-Sastre, A. and Levy, D.E. (2000) Type I IFN Modulates Innate and Specific Antiviral Immunity. Journal of Immunology, 164, 4220-4228. http://dx.doi.org/ 10.4049/jimmunol.164.8.4220
56. Ito, T., Wang, Y.H. and Liu, Y.J. (2005) Plasmacytoid Dendritic Cell Precursors/Type I Interferon-Producing Cells Sense Viral Infection by Toll-Like Receptor (TLR) 7 and TLR9. Springer Seminars in Immunopathology, 26, 221-229. http://dx.doi.org/10.1007/ s00281-004-0180-4
57. Jung, A., Kato, H., Kumagai, Y., Kumar, H., Kawai, T., Takeuchi, O. and Akira, S. (2008) Lymphocytoid Choriomeningitis Virus Activates Plasmacytoid Dendritic Cells and Induces a Cytotoxic T-Cell Response via MyD88. Journal of Virology, 82, 196-206. http://dx.doi.org/10.1128/JVI.01640-07
58. Szabo, G. and Dolganiuc, A. (2008) The Role of Plasmacytoid Dendritic Cell Derived IFNα in Antiviral Immunity. Critical Reviews in Immunology, 28, 61-94. http://dx.doi.org/ 10.1615/CritRevImmunol.v28.i1.40
59. Pedersen, L.G., Castelruiz, Y., Jacobsen, S. and Aasted, B. (2002) Identification of Monoclonal Antibodies that Cross- React with Cytokines from Different Animal Species. Veterinary Immunology and Immunopathology, 88, 111-122. http://dx.doi.org/ 10.1016/S0165-2427(02)00139-3
60. Paillot, R., Ellis, S.A., Daly, J.M., Audonnet, J.C., Minke, J.M., Davis-Poynter, N., Hannant, D. and Kydd, J.H. (2006) Characterisation of CTL and IFN-γ Synthesis in Ponies Following Vaccination with a NYVAC-Based Construct Coding for EHV-1 Immediate Early Gene, Followed by Challenge Infection. Vaccine, 24, 1490-1500. http://dx.doi.org/10. 1016/ j.vaccine.2005.10.019
61. Breathnach, C.C., Soboll, G., Suresh, M. and Lunn, D.P. (2005) Equine Herpesvirus-1 Infection Induces IFN-γ Pro- duction by Equine T Lymphocyte Subsets. Veterinary Immunology and Immunopathology, 103, 207-215. http://dx.doi.org/10. 1016/j. vetimm.2004.09.024
62. Luce, R., Shepherd, M., Paillot, R., Blacklawst, B., Wood, J.L. and Kydd, J.H. (2007) Equine Herpesvirus-1-Specific Intereferon Gamma (IFNγ) Synthesis by Peripheral Blood Mononuclear Cells in Thoroughbred Horses. Equine Veterinary Journal, 39, 202-209. http://dx.doi.org/10.2746/042516407X174216
63. Wagner, B., Hillegas, J.M., Brinker, D.R., Horohov, D.W. and Antczak, D.F. (2008) Characterization of Monoclonal Antibodies to Equine Interleukin-10 and Detection of T Regulatory 1 Cells in Horses. Veterinary Immunology and Immunopathology, 122, 57-64. http://dx.doi.org/10.1016/j.vetimm.2007.10.012
64. Wagner, B., Hillegas, J.M., Flaminio, M.J.B.F. and Wattrang, E. (2008) Monoclonal Antibodies to Equine Interferon-α (IFN-α): New Tools to Neutralize IFN Activity and to

Detect Secreted IFN-α. Veterinary Immunology and Immunopathology, 125, 315-325. http://dx.doi.org/10.1016/j.vetimm.2008.05.016
65. Wagner, B. and Freer, H. (2009) Development of a Bead-Based Multiplex Assay for Simultaneous Quantification of Cytokines in Horses. Veterinary Immunology and Immunopathology, 127, 242-248. http://dx.doi.org/10.1016/j.vetimm.2008.10.313
66. Soboll, G., Horohov, D.W., Aldridge, B.M., Olsen, C.W., Mcgregor, M.W., Drape, R.J., Macklin, M.D., Swain, W.F. and Lunn, D.P. (2003) Regional Antibody and Cellular Immune Responses to Equine Influenza Virus Infection, and Particle Mediated DNA Vaccination. Veterinary Immunology and Immunopathology, 94, 47-62. http://dx.doi.org/10.1016/S0165-2427(03)00060-6
67. Breathnach, C.C., Rudersdorf, R. and Lunn, D.P. (2004) Use of Recombinant Modified Vaccinia Ankara Viral Vectors for Equine Influenza Vaccination. Veterinary Immunology and Immunopathology, 98, 127-136. http://dx.doi.org/10.1016/j.vetimm.2003.11.004
68. Breathnach, C.C., Clark, H.J., Clark, R.C., Olsen, C.W., Townsend, H.G. and Lunn, D.P. (2006) Immunization with Recombinant Modified Vaccinia Ankara (rMVA) Constructs Encoding the HA or NP Gene Protects Ponies from Equine Influenza Virus Challenge. Vaccine, 24, 1180-1190. http://dx.doi.org/10.1016/j.vaccine.2005.08.091
69. Fraser, D.G., Oaks, J.L., Brown, W.C. and Mcguire, T.C. (2002) Identification of Broadly Recognized, T Helper 1 Lymphocyte Epitopes in an Equine Lentivirus. Immunology, 105, 295-305. http://dx.doi.org/10.1046/j.0019-2805.2001.01370.x
70. Fraser, D.G., Mealey, R.H. and Mcguire, T.C. (2003) Selecting Peptides to Optimize Th1 Responses to an Equine Lentivirus Using HLA-DR Binding Motifs and Defined HIV-1 Th Peptides. Immunogenetics, 55, 508-514. http://dx.doi.org/10.1007/s00251-003-0600-y
71. Lopez, A.M., Hines, M.T., Palmer, G.H., Alperin, D.C. and Hines, S.A. (2002) Identification of Pulmonary T-Lym- phocyte and Serum Antibody Isotype Responses Associated with Protection against Rhodococcus equi. Clinical and Diagnostic Laboratory Immunology, 9, 1270-1276.
72. Pusterla, N., Wilson, W.D., Conrad, P.A., Barr, B.C., Ferraro, G.L., Daft, B.M. and Leutenegger, C.M. (2006) Cytokine Gene Signatures in Neural Tissue of Horses with Equine Protozoal Myeloencephalitis or Equine Herpes Type 1 Myeloencephalopathy. Veterinary Record, 159, 341-345. http://dx.doi.org/10.1136/vr.159.11.341
73. Wagner, B., Burton, A. and Ainsworth, D.M. (2010) Interferon-Gamma, Interleukin-4 and Interleukin-10 Production by T Helper Cells Reveals Intact Th1 and Regulatory TR1 Cell Activation and a Delay of the Th2 Cell Response in Equine Neonates and Foals. Veterinary Research, 41, 47. http://dx.doi.org/10.1051/vetres/2010019
74. Soboll, G., Breathnach, C.C., Kydd, J.H., Hussey, S.B., Mealey, R.M. and Lunn, D.P. (2010) Vaccination of Ponies with the IE Gene of EHV-1 in a Recombinant Modified Live Vaccinia Vector Protects against Clinical and Virological Disease. Veterinary Immunology and Immunopathology, 135, 108-117. http://dx.doi.org/10.1016/j.vetimm.2009.11.009
75. Dutta, S.K., Myrup, A. and Bumgardner, M.K. (1980) Lymphocyte Responses to Virus and Mitogen in Ponies during Experimental Infection with Equine Herpesvirus 1. American Journal of Veterinary Research, 41, 2066-2068.
76. Hannant, D., O'Neill, T., Jessett, D.M. and Mumford, J.A. (1991) Evidence for Non-Specific Immunosuppression during the Development of Immune Responses to Equid Herpesvirus-1. Equine Veterinary Journal, 23, 41-45.
77. Chong, Y.C. and Duffus, W.P. (1992) Immune Responses of Specific Pathogen Free Foals to EHV-1 Infection. Veterinary Microbiology, 32, 215-239. http://dx.doi.org/10.1016/0378-1135(92)90146-K
78. Bumgardner, M.K., Dutta, S.K., Campbell, D.L. and Myrup, A.C. (1982) Lymphocytes from Ponies Experimentally Infected with Equine Herpesvirus 1: Subpopulation Dynamics and Their Response to Mitogens. American Journal of Veterinary Research, 43, 1308-1310.
79. Charan, S., Palmer, K., Chester, P., Mire-Sluis, A.R., Meager, A. and Edington, N. (1997) Transforming Growth Factor-β Induced by Live or Ultraviolet-Inactivated Equid Herpes Virus Type-1 Mediates Immunosuppression in the Horse. Immunology, 90, 586-591. http://dx.doi.org/10.1046/j.1365-2567.1997.00202.x

80. Van Der Meulen, K.M., Favoreel, H.W., Pensaert, M.B. and Nauwynck, H.J. (2006) Immune Escape of Equine Herpesvirus 1 and Other Herpesviruses of Veterinary Importance. Veterinary Immunology and Immunopathology, 111, 31- 40. http://dx. doi.org/10.1016/j.vetimm.2006.01.006
81. Kydd, J.H., Hannant, D. and Mumford, J.A. (1996) Residence and Recruitment of Leucocytes to the Equine Lung after EHV-1 Infection. Veterinary Immunology and Immunopathology, 52, 15-26. http://dx.doi.org/10.1016/0165-2427(95)05533-9
82. Lunn, D.P., Holmes, M.A. and Duffus, W.P. (1991) Three Monoclonal Antibodies Identifying Antigens on All Equine T Lymphocytes, and Two Mutually Exclusive T-Lymphocyte Subsets. Immunology, 74, 251-257.
83. Allen, G.P., Kydd, J.H., Slater, J.D. and Smith, K.C. (1999) Advances in Understanding of the Pathogenesis, Epidemiology, and Immunological Control of Equid Herpesvirus Abortion. In: Wernery, U., Wade, J.F., Mumford, J.A. and Kaaden, O.R., Eds., Equine Infectious Diseases VIII. Proceedings of the 8th International Conference, R & W Publications, Newmarket, 129-146.
84. O'Neill, T., Kydd, J.H., Allen, G.P., Wattrang, E., Mumford, J.A. and Hannant, D. (1999) Determination of Equid Herpesvirus 1-Specific, CD^{8+}, Cytotoxic T Lymphocyte Precursor Frequencies in Ponies. Veterinary Immunology and Immunopathology, 70, 43-54. http://dx.doi.org/10.1016/S0165-2427(99)00037-9
85. Siedek, E.M., Whelan, M., Edington, N. and Hamblin, A. (1999) Equine Herpesvirus Type 1 Infects Dendritic Cells in Vitro: Stimulation of T Lymphocyte Proliferation and Cytotoxicity by Infected Dendritic Cells. Veterinary Immunology and Immunopathology, 67, 17-32. http://dx.doi.org/10.1016/S0165-2427(98)00203-7
86. Hannant, D., O'Neill, T., Ostlund, E.N., Kydd, J.H., Hopkin, P.J. and Mumford, J.A. (1999) Equid Herpesvirus-In- duced Immunosuppression Is Associated with Lymphoid Cells and Not Soluble Circulating Factors. Viral Immunology, 12, 313-321. http://dx.doi.org/10.1089/vim.1999.12.313
87. O'Neill, T. (1995) T Lymphocyte Responses to Equid Herpesviruses-1 and -4 in Horses. Ph.D. Thesis, Open University, London.
88. Scott, J.C., Dutta, S.K. and Myrup, A.C. (1983) In Vivo Harboring of Equine Herpesvirus-1 in Leukocyte Populations and Subpopulations and Their Quantitation from Experimentally Infected Ponies. American Journal of Veterinary Research, 44, 1344-1348.
89. Dutta, S.K. and Myrup, A.C. (1983) Infectious Center Assay of Intracellular Virus and Infective Virus Titer for Equine Mononuclear Cells Infected in Vivo and in Vitro with Equine Herpesviruses. Canadian Journal of Comparative Medicine, 47, 64-69.
90. Van Der Meulen, K.M., Nauwynck, H.J., Buddaert, W. and Pensaert, M.B. (2000) Replication of Equine Herpesvirus Type 1 in Freshly Isolated Equine Peripheral Blood Mononuclear Cells and Changes in Susceptibility Following Mitogen Stimulation. Journal of General Virology, 81, 21-25.
91. Gryspeerdt, A.C., Vandekerckhove, A.P., Garre, B., Barbe, F., Van De Walle, G.R. and Nauwynck, H.J. (2010) Differences in Replication Kinetics and Cell Tropism between Neurovirulent and Non-Neurovirulent EHV1 Strains during the Acute Phase of Infection in Horses. Veterinary Microbiology, 142, 242-253. http://dx.doi.org/10. 1016/j.vetmic.2009.10.015
92. Goodman, L.B., Loregian, A., Perkins, G.A., Nugent, J., Buckles, E.L., Mercorelli, B., Kydd, J.H., Palù, G., Smith, K.C., Osterrieder, N. and Davis-Poynter, N. (2007) A Point Mutation in a Herpesvirus Polymerase Determines Neuropathogenicity. PLoS Pathogens, 3, e160. http://dx.doi.org/10.1371/journal.ppat.0030160
93. Wilsterman, S., Soboll-Hussey, G., Lunn, D.P., Ashton, L.V., Callan, R.J., Hussey, S.B., Rao, S. and Goehring, L.S. (2011) Equine Herpesvirus-1 Infected Peripheral Blood Mononuclear Cell Subpopulations during Viremia. Veterinary Microbiology, 149, 40-47. http://dx.doi.org/10.1016/j.vetmic.2010.10.004
94. Ambagala, A.P.N., Gopinath, R.S. and Srikumaran, S. (2004) Peptide Transport Activity of the Transporter Associated with Antigen Processing (TAP) Is Inhibited by an Early Protein of Equine Herpesvirus-1. Journal of General Virology, 85, 349-353. http://dx.doi.org/10.1099/vir.0.19563-0

95. Favoreel, H.W., Van Minnebruggen, G., Van De Walle, G.R., Ficinska, J. and Nauwynck, H.J. (2006) Herpesvirus Interference with Virus-Specific Antibodies: Bridging Antibodies, Internalizing Antibodies, and Hiding from Antibodies. Veterinary Microbiology, 113, 257-263. http://dx.doi.org/10.1016/j.vetmic.2005.11.003
96. Van De Walle, G.R., May, M.L., Sukhumavasi, W., Von Einem, J. and Osterrieder, N. (2007) Herpesvirus Chemokine- Binding Glycoprotein G (gG) Efficiently Inhibits Neutrophil Chemotaxis in Vitro and in Vivo. Journal of Immunology, 179, 4161-4169. http://dx.doi.org/10.4049/jimmunol.179.6.4161
97. Van De Walle, G.R., Goupil, R., Wishon, C., Damiani, A., Perkins, G.A. and Osterrieder, N. (2009) A Single-Nucleo- tide Polymorphism in a Herpesvirus DNA Polymerase Is Sufficient to Cause Lethal Neurological Disease. Journal of Infectious Diseases, 200, 20-25.
98. Van De Walle, G.R., Jarosinski, K.W. and Osterrieder, N. (2008) Alphaherpesviruses and Chemokines: Pas De Deux Not Yet Brought to Perfection. Journal of Virology, 82, 6090-6097. http://dx.doi.org/10.1128/JVI.00098-08
99. Stokes, A. and Wardley, R.C. (1988) ADCC and Complement-Dependent Lysis as Immune Mechanisms against EHV-1 Infection in the Horse. Research in Veterinary Science, 44, 295-302.
100. Huemer, H.P., Nowotny, N., Crabb, B.S., Meyer, H. and Hübert, P.H. (1995) gp13 (EHV-gC): A Complement Receptor Induced by Equine Herpesviruses. Virus Research, 37, 113-126. http://dx.doi.org/10.1016/0168-1702(95)00027-N
101. Rappocciolo, G., Birch, J. and Ellis, S.A. (2003) Down-Regulation of MHC Class I Expression by Equine Herpesvirus-1. Journal of General Virology, 84, 293-300. http://dx.doi.org/10.1099/vir.0.18612-0
102. Nugent, J., Birch-Machin, I., Smith, K.C., Mumford, J.A., Swann, Z., Newton, J.R., Bowden, R.J., Allen, G.P. and Davis-Poynter, N. (2006) Analysis of Equid Herpesvirus 1 Strain Variation Reveals a Point Mutation of the DNA Polymerase Strongly Associated with Neuropathogenic versus Nonneuropathogenic Disease Outbreaks. Journal of Virology, 80, 4047-4060. http://dx.doi.org/10.1128/JVI.80.8.4047-4060.2006
103. Van Der Meulen, K.M., Nauwynck, H.J. and Pensaert, M.B. (2003) Absence of Viral Antigens on the Surface of Equine Herpesvirus-1-Infected Peripheral Blood Mononuclear Cells: A Strategy to Avoid Complement-Mediated Lysis. Journal of General Virology, 84, 93-97. http://dx.doi.org/10.1099/vir.0.18864-0
104. Crabb, B.S. and Studdert, M.J. (1996) Equine Rhinopneumonitis (Equine Herpesvirus 4) and Equine Abortion (Equine Herpesvirus 1). In: Studdert, M.J., Ed., Virus Infections of Equines, Elsevier Sciences, 11-37.
105. Slater, J. and Hannant, D. (2000) Equine Immunity to Viruses. Veterinary Clinics of North America, 16, 49-68.
106. Mumford, J.A., Rossdale, P.D., Jessett, D.M., Gann, S.J., Ousey, J. and Cook, R.F. (1987) Serological and Virological Investigations of an Equid Herpesvirus 1 (EHV-1) Abortion Storm on a Stud Farm in 1985. Journal of Reproduction and Fertility Supplement, 35, 509-518.
107. Rosas, C.T., Goodman, L.B., Von Einem, J. and Osterrieder, N. (2006) Equine Herpesvirus Type 1 Modified Live Virus Vaccines: Quo vaditis? Expert Review of Vaccines, 5, 119-131. http://dx.doi.org/10.1586/14760584.5.1.119

CHAPTER 13

Differential expression of Toll-like receptors and inflammatory cytokines in ovine interdigital dermatitis and footrot

Rebecca Davenport[a], Christopher Heawood[a], Kate Sessford[a], Melissa Baker[a], Kerstin Baiker[a], Barbara Blacklaws[b], Jasmeet Kaler[a], Laura Green[c], Sabine Tötemeyer[a,*]

[a] *School of Veterinary Medicine and Science, University of Nottingham, Loughborough LE12 5RD, UK*

[b] *Department of Veterinary Medicine, University of Cambridge, Madingley Road, Cambridge CB3 0ES, UK*

[c] *School of Life Sciences, University of Warwick, Coventry CV4 7AL, UK*

ABSTRACT

Footrot is a common inflammatory bacterial disease affecting the health and welfare of sheep worldwide. The pathogenesis of footrot is complex and multifactorial. The primary causal pathogen is the anaerobic bacterium *Dichelobacter nodosus*, with *Fusobacterium necrophorum* also shown to play a key role in disease. Since immune-mediated pathology is implicated, the aim of this research was to investigate the role of the host response in interdigital dermatitis (ID) and footrot. We compared the expression of

Toll-like receptors (TLRs) and pro-inflammatory cytokines and the histological appearance of clinically normal in comparison to ID and footrot affected tissues. Severe ID and footrot were characterised by significantly increased transcript levels of pro-inflammatory cytokines TNFα and IL1β and the pattern recognition receptors TLR2 and TLR4 in the interdigital skin. This was reflected in the histopathological appearance, with ID and footrot presenting progressive chronic-active pododermatitis with a mixed lymphocytic and neutrophilic infiltration, gradually increasing from a mild form in clinically normal feet, to moderate in ID and to a focally severe form with frequent areas of purulence in footrot. Stimulation with *F. necrophorum* and/or *D. nodosus* extracts demonstrated that dermal fibroblasts, the resident cell type of the dermis, also contribute to the inflammatory response to footrot bacteria by increased expression of TNFα, IL1β and TLR2. Overall, ID and footrot lead to a local inflammatory response given that expression levels of TLRs and IL1β were dependent on the disease state of the foot not the animal.

Keywords: Footrot Sheep, Toll-like receptor, Inflammation, Histopathology, Cytokines

1. INTRODUCTION

Footrot is a bacterial infection of the interdigital skin of the sheep foot resulting in lameness, and is the greatest welfare and economic concern for sheep farmers and veterinary surgeons in the UK (Goddard et al., 2006). In England, more than 95% of sheep flocks have footrot (Kaler and Green, 2008). In the UK current vaccination strategies, while recommended as part of a comprehensive lameness management programme (EBLEX; FAI), are on their own neither long lasting nor efficacious (Duncan et al., 2012).

The pathogenesis of footrot is complex and multifactorial. Physical damage to the interdigital skin that is caused by, for example, prolonged exposure to moisture is required to initiate disease. Bacterial replication in this damaged skin leads to interdigital dermatitis (ID) where the superficial epidermal layers are inflamed, damaged and slough off irregularly. Disease may progress to footrot with separation of the hoof horn capsule from the underlying sensitive tissue. Hoof horn separation does not occur without the involvement of *Dichelobacter nodosus*, an obligate anaerobe bacterium. A second bacterium, *Fusobacterium necrophorum*, may also play a role in the pathogenesis of footrot. It either facilitates disease development by increasing the damage to the interdigital skin and promoting ID that subsequently permits replication of *D. nodosus* (Egerton et al., 1969;

Roberts and Egerton, 1969) or *F. necrophorum* is secondary to *D. nodosus* but may exacerbate the severity and persistence of footrot (Beveridge, 1941; Kennan et al., 2010, 2011; Witcomb et al., 2014).

We hypothesise that the pathology of footrot is a host mediated over-expression of local immune responses leading to acute severe inflammation in the foot that can progress to hoof horn separation from underlying tissues. The host response to bacterial invasion is characterised by the recruitment of large numbers of neutrophils into the epidermis. This causes inflammation and pressure in the hoof horn capsule resulting in its separation from the underlying tissue. However, not all sheep exposed to *D. nodosus* get footrot and fewer than half of cases progress to separation of the hoof horn (Wassink et al., 2010). Early histopathological observations of tissue sections from footrot lesions described little inflammatory response in areas with *D. nodosus*, but severe inflammation in response to invasion by *F. necrophorum* (Egerton et al., 1969). Infiltration by polymorphonuclear leucocytes and a dense population of filamentous bacteria, visually identified as *F. necrophorum* were observed in ID sections (Parsonson et al., 1967). It is, therefore, beneficial to elucidate the basis of the ovine inflammatory response in the interdigital skin to these pathogens and, thus, contribute to the more targeted development of novel vaccines and their adjuvants.

Key components of the host response to bacterial infection are the innate immune pathogen recognition receptors, such as the Toll-like receptors (TLRs) that are expressed in a wide range of cell types (Hancock et al., 2012). TLRs 1–10, which recognise a range of pathogen associated molecular patterns, have been cloned and characterised in sheep (Chang et al., 2009). Footrot is a mixed bacterial infection and, therefore, the TLRs likely to be involved in their recognition are TLR2, TLR2/TLR1 and TLR2/TLR6 heterodimers (which recognise bacterial lipoproteins, lipoteichoic acid and atypical LPS, respectively), TLR4 (which is activated by bacterial LPS), TLR5 (which recognises bacterial flagellin) and TLR9 (which recognises bacterial DNA). *D. nodosus* and *F. necrophorum* are both Gram-negative bacteria but, interestingly, *F. necrophorum* may have an atypical lipopolysaccharide (LPS) as shown for a closely related bacterium, *Fusobacterium nucleatum* (Kikkert et al., 2007) and hence may be signalling via TLR2. The activation of TLRs initiates a complex signalling network, leading to the expression of a wide range of inflammatory mediators such as nitric oxide, cytokines and chemokines (Carpenter and O'Neill, 2007; Takeda and Akira, 2004). Key pro-inflammatory cytokines of the skin include IL-1β and TNF-α (Nestle et al., 2009).

The aim of this research was to investigate this inflammatory response by focussing on histological appearance and the expression of TLRs and pro-inflammatory cytokines in clinically normal, ID and footrot tissues and cultured ovine dermal fibroblasts stimulated with bacterial extracts.

2. MATERIALS AND METHODS

2.1. Ovine interdigital skin biopsies

Ovine feet, obtained from an abattoir, were cleaned, scored for ID and footrot using a scoring system adapted from Parsonson et al. (1967) and classified into one of four categories; clinically healthy, mild ID (slight ID +/- foetid smell, <5% of area affected), severe ID (moderate - severe ID with a foetid smell, >5% of area affected) and footrot (active footrot lesion with under-running of the heel and/or sole area of the digit). Six millimetre biopsies from healthy and a range of severities of diseased feet were sampled from the skin/hoof interface and stored in RNAlater (Sigma, UK) for RNA isolation or fixed in 10% neutral buffered formalin (NBF).

2.2. Histopathology

For histopathology, the formaldehyde fixed biopsies (see above) from 10 normal, 7 ID and 8 footrot affected ovine feet were processed and embedded in paraffin wax. Upon wax embedding, biopsies were orientated to ensure a cross section of the skin-hoof interface was obtained. Cutting these wax blocks was facilitated by soaking the blocks in ice cold 10% ammoniated water prior to cutting. Sections (5 μm) were cut and stained with haematoxylin and eosin (H&E). Microscopy (Leica model DM5000B, software Leica 2000) was used to analyse the tissue sections.

2.3. Cell stimulation

Ovine fibroblasts from five individual Finn Dorset crossed sheep (B1123, 1378X, 1211A, 1220A and 1222A) were isolated and grown as described by Bird et al. (1993). For stimulation studies, they were cultured in Dulbecco's Modified Eagle's Medium supplemented with 10% foetal bovine serum, 2 mM L-glutamine, 5 μg/ml penicillin/streptomycin (Gibco), 0.63 μg/ml fungizone (Lonza) and 100 μg/ml gentamicin (Sigma). Cells were stimulated with *Escherichia coli* LPS (1 μg/ml) or heat killed *D. nodosus, F. necrophorum* or both (10 μg/ml) as established by preliminary experiments with 1–100 μg/ml bacterial extracts (data not shown). To capture the early host response, the cells were stimulated for 4 h (Widdison et al., 2011). For subsequent RNA isolation, cells were lysed with 350 μl of RNA lysis buffer (Nucleospin® RNA isolation kits, Machery-Nagel, UK).

2.4. RNA isolation and cDNA synthesis

Biopsies were homogenised using a 5 mm stainless steel ball-bearing in a Retsch® Bead Mill MM 301 for 4 min, at 30 oscillations per second. Total RNA from biopsy homogenate and lysed fibroblasts was isolated using Nucleospin®RNA isolation kits (Machery-Nagel, UK) following the manufacturer's instructions. RNA concentration was measured using a NanoDrop™ (ND-1000, ThermoScientific, UK). RNA was diluted in water to 100 ng/μl, and cDNA synthesised using M-MLV Reverse Transcriptase and random hexamers (Promega, Madison, USA) according to manufacturer's instructions. To identify any residual genomic DNA contamination, samples with and without reverse transcription were PCR amplified for GAPDH, an abundant transcript, using forward (5′-CCACCAACTGCTTGGCCCCC-3′) and reverse (5′-GGACACGTTGGGGGTGGGGA-3′) primers. This was performed using DreamTaq Polymerase (Fermentas Life Sciences, York, UK), 10 μM of each primer, 10 μM dNTP mix (Promega, Madison, USA) and 25 ng of cDNA. The PCR program consisted of a 95 °C denaturation for 3 min, 25 cycles at 95 °C for 10 s, 55 °C for 1 min, 72 °C for 1 min, and a final extension at 72 °C for 10 min.

2.5. Quantitative real time PCR

All quantitative real time PCR (qPCR) experiments were designed and performed to comply with the quality controls detailed in the MIQE guidelines (Bustin et al., 2010). Primers for TLR2 (forward 5′-CATCTTTGTGCTTTCGGAGA-3′, reverse 5′-AAGAGACGGAAGTGGGAGAA-3′, 78 bp product, AM981300.1) and TLR4 (forward: 5′-AGAAACCT CCGCT ACCTTGA, reverse: 5′-CAGGGAGCAAGTTGTTCTGA-3′, 130 bp, NM_00 1135930.1) were designed and assessed using NCBI PrimerBLAST. In addition qPCR primers for ovine β-actin, GAPDH (Hughes et al., 2011), PPIA (Lloyd et al., 2012), β2-microglobulin, α-tubulin, 18S, TLR1 (Taylor et al., 2008), TLR6 (Plain et al., 2010), TNFα (Sow et al., 2009) and IL-1β (Darlay et al., 2011) were used. The mRNA expression was measured using qPCR on a LightCycler® 480 (Roche Applied Science, UK). Reactions contained 5 μl of diluted sample cDNA (1/10 dilutions apart from 18S and β-actin, where 1/100 dilution was used) in 1× SYBR green qPCR master mix (Roche Applied Science, UK) with 1 μM forward and reverse primers. The following cycle conditions were used: 95 °C for 10 min followed by 45 cycles at 95 °C for 10 s, 60 °C for 50 s, 72 °C for 1 min and a final dissociation gradient up to 97 °C to obtain a melt curve. Standard curves were generated using ovine lymph node or fibroblast cDNA to assess primer efficiency (Table S1).

To facilitate mRNA expression studies, the six housekeeping genes were investigated to detemine a stable reference gene transcript. geNORM (Vandesompele et al., 2002) analysis for stability of expression was performed following qPCR analysis of biopsies from 16 clinically normal, eight ID, and four footrot ovine feet. Stability of a gene is defined as the consistancy of expression between samples. The geNorm recommended minimum (M) value for stability is 0.5. For normalisation, β-actin was used as the housekeeping gene (M-value of 0.49).

C_t values for each sample and transcript were calculated from the mean of triplicate reactions. Normalised expression of each gene was calculated using the following formulae (Hughes et al., 2007):

$$\text{Differentiation factor} = \frac{\text{overall mean of } 40 - \text{CP value for housekeeping gene}}{40 - \text{CP value of housekeeping gene of thatsample}}$$

$$\text{Normalised expression} = \frac{\text{Mean}(40 - \text{CP value for the sample}) \times \text{target primer slope}}{\text{Differentiation factor for that sample} \times \text{housekeeping gene}}$$

Fold change = $2^{(T-C)}$, whereby T = normalised expression level of treated samples; C = s normalised expression level of control samples (Hughes et al., 2007). Expression data are presented as box and whisker plots showing minimum, lower quartile, median, upper quartile and maximum.

2.6. Statistical analysis

The biopsy data were modelled in a mixed effect two level model (Dohoo et al., 2009), which incorporated autocorrelation of feet within sheep. The model took the form:

$$Y_{ij} = \alpha + \beta_1 X_{ij} + v_j + e_{ij}$$

$$v_j \sim N(0, \Omega_v^2)$$

$$e_{ij} \sim N(0, \Omega_e^2)$$

where Y is the immune response, the subscripts i and j denote the ith foot in the jth sheep respectively, α the regression intercept, X_{ij} the vector of covariates associated with each observation, β_1 the coefficients for covariates X_{ij}, v_j a random effect to reflect residual variation between sheep (mean = 0 and variance Ωv2), e_{ij} a random effect to reflect residual variation between feet restricted to following a binomial distribution. For detailed model results with all confidence intervals (CI) see Supplementary Table S2.

The stimulated fibroblast data were analysed using One-way Analysis of Variance (ANOVA) with repeated measures followed by a Dunnett post-test.

3. RESULTS

3.1. *TLR2*, *TLR4* and *IL-1β* expression was increased in severe ID and footrot

Expression of transcripts of TLR1, 2, 4, and 6 and the inflammatory mediators IL-1β and TNFα were analysed in biopsies from the skin/hoof interface of feet that were either clinically normal, had mild ID, severe ID or footrot (Fig. 1). Expression of *TLR2*, *TLR4* and *IL-1β* was significantly higher in severe ID (CI: TLR2 1.41–4.45, TLR4 1.41–4.8, IL-1β 1.35–6.27) and footrot (CI: TLR2 1.71–4.36, TLR4 4.65–4.63, IL-1β 4.37–8.68) (Fig. 1B, C and E). Expression levels were dependent on the disease state of the foot not the animal, i.e. in animals that had one or more foot affected with ID or footrot, the clinically normal feet had the same expression levels as normal feet from animals with no diseased feet (Fig. S1). In contrast, *TLR6* expression was lower in mild ID (CI: (−2.81) – (−0.22)) but not in severe ID and footrot (Fig. 1D). Expression of *TLR1* and *TNFα* was similar in all samples (Fig. 1A and F). A strong correlation between *TLR2* and *TLR4* expression was observed (Fig. 1G, $R^2 = 0.81$). *TLR1* and *TLR6* expression did not mirror that of *TLR2*, suggesting a mixture of TLR1/TLR2 and TLR6/TLR2 heterodimers may be being expressed (Fig. 1A, B and D). However, *TLR6* was expressed at significantly higher levels than *TLR1* ($p < 0.0001$).

3.2. ID and footrot lead to progressive pododermatitis with mixed neutrophil and lymphocyte infiltration

Tissue sections from clinically normal feet had no obvious damage to the interdigital skin. The epidermis had a prominent granular layer and the dermal–epidermal junction had a low number of infiltrating leukocytes. The papillary dermis showed minimal to focally mild perivascular infiltration of lymphocytes and neutrophils (Fig. 2A). Thus, clinically normal feet had a minimal to mild chronic-active pododermatitis with a mixed lymphocytic and neutrophilic infiltration.

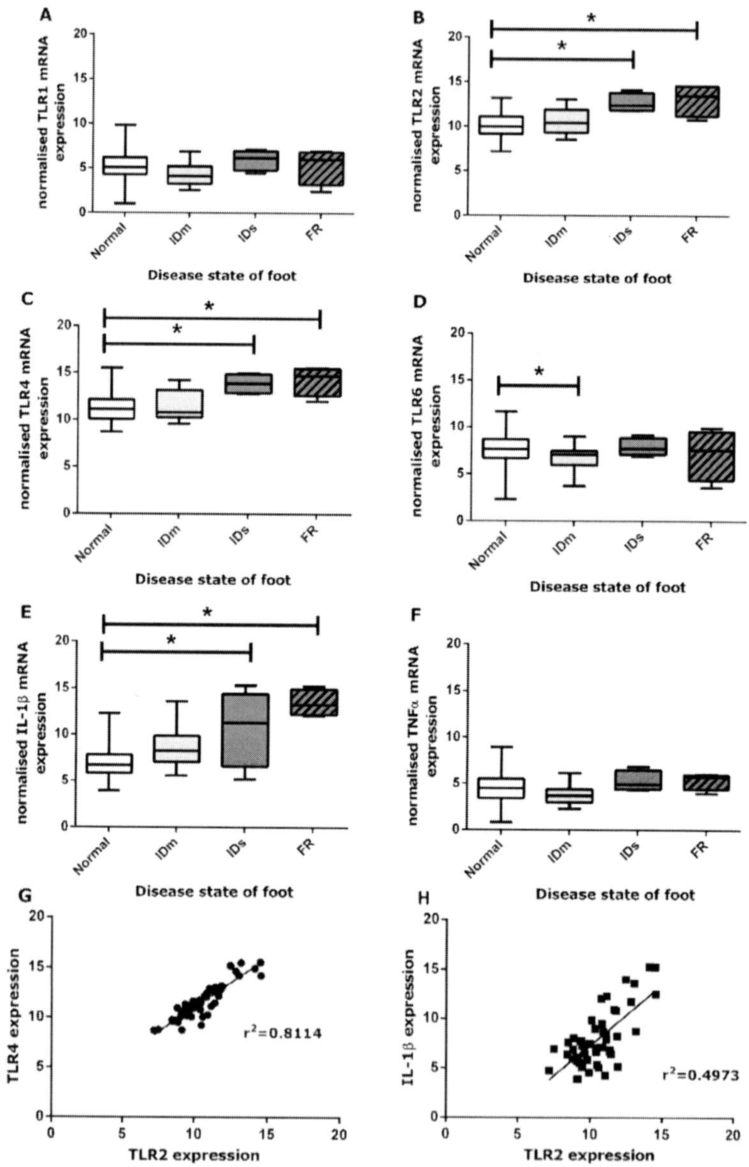

Figure. 1. TLR and inflammatory mediator expression in the interdigital space. TLR1 (A), TLR2 (B), TLR4 (C), TLR6 (D), IL-1β (E) and TNFα (F) transcript levels in clinically normal, mild ID (IDm), severe ID (IDs) and footrot (FR) biopsies of the skin/hoof interface. Data are presented as box and whisker plots showing minimum, lower quartile, median, upper quartile and maximum. Correlation of transcript expression of TLR2 with TLR4 (G) and IL-1β (H) for all samples.

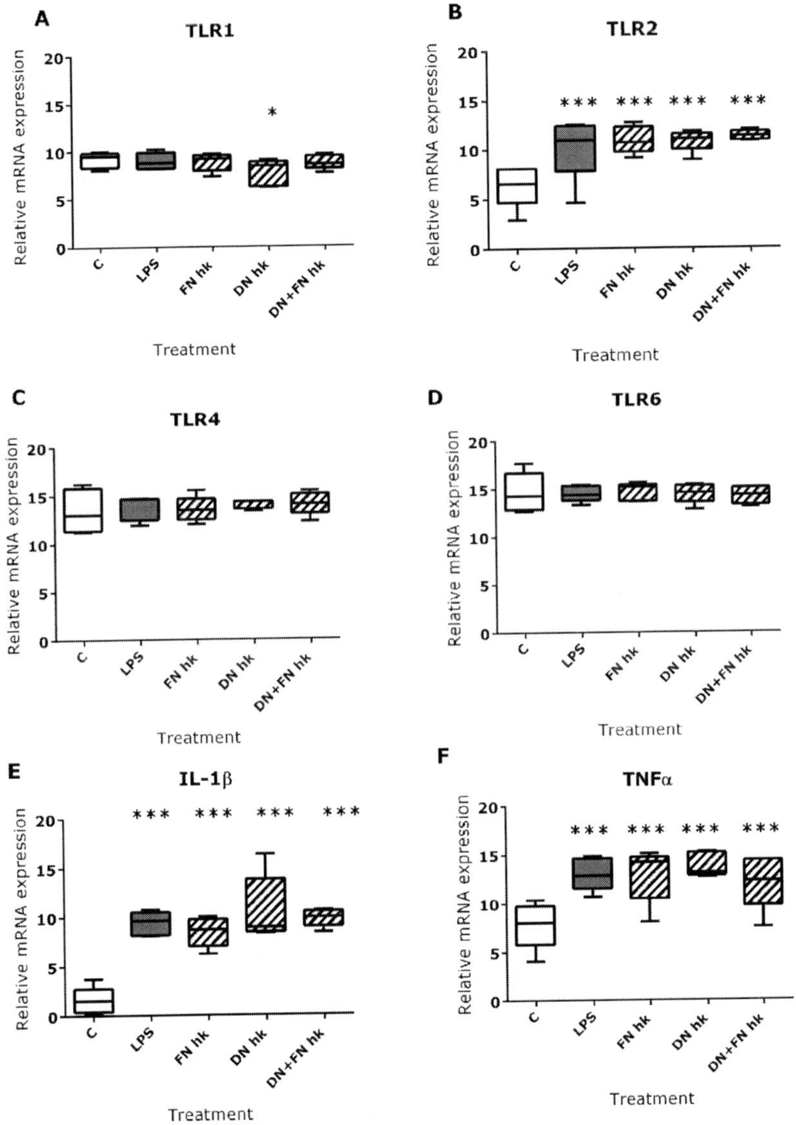

Supplementary Fig. S1. TLR2, TLR4 and IL-1β transcript levels for healthy, ID and footrot (FR) samples, pooled by foot disease state for all samples and by foot disease state for each animal disease group (healthy, ID, FR), whereby clinically healthy animals only had healthy feet, ID animals had healthy and ID feet and FR animals had healthy, ID and FR feet.

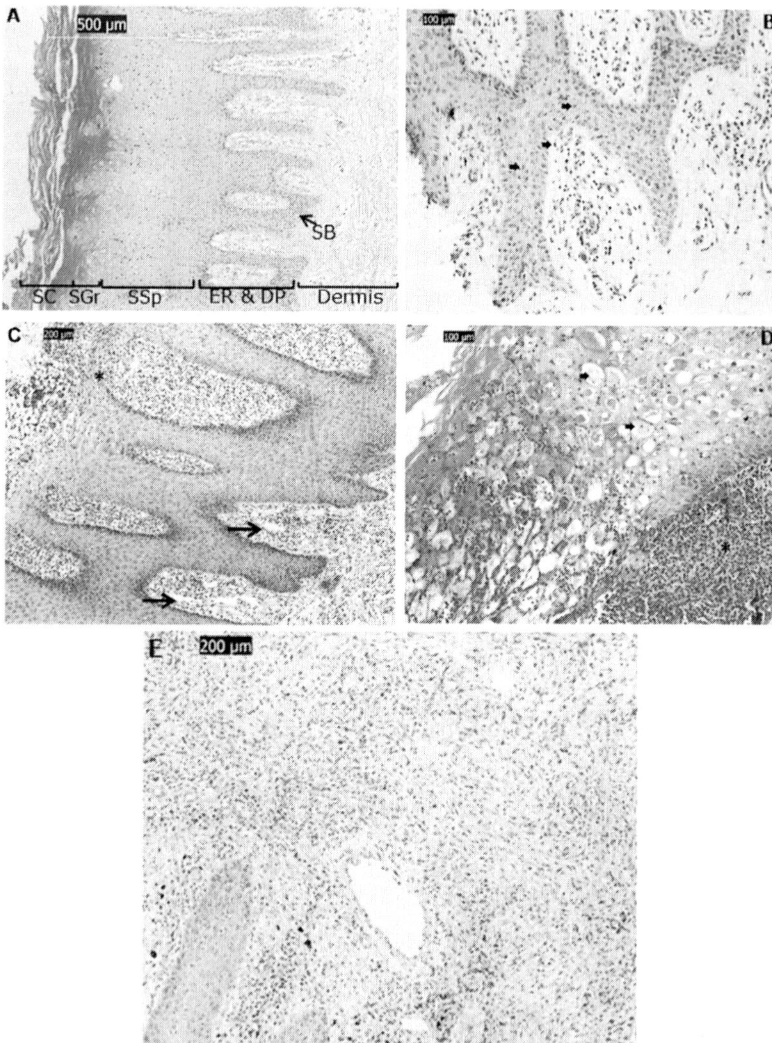

Figure. 2. Histological features of clinically healthy, ID and footrot samples. H&E stain of clinically healthy (A): overview, with SC, stratum corneum; SGr, stratum granulosum; SSp, stratum spinosum; ER, epidermal ridges; DP, dermal papillae; SB, stratum basale; ID (B), superficial mild to moderate dermatitis, exocytosis, granulated neutrophils (➡); FR (C) marked dermatitis, oedema (arrow), congested vessels (*); FR (D) epidermal degeneration (ballooning with nuclear pyknosis) (➡) and large numbers of degenerating neutrophils (purulence) (*) underneath the stratum basale; FR (E) dermal fibrosis.

ID is characterised by erythema (redness of the skin) in the interdigital skin, which is a direct result of dermal blood vessel dilation and hyperaemia. Perivascular immune cell infiltration was seen surrounding dilated superficial dermal blood vessels and resulted in a cell dense dermal–epidermal junction, predominantly involving lymphocytes, neutrophils and a few plasma cells (Fig. 2B). Migrating leukocytes in the stratum basale and stratum spinosum represented cells that had moved through the dermal–epidermal junction (exocytosis) (Fig. 2B), with neutrophilic degranulation observed in the superficial epidermis. Overall, ID presented as a moderate chronic-active pododermatitis with a mixed lymphocytic and neutrophilic infiltration with mild epidermal exocytosis.

In footrot, in addition to erythema, under running of the hoof horn interface was observed indicating a loss of tissue integrity. The dermis showed large numbers of perivascular lymphocytes, neutrophils and fewer plasma cells leading to a cell dense migration through the papillary dermis and epidermis (Fig. 2C). Lymphocytes and neutrophils accumulated uniformly in all layers of the epidermis. Purulence, composed of large numbers of non-viable neutrophils, necrotic debris and plasma proteins, was seen frequently in areas of epidermal degeneration and necrosis and in areas of epidermal–dermal clefts (Fig. 2D). Epidermal ballooning, characteristic of hydropic cell degeneration, was noted with marked swelling of the cytoplasm and additional condensation (pyknosis) of the nucleus (Fig. 2D). Increased fibrous tissue proliferation (superficial dermal scarring) occurred in some cases indicating a chronic reaction due to the loss of tissue integrity at the skin hoof interface (Fig. 2E). The papillary dermis showed vascular congestion and dilated lymphatic vessels with dermal oedema (Fig. 2C), as a consequence of the intense epidermal and dermal inflammation. In summary, footrot presented as moderate to focally severe chronic-active pododermatitis with a mixed lymphocytic and neutrophilic infiltration, pus formation, horn clefting in some cases.

3.3. *TLR2, IL-1β, TNFα* but not *TLR4* expression is increased in dermal fibroblasts in response to *D. nodosus* and *F. necrophorum*

Since bacteria were observed in the dermis in tissue sections from feet with footrot, we investigated *TLR1, TLR2, TLR4, TLR6, IL-1β* and *TNFα* expression in ovine dermal fibroblasts, a resident dermal cell type, in response to LPS and heat-killed extracts of *D. nodosus* and *F. necrophorum* (Fig. 3). In response to LPS and bacterial extract stimulation, *TLR2, IL-1β* and *TNFα* transcript expression was significantly increased in dermal fibroblasts (Fig. 3B, E and F); no effect was seen on *TLR4* and *TLR6* expression (Fig. 3C and D).

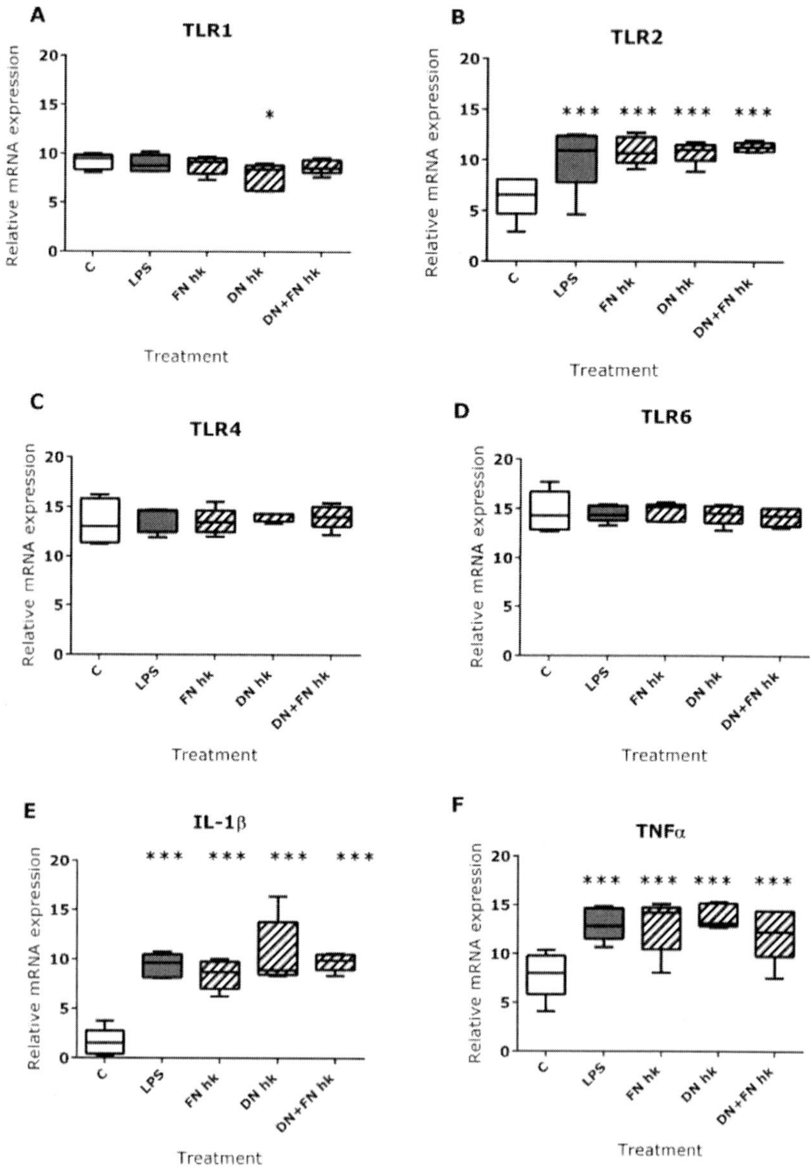

Figure. 3. TLR and inflammatory mediator expression in ovine dermal fibroblasts. TLR1 (A), TLR2 (B), TLR4 (C), TLR6 (D), IL-1β (E) and TNFα (F) transcript levels in ovine dermal fibroblasts stimulated with 1 mg/ml LPS or 10 mg/ml heat killed *D. nodosus* (DN), *F. necrophorum* (FN) or both (DN + FN) for 4 h. Data are presented as box and whisker plots showing minimum, lower quartile, median, upper quartile and maximum. One-way ANOVA, repeated measures, Dunnett post-test, $*p \leq 0.05$, $***p \leq 0.001$ to control.

TLR1 transcript expression was unchanged except for a small but significant decrease in response to *D. nodosus* (Fig. 3A). As seen for the foot biopsies, *TLR1* and *TLR6* expression did not mirror that of TLR2, suggesting expression of *TLR1/TLR2* and *TLR6/TLR2* heterodimers. However, *TLR6* was expressed at significantly higher levels than *TLR1* (Fig. 3A, B and D, $p < 0.0001$).

4. DISCUSSION

This study is the first to report changes in expression of TLRs and pro-inflammatory cytokines in ovine interdigital skin samples from clinically normal and feet diseased with ID or footrot. In interdigitial skin, *TLR2* and *TLR4* transcripts were the most abundant, and higher levels of *TLR6* than *TLR1* were expressed, this is similar to clinically healthy ovine skin from the flank (Nalubamba et al., 2007). Both flank and interdigital skin are exposed to a range of commensal microorganisms, however, there is less potential for tissue damage due to environmental conditions in the flank area compared with the interdigital skin, where damage and subsequent opportunistic infection of pathogens or commensal skin flora are common. Enhanced expression of *TLR2* and *TLR4* in response to infection has also been observed in another ovine infectious disease, Johne's disease, caused by *Mycobacterium avium* subspecies *paratuberculosis* (MAP) (Taylor et al., 2008). TLR1 and TLR6 can both form heterodimers with TLR2, recognising bacterial triacyl lipopeptides and diacyl lipopeptides respectively (Farhat et al., 2008). In MAP infections, TLR2/6 heterodimers are involved in the recognition of mycobacterial antigen (Bulut et al., 2001; Gomes et al., 2008), and in MAP-infected ileum both *TLR2* and *TLR6* expression were significantly increased, supporting a role for this heterodimer in the pathogenesis of Johne's disease (Plain et al., 2010). In contrast, our study of severe ID and footrot did not result in an expression profile indicative of a similarly predominant TLR2/6 heterodimer. Since damaged interdigital skin is exposed to a diverse opportunistic skin and environmental flora with at least 27 bacterial genera (Calvo-Bado et al., 2011), having a more diverse array of pathogen receptors, such as both, TLR1/2 and TLR6/2 heterodimers, may be an advantage to the host. The ovine inflammatory response associated with ID and footrot is also reflected in the increased expression of the pro-inflammatory cytokine *IL-1β*. Interestingly, expression of *TLR2*, *TLR4* and *IL-1β* was dependent on the disease states of the individual feet of an animal. This suggests that the disease has a local focus in the foot with little or no systemic innate immune response. This is different from the humoral response, where foot scoring methods that summed up the lesions on all the feet and weighted underrun lesions relative to non-underrun lesions provided the most accurate association with serum antibody levels to *D. nodosus* (Whittington and Nicholls, 1995).

The results of our study illustrate that ID and footrot lead to progressive chronic-active pododermatitis with a mixed lymphocytic and neutrophilic infiltration. This is mild in clinically normal feet, moderate in feet with ID and focally severe in footrot, with frequent areas of purulence and lytic necrosis. The mild inflammatory phenotype seen in apparently healthy ovine feet is likely to be dependent on many factors, including the environmental conditions and the variability in levels of bacterial exposure in individual sheep and host resistance. Similar observations have been made in a previous study (Parsonson et al., 1967), apparently describing a mild ID, but with no comparison to the histological appearance of clinically normal interdigital skin. Hence it might be that clinically normal skin has a constantly active immune response due to repeated exposure to commensal and pathogenic microbes. The gross footrot lesions in this study were similar to those described previously with intense inflammation, necrotic tissue and separation of the hoof-horn interface throughout the whole interdigital space (Egerton et al., 1969).

In addition, the inflammatory response, seen at the transcript level, in severe ID and footrot potentially resulted in the infiltration of immune cells into the epidermis. Bacteria, including *D. nodosus*, were morphologically identified in both, epidermal and dermal layers of the skin (Egerton et al., 1969), hence it is important to understand the role that individual cellular components of the skin play in the pathogenesis of this disease. One of the main cell types of the dermis is the fibroblast, and although originally thought to only play a structural role in the skin, it has subsequently been shown that these cells have an important role in the innate immune response, providing the supporting extracellular matrix and, on activation, produce cytokines, chemokines and prostanoids (Buckley et al., 2001). The release of pro-inflammatory cytokines and chemokines such as IL-1β, IL-6 and CXCL8 is fundamental in the cross talk between the innate and adaptive immune systems, increasing diversity of responses and hopefully protection against re-infection (Buckley et al., 2001). Our results showed enhanced *TLR2*, *IL-1β* and *TNFα* expression in cultured dermal ovine fibroblasts in response to defined (LPS) and mixed bacterial ligands (heat-killed *D. nodosus* and *F. necrophorum*), which demonstrates the ability of dermal fibroblasts to mount a pro-inflammatory response, thus contributing to the recruitment of innate immune cells.

In summary, we present here the first study of innate immune responses in ovine ID and footrot, linking immunopathology and inflammatory mediator expression localised to the interdigital skin and dermal fibroblasts. The differential expression of TLRs and proinflammatory cytokines correlates with the disease state of the sheep foot and strongly demonstrates the importance of the local immune response with little or no impact of the systemic innate immune response. Hence understanding of those local responses, particularlythe role of TLRs, can contribute to developing different approaches to vaccination.

ACKNOWLEDGEMENTS

This work was funded by the University of Nottingham and RD was awarded a Wellcome Trust CVRT vacation award and an MSD Connect award. We would like to thank Mohd Muzafar and Prof E. Wellington, University of Warwick for bacterial extracts. We would like to thank Dr. J. Daly and Dr. T. Coffey for critical reading of the manuscript.

REFERENCES

1. Beveridge, W.I.B., 1941. Foot-rot in sheep: a transmissible disease due toinfection with Fusiformis nodosus (n. sp.). Studies on its cause, epi-demiology and control. CSIRO Aust. Bull. 140, 1–56.
2. Bird, P., Blacklaws, B., Reyburn, H.T., Allen, D., Hopkins, J., Sargan, D.,McConnell, I., 1993. Early events in immune evasion by the lentivirusmaedi-visna occurring within infected lymphoid tissue. J. Virol. 67,5187–5197.
3. Buckley, C.D., Pilling, D., Lord, J.M., Akbar, A.N., Scheel-Toellner, D., Salmon,M., 2001. Fibroblasts regulate the switch from acute resolving tochronic persistent inflammation. Trends Immunol. 22, 199–204.
4. Bulut, Y., Faure, E., Thomas, L., Equils, O., Arditi, M., 2001. Cooperation ofToll-like receptor 2 and 6 for cellular activation by soluble tuberculosisfactor and Borrelia burgdorferi outer surface protein A lipoprotein: roleof Toll-interacting protein and IL-1 receptor signaling molecules inToll-like receptor 2 signaling. J. Immunol. 167, 987–994.
5. Bustin, S.A., Beaulieu, J.F., Huggett, J., Jaggi, R., Kibenge, F.S., Olsvik, P.A.,Penning, L.C., Toegel, S., 2010. MIQE precis: practical implementationof minimum standard guidelines for fluorescence-based quantitativereal-time PCR experiments. BMC Mol. Biol. 11, 74.
6. Calvo-Bado, L.A., Oakley, B.B., Dowd, S.E., Green, L.E., Medley, G.F., Ul-Hassan, A., Bateman, V., Gaze, W., Witcomb, L., Grogono-Thomas, R.,Kaler, J., Russell, C.L., Wellington, E.M., 2011. Ovine pedomics: thefirst study of the ovine foot 16S rRNA-based microbiome. ISME J. 5,1426–1437.
7. Carpenter, S., O'Neill, L.A.J., 2007. How important are Toll-like receptorsfor antimicrobial response? Cell. Microbiol. 9, 1891–1901.
8. Chang, J.S., Russell, G.C., Jann, O., Glass, E.J., Werling, D., Haig, D.M., 2009.Molecular cloning and characterization of Toll-like receptors 1–10 insheep. Vet. Immunol. Immunopathol. 127, 94–105.
9. Darlay, R.J., McCarthy, A.J., Illot, N.E., Smith, J.E., Shaw, M.A., 2011. Novelpolymorphisms in ovine immune response genes and their associa-tion with abortion. Anim. Genet. 42, 535–543.
10. Dohoo, I.R., Martin, W., Stryhn, H., 2009. Veterinary EpidemiologicResearch, 2nd ed. VER Inc., Charlottetown.
11. Duncan, J.S., Grove-White, D., Moks, E., Carroll, D., Oultram, J.W., Phythian,C.J., Williams, H.W., 2012. Impact of footrot vaccination and antibiotictherapy on footrot and contagious ovine digital dermatitis. Vet. Rec.170, 462–466.

12. EBLEX Reducing lameness for better returns, Sheep BRP Manual 7http:// www.eblex.org.uk/wp/wp-content/uploads/2013/07/brp lsheepSBRP Manual 7 reducing lameness170713.pdf (accessed09.02.14).
13. Egerton, J.R., Roberts, D.S., Parsonson, I.M., 1969. The aetiology and patho-genesis of ovine foot-rot. I. A histological study of the bacterialinvasion. J. Comp. Pathol. 79, 207–215.
14. FAI The Five Point Plan for the reduction of lameness in sheep.http:// www.fwi.co.uk/assets/getasset.aspx?itemid=5244754(accessed 09.02.14).
15. Farhat, K., Riekenberg, S., Heine, H., Debarry, J., Lang, R., Mages, J., Buwitt-Beckmann, U., Roschmann, K., Jung, G., Wiesmuller, K.H., Ulmer, A.J.,2008. Heterodimerization of TLR2 with TLR1 or TLR6 expands theligand spectrum but does not lead to differential signaling. J. Leukoc.Biol. 83, 692–701.
16. Goddard, P., Waterhouse, T., Dwyer, C., Stott, A., 2006. The perceptionof the welfare of sheep in extensive systems. Small Rumin. Res. 62,215–225.
17. Gomes, M.S., Sousa Fernandes, S., Cordeiro, J.V., Silva Gomes, S., Vieira,A., Appelberg, R., 2008. Engagement of Toll-like receptor 2 inmouse macrophages infected with Mycobacterium avium induces non-oxidative and TNF-independent anti-mycobacterial activity. Eur. J.Immunol. 38, 2180–2189.
18. Hancock, R.E.W., Nijnik, A., Philpott, D.J., 2012. Modulating immunity as atherapy for bacterial infections. Nat. Rev. Microbiol. 10, 243–254.
19. Hughes, J., Kwong, W.Y., Li, D., Salter, A.M., Lea, R.G., Sinclair, K.D., 2011.Effects of omega-3 and -6 polyunsaturated fatty acids on ovine follicu-lar cell steroidogenesis, embryo development and molecular markersof fatty acid metabolism. Reproduction 141, 105–118.
20. Hughes, S., Poh, T.Y., Bumstead, N., Kaiser, P., 2007. Re-evaluation ofthe chicken MIP family of chemokines and their receptors suggeststhat CCL5 is the prototypic MIP family chemokine, and that dif-ferent species have developed different repertoires of both the CCchemokines and their receptors. Dev. Comp. Immunol. 31, 72–86.
21. Kaler, J., Green, L.E., 2008. Naming and recognition of six foot lesions ofsheep using written and pictorial information: a study of 809 Englishsheep farmers. Prev. Vet. Med. 83, 52–64.
22. Kennan, R.M., Han, X., Porter, C.J., Rood, J.I., 2011. The pathogenesis ofovine footrot. Vet. Microbiol. 153, 59–66.
23. Kennan, R.M., Wong, W., Dhungyel, O.P., Han, X., Wong, D., Parker, D.,Rosado, C.J., Law, R.H., McGowan, S., Reeve, S.B., Levina, V., Powers,G.A., Pike, R.N., Bottomley, S.P., Smith, A.I., Marsh, I., Whittington, R.J.,Whisstock, J.C., Porter, C.J., Rood, J.I., 2010. The subtilisin-like proteaseAprV2 is required for virulence and uses a novel disulphide-tetheredexosite to bind substrates. PLoS Pathog. 6, e1001210.
24. Kikkert, R., Laine, M.L., Aarden, L.A., van Winkelhoff, A.J., 2007. Activationof Toll-like receptors 2 and 4 by Gram-negative periodontal bacteria.Oral Microbiol. Immunol. 22, 145–151.
25. Lloyd, L.J., Foster, T., Rhodes, P., Rhind, S.M., Gardner, D.S., 2012. Protein-energy malnutrition during early gestation in sheep blunts fetal renalvascular and nephron development and compromises adult renalfunction. J. Physiol. 590, 377–393.
26. Nalubamba, K.S., Gossner, A.G., Dalziel, R.G., Hopkins, J., 2007. Differen-tial expression of pattern recognition receptors in sheep tissues andleukocyte subsets. Vet Immunol Immunopathol 118, 252–262.
27. Nestle, F.O., Di Meglio, P., Qin, J.Z., Nickoloff, B.J., 2009. Skin immune sen-tinels in health and disease. Nat Rev Immunol 9, 679–691.
28. Parsonson, I.M., Egerton, J.R., Roberts, D.S., 1967. Ovine interdigital der-matitis. J. Comp. Pathol. 77, 309–313.
29. Plain, K.M., Purdie, A.C., Begg, D.J., de Silva, K., Whittington, R.J., 2010.Toll-like receptor (TLR)6 and TLR1 differentiation in gene expres-sion studies of Johne's disease. Vet. Immunol. Immunopathol. 137,142–148.

30. Roberts, D.S., Egerton, J.R., 1969. The aetiology and pathogenesis of ovinefoot-rot. II. The pathogenic association of Fusiformis nodosus and F.necrophorus. J. Comp. Pathol. 79, 217–227.
31. Sow, F.B., Gallup, J.M., Meyerholz, D.K., Ackermann, M.R., 2009. Gene pro-filing studies in the neonatal ovine lung show enhancing effects ofVEGF on the immune response. Dev. Comp. Immunol. 33, 761–771.Takeda, K., Akira, S., 2004. Microbial recognition by Toll-like receptors. J.Dermatol. Sci. 34, 73–82.
32. Taylor, D.L., Zhong, L., Begg, D.J., de Silva, K., Whittington, R.J., 2008.Toll-like receptor genes are differentially expressed at the sites ofinfection during the progression of Johne's disease in outbred sheep.Vet. Immunol. Immunopathol. 124, 132–151.
33. Vandesompele, J., De Preter, K., Pattyn, F., Poppe, B., Van Roy, N., De Paepe,A., Speleman, F., 2002. Accurate normalization of real-time quantita-tive RT-PCR data by geometric averaging of multiple internal controlgenes. Genome Biol. 3, RESEARCH0034.
34. Wassink, G.J., King, E.M., Grogono-Thomas, R., Brown, J.C., Moore, L.J.,Green, L.E., 2010. A within farm clinical trial to compare two treat-ments (parenteral antibacterials and hoof trimming) for sheep lamewith footrot. Prev. Vet. Med. 96, 93–103.
35. Whittington, R.J., Nicholls, P.J., 1995. Grading the lesions of ovine footrot.Res. Vet. Sci. 58, 26–34.
36. Widdison, S., Watson, M., Coffey, J.T., 2011. Early response of bovine alveo-lar macrophages to infection with live and heat-killed Mycobacteriumbovis. Dev. Comp. Immunol. 35, 580–591.
37. Witcomb, L.A., Green, L.E., Kaler, J., Ul-Hassan, A., Calvo-Bado, L.A., Medley,G., Grogono-Thomas, R., Wellington, E.M., 2014. A longitudinal studyof the role of Dichelobacter nodosus and Fusobacterium necrophorumload in initiation and severity of footrot in sheep. Prev. Vet. Med. 115,48–55.

CHAPTER 14

Comparison of the Serum Proteins and Immune Responses of Velogenic Newcastle Disease Virus Infected Chickens and Ducks

Chekwube Paul Eze[1], Vincent S. O. Shoyinka[1], John Osita Arinze Okoye[1], Wilfred Sunday Ezema[1], Innocent Okonkwo Ogbonna[2], Didacus Chukwuemeka Eze[1], Emmanuel Chukwudi Okwor[1], Ogbu Kenneth Ikejiofor[3]

[1]*Department of Veterinary Pathology and Microbiology, University of Nigeria, Nsukka, Nigeria*

[2]*Microbiology Unit, Department of Biological Sciences, University of Agriculture, Makurdi, Nigeria*

[3]*National Veterinary Research Institute, Vom, Nigeria*

ABSTRACT

The effect of velogenic Newcastle disease virus (vNDV) on the immune responses and serum proteins was investigated in six-week-old ducks and chickens. Results showed that weight loss was markedly significant ($p < 0.05$) from days 3 - 21 (PI) in chickens and mild ($p < 0.05$) on days 3 and 15 PI in ducks. The antibody response obtained showed significant ($p < 0.05$) increase in infected chickens (IC) than those of the infected ducks (ID).

While the total serum protein and serum globulin increased significantly ($p < 0.05$) in IC on days 7 and 14 PI, they decreased significantly ($p < 0.05$) in ID only on day 21 PI. The immune responses and serum protein values in this experiment X-ray showed less susceptibility of ducks when compared with the chickens. This may be related to marked anorexia and severe dehydration observed in the latter consequent upon serum concentration. Ducks could be maintaining the endemicity of Newcastle disease (ND) as reservoir host.

Keywords: Velogenic Newcastle Disease, Experimental Infection, Immune Response, Serum Proteins, Ducks, Chickens

1. INTRODUCTION

Proteins are the most abundant compound in the serum comprising amino-acid building blocks and are in turn, building blocks for cells and tissues [1]. They are vital components of enzymes, hormones, antibodies and clotting agents [1] - [3]. Some proteins function as transporters, osmotic pressure and acid-alkaline balancers, and reserved energy sources. Due to these multiple functions, any condition that alters the serum proteins extends its tentacles on these roles played by proteins. Newcastle disease (ND) is a highly contagious and infectious disease that affects almost all avian species including poultry, cage and wild life bird species [4] -[6] . It is widespread affecting many continents of the world namely: Asia, Africa, Europe and America [7]. ND causes considerable economic losses, not only due to high flock mortality but also through the economic impact arising from trade restrictions and embargoes [8] [9] . It has been reported to affect other birds such as guinea fowls, quails, turkeys, pheasants and peacock [4] . Mortality due to ND ranges from negligible to as high as 100% depending on the form or pathogenicity of the virus [10] . Kudu 113, (Kuru duck 113) a strain of NDV isolated from apparently healthy ducks and characterized by Echeonwu et al. [11] in Nigeria, has been studied in chickens and guinea fowls [12] . Anseriformes are suspected to maintain the endemicity of ND around the world [7] . In view of these facts, a comparative study of the serum proteins and immune response of vNDV infected ducks and chickens was done.

2. MATERIALS AND METHODS

2.1. Flock History

One hundred and seventy day old birds hatched the same day were obtained comprising 70 ducklings from poultry section of National Veterinary Research Institute NVRI Vom and 100 cockerel chicks from Zartech Hatchery. Brooding was done separately for the chicks and ducklings on deep litters under the same environmental conditions. The cockerels received IBD vaccine by intraocular route at days 10 and 24 post hatch (PH). Chicks' mash was given ad libitum to the birds from day old to 8 weeks PH. Growers' mash was given ad libitum also from 9 weeks PH until the end of the experiment. Water was allowed free choice.

2.2. NDV Inoculum

The VNDV strain, (Kudu-113) was acquired and used in the challenge experiment.

2.3. Experimental Design

At six weeks of age the chicks and ducklings were each randomly assigned into two groups of infected chicks (IC), uninfected chicks (UC) and infected ducks (ID), uninfected ducks (UD). The inoculum was reconstituted to give embryo lethal dose (ELD_{50}) titre of $10^{6.36}$ per ml. The chicks and ducks in groups IC and ID were inoculated intramuscularly (IM) with 0.2 ml of the inoculum (infected groups). The chicks and ducks in groups UC and UD received 0.2 ml of phosphate buffered solution IM (control groups). The infected and control groups were housed at different locations and maintained on deep litter system.

2.4. Live Body Weight

At day 3 PI, 10 birds from each group were randomly selected, marked, weighed and the weight recorded as live body weight (LBW) of each group. The marked birds were re-weighed at days 6, 9, 12, 15 and 21 PI.

2.5. Serology

Blood samples (3ml/bird) were collected from 10 birds in each group on days 0, 7, 14, and 21 PI through the wing vein, using sterile syringe. The sample bottles were stoppered, laid on near horizontal position, and allowed to clot. The serum samples were harvested into 2 ml vials. The duck sera were inactivated at 56°C in water bath for 30 minutes and with the chicken sera, use in the serological analysis.

2.5.1. Washing of Erythrocytes

Two ml of ND sero-negative chicken blood was collected from adult bird in a test tube containing EDTA and washed according to standard procedure. Using the formula CV = RxV/O, where CV = calculated volume (ml) of washed RBC, R = required % of RBC (0.5 %), V = volume (ml) of PBS intended to be used in the dilution and O = original PCV (%) of the RBC after washing, 0.5% of the washed chicken RBC was prepared [13]

2.5.2. Haemagglutination (HA) Test

The standard HA technique as described by Beard, [13] was used and two hundred doses of La Sota ND vaccine was diluted in 10 ml of PBS (pH 7.0). The titre was then adjusted to 4 hemagglutinin units (HAU) for the HI test.

2.5.3. Haemagglutination Inhibition (HI) Test

The standard HI technique as described by Beard, [13] was used. The arithmetic method of geometric mean titre (GMT) was calculated mathematically for 2 fold dilution with the following formula: GMT = antilog[(A − 1) × logB + logC] where A= average end point well number, B = the factor (2.0) and C = the reciprocal of first dilution (2.0) [14].

2.6. Serum Proteins

At days 0, 7, 14 and 21 PI, 10 birds from each group were randomly selected and at least 3 ml blood samples collected through the wing vein, into plain test tubes and serum samples harvested from the clotted blood into 2 ml vials and stored at −40°C for serum protein studies.

2.6.1. Determination of Total Serum Protein (TSP)

The TSP was determined using direct Biuret method. Biuret reagent containing NaOH, potassium iodide, copper (II) sulphate and sodium—potassium tartarate and Standard containing aqueous solution of protein, equivalent to 5 g/dL (50 g/L) were used [15].

The total proteins were calculated as follows: Total protein (g/dL) = absorbance of sample × 5/absorbance of standard.

2.6.2. Determination of Serum Albumin Level (SAL)

The SAL was determined using bromocresol green method. Bromocresol green reagent containing bromocresol green, succinate buffer (pH 4.2), surfactants, preservatives and stabilizers, and standard containing aqueous solution equivalent to 5 g/dL (50 g/L) of albumin were used [15]

The serum albumin was calculated as follows:

Serum albumin (g/dL) = absorbance of sample × 5/absorbance of standard.

2.6.3. Calculation of Serum Globulin Fractions (SGF)

The globulin fraction was calculated as the difference between total serum proteins and serum albumin level; SGF (g/dL) = TSP – SAL.

2.7. Data Analysis

The LBW, HI and serum protein values between and within groups were subjected to statistical analysis using independent sample t-test and the level of significance was determined and accepted at $p \leq 0.05$ for all the results using statistical product for service and solution (SPSS) version 16.0 computer software. The mean ± standard error of mean (SEM) of the results obtained in the experiment were calculated and presented in tables, graphs and charts.

3. RESULTS

3.1. Live Body Weight (LBW)

The result of live body weights of both species are shown on Table 1. The mean weights of IC were significantly lower ($p < 0.05$) than those of the control on days 3, 12, 15 and 21 PI, while the mean weights of ID were significantly lower ($p < 0.05$) than those of the control on days 3 and 15 PI.

3.2. Hemagglutination Inhibition (HI)

The PI HI values were determined and the results are shown in Table 2. On day 0 PI all the groups recorded zero HI titre. The HI titers of IC sera were significantly higher ($p < 0.05$) when compared with ID at days 7— (105.60 ± 20.27 and 22.40 ± 4.89), 14—(537.60 ± 59.73 and 32.00 ± 0.00), and 21—

(119.20 ± 17.64 and 36.80 ± 6.33), PI respectively. Control chickens and UD had zero titre throughout the period of the experiment.

Table 1. Mean live body weight (g) ± SEM of chickens and ducks.

Days PI	Infected chickens	Control chickens	Infected ducks	Control ducks
0	680.00 ± 23.81	680 ± 23.805	314.00 ± 9.91	314.00 ± 09.91
3	632.00 ± 18.61*	746.00 ± 17.59	283.00 ± 12.65*	349.50 ± 19.30
6	679.00 ± 28.61	745.00 ± 20.12	386.00 ± 24.13	356.00 ± 16.61
9	763.00 ± 51.45	835.00 ± 23.63	360.00 ± 20.98	384.50 ± 16.34
12	763.00 ± 46.60*	996.00 ± 31.70	363.00 ± 19.22	386.00 ± 19.56
15	795.00 ± 47.40*	1030.00 ± 41.95	358.00 ± 27.36*	414.50 ± 20.42
21	867.00 ± 38.73*	1044.00 ± 34.13	450.00 ± 19.72	448.50 ± 20.17

*Means values significantly different at $p < 0.05$ along the same row.

Table 2. Hemagglutination inhibition in chickens and ducks.

	Geometric mean titre ± SEM			
Days	IC	UC	ID	UD
D0 PI	000.00 ± 00.00	000.00 ± 00.00	00.00 ± 0.00	000.00 ± 00.00
D7 PI	105.60 ± 20.27	000.00 ± 00.00	22.40 ± 4.89*	000.00 ± 00.00
D14 PI	537.60 ± 59.73	000.00 ± 00.00	32.00 ± 0.00*	000.00 ± 00.00
D21 PI	119.20 ± 17.64	000.00 ± 00.00	36.80 ± 6.33*	000.00 ± 00.00

*Means values significantly different at $p < 0.05$.

3.3. Serum Proteins

The results of the serum proteins are shown in Table 3. The total serum proteins in infected chickens were significantly higher ($p < 0.05$) than the control on day 7 and 14 PI. By the day 21 PI, the infected chickens presented significantly lower TSP ($p < 0.05$) than the control group. In the infected ducks, the TSP were significantly lower ($p > 0.05$) than the control on days 7 and 21 PI.

The results of the SGF are shown in Table 4. The serum globulin protein fraction in infected chickens were significantly higher ($p < 0.05$) than the control on days 7 and 14 PI, but was significantly lower ($p > 0.05$) than the control on day 21 PI. In the infected ducks, the SGF was significantly lower (p > 0.05) than the control only on day 21 PI.

The results of the SAL are shown in Table 5. The serum albumin level in infected chickens was not significantly different (p > 0.05) when compared with the control throughout the experimental period. In the infected ducks, the SAL was significantly lower (p > 0.05) than the control only on day 7 PI.

Table 3. Total serum proteins ± SEM (g/dL × 10).

Days	IC	UC	ID	UD
0	4.09 ± 0.17	4.09 ± 0.17	5.25 ± 1.66	5.25 ± 1.66
7	6.00 ± 0.14*	5.34 ± 0.11	5.44 ± 0.12	5.76 ± 0.09
14	3.13 ± 0.12*	2.55 ± 0.40	3.81 ± 0.25	3.37 ± 0.13
21	2.55 ± 0.09*	3.61 ± 0.14	3.13 ± 0.05*	4.35 ± 0.13

*Means values significantly different at p < 0.05.

Table 4. Serum globulin proteins ± SEM (g/dL × 10).

Days	IC	UC	ID	UD
0	2.63 ± 0.19	2.63 ± 0.19	3.44 ± 0.11	3.44 ± 0.11
7	3.96 ± 0.27*	3.29 ± 0.21	4.10 ± 0.99	3.93 ± 0.17
14	1.63 ± 0.20*	1.13 ± 0.05	2.41 ± 0.24	1.80 ± 0.22
21	1.03 ± 0.12*	1.75 ± 0.21	1.42 ± 0.05*	2.61 ± 0.19

*Means values significantly different at p < 0.05.

Table 5. Serum albumin proteins ± SEM (g/dL × 10).

Days	IC	UC	ID	UD
0	1.46 ± 0.10	1.46 ± 0.10	1.81 ± 0.10	1.81 ± 0.10
7	2.04 ± 0.16	2.03 ± 0.17	1.30 ± 0.07*	1.81 ± 0.10
14	1.50 ± 0.11	1.42 ± 0.56	1.40 ± 0.06	1.59 ± 0.08
21	1.49 ± 0.06	1.87 ± 0.19	1.68 ± 0.29	1.74 ± 0.12

*Means values significantly different at p < 0.05.

4. DISCUSSION

Since maternal antibody could be detected in chickens up to 3 weeks of age [4] , the choice of 6 weeks as the age of inoculation in this experiment was to ensure that the antibodies did not interfere with the susceptibility of the birds in these two species [9] . The results of the HI tests showed that the virus elicited a progressive serological response and sero-conversion detected earlier from day 7 PI which is different from day 10 PI first response recorded by Piacenti et al. [16] and Igwe [9] . Significantly higher antibody titers were observed in the chickens when compared with the ducks, indicating that the immune system in chickens may be more responsive, but were more stable in ducks than that in the chickens since there was rapid antibody decay in the latter. However, both species presented protective levels of antibodies ($>2^3$) at day 21 PI as reported by Villagas and Purchase [14] . The results underscore variations between both species and may be due to dehydrations suffered by chickens. The response was in tandem with the results of the study of vNDV in chickens by Mishra et al. [17] , Okoye et al. [12] and Piacenti et al. [16] who recorded the highest antibody response on days 15 and 21 PI. Contrary to this result, Oladele et al. [18] detected the highest HI titers by day 4 PI using Kudu-113 strain. The variations may be due to differences in the immune status of the chickens [19] [20] and the laboratory procedures used in Table 3. Total serum proteins ± SEM (g/dL × 10). the HI test. One of the mechanisms by which antibodies fight against pathogens is neutralization, particularly viruses [21] . Neutralized viruses are unable to attach to surface receptors of target cells and are thus prevented from replication [22] [23] .

Chekwube Paul Eze, Vincent S. O. Shoyinka, John Osita Arinze Okoye, Wilfred Sunday Ezema, Innocent Okonkwo Ogbonna, Didacus Chukwuemeka Eze, Emmanuel Chukwudi Okwor, Ogbu Kenneth Ikejiofor (1,28)">1] [3] [24] or hemo-concentration associated with the dehydration occasioned by the infection and reduced feed and water intake [25] . Serum proteins have been known to play important roles in infection following invasion of the body by pathogens such as virus, bacteria etc. [1] [2] [18] and as vital substrate for antibody formation [26] . Low level of albumin in serum in livestock may be due to heavy loss in urine or loss due to enteropathy (ulcers and diarrhea), or decreased production by the liver probably because of insufficient intake of protein in diet [1] [25] [27] . Serum albumin proteins which are the most abundant proteins in the clear fluid portion of the blood, typically decrease during inflammation [28] [29] and in a condition of anorexia, albumin is said to depress leading to osmotic disequilibrium and dehydration [29] [30] . There is a strong indication that ducks are far less susceptible to ND than chickens and may likely be maintaining the disease in poultry.

REFERENCES

1. Kaslow, E.J. (2011) Serum Proteins and Functions. Califonia (800), 633-2322. www.mbc.ca.gov
2. Eckersall, P.D. (2008) Proteins Proteomics and the Dysproteinemias. In: Kaneko, J.J., Harvey, J.W. and Bruss, M.L., Eds., Clinical Biochemistry of Domestic Animals, 6th Edition, Academic Press, San Diego, 117-155.
3. Schmidt, E.M.S., Paulillo, A.C., Locatelli-Dittrich, R., Beltrame, O. and Denadai, J. (2009) Serum Protein Profiles of Ring-Necked Pheasants Vaccinated or Not against Newcastle Disease. International Journal of Poultry Science, 8, 359-362. http://dx.doi.org/10.3923/ijps.2009.359.362
4. Alexander, D.J. and Gough, R.E. (2003) Newcastle Disease and Other Avian Paramyxovirus Infections. In: Saif, Y.M., Barnes, H.J., Glisson, J.R., Fadly, A.M., McDougal, L.R. and Swayne, D.E., Eds., Disease of Poultry, 11th Edition, Iowa State University Press, Ames, 63-87.
5. Terregino, C., Cattoli, G., Grossele, B., Bertoli, E., Tisato, E. and Capua, I. (2003) Characterization of Newcastle Disease Virus Isolates Obtained from Eurasian Collared Doves (Streptopelia decaocto) in Italy Short Communication. Avian Pathology, 32, 63-68. http:// dx.doi.org/10.1080/0307 945021000070732
6. Vidanovic, D., Sekler, M., Asanin, R., Milic, N., Nisavic, J., Petrovic, T. and Savic, V. (2011) Characterization of Ve-logenic Newcastle Disease Viruses Isolated from Dead Wild Birds in Serbia during 2007. Journal of Wildlife Diseases, 47, 433-444. http://dx.doi.org/10.7589/0090-3558-47.2.433
7. Office International des Epizooties (OIE) (2009) Manual of Diagnostic Tests and Vaccines for Terrestial Animals. Newcastle Disease. OIE Standard Commission Publication, Office International Epizooties, 2009 Version, Part 2, Section 2.1, Chapter 2.3.14, 576-589.
8. Leslie, J. (2000) Newcastle Disease: Outbreak Losses and Control Policy Costs. Veterinary Record (Journal of British Veterinary Association), 146, 603-606. http://dx.doi.org/10.1136/vr.146.21.603
9. Igwe, A.O. (2009) Comparative Study of the Pathogenicity and Pathogenesis of a Local Nigerian Velogenic Newcastle Disease Virus in Guinea Fowls and Chickens. MSc Dissertation, University of Nigeria, Nsukka, 1-112.
10. Alexander, D.J., Bell, J.G. and Alders, R.G. (2004) Newcastle Disease A Technology Review: With Special Emphasis on Its Effects on Village Chickens FAO Animal Production and Health. Paper 161 Food and Agriculture Organization of the United Nations, Rome.
11. Echeonwu, G.O.N., Ireogbu, C.U. and Emeruwa, A.C. (1993) Recovery of Velogenic Newcastle Disease Virus from Dead and Healthy Free Roaming Birds in Nigeria. Avian Pathology, 22, 383-387. http://dx.doi.org/10.1080/03079459308418928
12. Okoye, J.O.A., Agu, A.O., Chineme, C.N. and Echeonwu, G.O.N. (2000) Pathological Characterization in Chicken of a Velogenic Newcastle Disease Virus Isolate from Guinea Fowl. Revue d'élevage et de Médecine Vétérinaire des pays Tropicaux, 53, 325-330.
13. Beard, C.W. (1980) Isolation and Identification of Avian Pathogens. 2nd Edition, American Association of Avian Pa-thologists, 129-135.
14. Villegas , P. and Purchase, H.G. (1989) Titration of Biological Suspension. In: A Laboratory Manual for the Isolation and Identification of Avian Pathogens, USA Kendal Hunt American Association of Avian Pathologists, Iowa, 186-190
15. Dumas, B.T. (1971) Diagnostic Reagent Kit for in Vitro Determination of Total Protein and Albumin in Serum (Code No25931). Clinical Chemistry Acta, 31, 87-96.
16. Piacenti, A.M., King, D.J., Seal, B.S., Zhang, J. and Brown, C.C. (2006) Pathogenesis of Newcastle Disease in Com-mercial and Specific Pathogen-Free Turkeys Experimentally Infected with Isolates of Different Virulence. Journal of Veterinary Pathology, 43, 168-178. http://dx.doi.org/10.1354/vp.43-2-168
17. Mishra, S., Kataria, J.M., Verma, K.C. and Sah, R.L. (2000) Response of Chickens to Infection with Newcastle Disease Virus Isolated from a Guinea Fowl. Tropical Animal Health and Production, 32, 276-284.

18. Oladele, S.B., Abdu, P., Nok, A.J., Esievo, K.A.N. and Useh, N.M. (2005) Haemagglutination Inhibition Antibodies rectal Temperature and Total Protein of Chicken Infected with a Local Nigerian Isolate of Velogenic Newcastle Disease Virus. Veterinary Research Communication, 29, 171-179. http://dx.doi.org/10.1023/B:VERC.0000047495.03341.2b
19. Scott, T.R. (2004) Our Current Understanding of Humoral Immunity of Poultry. Poultry Science, 83, 574-579. http://dx.doi.org/10.1093/ps/83.4.574
20. Dortmans, J.C.F.M., Koch, G., Rottier, P.J.M. and Peeters, B.P.H. (2011) Virulence of Newcastle Disease Virus: What Is Known So Far? Veterinary Research, 42, 122.
21. Grogan, K.B., Fernandez, R.J., Barranon, R.F.J. and Espinosa, H.G. (2008) Avian Immune System: A Brief Review. Merial Select, Gainesville, 1-12.
22. Nester, W.E., Anderson, G.D., Evans, C.R., Pearsall, N.N., Nester M.T. and Hurley, D. (2004) Microbiology: A Human Perspective. McGraw Hill, New York.
23. Talaro, K.P. (2005) Foundation in Microbiology. 5th Edition, The McGraw Hill Companies, New York.
24. Filipovic, N., Stojevic, Z., Milinkovic-Tur, S., Ljubic, B.B. and Zdelar-Tuk, M. (2007) Changes in Concentration and Fractions of Blood Serum Proteins of Chickens during Fattening. Veterinarski Archiv, 77, 319-326.
25. Ihedioha, J.I. and Chineme, C.N. (2005) Fundamental of Systemic Veterinary Pathology. Vol. 2, Great A P Express, Nsukka.
26. Nnadi, P.A., Eze, P.C. and Ezema, W.S. (2010) Influence of Delayed Feeding on the Performance Development and Response of Immune System to Newcastle Disease Vaccination in Chickens. International Journal of Poultry Science, 9, 669-674. http://dx.doi.org/10.3923/ijps.2010.669.674
27. Daramola, J., Adeboye, A., Fatoba, T. and Soladoye, A.O. (2003) Hematological and Biochemical Parameters of the West African Dwarf Goat. Department of Animal Production University of Ilorin Nigeria, Longman, Singapore City.
28. Lumeij, J.T. (2008) Avian Clinical Biochemistry. In: Kaneko, J.J., Harvey, J.W. and Bruss, M.L., Eds., Clinical Bio-chemistry of Domestic Animals, 6th Edition, Academic Press, San Diego, 839-872. http://dx.doi.org/10.1016/B978-0-12-370491-7.00030-1
29. Petersen, H.H., Nielsen, J.P. and Heegaard, P.M. (2004) Application of Acute Phase Protein Measurements in Veteri-nary Clinical Chemistry. Veterinary Research, 35, 163-187. http://dx.doi.org/10.1051/vetres:2004002
30. Thomson, R.G. (1984) General Veterinary Pathology.

CHAPTER 15

A new tetravalent canine leptospirosis vaccine provides at least 12 months immunity against infection

H.L.B.M. Klaasen[a,*], M. van der Veen[a,1], D. Sutton[b,2], M.J.C.H. Molkenboer[c,3]

[a]*Department of Microbiological R&D, MSD Animal Health, Boxmeer, The Netherlands*

[b]*Global Marketing Department, MSD Animal Health, Milton Keynes, UK*

[c]*Department of Global Regulatory Affairs Immunologicals, MSD Animal Health, Boxmeer, The Netherlands*

ABSTRACT

A key success factor in the vaccination of dogs against leptospirosis is long term protection against establishment of the renal carrier state, in order to protect other dogs, as well as humans, against this re-emerging zoonotic disease. In this paper, we describe the ability of a new European tetravalent vaccine containing antigen from *Leptospira interrogans* (*sensu lato*) serogroups Icterohaemorrhagiae, Canicola, Grippotyphosa and Australis to control infection and renal excretion in dogs at 12 months after vaccination.

In order to demonstrate the efficacy of all four vaccine components, four separate challenge studies were performed. For each study two groups of dogs were used (a group receiving the leptospirosis vaccine and a control group). Twelve months after the second vaccination all dogs in the vaccine and control groups were challenged, both intraperitoneally and conjunctivally, using a pathogenic challenge strain from one of four serogroups. Parameters recorded post-challenge were: clinical signs of disease, change in body temperature, total leucocyte count, thrombocyte count, presence of challenge organisms in blood, urine and kidney tissue, and evidence of interstitial nephritis at necropsy four weeks after challenge.

The vaccine was able to either prevent or significantly reduce infection following challenge with the strains of all four serogroups. The vaccine was also able to prevent or significantly reduce renal infection following Canicola and Icterohaemorrhagiae challenge, and there was a trend of reduction of renal infection with Australis (serovar Bratislava). In the case of the Grippotyphosa study, challenge led to no detectable renal infection in any dog of the control group.

In conclusion, in this study significant protective immunity was achieved in dogs 12 months after a basic vaccination schedule of two doses against strains of serogroups Canicola, Icterohaemorrhagiae, Grippotyphosa and Australis.

Keywords: *Leptospira*; Vaccine; Dog; Immunity; Renal

1. INTRODUCTION

Although canine *Leptospira* vaccines are strictly regarded as 'non-core', many dogs are at risk of disease and thus vaccination is widespread in many countries. Traditionally, leptospirosis in dogs has been associated with serovars from serogroups Canicola and Icterohaemorrhagiae and bivalent vaccines containing strains from these serogroups have been used for the last 50 years. In recent years however there has been an increasing recognition of disease associated with serovars from other serogroups; in the USA from serogroups Grippotyphosa and Pomona and in Europe predominantly from serogroups Grippotyphosa and Australis (Ellis, 2010). As a consequence, in the USA, there are now a number of tetravalent canine leptospirosis vaccines available containing, in addition to traditional Icterohaemorrhagiae and Canicola antigens, strains from serogroups Grippotyphosa and Pomona. In Europe the vaccines are still predominantly bivalent, albeit with one recently-introduced trivalent product (serogroups Icterohaemorrhagiae–Canicola–Grippotyphosa). Most of the available vaccines, whether bivalent, trivalent or tetravalent, are regarded as being effective at controlling clinical disease and preventing mortality but only a few claim to be able to reduce infection or renal excretion following challenge; an important property in reducing the spread of this zoonotic disease (Feigin et al., 1973). Additionally, concerns have been raised about whether vaccine immunity persists for a full 12 months or whether more frequent re-vaccination is necessary.

A new European tetravalent vaccine containing antigen from *Leptospira interrogans* (*sensu lato*) serogroups Icterohaemorrhagiae, Canicola, Grippotyphosa and Australis has recently been developed (Nobivac® L4 – MSD Animal Health) which has been shown to reduce infection and/or renal excretion following challenge with specific serovars of these four serogroups shortly after vaccination (Klaasen et al., 2013). The following studies demonstrate the ability of this new vaccine to control infection and renal excretion in dogs at 12 months after vaccination.

2. MATERIALS AND METHODS

2.1. Animals

Six-week-old conventional beagle dogs without detectable agglutinating serum antibodies against *Leptospira* serogroups Canicola, Icterohaemorrhagiae, Grippotyphosa and Australis were provided by a commercial supplier. In each of the four studies, treatment groups (with nine dogs per group) consisted of pups of both sexes and pups derived from different litters in order to prevent gender and litter effects interfering with treatment effects. The selected dogs were free of clinical abnormalities or disease prior to inclusion in these studies. Husbandry was the same in each study; during the first part of the study (pre-challenge, up to 64 weeks of age) the dogs were housed in the dog facilities of the supplier; at the age of eight weeks the pups were weaned and for the challenge phase of the study the dogs were transferred to the animal facilities of MSD Animal Health, where, after being allowed to acclimatise for seven days, they were challenged at the age of 65 weeks. All housing systems used in these studies fully complied with the requirements of the Federation of European Laboratory Animal Science Associations (FELASA). The animal studies described in this paper were conducted after prior written approval by the responsible ethics review committee and thus this work follows international, national and institutional guidelines for humane animal treatment and complies with relevant legislation.

2.2. Study design

In order to demonstrate the efficacy of all four vaccine components, four separate challenge studies were undertaken using the same basic protocol. For each study two groups of dogs were used. One group (vaccine group) was vaccinated subcutaneously, twice with Nobivac® DHPPi* reconstituted in Nobivac® L4 at the ages of 6 and 10 weeks, and once (at the age of 6 weeks) with Nobivac® KC** intranasally. The second group (control group) was vaccinated twice with Nobivac® DHPPi reconstituted in Nobivac Solvent subcutaneously at 6 and 10 weeks of age, and once (at the age of 6 weeks) with Nobivac KC intranasally. Nobivac Solvent does not affect the immune response to the DHPPi vaccine, because it is a buffered salt solution, and Nobivac L4 licensing studies (in 2012 approved by the Committee for Medicinal Products for Veterinary Use of the European Medicines Agency) demonstrated that there was no effect of Nobivac DHPPi or Nobivac KC on the immune response in dogs to Nobivac L4. For the next year the dogs were housed under strict infection barrier conditions which prevented the possibility of exposure to field infection. Twelve months after the second vaccination all dogs in the vaccine and control groups were challenged, both intraperitoneally and conjunctivally, using a pathogenic challenge strain from one of four serogroups. Details of the grouping, vaccination schedules and challenge are shown in Table 1. For challenge with all four strains the method described in a recent publication (Klaasen et al., 2013) was used. However, to reduce the risk of a failing challenge, for the Bratislava strain two challenges on two consecutive days were performed.

2.3. Sample collection and parameters

Post-challenge the dogs were monitored for four weeks for any clinical signs of disease and change in body temperature. Samples of blood, serum and urine were collected at intervals during the four weeks following challenge and were evaluated for total leucocyte count, thrombocyte count and for the presence of challenge organisms or leptospiral DNA by culture and PCR, respectively (Klaasen *et al.*, 2003 ; Ahmed *et al.*, 2009). Four weeks after challenge the dogs were euthanized and a detailed post-mortem examination was undertaken. In addition a sample of kidney cortex was taken aseptically for leptospiral culture. In this study it was crucial to differentiate between dogs in which direct or indirect evidence was only found for leptospiraemia (early phase of the infection) and dogs in which renal infection (subsequent phase of the infection) was demonstrated. In accordance with the European Pharmacopoeia Monograph 0447 ["Canine Leptospirosis vaccine (Inactivated)"] the somewhat arbitrary term infection (referring to the bacteraemic phase of the infection) and the term renal infection (referring to urinary tract infection and excretion) are used here to assess per challenge strain the numbers of dogs with infection and renal infection in vaccinated groups versus control groups. The following criteria were used to define a dog positive for infection and renal infection:

Table 1. Details of grouping, vaccination and challenge.

Study	Group	No. of dogs	1st vaccination		2nd vaccination		*Leptospira* challenge		
			Age (weeks)	Vaccines	Age (weeks)	Vaccines	Age (weeks)	Serogroup	Serovar
1	Vaccine	9	6	L_4, DHPPi[a], KC[b]	10	L_4, DHPPi	65	Canicola	Canicola
	Control	9	6	DHPPi, KC	10	DHPPi	65	Canicola	Canicola
2	Vaccine	9	6	L_4, DHPPi, KC	10	L_4, DHPPi	65	Icteroh.	Copenhageni
	Control	9	6	DHPPi, KC	10	DHPPi	65	Icteroh.	Copenhageni
3	Vaccine	9	6	L_4, DHPPi, KC	10	L_4 + DHPPi	65	Gripp.	Bananal/Liangguang
	Control	9	6	DHPPi, KC	10	DHPPi	65	Gripp.	Bananal/Liangguang
4	Vaccine	9	6	L_4, DHPPi, KC	10	L_4, DHPPi	65	Australis	Bratislava
	Control	9	6	DHPPi, KC	10	DHPPi	65	Australis	Bratislava

[a] Nobivac DHPPi – live, freeze-dried vaccine containing canine distemper virus, canine adenovirus type 2, canine parvovirus and canine parainfluenza virus for subcutaneous inoculation following reconstitution.

[b] Nobivac KC – live, freeze-dried vaccine containing *Bordetella bronchiseptica* and canine parainfluenza virus for intranasal inoculation following reconstitution.

- A *dog positive for infection* (early stage of leptospirosis in which leptospiraemia plays a central role) is a dog with at least two positive samples of blood or serum or urine/kidney on different days or a dog with challenge-induced nephritis or clinical or haematological evidence for leptospirosis.In general, an initial phase of bacteraemia (i.e., leptospiraemia) has occurred in each dog in which infection, even in the absence of positive blood samples, or renal infection has been demonstrated, since: (i) challenge-induced clinical signs or haematological changes in the first two weeks are associated with bacteraemia, because these are effects of leptospires in the bloodstream (Faine, 1994a); (ii) in any dog in which challenge organisms are isolated from urine or kidney and in any dog with challenge-induced nephritis a preceding phase of bacteraemia has occurred: the bacteria are disseminated from the bloodstream, and not via another route, to the kidneys (Faine, 1994a ; Faine, 1998).
- A *dog positive for renal infection* is a dog with at least one positive sample of urine/kidney from day 14 post-challenge onwards or challenge-induced nephritis (demonstrated by histopathological examination).This definition is considered valid, since: (i) the presence of leptospires in urine or kidney tissue from day 14 post-challenge onwards is considered as evidence for an active renal infection based on generally known scientific data on renal disease and patterns of urinary excretion in canine leptospirosis (Faine, 1998 ; Levett, 2001); (ii) challenge-induced nephritis is a result of ischaemia and inflammation caused by the prolonged presence of leptospires in the blood vessels of the kidney (Faine, 1994a ; Levett, 2001).The numbers of dogs which were scored as positive for infection and renal infection in each of the four challenge studies were recorded and then compared to identify significant differences using a two-sided Fisher's Exact test.

3. RESULTS AND DISCUSSION

Although in the USA serovar Bratislava has been reported as a pathogen in dogs (Adin and Cowgill, 2000), it is mainly in Europe where this serovar has been recognised as a dog pathogen (Ellis, 2010). In the present study the Bratislava strain was able to persist in the bloodstream of non-vaccinated control dogs for up to four days and to appear in the urine on days 3, 16 and 22 post-challenge, and caused interstitial nephritis in two control dogs (results not shown). These results clearly demonstrate the dog-pathogenicity of Bratislava.

The results are summarised in Table 2. As can be seen the vaccine was able to either prevent or significantly reduce infection following challenge one year after vaccination in all four studies. The vaccine was also able to prevent or significantly reduce renal infection following Canicola and Icterohaemorrhagiae challenge one year after vaccination. In the Canicola challenge study one out of nine vaccinated dogs had a positive urine culture on three out of five post-challenge sampling points, but without any evidence of leptospiral infection in the kidney (negative on renal culture and no signs of interstitial nephritis) at the end of the study. Nonetheless this dog was classified as "positive for infection" and "positive for renal infection" according to the criteria described. None of the other vaccinated dogs had any positive urine, whereas in all control dogs convincing evidence of renal infection was found (multiple positive urine cultures, interstitial nephritis). In the case of Australis challenge the difference between vaccinated and control dogs with renal infection was too small to reach statistical significance ($P = 0.0824$), due to a low number of positive control dogs (four out of nine control dogs). However, there was a clear tendency of reduction of urinary shedding by vaccination. With one additional control dog being positive the difference would have been statistically significant ($P = 0.0294$). Apart from the four positive control dogs, a fifth control dog had a positive urine sample on day 3 post-challenge but the other urine samples were negative, so that this dog was classified as negative for renal infection (results not shown). In the case of the Grippotyphosa study, challenge did not lead to detectable renal infection in any of the control group. In contrast, a similar challenge study in our laboratory using the same challenge (strain, method and dose) carried out in puppies at 13 weeks of age induced signs of renal infection in seven out of eight control dogs (Klaasen et al., 2013). Apart from the Bratislava study where a repeated challenge on two consecutive days was performed, it was observed that fewer control dogs had positive blood and or urine cultures in the present studies involving challenge in adult dogs than in the previously published studies with challenge in pups. This discrepancy can be explained by the fact that adult animals are known to have a higher resistance to infection with pathogenic *Leptospira* bacteria compared to young animals (Faine, 1994b). In the present Grippotyphosa challenge study there was also one vaccinated dog with signs of interstitial nephritis at post-mortem but with no evidence of the presence of leptospires (by either culture or PCR) and no clinical signs or evidence of thrombocytopaenia (results not shown). Although this case, according to the definition of a dog positive for renal infection, has been scored as a positive case it is likely that the interstitial nephritis could have resulted from some other cause.

Table 2. Numbers of dogs positive for infection or renal infection post-challenge.

Study (challenge)	Group	No. of dogs/group positive for:			
		Infection		Renal infection	
		No. of dogs	Significant?	No. of dogs	Significant?
1 (Canicola)	Vaccine	1/9	S ($P = 0.0004$)	1/9[a]	S ($P = 0.0004$)
	Control	9/9		9/9	
2 (Ictероh.)	Vaccine	0/9	S ($P = 0.0091$)	0/9	S ($P = 0.0091$)
	Control	6/9		6/9	
3 (Gripp.)	Vaccine	1/8[b]	S ($P = 0.0152$)	1/8[b]	NS ($P = 0.4706$)
	Control	7/9		0/9	
4 (Australis)	Vaccine	2/8	S ($P = 0.0152$)	0/8	NS ($P = 0.0824$)
	Control	8/9		4/9	

[a] Dog had three positive urine samples; no clinical signs, no thrombocytopania, no positive kidney, no interstitial nephritis.
[b] Dog had interstitial nephritis; no positive culture/PCR results, no clinical signs, no thrombocytopenia.

In these studies we were able to reproduce transient leptospiraemia and urinary shedding of the challenge organisms in non-vaccinated control dogs following challenge with pathogenic strains from serogroups Canicola, Icterohaemorrhagiae and Bratislava, and transient leptospiraemia for Grippo typhosa. With this vaccine urinary shedding of leptospires is reduced, which implies that the vaccine helps prevent transmission of the infection to other animals and to humans and, therefore, is an aid in preventing these zoonotic infections (Feigin et al., 1973).

4. CONCLUSION

In this study significant protective immunity was achieved in dogs 12 months after a basic vaccination schedule of two doses against strains of serogroups Canicola, Icterohaemorrhagiae, Grippotyphosa and Australis. Reduction of infection (leptospiraemia) was demonstrated against a Grippotyphosa strain, and reduction of infection, renal infection and urinary shedding was demonstrated against strains of serogroups Canicola, Icterohaemorrhagiae and Australis.

ACKNOWLEDGEMENT

The authors would like to thank Karel van Stokkom for his technical assistance in performing the animal studies.

REFERENCES

1. Adin, C.A., Cowgill, L.D., 2000. Treatment and outcome of dogs with leptospirosis: 38 cases (1990–1998). J. Am. Vet. Med. Assoc. 216, 371–375.
2. Ahmed, A., Engelberts, M.F.M., Boer, K.R., Ahmed, N., Hartskeerl, R.A., 2009. Development and validation of a real-time PCR for detection of pathogenic Leptospira species in clinical materials. PLoS ONE 4 (9), e7093.
3. Ellis, W.A., 2010. Control of canine leptospirosis in Europe: time for a change? Vet. Rec. 167, 602–605.
4. Faine, S., 1994a. Leptospira and Leptospirosis. CRC Press Inc., Boca Raton, FL, USA, pp. 155–159, 199–200.
5. Faine, S., 1994b. Leptospira and Leptospirosis. CRC Press Inc., Boca Raton, FL, USA, 146, 172.
6. Faine, S., 1998. Leptospirosis. In: Collier, L., Balows, A., Sussman, M. (Eds.), Topley and Wilson's Microbiology and Microbial Infections, volume 3. Bacterial Infections (Hausler, W.J., Sussman, M.). , 9th ed. Arnold, London, UK, pp. 850–856.
7. Feigin, R.D., Lobes, L.A., Anderson, D., Pickering, L., 1973. Human leptospirosis from immunized dogs. Ann. Intern. Med. 79, 777–785.

8. Klaasen, H.L.B.M., Molkenboer, M.J.C.H., Vrijenhoek, M.P., Kaashoek, M.J.,\ 2003. Duration of immunity in dogs vaccinated against leptospirosis with a bivalent inactivated vaccine. Vet. Microbiol. 95, 121–132.
9. Klaasen, H.L.B.M., van der Veen, M., Molkenboer, M.J.C.H., Sutton, D., 2013. A novel tetravalent Leptospira bacterin protects against infection and shedding following challenge in dogs. Vet. Rec. 172, 181–186.
10. Levett, P.N., 2001. Leptospirosis. Clin. Microbiol. Rev. 14, 296–326.

CHAPTER 16

Genes controlling vaccine responses and disease resistance to respiratory viral pathogens in cattle

Elizabeth J. Glass*, Rebecca Baxter, Richard J. Leach, Oliver C. Jann

The Roslin Institute and Royal (Dick) School of Veterinary Studies, University of Edinburgh, Easter Bush, Midlothian, EH25 9RG, UK

ABSTRACT

Farm animals remain at risk of endemic, exotic and newly emerging viruses. Vaccination is often promoted as the best possible solution, and yet for many pathogens, either there are no appropriate vaccines or those that are available are far from ideal. A complementary approach to disease control may be to identify genes and chromosomal regions that underlie genetic variation in disease resistance and response to vaccination. However, identification of the causal polymorphisms is not straightforward as it generally requires large numbers of animals with linked phenotypes and genotypes. Investigation of genes underlying complex traits such as resistance or response to viral pathogens requires several genetic approaches including candidate genes deduced from knowledge about the cellular pathways leading to protection or pathology, or unbiased whole genome scans using markers spread across the genome.

Evidence for host genetic variation exists for a number of viral diseases in cattle including bovine respiratory disease and anecdotally, foot and mouth disease virus (FMDV). We immunised and vaccinated a cattle cross herd with a 40-mer peptide derived from FMDV and a vaccine against bovine respiratory syncytial virus (BRSV). Genetic variation has been quantified. A candidate gene approach has grouped high and low antibody and T cell responders by common motifs in the peptide binding pockets of the bovine major histocompatibility complex (BoLA) DRB3 gene. This suggests that vaccines with a minimal number of epitopes that are recognised by most cattle could be designed. Whole genome scans using microsatellite and single nucleotide polymorphism (SNP) markers has revealed many novel quantitative trait loci (QTL) and SNP markers controlling both humoral and cell-mediated immunity, some of which are in genes of known immunological relevance including the toll-like receptors (TLRs).

The sequencing, assembly and annotation of livestock genomes and is continuing apace. In addition, provision of high-density SNP chips should make it possible to link phenotypes with genotypes in field populations without the need for structured populations or pedigree information. This will hopefully enable fine mapping of QTL and ultimate identification of the causal gene(s). The research could lead to selection of animals that are more resistant to disease and new ways to improve vaccine efficacy.

Keywords: Cattle Genetics, Vaccine response, Bovine respiratory, syncytial, virus Foot and mouth disease, virus Quantitative trait, loci, Whole genome scan Polymorphism Bovine major histocompatibility complex BoLA Toll like receptor

1. GENETIC VARIATION IN DISEASE RESISTANCE AND VACCINE RESPONSE

Endemic and exotic pathogens continue to have a significant effect on livestock with high economic costs and welfare implications worldwide. In addition emerging pathogens are increasingly recognised as a threat to biosecurity and zoonotic pathogens and chemical usage still cause food safety hazards. These concerns remain despite increasing regulation and control measures such as good management and the availability of vaccines at least to some pathogens. Globalisation of trade and climate change bring new urgency to find ways to reduce the impact of infectious disease which remains a barrier to further improvement in livestock efficiency in both the developed and developing world.

One potential solution is to breed livestock that are inherently more resistant or tolerant to prevailing pathogens. However it is not easy to identify markers and genes that would enable selection of such desirable traits, partly because of the cost and logistics of collecting appropriate large-scale data either from the field or experimental challenges. Thus there is a phenotype gap. Furthermore resistance or tolerance traits are likely to be underpinned by many genes of small effect, making it more difficult to identify them. Another solution would be to design more effective vaccines. After the foot and mouth disease virus (FMDV) outbreak in the U.K. in 2001, the Royal Society stated that research should concentrate on developing an ideal vaccine that would be "safe, synthetic, induce sterile immunity, in one shot, be cost effective and protect all animals" (Royal Society, 2002). It is debatable whether any livestock vaccine would meet all of these criteria. In addition for many diseases there are no appropriate vaccines.

Generally, most research into vaccines has focused on the pathogens and the identification of pathogen-encoded vaccine candidates, and less on the host response. Yet, in most vaccine studies, evidence of variation in response has been reported, with varying proportions of non-responders, even after several immunisations, as well as individuals with adverse reactions. However this variation has mainly been ignored, or considered as inevitable animal-to-animal variation. On the other hand this suggests that at least some of the observed variation may be genetic and understanding what leads to non- or low-response, or pathology at the other extreme might identify new targets both for immunomodulators as well as for immunotherapeutics. These genes may also highlight new pathways that regulate the response to pathogens and be of more general importance. Genetic variability in response to vaccination is likely to become an even more significant factor in designing ideal "safe, synthetic" vaccines. The genes identified might also be important for disease resistance traits, and could potentially provide the tools to select "good responders", as originally suggested by Wilkie and Mallard (1999) and more recently by Gay et al. (2007). However, the genes underlying variation in vaccine response have not been greatly explored in either humans (Poland et al., 2008) or livestock (Glass, 2004).

Probably the most complex issue in aims to identify genes controlling disease resistance or vaccine response relates to the choice of phenotype(s). Depending on the pathogen, the goal may be to prevent infection completely or more likely to reduce the consequences of infection, particularly those that impact on performance traits such as growth, milk and meat quality. Although it has been reported that selecting for performance has had a detrimental effect on resistance to infectious disease, for example milk yield and mastitis (Heringstad et al., 2005), the relationship between performance and disease resistance can show beneficial correlations. Susceptibility to *Mycobacterium bovis* is negatively genetically correlated with milk yield (Brotherstone et al., 2010) as well as survival (Bermingham et al., 2010) and susceptibility to *M. avium* subspecies *paratuberculosis* (MAP) is negatively correlated with several productivity traits including a combined productivity score, Net Merit (Attalla et al., 2010). Although these results need confirmation in other populations, they would suggest that selection for disease resistance and improved performance is a realistic goal. Furthermore the livestock industry routinely selects for a number of different criteria using a selection index of weighted traits (for example, Haile-Mariam et al., 2010). Thus, it should be possible to incorporate both disease resistance and performance as goals for selection. However, identifying relevant phenotypes for both infectious disease resistance and vaccine response is not straightforward, as in many cases the correlates of protection are unclear and/or difficult to measure.

We have undertaken a number of approaches to begin to address the hypothesis that small sequence changes (polymorphisms) in key host immune response genes, results in much greater changes in immune outcome. The first steps in trying to identify these genes is to determine if there is in fact any variation in phenotype – in this case in vaccine or immune responsiveness – and if so, to determine if any phenotypic variance can be accounted for by genetics. Usually this involves a large number of animals and linked phenotypes, in which the observed variation is partitioned into "environmental effects" (i.e. year of sampling, farm, etc.) and genetic effects (i.e. sire, line, breed, etc.) using multivariate statistical methods such as restricted maximum likelihood (REML) methods. The proportion of variance in the population being tested attributed to genetic factors is referred to as "heritability" or "h^2". In general, most disease resistance studies based on field studies report relatively low heritability e.g. mastitis studies typically report $h^2 \sim 0.05$ (Rupp and Boichard, 2003). However, these are likely to be underestimates as they are often based on treatment records or outcome at slaughter and are thus not very accurate. They are also based on all animals in a study and it is often unknown whether all the animals were exposed or not. This is particularly important as in many cases the resistance or tolerance phenotype is only expressed following exposure. Thus low estimates of heritability should not necessarily be viewed as a stumbling block to further investigation (Bishop and Woolliams, 2010). In addition heritability depends on the population and its environment, and does not describe the nature of the underlying genes, their number or impact on phenotype.

Generally, once variation has been established, two main types of study in which the phenotype is correlated with the genotype are undertaken. These involve linkage (in which the phenotypes are linked to markers inherited within family groups) or association (in which unrelated individuals are divided according to phenotype and the frequencies of markers within each group are compared). Investigation into the complexities of immune or vaccine responsiveness can take complementary approaches: either candidate genes previously determined by knowledge of mechanisms leading to protection or pathology, or whole genome approaches which do not rely on a priori discoveries and use markers distributed across the entire genome. In the past, association studies have mainly relied on candidate gene polymorphisms, whereas whole genome studies have employed linkage analysis with microsatellite markers to identify chromosomal regions that affect the phenotype: so called quantitative trait loci (QTL). Only a limited number of studies in livestock have identified QTL for disease resistance and even fewer for immune/vaccine response. Until recently, these have mainly involved a relatively limited number of microsatellite markers, and the chromosomal regions identified as having significant impact on the traits have been very broad and contained large numbers of candidate genes. However with the advent of livestock genome sequences and identification of very large numbers of single nucleotide polymorphisms (SNPs), it is now possible to conduct association analyses with thousands of SNPs as markers spread across genomes which should make it easier to identify causal genes (reviewed by Fan et al., 2010). For example, using the Illumina Bovine 50 Bead chip which contains approximately 50,000 SNPs, several studies have recently reported multiple loci controlling infection with or antibody response to MAP (Settles et al., 2009; Minozzi et al., 2010; Pant et al., 2010; Kirkpatrick et al., 2011). However, there is only limited concordance between the studies in terms of identified loci, some of which may relate to differences in MAP trait definitions. Higher density SNP arrays are now available for cattle containing over ten times the number of SNPs in the Illumina Bovine 50 Bead chip which in theory means that details of pedigree structures are probably no longer necessary, and should make genetic studies more tractable for livestock. Nonetheless although there has been considerable investment in the technology and resources for large-scale genotyping in cattle and other domestic species, the issues surrounding the "phenotype" gap remain to be solved.

Control of viral pathogens poses particular problems for livestock management because of their very nature. Their ability to proliferate, mutate and to modulate the host immune response might suggest that host species could not evolve counter-measures fast enough to become resistant, yet there is considerable evidence indicating that genes involved in host defence are extremely diverse both within and between species (Barreiro and Quintana-Murci, 2010). Evidence for host genetic variation exists for viral diseases in many host species including cattle (Glass et al., 2010) and the remainder of this article will focus on the genetics of bovine respiratory disease and vaccine responses in cattle.

2. BOVINE RESPIRATORY DISEASE

Bovine respiratory disease (BRD) is a major welfare and economic burden affecting both beef and dairy cattle, and costing around $750 million per annum in the United States (Snowder et al., 2006; Miles, 2009). BRD is a complex of diseases with many viral and bacterial infectious agents. The most common viral pathogens which are known to contribute to the development of BRD include bovine respiratory syncytial virus (BRSV), parainfluenza virus (PIV)-3, bovine herpes virus (BHV)-1 and bovine viral diarrhoea virus (BVDV) (Fulton, 2010). Often infection with viral respiratory pathogens results in secondary bacterial infections, including *Mannheimia haemolytica*, *Histophilus somni* and *Pasteurella multocida* (Griffin et al., 2010). Currently the control methods for BRD include good management, antibiotics and vaccination against the viral pathogens, BHV-1, BVDV, BRSV and PIV-3 (Bowland and Shewen, 2000) as well as vaccination against the bacterial pathogens, *M. haemolytica* and *P. multocida* associated with BRD (Fulton, 2010), although the efficacy of these vaccines has mainly not been subject to controlled challenge studies (Bowland and Shewen, 2000). The pathogens involved in BRD have co-evolved with their hosts and developed strategies for manipulating and evading the host immune response and this enables these pathogens to persist, and may be at least partly responsible for the failure of current vaccines to control BRD (Srikumaran et al., 2007).

A number of studies have indicated that different breeds of cattle have different degrees of susceptibility to BRD and heritability has been estimated to be around 0.04–0.08 (Muggli-Cockett et al., 1992; Snowder et al., 2005, 2006, 2007; Heringstad et al., 2007). It is possible that susceptibility to BRD is greater than estimated for the reasons described above, especially as the specific pathogens causing BRD are often unknown. In any case the genes underlying any genetic differences between animals or breeds are unknown.

3. BOVINE RESPIRATORY SYNCYTIAL VIRUS

One of the major pathogens causing BRD as well as bovine shipping fever is considered to be BRSV which is a large enveloped, negative sense single stranded RNA *Pneumovirus* of the *Paramyxoviridae* family. It is ubiquitous in both dairy and beef cattle worldwide. Different strains of RSV also infect sheep, goats and humans. The clinical signs of BRSV include severe infection of the lower respiratory tract, resulting in coughing and nasal discharge and abnormal breathing sounds (Antonis et al., 2003). However, its effect on the host is variable from mild to severe; BRSV causes high morbidity in young animals and is the most important cause of lower respiratory tract infection in young calves (Valarcher and Taylor, 2007). By nine months of age over 70% of calves are estimated to have been infected with BRSV. Intensification of farming has probably increased the prevalence of BRSV in livestock. Thus this pathogen alone has high economic impact.

Currently the control of BRSV infection is mainly through management practices that reduce the level of circulating pathogen as well vaccination with modified live virus (MLV) or killed virus vaccines (Meyer et al., 2008). However, there is little concensus about which vaccines are the most efficacious as most studies conducted have been experimental trials and there are few field studies demonstrating clinical protection or reduction in clinical disease with any BRSV vaccine (Meyer et al., 2008). One of the most common MLV vaccines used in Europe is Rispoval RS (Pfizer) which is based on strain RB-94 (1969 Belgian isolate) (Zygraich, 1982).

Infection with BRSV, as with its human counterpart, HRSV, does not induce long lasting immunity (Meyer et al., 2008) and often calves are repeatedly infected before they become essentially immune to further infection. Similarly BRSV vaccines have limited protective capacity. Furthermore calves are particularly at risk as maternal antibody begins to wane. Yet, it is clear that this passively acquired antibody inhibits responses induced by vaccination (O'Neill et al., 2007; Meyer et al., 2008) and clearance of antibody is variable and at least in part genetically determined (O'Neill et al., 2006), making the timing of effective vaccination difficult to achieve. It is generally considered that neutralising antibody is the most important protective mechanism, and particularly mucosal antibody, but the precise role of different subclasses of immunoglobulin, in clearance of virus, maintenance of protection and induction of pathology is less clear (Meyer et al., 2008). In cattle, as in mice and humans, it appears that protection is associated with both a Th1 type response involving interferon-γ (IFN-γ) as well as the corresponding IgG subclass, IgG2 (Mapletoft et al., 2006), and a Th2 type response involving IgG1 (Kalina et al., 2005). BRSV infection induces bovine major histocompatibility complex (MHC) (BoLA) class II restricted $CD4^+$ and BoLA class I restricted $CD8^+$ T cells (Taylor et al., 1995; Gaddum et al., 1996; Fogg et al., 2001) and depletion of T cells in the respiratory tract leads to delayed BRSV clearance in calves (Taylor et al., 1995), suggesting that cell mediated immunity must also be important for protection in cattle.

Although BRSV is cytopathic *in vivo* and is associated with pathology involving innate and adaptive immune cells, *in vitro*, cell death does not necessarily occur and depends on the cell type infected (Valarcher and Taylor, 2007). Furthermore, pathology does not correlate with viral load (Gershwin, 2007). These findings suggest that the pathology seen in cattle is related more to the host immune response than to the virus (Gershwin, 2007; Ellis, 2009).

There are no licensed vaccines for human RSV, partly because formalin inactivated vaccines administered to babies, resulted in higher levels of pathology following natural infection with HRSV (Meyer et al., 2008). It has been suggested that these vaccines induced a non-protective Th2 biased response and that this was then invoked by natural infection and caused the observed pathology (Meyer et al., 2008). Similarly, experimental administration of inactivated BRSV vaccines can result in pathology following challenge with live BRSV, which appears to be mediated by an exacerbated Th2 mediated response including eosinophilia and IgE and suppressed IFN-γ (Th1) response (Gershwin, 2007). Nonetheless, these type of vaccines for cattle have been in use for many years and generally not found to result in such disease enhancement (Meyer et al., 2008). Thus it currently remains difficult to draw firm conclusions about the likelihood of inactivated BRSV vaccines inducing protective or pathogenic responses to natural BRSV infection. As with all BRSV vaccine responses, it seems likely that both vaccine and host factors such as strain of BRSV, dose, differences in adjuvants, maternal antibody and age, may account for some of the variation between animals and experimental studies. However, the role of host genetics in BRSV vaccine induced pathology and protection in cattle has up to now not been explored.

4. THE ROLE OF GENETICS IN DETERMINING THE OUTCOME OF BRSV INFECTIONS AND VACCINATION

Although a role for genetics in variability in clinical outcome for RSV infection has been evinced in human studies (Janssen et al., 2007; Miyairi and DeVincenzo, 2008; Glass et al., 2010), such evidence is lacking for cattle and BRSV. However as the pathology in BRSV and HRSV shows many similarities, it seems likely that genetics must play a role (Glass et al., 2010). Furthermore, the evidence of genetics playing a role in BRD (see above) and the strong correlation of seropositivity for BRSV with BRD might also indicate a role for genetics in the bovine response to natural infection with BRSV. In addition, variation in the clearance of circulating maternal antibody was also shown to have a genetic component (O'Neill et al., 2006).

In contrast to the genetics of natural infection with BRSV, the response to vaccination with a MLV BRSV vaccine (Rispoval RS) has been linked to genetics. BRSV-specific IgG1 and IgG2 levels were measured prior to and following vaccination of approximately 500 young calves, and found to be highly variable between animals. These parameters were chosen as potential indicators of Th2 and Th1 responses in cattle respectively (Estes and Brown, 2002) as it has been hypothesised that genetic control of these pathways may underlie susceptibility and resistance to many pathogens. Furthermore both isotypes appear to be important for protection against BRSV (Kalina et al., 2005; Mapletoft et al., 2006). Heritabilities for serum IgG1 and IgG2 responses to vaccination were estimated to be as high as 0.36 at day 35 following vaccination, and significant sire effects were also discovered (O'Neill et al., 2006). This study exploited an experimental second generation cross population of dairy (Holstein) and beef (Charolais) cattle for which pedigrees were established. This population was also genotyped for 165 microsatellite markers and also for BoLA class II DRB3 alleles by a modified sequence based typing method of the polymorphic second exon of DRB3 (Baxter et al., 2008). QTL studies on production related traits on this population have revealed many QTL (Gutierrez-Gil et al., 2007, 2008a,b, 2009, 2010), and in addition several immune-related studies have now been carried out. In addition to the BRSV study (O'Neill et al., 2006), a genetic component to the T cell response to a mastitis causing pathogen, *Staphylococcus aureus*, has been reported (Young et al., 2005). Furthermore, following immunisation of this herd with a 40-mer peptide derived from foot-and-mouth disease virus (FMDV), a large number of QTL (Leach et al., 2010) and BoLA DRB3 polymorphisms (Baxter et al., 2009) have been associated with antibody responses to this peptide.

We have now found that, as with the response to the FMDV peptide (Leach et al., 2010), many regions of the bovine genome are strongly associated with the levels pre- and post-vaccination with the BRSV vaccine (summarised in Table 1). The variance in the phenotypic traits (total BRSV specific IgG, BRSV specific IgG1 and BRSV specific IgG2) explained by each QTL was around 2–3%. It can be seen that some chromosomal regions (on BTA 10 and 15) only influenced the clearance of maternally derived antibody and not the response to vaccination whereas others (on BTA 7, 8, 9, 14, 18, 23 and 24) only influenced the response to vaccination and not the level of antibody prior to vaccination. Only two chromosomes (BTA 2 and 17) harboured genes that influenced the level of antibody both pre- and post-vaccination. Interestingly, the level of variation of IgG2 at 14 days post vaccination appears to be the most polygenic trait, with six different chromosomal regions significantly associated with this trait, in contrast to the variation in the IgG1 levels at the same time point, where only a single QTL was detected. Since IgG1 and IgG2 are believed to be controlled by Th2 and Th1 cells, respectively (Estes and Brown, 2002), these results suggest that these pathways are influenced by different genes. It is also apparent that the early and later responses to vaccination are controlled by different genes.

Table 1. The sequential influence of significant quantitative trait loci (QTL) associated with BRSV specific antibody levels pre- and post-vaccination across time.

Day[a]/Trait[b]	Chromosome number (BTA[c])											
	3	17	2	10	15	8	18	7	9	23	24	14
Pre-vaccination												
−14	X[d]	XX	X									
0: Total[e] IgG			X									
0: IgG1			X		X							
0: IgG2				X								
Post-vaccination												
0–14			X			X	X					
14: Total IgG		X						X				
I4: IgG1		X										
14: IgG2	X						X		X	X	X	X
0–49			X									
49: Total IgG						X	X					X
49: IgG1							X					
Overall[f] IgG1						X						X
Overall IgG2										X		

[a]Day is relative to vaccination at day 0.

[b]Trait = BRSV specific antibody level, measured by ELISA as detailed in (O'Neill et al., 2006).

[c]BTA = *Bos taurus* autosome. The order shown attempts to reflect the influence of different QTL across the time course, with those loci which influence the response at the earliest time point being shown first.

[d]Each X represents the presence of a significant QTL ($p < 0.05$); XX represents two QTL on the same chromosome.

[e]Total refers to IgG1 and IgG2 levels.

[f]Overall refers to the total amount of IgG isotype generated across time (calculated as "area under the curve" by the Trapezoidal rule). All QTL are at least 5% chromosome wide significant

It is worth noting that the QTL on BTA23 whilst they have broad confidence intervals, do cover the bovine MHC region suggesting that the bovine MHC may be involved in determining the level of antibody response to the BRSV vaccine (and see below).

In almost every immune-related study for which there is evidence for a genetic component, the MHC has been identified as playing a role (Glass, 2004). Given the likely important role for cell mediated immunity in BRSV infection, including $CD4^+$ and $CD8^+$ Tells (Taylor et al., 1995), it would seem likely that polymorphisms in MHC genes would affect both the outcome from infection and also vaccination. However, although many candidate genes have been associated with outcome of HRSV infection in human studies, none have reported MHC associations for human or bovine RSV infection or vaccination (Janssen et al., 2007; Miyairi and DeVincenzo, 2008; Glass et al., 2010), although a very recent study in congenic strains of mice has shown that MHC haplotypes play a role in the susceptibility of neonatal mice (Tregoning et al., 2010). We have now discovered a number of significant associations between BoLA class II DRB3 polymorphisms and both the circulating level of antibody prior to vaccination as well as the antibody response to the BRSV vaccine (Tables 2 and 3). Although a few DRB3 alleles were significantly associated with both pre-vaccination and post-vaccination levels of anti-BRSV antibody (Table 2), there were very few animals in the herd that carried these alleles and the results must be treated with caution, and require verification in other populations. However, the majority of polymorphisms found in MHC class II DRB genes are concentrated in exon 2 and particularly in the sequences encoding the peptide binding pockets. These are the major binding anchors for presented peptides. Thus BoLA DRB3 alleles can be grouped into a series of peptide binding pocket motifs (Sharif et al., 2000; Baxter et al., 2009). Several individual positions within three of these pockets were significantly associated with the pre- and post-vaccination antibody levels to BRSV (Table 3). It is striking that none of the associations were significant for both the pre- and post-vaccination BRSV specific antibody levels. Perhaps not surprisingly this suggests that the rate of clearance of maternal antibody is controlled by different mechanisms from those operating in the induction of an adaptive response. The relative paucity of significant associations reported here are in marked contrast to those found with the FMDV peptide (Baxter et al., 2009) where several alleles and pocket positions were highly significantly associated with response. There are likely to be many reasons for this, not least that the FMDV 40-mer peptide has a limited number of epitopes whereas BRSV is likely to have many different epitopes recognised by the host immune system. In addition, the BRSV antibody levels are confounded with the pre-vaccination levels of circulating antibody. Nonetheless it is interesting that the pocket with the greatest associations was pocket 4 and this mirrors the findings with the

FMDV peptide response (Baxter et al., 2009). Pocket 4 is located centrally within the peptide binding groove of DR molecules and has been implicated as playing a significant role in human immune responsiveness (Ou et al., 1998). This might indicate that in terms of vaccine design for any pathogen, ensuring that the vaccine includes epitopes that bind strongly to the major pocket 4 variants of DRB3 would ensure that all animals within herds would be protected.

Table 2. Significant *BoLA DRB3* alleles associated with BRSV-specific IgG levels pre- and post-vaccination.

DRB3 allele[a]	**Pre-vaccination**	**Post-vaccination**
*0801	<0.05[b]	N.S.
*0901	N.S.[c]	<0.05
*1002	N.S.	<0.05
*1701	N.S.	<0.05

[a]DRB3 alleles determined by a sequence based typing method of the 2nd exon of BoLA DRB3, as described in Baxter et al. (2008).

[b]*p* values determined by Wald test.

[c]N.S. = not significant.

Table 3. Peptide binding pockets significantly associated with BRSV-specific IgG levels pre- and post-vaccination.

DRB3 Pocket[a]	**Pocket position**	**Pre-vaccination**	**Post-vaccination**
4	β13	<0.05[b]	N.S.[c]
4	β70	N.S.	<0.05
4	β74	N.S.	<0.001
7	β28	N.S.	<0.05
7	β30	N.S.	<0.05
9	β37	<0.05	N.S.

[a]DRB3 Pockets determined as described in Baxter et al. (2009).

[b]*p* values determined by Wald test.

[c]N.S. = not significant.

Of course in terms of candidate genes for the vaccine response to BRSV, it is unlikely that only the MHC region would be important. As we (Clapperton et al., 2009; Glass et al., 2010) have argued elsewhere, genes expressed in the innate immune system are strong candidate genes for infectious disease associated traits in livestock as (a) activation of the innate immune system is essential and indeed determines the nature of the acquired immune response, (b) the genes involved are present in the germ-line, are highly polymorphic and code for molecules involved in the initial interactions of pathogens and the host defence system and (c) these molecules include pattern recognition receptors and soluble molecules (PRR and PRM) that sense a wide range of pathogens. PRMs and PRRs interact with pathogen associated molecular patterns (PAMPs) which are conserved on pathogens but generally not found in host species. The most well known PRRs are the Toll-like receptors (TLRs), but more recently a number of other PRR families have been described including the RNA helicase-type retinoic acid-inducible gene-1 receptors (RIG-1-like helicases) and NOD-like receptors which appear to be relevant for host detection of viruses (reviewed by Brennan and Bowie, 2010). Many candidate gene variants have been associated with severe humans RSV infections including PRMs – collectins, surfactant proteins A–D and mannose binding lectin (MBL), and the PRRs – TLR4 and TLR8 (Janssen et al., 2007; Miyairi and DeVincenzo, 2008; Glass et al., 2010). Although TLR2, TLR3 and RIG-1 have all been implicated in host interactions with RSV, no genetic associations in human studies have been reported. In cattle, a number of distinct collectin genes form a cluster on BTA28 (Gjerstorff et al., 2004), and some have been associated with viral pathogen interactions (Lillie et al., 2005). Polymorphisms have been described in bovine MBL (Wang et al., 2011) and bovine conglutinin levels are heritable (Holmskov et al., 1998). Similarly, bovine viral recognition *TLRs* (Cargill and Womack, 2007), and *NOD2* (also known as *CARD15*) harbour polymorphisms (Taylor et al., 2006) and a single SNP has been reported in bovine *RIG-1* (Cargill et al., 2006). Some associations of TLR variants with bacterial but not viral diseases have been reported in cattle. We have argued that TLRs and their intracellular down-stream signalling components may be the most likely candidates for many infectious disease traits in livestock species (Jann et al., 2009; Werling et al., 2009). Furthermore, as adjuvants are essential to invoke responses to vaccine components by triggering innate immunity, we suggest that gene variants encoding innate immune system proteins may also influence vaccine responsiveness. Although many SNPs in candidate genes have been proposed, very few studies in livestock have reported significant associations and this may in some cases be because the SNPs have no functional consequences. Determining whether individual SNPs are functionally relevant requires complex molecular biology and we have proposed that a simpler approach to narrow down the candidates is to consider only those SNPs that are both non-synonymous and positively selected (Jann et al., 2008). Of course this excludes SNPs that are non-coding and we accept that it is likely that SNPs in promoter regions can change expression levels and therefore could also have functional consequences. However even in human populations the relative importance of coding SNPs and non-coding SNPs in determining complex disease is as yet unknown, although it is clear that there is considerable inter-individual variation in gene expression across the human genome (Cookson et al., 2009; Skelly et al., 2009). With the advent of massive parallel sequencing for transcriptomics the contribution of inherited variation in gene expression to complex traits should become clearer (Majewski and Pastinen, 2011).

However we decided to test whether our proposal has merit and investigated whether any non-synonymous and positively selected SNPs in bovine TLRs were associated with the response to the BRSV vaccine and also to the FMDV peptide. In addition, the maximum number of microsatellite markers per chromosome in the QTL study was six, and thus as discussed above large confidence intervals were obtained which contain hundreds of candidate genes. Clearly this limits both the power and accuracy of the study, and in order to narrow the regions, we added additional SNP markers within some of the QTL. We found that a number of SNPs on different chromosomes were significantly associated with the response to the BRSV vaccine (Table 4) and also to the FMDV peptide (results not shown). The variance of the phenotypic trait accounted for by individual SNPs was in general much higher than for the QTL study, with some explaining over 10% of the phenotypic variance. However the majority of SNPs were not on the same chromosomes as the QTL, and this may well reflect the size of the study (~500 animals) and the fact that the phenotypes were measured on a F2 generation which means that little recombination will have occurred. However as with the QTL study it is clear that both time and isotype of antibody must be under different genetic control. Although we tested for association of bovine SNPs in all ten TLRs, we only found two associated with response to the BRSV vaccine. Of particular interest, one of the most significant SNPs was on BTA 8, which influenced the IgG1 and IgG2 response to the BRSV vaccine and was located within the TLR4 gene. Further analysis determined that this SNP was non-synonymous and positively selected. Two further highly significant non-synonymous and positively selected SNPs were located in TLR8 and influenced the response to both BRSV vaccine and also the FMDV peptide. It is possible that both TLR4 and TLR8 are involved in the recognition of BRSV as there is evidence that HRSV F protein binds to TLR4 (Kurt-Jones et al., 2000) and BRSV induces a TLR4-dependent NFkB response in bovine TLR transfected cells (Lizundia et al., 2008). In addition TLR8 senses ssRNA and RSV is an RNA virus although no direct interactions between TLR8 and RSV have been reported. In humans polymorphisms in TLR4 and TLR8 have been associated with severity of BRSV infection (Janssen et al., 2007).

Table 4. The sequential influence of significant SNPs associated with BRSV specific antibody levels at vaccination day and post-vaccination.

Trait[a]	**Day[b]**	**Chromosome number (BTA[c])**												
		12	**16**	**22**	**24**	**6**	**7**	**19**	**25**	**X**	**21**	**27**	**29**	**8**
IgG2	0	X[d]	X	X	X									
	14					X	X							
	35							X	X	X				
	49								X		X			
IgG1	14			X					X			X	X	
	35						X							X
	49													
Overall[e] IgG1												X		
Overall IgG2														X

[a]Trait = BRSV specific antibody level, measured by ELISA as detailed in (O'Neill et al., 2006).

[b]Day is relative to vaccination at day 0.

[c]BTA = *Bos taurus* autosome. The order shown attempts to reflect the sequential influence of different SNP across the time course, with those loci which influence the response at the earliest time point being shown first.

[d]Each X represents the presence of a significant SNP ($p < 0.05$).

[e]Overall refers to the total amount of IgG isotype generated across time (calculated as "area under the curve" by the Trapezoidal rule).

5. CONCLUDING REMARKS

In conclusion, genetics clearly influences both the response to vaccination and also to infection. In addition, genetics also plays a key role in the rate of clearance of maternal antibody. Although we have identified a number of candidate variants, which include BoLA DRB3 and TLRs 4 and 8 as playing a significant role, we have also identified many regions that may harbour novel genes that are important in determining the response to vaccination. Because of the strong linkage between genes in the BoLA region, we currently cannot disentangle whether our data supports the role of BoLA DRB3 or is linked to polymorphisms in other classical class II genes (DQA and/or DQB) or indeed class I genes.

However, the pocket 4 associations would strongly suggest that BoLA DRB3 is playing an important role in the observed variation in antibody response. We and others have demonstrated the importance and complexity of DQ molecules in T cell responses in cattle (Glass et al., 2000; Norimine and Brown, 2005; Gerner et al., 2009), and further exploration of the role of DQ in bovine responses to BRSV is warranted.

We found three non-synonymous, positively selected SNPs in two TLRs providing some support to our hypotheses relating to the involvement of innate immune genes in disease resistance and vaccine responsiveness. The lack of evidence for association with SNPs in other candidate genes does not of course rule out a role for them, as it could simply be that the study was too small to detect their influence and/or other polymorphisms may not be represented in the study herd. Although the uncertainty surrounding the correlates of protection for BRSV remains to be solved, our initial studies suggest that another significant factor in observed variation in protection and pathology following natural infection and vaccination in cattle may be host genetics. However, in order to investigate further, the QTL and SNPs require validation in new and larger populations as well as more detailed molecular studies, for example cells transfected with different TLR variants may reveal the underlying mechanisms.

Future studies that improve our understanding of the role of genetics in immune responsiveness and disease resistance have the potential to improve vaccine efficacy as well as provide selectable markers that breeders could use to improve livestock disease resistance as well as vaccine responsiveness (Fig. 1).

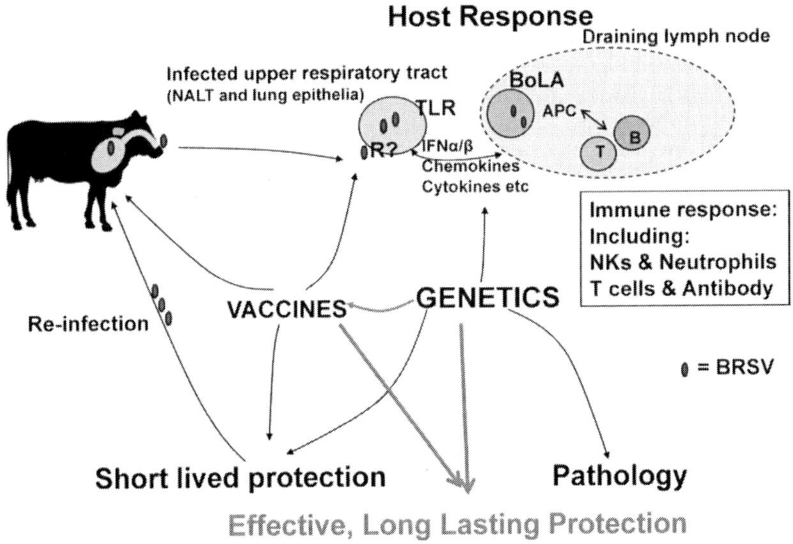

Figure. 1. Schematic diagram illustrating the potential role of genetics in the variation in outcome following natural infection or vaccination with BRSV. Understanding the relationship that genetics has with outcome may lead to improvements in host disease resistance and/or vaccine efficacy.

ACKNOWLEDGEMENTS

We thank many current and past colleagues at Roslin Institute including John Williams, now at Parco Tecnologico Padano, Italy, who was instrumental in setting up the cattle cross herd, John Woolliams who contributed to some of the initial genetic and QTL studies, and the farm staff who looked after the welfare of the herd and conducted many of the immunisations and bleeding required. All animal experimentation was conducted under the auspices of the UK Animals (Scientific Procedures) Act, 1986. In addition the BRSV ELISA phenotypes were collected by Ronan O'Neill, now of The Department of Agriculture, Fisheries and Food, Republic of Ireland. We also acknowledge the input contributed by Julie Fitzpatrick, now Director of the Moredun Research Institute (MRI), UK, and Sara Knott, the Institute of Evolutionary Biology, University of Edinburgh, UK, as well as the TLR consortium which consisted of David Haig, University of Nottingham, UK, Dirk Werling, Royal Veterinary College, UK, George Russell, MRI, UK, Tracey Coffey and Jayne Hope, Institute for Animal Health, UK. This work was supported by the Biotechnology and Biological Sciences Research Council (BBSRC) (Institute Strategic Programme Grant funding), a

Genesis Faraday (now KTN Biosciences) facilitated project grant BB/D524040/1 jointly funded by BBSRC, The Scottish Government Rural and Environment Research and Analysis Directorate (RERAD) and Pfizer Animal Health (PAH), a BBSRC CASE studentship with EBLEX, Hybu Cig Cymru/Meat Promotion Wales (HCC), Quality Meat Scotland (QMS), DairyCo, the UK Department for Environment, Food and Rural Affairs and RERAD, and a PAH CASE studentship.

REFERENCES

1. Antonis, A.F.G., Schrijver, R.S., Daus, F., Steverink, P.J.G.M., Stockhofe, N., Hensen, E.J., Langedijk, J.P.M, van der Most, R., 2003. Vaccine-induced immunopathology during bovine respiratory syncytial virus infection: exploring the parameters of pathogenesis. J. Virol. 77, 12067–12073.
2. Attalla, S.A., Seykora, A.J., Cole, J.B., Heins, B.J., 2010. Genetic parameters of milk ELISA scores for Johne's disease. J. Dairy Sci. 93, 1729–1735.
3. Barreiro, L.B., Quintana-Murci, L., 2010. From evolutionary genetics to human immunology: how selection shapes host defence genes. Nat. Rev. Genet. 11, 17–30.
4. Baxter, R., Hastings, N., Law, A., Glass, E.J., 2008. A rapid and robust sequence based genotyping method for BoLA-DRB3 alleles in large numbers of heterozygous cattle. Anim. Genet. 39, 561–563.
5. Baxter, R., Craigmile, S.C., Haley, C., Douglas, A.J., Williams, J.L., Glass, E.J., 2009. BoLA-DR peptide binding pockets are fundamental for foot-andmouth disease virus vaccine design in cattle. Vaccine 28, 28–37.
6. Bermingham, M.L., More, S.J., Good, M., Cromie, A.R., Higgins, I.M., Berry, D.P., 2010. Genetic correlations between measures of Mycobacterium bovis infection and economically important traits in Irish Holstein-Friesian dairy cows. J. Dairy Sci. 93, 5413–5422.
7. Bishop, S.C., Woolliams, J.A., 2010. On the genetic interpretation of disease data. PLoS ONE 5 (1), e8940, doi:10.1371/journal.pone.0008940.
8. Bowland, S.L, Shewen, P.E., 2000. Bovine respiratory disease: commercial vaccines currently available in Canada. Can. Vet. J. 41, 33–48.
9. Brennan, K., Bowie, A.G., 2010. Activation of host pattern recognition receptors by viruses. Curr. Opin. Microbiol. 13, 503–507.
10. Brotherstone, S., White, I.M.S., Coffey, M., Downs, S.H., Mitchell, A.P., Clifton-Hadley, R.S., More, S.J., Good, M., Woolliams, J.A., 2010. Evidence of genetic resistance of cattle to infection with Mycobacterium bovis. J. Dairy Sci. 93, 1234–1242.
11. Cargill, E.J., Paetzold, L., Womack, J.E., 2006. Radiation hybrid mapping and comparative sequence analysis of bovine RIG-I and MAVS genes. DNA Sequence 17, 314–318.
12. Cargill, E.J., Womack, J.E., 2007. Detection of polymorphisms in bovine toll-like receptors 3, 7, 8, and 9. Genomics 89, 745–755.
13. Clapperton, M., Diack, A.B., Matika, O., Glass, E.J., Bishop, S.C., 2009. Traits associated with innate and adaptive immunity in pigs: heritability and associations with performance under different health status conditions. Genet. Sel. Evol. 41, 54.
14. Cookson, W., Liang, L., Abecasis, G., Moffatt, M., Lathrop, M., 2009. Mapping complex disease traits with global gene expression. Nat. Rev. Genet. 10, 184–194.
15. Ellis, J.A., 2009. Update on viral pathogenesis in BRD. Anim. Health Res. Rev. 10, 149–153.
16. Estes, M.D., Brown, W.C., 2002. Type 1 and Type 2 responses in regulation of Ig isotype expression in cattle. Vet. Immun. Immunopathol. 90, 1–10.

17. Fan, B., Du, Z.Q., Gorbach, D.M., Rothschild, M.F., 2010. Development and application of high-density SNP arrays in genomic studies of domestic animals. Asian-Australasian J. Anim. Sci. 23, 833–847.
18. Fogg, M.H., Parsons, K.R., Thomas, L.H., Taylor, G., 2001. Identification of CD4(+) T cell epitopes on the fusion (F) and attachment (G) proteins of bovine respiratory syncytial virus (BRSV). Vaccine 19, 3226–3240.
19. Fulton, R.W., 2010. Bovine respiratory disease research (1983–2009). Anim. Health Res. Rev. 10, 131–139.
20. Gaddum, R.M., Cook, R.S., Thomas, L.H., Taylor, G., 1996. Primary cytotoxic T-cell responses to bovine respiratory syncytial virus in calves. Immunology 88, 421–427.
21. Gay, C.G., Zuerner, R., Bannantine, J.P., Lillehoj, H.S., Zhu, J.J., Green, R., Pastoret, P.P., 2007. Genomics and vaccine development. Rev. Sci. Technol. Off. Int. Epiz. 26, 49–67.
22. Gerner, W., Hammer, S.E., Wiesmuller, K.H., Saalmuller, A., 2009. Identification of major histocompatibility complex restriction and anchor residues of foot-and-mouth disease virus-derived bovine T-cell epitopes. J. Virol. 83, 4039–4050.
23. Gershwin, L.J., 2007. Bovine respiratory syncytial virus infection: immunopathogenic mechanisms. Anim. Health Res. Rev. 8, 207–213.
24. Gjerstorff, M., Hansen, S., Jensen, B., Dueholm, B., Horn, P., Bendixen, C., Holmskov, U., 2004. The genes encoding bovine SP-A, SP-D, MBL-A, conglutinin, CL-43 and CL-46 form a distinct collectin locus on Bos taurus chromosome 28 (BTA28) at position q.1.8-1.9. Anim. Genet. 35, 333–337.
25. Glass, E.J., 2004. Genetic variation and responses to vaccines. Anim. Health Res. Rev. 5, 197–208.
26. Glass, E.J., Oliver, R.A., Russell, G.C., 2000. Duplicated DQ-haplotypes increase the complexity of restriction element usage in cattle. J. Immunol. 165, 134–138.
27. Glass, E.J., Baxter, R., Leach, R., Taylor, G., 2010. Bovine viral diseases – the role of host genetics. In: Bishop, S.C., Axford, R., Owen, J., Nicholas, F. (Eds.), Breeding for Disease Resistance in Farm Animals. , 3rd edition. CAB International, Nosworthy Way, Wallingford OX10 8DE, United Kingdom.
28. Griffin, D., Chengappa, M.M., Kuszak, J., McVey, D.S., 2010. Bacterial pathogens of the bovine respiratory disease complex. Vet. Clin. N. Am. Food Anim. Pract. 26, 381–394.
29. Gutierrez-Gil, B., Wiener, P., Williams, J.L., 2007. Genetic effects on coat colour in cattle: dilution of eumelanin and phaeomelanin pigments in an F2-Backcross Charolais × Holstein population. BMC Genet. 8, 56.
30. Gutierrez-Gil, B., Wiener, P., Nute, G.R., Burton, D., Gill, J.L., Wood, J.D., Williams, J.L., 2008a. Detection of quantitative trait loci for meat quality traits in cattle. Anim. Genet. 39, 51–61. Gutierrez-Gil, B., Ball, N., Burton, D., Haskell, M., Williams, J.L., Wiener, P., 2008b. Identification of quantitative trait loci affecting cattle temperament. J. Heredity 99, 629–638.
31. Gutierrez-Gil, B., Williams, J.L., Homer, D., Burton, D., Haley, C.S., Wiener, P., 2009. Search for quantitative trait loci affecting growth and carcass traits in a cross population of beef and dairy cattle. J. Anim. Sci. 87, 24–36.
32. Gutierrez-Gil, B., Wiener, P., Richardson, R.I., Wood, J.D., Williams, J.L., 2010. Identification of QTL with effects on fatty acid composition of meat in a Charolais × Holstein cross population. Meat Sci. 85, 721–729.
33. Haile-Mariam, M., Bowman, P.J., Goddard, M.E., 2010. Calculation of lifetime net income per year (LTNI/year) of Australian Holstein cows to validate the Australian profit ranking of their sires, 2. Validation of the Australian profit ranking of sires and test for genotype by environment interaction based on LTNI/year. Anim. Prod. Sci. 50, 767–774.
34. Heringstad, B., Chang, Y.M., Gianola, D., Klemetsdal, G., 2005. Genetic association between susceptibility to clinical mastitis and protein yield in Norwegian dairy cattle. J. Dairy Sci. 88, 1509–1514.
35. Heringstad, B., Chang, Y.M., Gianola, D., Osteras, O., 2007. Short Communication: Genetic analysis of respiratory disease in Norwegian Red calves. J. Dairy Sci. 91, 367–370.

36. Holmskov, U., Jensenius, J.C., Tornoe, I., Lovendahl, P., 1998. The plasma levels of conglutinin are heritable in cattle and low levels predispose to infection. Immunology 93, 431–436.
37. Jann, O.C., Werling, D., Chang, S., Haig, D., Glass, E.J., 2008. Molecular evolution of bovine TLR2 suggests substitutions of functional relevance. BMC Evol. Biol. 8, 288.
38. Jann, O.C., King, A., Corrales, N.L., Anderson, S., Jensen, K., Ait-ali, T., Tang, H., Wu, C., Archibald, A.L., Cockett, N.E., Glass, E.J., 2009. Comparative genomics of Toll-like receptor signalling in five species. BMC Genomics 10, 219.
39. Janssen, R., Bont, L., Siezen, C.L., Hodemaekers, H.M., Ermers, M.J., Doornbos, G., vant Slot, R., Wijmenga, C., Goeman, J.J., Kimpen, J.L., van Houwelingen, H.C., Kimman, T.G., Hoebee, B., 2007. Genetic susceptibility to respiratory syncytial virus bronchiolitis is predominantly associated with innate immune genes. J. Infect. Dis. 196, 826–834.
40. Kalina, W.V., Woolums, A.R., Gershwin, L.J., 2005. Formalin-inactivated bovine RSV vaccine influences antibody levels in bronchoalveolar lavage fluid and disease outcome in experimentally infected calves. Vaccine 23, 4625–4630.
41. Kirkpatrick, B.W., Shi, X., Shook, G.E., Collins, M.T., 2011. Whole-Genome association analysis of susceptibility to paratuberculosis in Holstein cattle. Anim. Genet. 42, 149–160.
42. Kurt-Jones, E.A., Popova, L., Kwinn, L., Haynes, L.M., Jones, L.P., Tripp, R.A., Walsh, E.E., Freeman, M.W., Golenbock, D.T., Anderson, L.J., Finberg, R.W., 2000. Pattern recognition receptors TLR4 and CD14 mediate response to respiratory syncytial virus. Nat. Immunol. 1, 398–401.
43. Leach, R.J., Craigmile, S.C., Knott, S.A., Williams, J.L., Glass, E.J., 2010. Quantitative trait loci for variation in immune response to a Footand- Mouth Disease Virus peptide. BMC Genet. 11, 107.
44. Lillie, B.N., Brooks, A.S., Keirstead, N.D., Hayes, M.A., 2005. Comparative genetics and innate immune functions of collagenous lectins in animals. Vet. Immunol. Immunopathol. 108, 97–110.
45. Lizundia, R., Sauter, K.S., Taylor, G., Werling, D., 2008. Host species-specific usage of the TLR4-LPS receptor complex. Innate Immun. 14, 223–231.
46. Majewski, J., Pastinen, T., 2011. The study of eQTL variations by RNA-seq, from SNPs to phenotypes. Trends Genet. 27, 72–79.
47. Mapletoft, J.W., Oumouna, M., Townsend, H.G., Gomis, S., Babiuk, L.A., Littel-van Den Hurk, S.V., 2006. Formulation with CpG oligodeoxynucleotides increases cellular immunity and protection induced by vaccination of calves with formalin-inactivated bovine respiratory syncytial virus. Virology 353, 316–323.
48. Meyer, G., Deplanche, M., Schelcher, F., 2008. Human and bovine respiratory syncytial virus vaccine research and development. Comp. Immunol. Microbiol. Inf. Dis. 31, 191–225.
49. Miles, D.G., 2009. Overview of the North American beef cattle industry and the incidence of bovine respiratory disease (BRD). Anim. Health Res. Rev. 10, 101–103.
50. Minozzi, G., Buggiotti, L., Stella, A., Strozzi, F., Luini, M., Williams, J.L., 2010. Genetic loci involved in antibody response to Mycobacterium avium ssp. paratuberculosis in cattle. PLoS ONE 5, e11117.
51. Miyairi, I., DeVincenzo, J.P., 2008. Human genetic factors and respiratory syncytial virus disease severity. Clin. Microbiol Rev. 21, 686–703.
52. Muggli-Cockett, N.E., Cundiff, L.V., Gregory, K.E., 1992. Genetic-analysis of bovine respiratory-disease in beef-calves during the 1st year of life. J. Anim. Sci. 70, 2013–2019.
53. Norimine, J., Brown, W.C., 2005. Intrahaplotype and interhaplotype pairing of bovine leukocyte antigen DQA and DQB molecules generate functional DQ molecules important for priming CD4(+) T-lymphocyte responses. Immunogenetics 57, 750–762.
54. O'Neill, R.G., Woolliams, J.A., Glass, E.J., Williams, J.L., Fitzpatrick, J.L., 2006. Quantitative evaluation of genetic and environmental parameters determining antibody response induced by vaccination against bovine respiratory syncytial virus. Vaccine 24, 4007–4016.
55. O'Neill, R.G., Fitzpatrick, J.L., Glass, E.J., Williams, J.L., Woolliams, J.A., 2007. Optimising responses to respiratory virus vaccines in cattle. Vet. Rec. 161, 269–270.

56. Ou, D., Mitchell, L.A., Tingle, A.J., 1998. A new categorization of HLA DR alleles on a functional basis. Hum. Immunol. 59, 665–676.
57. Pant, S.D., Schenkel, F.S., Verschoor, C.P., You, Q.M., Kelton, D.F., Moore, S.S., Karrow, N.A., 2010. A principal component regression based genome wide analysis approach reveals the presence of a novel QTL on BTA7 for MAP resistance in Holstein cattle. Genomics 95, 176–182.
58. Poland, G.A., Ovsyannikova, N.G., Jacobson, R.M., 2008. Vaccine immunogenetics: bedside to bench to population. Vaccine 26, 6183–6188.
59. Royal Society, 2002. Infectious diseases in livestock. http://www.royalsoc.ac.uk/inquiry/.
60. Rupp, R., Boichard, D., 2003. Genetics of resistance to mastitis in dairy cattle. Vet. Res. 34, 671–688.
61. Settles, M., Zanella, R., Mckay, S.D., Schnabel, R.D., Taylor, J.F., Whitlock, R., Schukken, Y., Van Kessel, J.S., Smith, J.M., Neibergs, H., 2009. A whole genome association analysis identifies loci associated with Mycobacterium avium subsp. paratuberculosis infection status in US Holstein cattle. Anim. Genet. 40, 655–662.
62. Sharif, S., Mallard, B.A., Sargeant, J.M., 2000. Presence of glutamine at position 74 of pocket 4 in the BoLA-DR antigen binding groove is associated with occurrence of clinical mastitis caused by Staphylococcus species. Vet. Immunol. Immunopathol. 76, 231–238.
63. Skelly, D.A., Ronald, J., Akey, J.M., 2009. Inherited variation in gene expression. Ann. Rev. Genomics Hum. Genet. 10, 313–332.
64. Snowder, G.D., Van Vleck, L.D., Cundiff, L.V., Bennett, G.L., 2005. Influence of breed, heterozygosity, and disease incidence on estimates of variance components of respiratory disease in preweaned beef calves. J. Anim. Sci. 83, 1247–1261.
65. Snowder, G.D., Van Vleck, L.D., Cundiff, L.V., Bennett, G.L., 2006. Bovine respiratory disease in feedlot cattle: environmental, genetic, and economic factors. J. Anim. Sci. 84, 1999–2008.
66. Snowder, G.D., Van Vleck, L.D., Cundiff, L.V., Bennett, G.L., Koohmaraie, M., Dikeman, M.E., 2007. Bovine respiratory disease in feedlot cattle: phenotypic, environmental, and genetic correlations with growth, carcass, and longissimus muscle palatability traits. J. Anim. Sci. 85, 1885–1892.
67. Srikumaran, S., Kelling, C.L., Ambagala, A., 2007. Immune evasion by pathogens of bovine respiratory disease complex. Anim. Health Res. Rev. 8, 215–229.
68. Taylor, G., Thomas, L.H., Wyld, S.G., Furze, J., Sopp, P., Howard, C.J., 1995. Role of T-lymphocyte subsets in recovery from respiratory syncytial virus-infection in calves. J. Virol. 69, 6658–6664.
69. Taylor, K.H., Taylor, J.F., White, S.N., Womack, J.E., 2006. Identification of genetic variation and putative regulatory regions in bovine CARD15. Mam. Genome 17, 892–901.
70. Tregoning, J.S., Yamaguchi, Y., Wang, B., Mihm, D., Harker, J.A., Bushell, E.S.C., Zheng, M., Liao, G.C., Peltz, G., Openshaw, P.J.M., 2010. Genetic susceptibility to the delayed sequelae of neonatal respiratory syncytial virus infection is MHC dependent. J. Immunol. 185, 5384–5391.
71. Valarcher, J.F., Taylor, G., 2007. Bovine respiratory syncytial virus infection. Vet. Res. 38, 153–180.
72. Wang, C.F., Liu, M., Li, Q.L., Ju, Z.H., Huang, J.M., Li, J.B., Wang, H.M., Zhong, J.F., 2011. Three novel single-nucleotide polymorphisms of MBL1 gene in Chinese native cattle and their associations with milk performance traits. Vet. Immunol. Immunopathol. 139, 229–236.
73. Werling, D., Jann, O.C., Offord, V., Glass, E.J., Coffey, T.J., 2009. Variation matters: TLR structure and species-specific pathogen recognition. Trends Immunol. 30, 124–130.
74. Wilkie, B.N., Mallard, B.A., 1999. Genetic effects on vaccination. Adv. Vet. Med. 41, 39–51.
75. Young, F.J., Woolliams, J.L., Williams, J.L., Glass, E.J., O'Neill, R.G., Fitzpatrick, J.L., 2005. In vitro peripheral blood mononuclear cell proliferation in a crossbred cattle population. J. Dairy Sci. 88, 2643–2651.
76. Zygraich, N., 1982. Vaccination against RS virus – a five year experience with Rispoval. World Association for Buiatrics Proceedings, September 1982, Amsterdam, The Netherlands, 12th World Congress on Diseases of Cattle, pp. 189–195.

CHAPTER 17

Evaluation of the Immunoregulatory Capacities of Feed Microbial Materials in Porcine Intestinal Immune and Epithelial Cells

Naosuke Kumagae[1, 2*], Julio Villena[1, 3*], Yohsuke Tomosada[1*], Hisakazu Kobayashi[1], Paulraj Kanmani[1,4], Hisashi Aso[5], Takashi Sasaki[2], Motohiko Yoshida[2], Hiroshi Tanabe[2], Isao Shibata[2], Tadao Saito[1], Haruki Kitazawa[1#]

[1]*Food and Feed Immunology Group, Tohoku University, Sendai, Japan*

[2]*Scientific Feed Laboratory Co. Ltd., Tokyo, Japan*

[3]*Laboratory of Immunobiotechnology, Reference Centre for Lactobacilli (CERELA-CONICET), Tucuman, Argentina*

[4]*Japan Society for the Promotion of Science (JSPS), Tokyo, Japan*

[5]*Cell Biology Laboratory, Graduate School of Agricultural Science, Tohoku University, Sendai, Japan*

ABSTRACT

The establishment of drug-free feeding systems has been required for secure and healthy livestock production. Although functional feed materials containing microorganisms as alternatives to enhance intestinal immunity are expected to be beneficial for reducing diarrhoea caused by pathogens in weaned piglets, the effects of such materials on porcine intestinal cells have not been investigated in detail. Therefore, this work evaluated the immunoregulatory functions of microbial feed materials in porcine intestinal immune and epithelial cells. Porcine immune cells isolated from Peyer's patches and mesenteric lymph nodes were stimulated with six different feed materials containing microorganisms, and evaluated for lymphocyte mitogenicity and cytokine inductions. In addition, porcine intestinal epithelial cells were stimulated with the materials before treatment with heat-killed enterotoxigenic Escherichia coli (ETEC), and analyzed for the proinflammatory cytokine expressions. The material containing *Bifidobacterium thermophilum* significantly augmented lymphocytes' mitogenicity and also induced a high expression of IL-2, IL-6 and IFN-γ in immune cells, and inhibited ETEC-induced overexpression of IL-6 and IL-8 via regulation of Toll-like receptor signaling. These results suggest that this feed material stimulates intestinal epithelial and immune cells to exert immunoregulation, suggesting that this feed is expected to contribute to promoting the health of piglets without using antimicrobial feed materials.

Keywords: Feed Microbial Material; Immunoregulatory Effect; Porcine Intestinal Epitheliocytes; Porcine Immune Cells

1. INTRODUCTION

Weaning-associated intestinal inflammation occurs in various animal species including the pig. Following the withdrawal of sow's milk, young piglets are highly susceptible to enteric diseases partly as a result of the altered balance between developing beneficial microbiota and the establishment of intestinal bacterial pathogens. In addition to the changes in microbiota composition, the intestinal immune system of the newborn piglet undergoes a rapid period of maturation, expansion, and specialization that is not achieved before commercial weaning [1]. Then, intensification of the pig industry has brought increased risks of both clinical and sub-clinical enteric disease. Piglets are vulnerable to harmful microorganisms such as *Escherichia coli, Salmonella* spp., and *Clostridium perfringens* [2]. Antibiotics have been applied widely in animal husbandry to prevent and treat the gastrointestinal infection caused by pathogens [3]. However, the promiscuous use of antibiotics has resulted not only in the emergence and spread of resistant bacteria in humans but also in animals [4].

Various nutritional approaches for optimizing the weaning transition and minimizing gut inflammation and enteric diseases have been tested in the past decade. Among the novel dietary strategies investigated that are focused on improving gut health in pigs, probiotics and prebiotics are clear nutritional options. A growing body of evidence supports the therapeutic and preventive application of probiotics for several gastrointestinal disorders in pigs. Qiao *et al.* [5] conducted experiments to evaluate the effects of a complex Lactobacilli preparation on performance, resistance to *E. coli* infection and gut microbial flora of weaning pigs. The mix of four lactobacilli isolated from weaning pigs was able to reduce *E. coli* and anaerobe counts in the gut, and decrease diarrhoea. Additionally, lactobacilli treatment significantly improved average daily feed intake of pigs [5]. Another study from Herfel *et al.,* [6] examined the impact of a novel probiotic strain of *Bifidobacterium longum* AH1206 on the health, growth and development of neonatal pigs. Authors found that ileal IL-10 expression increased progressively with AH1206 supplementation, which indicated the potential for modulation of the inflammatory tone of the intestinal mucosa of suckling piglets. However, no differences were found between AH1206-treated and control piglets when comparing body weight gain, feed efficiency (gain:intake), and physiological modifications in the gut. Another recent study evaluated the effect of the co-administration of *Bacillus subtilis* RJGP16 and *Lactobacillus salivarius* B1 on intestinal immunity in piglets [7]. Authors demonstrated that probiotic administration increased the expression of IL-6, porcine beta-defensins and IgA producing cells in the intestine, clearly showing that co-administration of RJGP16 and B1 strains strongly enhances the intestinal mucosal immunity of piglets.

In addition, some recent studies have specifically evaluated the capacity of probiotics to improve the resistance of piglets against enterotoxigenic *Escherichia coli* (ETEC). It was shown that the probiotic strain *L. plantarum* CJLP243 may serve as a potential alternative to antibiotic supplementation to improve the growth and health performance of weaning pigs because of its capacity to reduce the severity of ETEC-induced diarrhoea [8]. Li *et al.,* [9] showed that pretreatment of piglets with *L. rhamnosus* ATCC7469 ameliorates $F4^+$ETEC-induced diarrhoea. In piglets exposed to $F4^+$ETEC, jejunal Toll-like receptor (TLR) 4 and IL-8 expression were increased; however, these increases were attenuated by administration of *L. rhamnosus.* Notably, expression of jejunal TLR2, ileal TLR9, NOD1 and TNF-α was upregulated in the ATCC7469-treated piglets after $F4^+$ETEC challenge [9].

Although these studies demonstrated that it is possible to modulate piglets' gut microbiota and immunity resulting in improvement of growth performance by using appropriate probiotics strains, the true efficacy of probiotics in livestock animals remains unclear because of incomplete studies of mechanisms. In this regard, some feed microbial materials

(FMAs) containing Aspergilli, Lactobacilli and Bifidobacteria have been used to successfully improve growth performance in livestock in Japan [10]. However, the exact cellular and molecular mechanisms involved in the beneficial effects have not been studied in detail.

Our laboratory has shown that probiotic microorganism with immunoregulatory functions (immunobiotics) can beneficially modulate the immune response in the gut by modulating the functions of intestinal epithelial cells (IECs) and immune cells [11] -[15]. Therefore, the aims of the present work were: 1) to evaluate the immunoregulatory capacities of common microbial feed materials used in Japan in porcine intestinal immune and epithelial cells and; 2) to advance in the knowledge of the cellular and molecular mechanisms involved in the immunoregulatory effect these materials by evaluating the role of TLRs and negative regulators of TLR.

2. MATERIALS AND METHODS

2.1. Feed Microbial Materials

The following feed microbial materials (FMAs) used as livestock feed in Japan were evaluated in this study. FMA1: Grinded freeze dry soymeal fermented with *Enterococcus faecium* and *Aspergillus oryzae*. FMA2: Freeze dry of *Candida utilis* cultures. FMA3: Freeze dry of *Saccharomyces cerevisiae* cultures. FMA4: Freeze dry of Enterococcus faecalis cultures. FMA5: Freeze dry of *Bifidobacterium thermophilum* cultures. FMA6: Freeze dry of *Leuconostoc mesenteroides* cultures.

Enterotoxigenic *Echerichia coli* (ETEC) strain 987P was kindly provided by the National Institute of Animal Health (Tsukuba, Japan). ETEC cells were plated into tryptic soy agar (TSA, Becton, Dickinson and Company) supplemented with 5% sheep blood (Nippon Biotest Laboratories Inc., Tokyo, Japan). After overnight incubation at 37°C, ETEC was transferred to tryptic soy broth (TSB, Becton) and grown for 5 days without shaking to form a pellicle containing the piliated phase. Then, ETEC cells were collected from the pellicle and transferred to TSB and cultured 20 hours at 37°C with shaking (200 rpm). Finally, the subculture of the ETEC strain was centrifuged at 1900 × g for 10 min at 4°C, washed with phosphate buffered saline (PBS) and heat-killed at 65°C for 30 min, and resuspended in Dulbecco's modified Eagle's medium (DMEM) (Invitrogen corporation, Carlsbad, CA) as described previously [12] [13].

2.2. Porcine Intestinal Epitheliocyte (PIE) Cells

Porcine intestinal epitheliocyte (PIE) cells were obtained as described before [16]. PIE cells were maintained in DMEM (Invitrogen Corporation, Carlsbad, CA) supplemented with 10% fetal calf serum (FCS), 100 mg/ml penicillin, and 100 U/ml streptomycin at 37°C in an atmosphere of 5% CO_2. PIE cells grow rapidly and are well adapted to culture conditions even without transformation or immortalization. However, the proliferative ability of PIE cells diminishes after 50 passages in culture. Therefore, we used PIE cells only between the 20th and 40th passages in experiments.

2.3. Immunocompetent Cells from Swine Peyer's Patches and Mesenteric Lymph Nodes

This study was carried out in strict accordance with the recommendations in the Guide for the Care and Use of Laboratory Animals of the Guidelines for Animal Experimentation of Tohoku University, Sendai, Japan. The present study was approved by the Institution Animal Care and Use Committee of Tohoku University with a permitted No. 2011-noudou-5. All efforts were made to minimize suffering. Immunocompetent cells from Swine peyer's patches (PPs) and mesenteric lymph nodes (MLNs) were prepared from adult swine intestine as described previously [11] [14]. Briefly, PPs or MLNs were cut into small fragments, which were gently pressed through a nylon mesh and washed three times in complete RPMI 1640 medium (Sigma, St. Louis, MO) supplemented with 10% FCS (Sigma). Residual erythrocytes were lysed by resuspending in hypotonic salt solution (0.2% NaCl). Next, harvested PPs or MLN cells were subjected to hypertonic rescue in an equal volume of 1.5% NaCl. Finally, immune cells were fractionated using Lympholyte-mammal (Cedarlane, Hornby, Ontario, Canada) density gradient centrifugation, and the isolated immune cells were suspended in complete DMEM (Invitrogen, Tokyo, Japan) supplemented with 10% FCS (Sigma) and 50 mg/ml penicillin-streptomycin (Life Tech., Carlsbad, CA, USA).

2.4. Mitogenicity Assay

The mitogenicity assay was performed as described previously [11]. Briefly, PPs or MLN immunocompetent cells were placed in a 96-well microplate (Costar, 2×10^5 cells/well) and incubated at 5% CO_2, 37°C for 48 h in complete RPMI 1640 medium (Sigma) supplemented with 2% FCS. Immune cells were stimulated with FMAs (0.1 - 1000 μg/mL) for 48 h. During the final hour of culture, the cells were radiolabeled with 9.25 kBq per well of [methyl-^{3}H]-Thymidine (GE Healthcare, Tokyo, Japan). The cells were then harvested with a glass fibre filter (PerkinElmer Japan, Kanagawa, Japan). The [methyl-^{3}H]-Thymidine incorporation was quantitated in a liquid scintillation counter (Beckman Instruments, Palo Alto, CA, USA). Results are presented as the Stimulation Index (SI), calculated with the following equation:

$$\frac{\left[(\text{counts per minute in treated cultures}) - (\text{counts per minute in background})\right]}{\left[(\text{counts per minute in control cultures}) - (\text{counts per minute in background})\right]} \quad 1$$

2.5. Immunomoregulatory Effect of FMAs in Immune Cells and PIE Cells

Evaluation of the immunoregulatory activity of FMAs was performed using mononuclear cells from PPs and MLNs or PIE cells. PP and MLM cells ware plated at a density of 4.0 × 10^6 cells/well in 12-well type I collagencoated plates (Iwaki, Tokyo, Japan). PIE cells were plated at a density of 3.0 × 10^4 cells/well in 12-well type I collagen-coated plates (Iwaki) and cultured for 3 days before stimulation of FAMs. FMAs were added into each well and cells were incubated for 48 h. Expression of IFN-γ, IL-2, IL-4, IL-6, IL-8, IL-10, IL-13 and MCP-1 was evaluated with quantitative real time RT-PCR as described below. In addition, expression levels of six TLR negative regulators, A20, Bcl-3, IRAK-M, MKP-1, SIGIRR, and Tollip, were evaluated in PIE cells as described previously [12] [13].

2.6. Anti-Inflammatory Effect of FMAs in PIE Cells

PIE cells were seeded at 3 × 10^4 cells/12-well plate on type I collagen-coated plates (Iwaki) and cultured for 3 days. After changing medium, FMAs were added; 48 h later, each well was washed vigorously with medium at least three times to eliminate all stimulants, and then cells were stimulated with ETEC (equivalent to 5 × 10^7 cells/ml) for 12 h. Expression of cytokines was evaluated with quantitative real time RT-PCR as described below. In blocking experiments, unlabeled anti-porcine TLR2- or TLR4-rabbit IgG (Biolegend, San Diego, CA) were used. Cultured porcine cells were incubated with unlabeled anti-TLR2 of TLR4 antibodies for 12 h before stimulation with FMAs.

2.7. Real-Time Quantitative RT-PCR

We performed two-step real-time quantitative PCR to characterize the expression of mRNAs in PIE cells and immune cells. Total RNA was isolated from each PIE or immune cell sample using TRIzol reagent (Invitrogen). All cDNAs were synthesized using a Quantitect reverse transcription (RT) kit (Qiagen, Tokyo, Japan) according to the manufacturer's recommendations. Real-time quantitative PCR was carried out using a 7300 real-time PCR system (Applied Biosystems, Warrington, UK) and the Platinum SYBR green qPCR SuperMix uracil-DNA glycosylase with 6-carboxyl-X-rhodamine (Invitrogen). The primers for IFN-γ, IL-2, IL-4, IL-6, IL-8, IL-10, IL- 13, MCP-1, A20, SIGIRR, Tollip, Bcl-3, MKP-1, and IRAK-M used in this study were previously described [11] [12]. The PCR cycling conditions were 5 min at 50°C, followed by 5 min at 95°C, and then 40 cycles of 15 s at 95°C, 30 s at 60°C, and 30 s at 72°C. The reaction mixtures contained 2.5 m/l sample cDNA and 7.5 m/l master mix, which included the sense and antisense primers. Expression of β-actin in each sample was assessed, and the β-actin data were used as an internal control to normalize differences between samples and to calculate normalized fold expression.

2.8. Western Blot Analysis

PIE cells were resuspended in 200 μl CelLytic M cell lysis reagent (Sigma) including protease and phosphate inhibitors (Complete Mini, PhosSTOP; Roche, Mannheim, Germany). Cells were transferred into Eppendorf tubes and boiled for 5 min at 95°C. Protein concentration was measured using the bicinchoninic acid protein assay kit (Pierce, Rockford, IL). Total protein samples (2 μg/sample) were loaded onto 10% SDS-polyacrylamide gels. Separated proteins were electrophoretically transferred to a nitrocellulose membrane. Phosphorylation of p38, and IκBα degradation were evaluated using anti-phosphorylated p38, and anti-IκB antibodies, respectively (Cell Signaling Technology, Beverly, MA). After detecting the phosphorylation, the membranes were stripped with Ten Minute Western Blot Re-Probe Kit (Jacksun Easy Biotech, Inc., New York, USA) for the detection of each total protein using p38 antibody (Cell Signaling Technology) and -actin antibody (Cell Signaling Technology). Anti-rabbit IgG, AP-linked antibody (Cell Signaling Technology) was used as secondary antibody. The optical protein bands were detected by ECF substrate (GE Healthcare Japan Co., Tokyo, Japan) and estimated from the peak area of densitogram by using Image J software (National Institute of Health, Bethesda, MD, USA).

2.9. Statistical Analysis

Statistical analyses were performed using the GLM and REG procedures available in the SAS computer program (SAS, 1994). Comparisons between mean values were carried out using one-way analysis of variance and Fisher's least-significant-difference test. For these analyses, P values of 0.05 were considered significant.

3. RESULTS

3.1. Immunoregulatory Activity of Feed Microbial Materials in Porcine Intestinal Immune Cells

We first evaluated the mitogenic capacity of the feed microbial materials (FMAs) in porcine intestinal immune isolated from Peyer's patches (PPs) or MLN (Figure 1).

All the FMAs were able to significantly increase the mitogenic activity of PPs and MLN cells compared with the control group, in a dose dependent manner. However, FMA5 was the most effective treatment to increase this parameter in MLN cells, being the doses of 100 μg/ml and 1000 μg/ml able to increase mitogenic activity more than 10 folds (Figure 1(a)). In addition, both FMA4 and FMA5 significantly increased mitogenic activities of immune isolated from PPs, being FMA5 in a dose of 100 μg/ml the most effective treatment to increase this parameter (Figure 1(b)).

We also evaluated the changes in cytokine expression in immune isolated from PPs or MLN after the treatments with the FMAs. As shown in Figure 2(a), all the treatments were able to upregulate the expression of IL- 4 and IL-6 in MLN cells. In addition, FMA5 significantly increased the expression of IL-2, IL-10 and IL-13 in those cells (Figure 2(a)). On the other hand, changes in the expression of cytokines in immune isolated from PPs were different for each FMA (Figure 2(b)). All the FMAs upregulated the expression of IFN-γ, being FMA5 and FMA6 the most effective treatments for improving this cytokine. In addition, FMA5 effectively increased the expression of IL-2 and IL-6 in PPs cells while FMA6 significantly increased IL-4 and IL-13 (Figure 2(b)).

3.2. Immunoregulatory Activity of Feed Microbial Materials in PIE Cells

The immunoregulatory capacities of FMAs were then evaluated in PIE cells. First, PIE cell were stimulated with the different FMAs and the expression of IL-6, IL-8 and MCP-1 was evaluated (Figure 3(a)).

FMA5 significantly increased the expression of IL-6 in PIE cells while FMA6 upregulated the expression of MCP-1 (Figure 3(a)).

In addition, the effect of FMAs in the response of PIE cell to heat-killed ETEC challenge was evaluated. As shown in Figure 3(b), ETEC challenge significantly increased the production of proinflammatory cytokines in PIE cells.

However, PIE cells pretreated with FMA5 before the challenge with ETEC showed significantly lower levels of IL-6, IL-8 and MCP-1 when compared to ETEC controls (Figure 3(b)). In addition, both FMA4 and FMA6 were able to reduce IL-8 expression.

We next evaluated whether FMA5 attenuated the ETEC-mediated proinflammatory response by modulating the NF-κB pathway (Figure 4(a)). Challenge of PIE cells with ETEC significantly reduced the levels of the counter-regulatory factor IκBα after 5 to 40 min, indicating activation of the NF-κB pathway. PIE cells previously stimulated with FMA5 or FMA1 showed significantly higher levels of IκBα when compared to control cells.

In fact, no significant degradation of IκBα was observed in FMAs-treated PIE cells, indicating an inhibitory effect of the NF-κB pathway (Figure 4(a)). We also examined the relationship between p38 MAPK activation and the regulation of proinflammatory cytokines in PIE cells by FMA5 (Figure 4(b)). PIE cells were stimulated with FMA1, FMA5, or control medium, and the activation profiles of p38 after ETEC challenge was compared. ETEC challenge induced the phosphorylation of p38 after 5 min. FMA5-treated PIE cells showed significantly lower levels of phosphorylated p38 than control cells (Figure 4(b)).

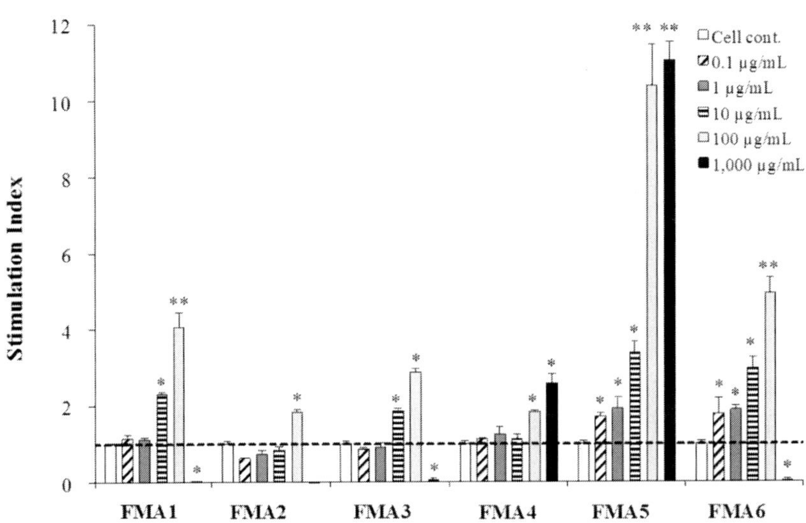

(a)

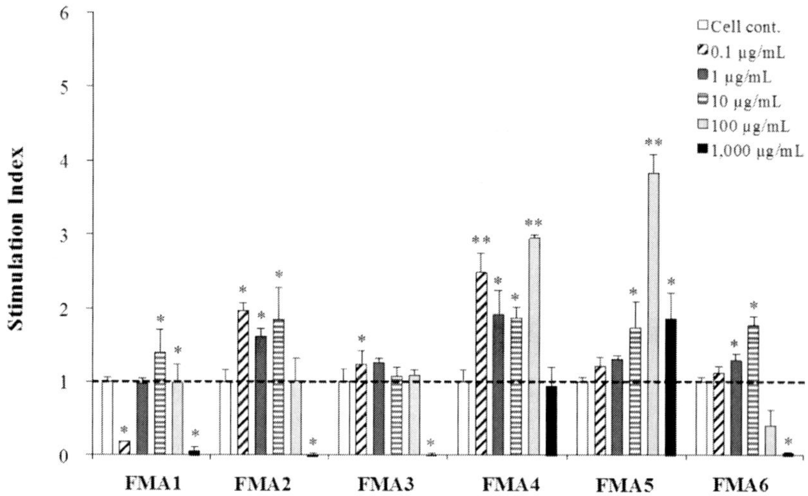

(b)

Figure 1. Mitogenic activity of functional feed materials (FMAs) on porcine mesenteric lymphoid nodes (MLN) (a) and Peyer's patches (PPs) (b) cells. FMA1: Grinded freeze dry soymeal fermented with Enterococcus faecium and Aspergillus oryzae. FMA2: Freeze dry of Candida utilis cultures. FMA3: Freeze dry of Saccharomyces cerevisiae cultures. FMA4: Freeze dry of Enterococcus faecalis cultures. FMA5: Freeze dry of Bifidobacterium thermophilum cultures. FMA6: Freeze dry of Leuconostoc mesenteroides cultures. Values represent means and error bars indicate the S.D. The results are representative of three independent experiments. $^{*}p < 0.05$, $^{**}p < 0.01$ against cell control.

3.3. Effect of Feed Microbial Materials on Negative Regulators of the TLR Signaling Pathway in PIE Cells

We also studied regulators that inhibit the TLR signaling pathway. The expression of SIGIRR, Tollip, A20, Bcl-3, MKP-1, and IRAK-M mRNAs in PIE cells was determined after the challenge with heat-killed ETEC (Figure 5). None of the treatments induced statistical significant changes in expression of Bcl-3 (Figure 5). However, all FMAs except FMA6 caused upregulation of SIGIRR. Additionally, FMA5 significantly increased the expression of Tollip and IRAK-M and A20 (Figure 5). FMA6 also upregulated A20 and IRAK-M while Bcl-3 was increased in PIE cells only with FMA1 (Figure 5).

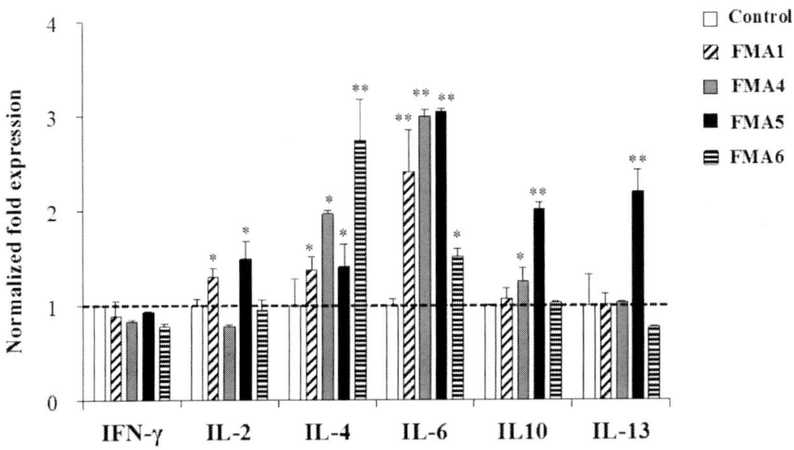

(a)

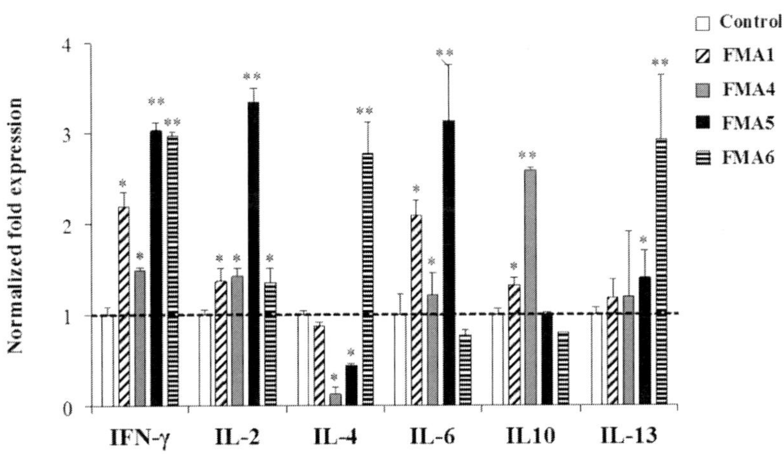

(b)

Figure 2. Expression of cytokines mRNA in porcine mesenteric lymphoid nodes (MLN) (a) and Peyer's patches (PPs) (b) cells stimulated with functional feed materials (FMAs). FMA1: Grinded freeze dry soymeal fermented with *Enterococcus faecium* and *Aspergillus oryzae.* FMA4: Freeze dry of *Enterococcus faecalis* cultures. FMA5: Freeze dry of *Bifidobacterium thermophilum* cultures. FMA6: Freeze dry of *Leuconostoc mesenteroides* cultures. Values represent means and error bars indicate the S.D. The results are representative of three independent experiments. $^{*}p < 0.05$, $^{**}p < 0.01$ against control.

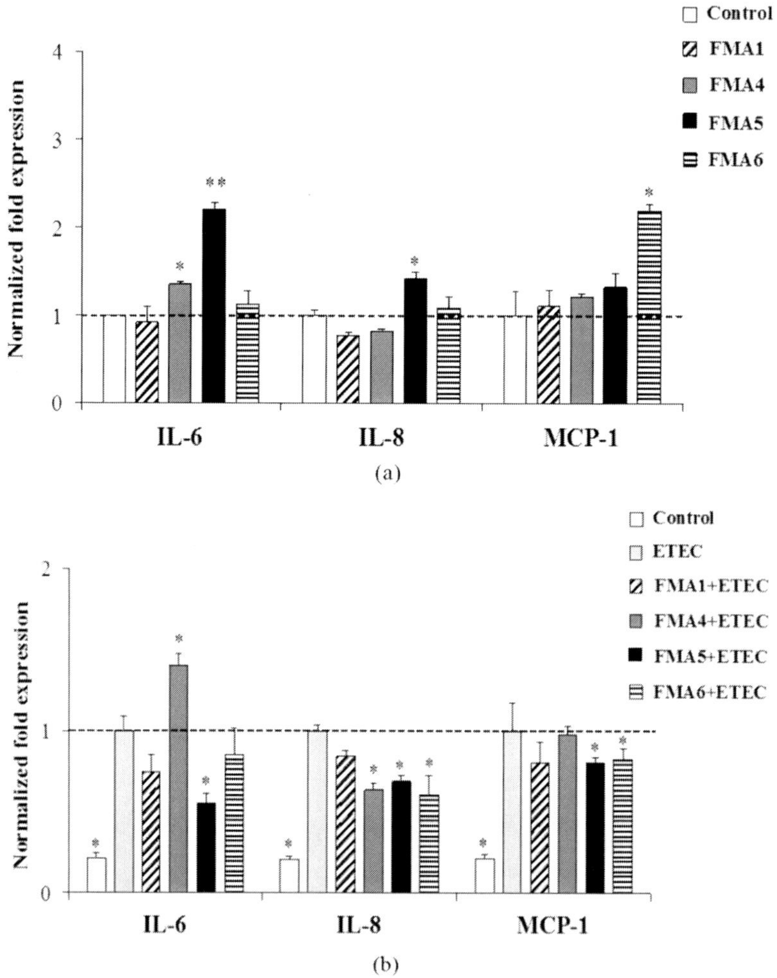

Figure 3. Effect of feed microbial materials (FMAs) on the expression of IL-6, IL-8 and MCP-1 mRNA in porcine intestinal epithelial (PIE) cells in non inflammatory conditions (a) and after inflammatory stimulation (b). (a) PIE cells were pre-treated with FMAs for 48 h or, (b) stimulated with FMAs for 48 h and then stimulated with heat-killed enterotoxigenic Escherichia coli (ETEC) for 12 h. PIE cells without pre-treatment (ETEC control group) and without ETEC challenge (control group) were used as controls. FMA1: Grinded freeze dry soymeal fermented with Enterococcus faecium and Aspergillus oryzae. FMA4: Freeze dry of Enterococcus faecalis cultures. FMA5: Freeze dry of Bifidobacterium thermophilum cultures. FMA6: Freeze dry of Leuconostoc mesenteroides cultures. Values represent means and error bars indicate the S.D. The results are representative of four independent experiments. *$p < 0.05$, **$p < 0.01$ against ETEC control.

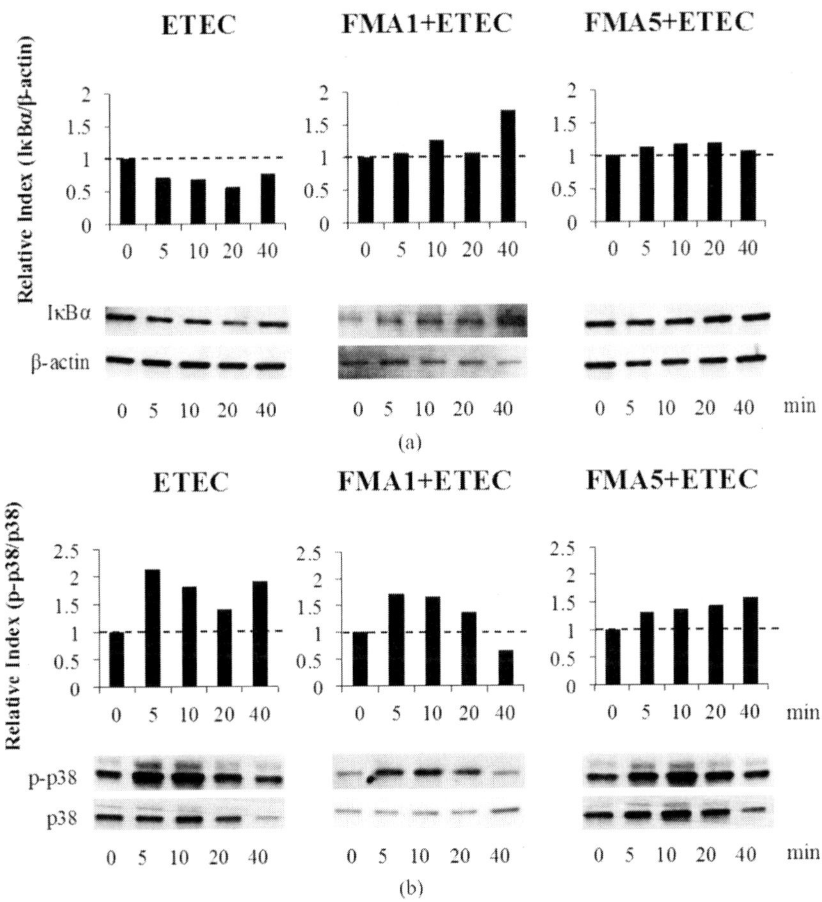

Figure 4. Effect of feed microbial materials (FMAs) on NF-B and p38 mitogen-activated protein kinase pathways in porcine intestinal epithelial (PIE) cells after inflammatory stimulation. PIE cells were pre-treated with FMAs for 48 h and then stimulated with heat-killed enterotoxigenic *Escherichia coli* (ETEC). The degradation of IκBα and phosphorylation of p38 was evaluated at different time intervals (0, 5, 10, 20, and 40 min). Intensities of protein bands were calculated from the peak area of densitogram by using image software and were expressed as bar graph of relative index with setting 1.0 at time 0. FMA1: Grinded freeze dry soymeal fermented with *Enterococcus faecium* and *Aspergillus oryzae*. FMA5: Freeze dry of *Bifidobacterium thermophilum* cultures.

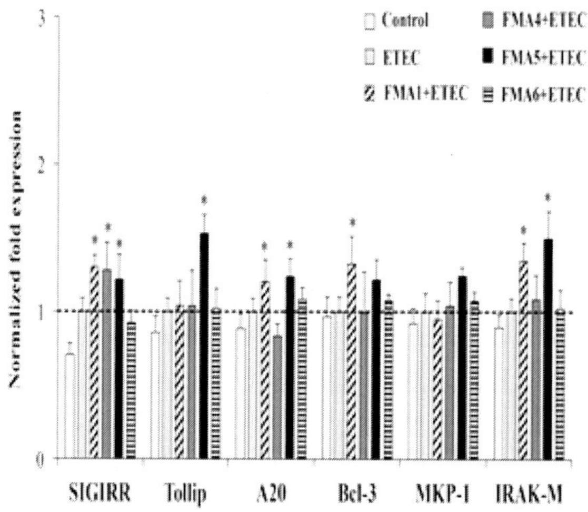

Figure 5. Effect of feed microbial materials (FMAs) on the expression of tolllike receptor negative regulators in porcine intestinal epithelial (PIE) cells. PIE cells were pre-treated with different FMAs for 48 h, stimulated with heat-killed enter otoxigenic Escherichia coli (ETEC) for 3 h, and the expression of SIGIRR, Tollip, A20, Bcl-3, MKP-1, and IRAK-M negative regulators was evaluated. FMA1: Grinded freeze dry soymeal fermented with Enterococcus faecium and Aspergillus oryzae. FMA4: Freeze dry of Enterococcus faecalis cultures. FMA5: Freeze dry of Bifidobacterium thermophilum cultures. FMA6: Freeze dry of Leuconostoc mesenteroides cultures. Values represent means and error bars indicate the S.D. The results are representative of three independent experiments. $^{*}p < 0.05$ against ETEC control.

3.4. Role of TLRs in the Immunoregulatory Activity of Feed Microbial Materials

To study the role of TLR2 and TLR4 in the immunomodulatory effect of FMA5, we next performed comparative studies with FMA5 and FMA1 and used anti-TLR2 and anti-TLR4 blocking antibodies, in PIE cells challenged with heat-killed ETEC (Figure 6). The use of anti-TLR2 antibodies did not induce any modification in the capacity of FMA5 to downregulate the expression of IL-6 or IL-8 in ETEC-challenged PIE cells, while the reduction of MCP-1 was abolished with anti-TLR2 antibodies (Figure 6). On the contrary, reduction of IL-6 and IL-8 induced by FMA5 was abolished with anti-TLR4 antibodies, while these blocking antibodies did not induce changes in the capacity of FMA5 to reduce MCP-1 (Figure 6).

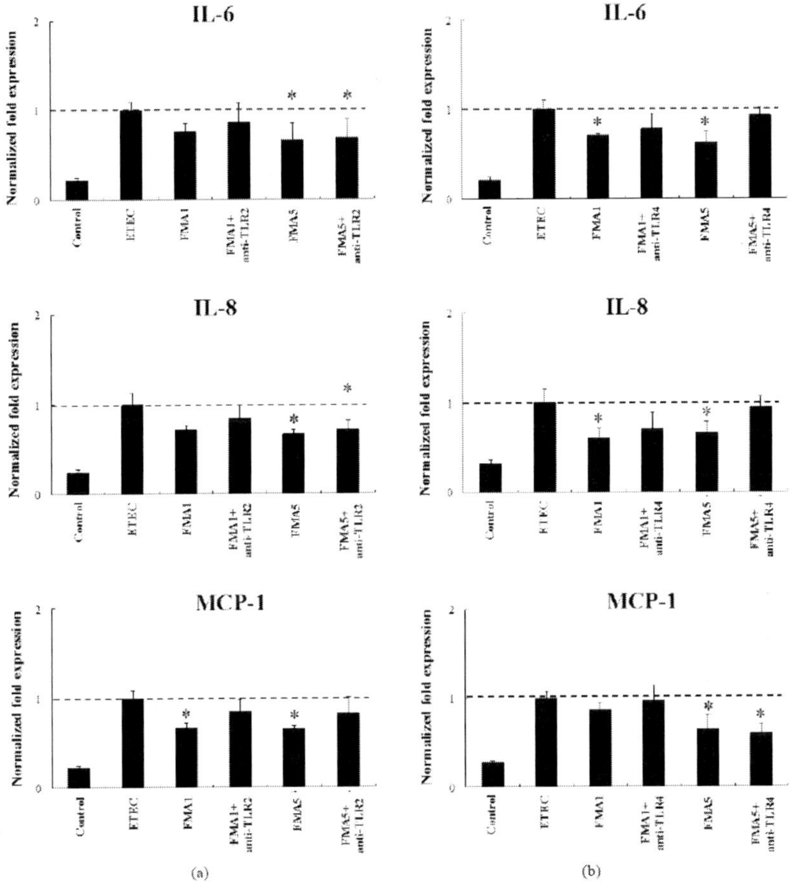

Figure 6. Role of Toll-like receptor (TLR)-2 and TLR4 s in the immunomodulatory effect of feed microbial materials (FMAs) in porcine intestinal epithelial (PIE) cells. PIE cells were pre-treated with FMAs in the presence or absence of anti-TLR2 or TLR4 antibodies. Untreated PIE cells were used as controls. PIE cells were stimulated with heat-killed enterotoxigenic *Escherichia coli* (ETEC) and then the expression of MCP-1, IL-6, and IL-8 mRNA were studied at 12 h post-stimulation. FMA1: Grinded freeze dry soymeal fermented with *Enterococcus faecium* and *Aspergillus oryzae*. FMA5: Freeze dry of *Bifidobacterium thermophilum* cultures. Values represent means and error bars indicate the S.D. The results are representative of three independent experiments. $^{*}p < 0.05$ against ETEC control.

4. DISCUSSION

The weaning transition is a complex period during which the piglets have to face abrupt changes. They are separated from their mother, mixed with other litters in a new environment, and switched from milk to a solid feed which involves a change from a highly digestible to a less-digestible and more-complex feed. In consequence, several physiological changes occur in the intestine of pigs during weaning which increase the susceptibility to uncontrolled inflammation and infections [reviewed in 17]. Probiotic microorganisms have been proposed to functionally modulate the microbiota and the immune system during the weaning transition in the pig, in order to avoid gut alterations and increase pig productivity.

In Japan, several FMAs are used as expected alternatives to antibiotics to improve pig health. However, the exact cellular and molecular mechanisms involved in the beneficial effects obtained in pigs have not been studied in detail. In this work we evaluated the immunoregulatory capacities of several FMAs using porcine in vitro systems, and we found that the freeze dry of B. thermo philum cultures (called FMA5 in this work) was notably able to modulate porcine immune cells as well as PIE cells.

Mitogenic activities are generally considered to be one of the most important properties of immune modulators, and have been used to evaluate the immunomodulating activities of potential probiotic microorganisms. Here, we used this methodology to evaluate the immunoregulatory effect of different FMAs in porcine immune cells. Results showed that FMA5 significantly increased mitogenic activity of both PPs and MLN cells. Moreover, FMA5 was able to increase the expression of IL-2, IL-6, IL-10 and IFN-γ in porcine immune cells indicating a clear immunostimulatory. In fact, early studies from Sasaki et al. [10] demonstrated that oral administration of B. *thermophilum* peptidoglycan to weaning piglets significantly improved intestinal humoral immunity and protected against post-weaning diarrhoea. Authors demonstrated that *B. thermop*hilum peptidoglycan administration augments IgA+ cells in pigs' intestine, which correlates with the augmented levels of IL-6 and IL-10 found in our experiments. Moreover, we extend these findings by demonstrating the capacity of B. thermophilum to improve intestinal Th1 immunity. This capacity of FMA5 to improve IgA and to stimulate both IL-10 and IFN-γ in immune cells could explain the beneficial effect in vivo because the administration of this FMA may result in an improvement in the adaptive immune response and protection from unproductive inflammation. In fact, evidence from in vivo trials of some strains indicates that immunobiotics can simultaneously improve both resistance against infection and protection from inflammatory tissue damage [17] -[19].

In addition, we showed that FMA5 is able to functionally modulate PIE cells. Previously, we demonstrated that the stimulation of PIE cells with ETEC significantly increases the levels of IL-6, IL-8, and MCP-1 [12] [16]. It was also found that damage to PIE cells correlates with the levels of proinflammatory cytokines produced after stimulation with heat-killed ETEC and LPS [12] , which is consistent with reports demonstrating that challenging human intestinal Caco-2 cells with ETEC causes strong up-regulation of proinflammatory mediators that lead to membrane damage [20] [21]. Then, this porcine in vitro system is a useful tool to evaluate the immunoprotective capacity of potential probiotic microorganisms. In this work, experiments with PIE cells allow us to conclude that FMA5 has an immunoprotective effect against ETEC challenge. The stimulation of PIE cells with ETEC significantly increased levels of proinflammatory cytokines in control cells. However, IL-6, IL-8 and MCP-1 levels in PIE cells stimulated with FMA5 were significantly lower than those in the control. Moreover we showed that this effect was related to the capacity of FMA5 to modulate NF-kB and p38 MAPK pathways in PIE cells. This is in line with reports demonstrating that probiotics can inhibit excessive NF-kB and MAPK induced by proinflammatory cytokine secreted from IECs. Immunobiotics can suppress TNF or *Salmonella enterica* serovar T*yphimurium*-induced IL-8 mRNA expression and secretion by IECs in an NF-kB-dependent manner [22]. Moreover, we demonstrated that some immunobiotic strains such as *L. jensenii* TL2937, *B. longum* BB536, *B. breve* M-16V and *B. breve* MCC-117 significantly downregulated levels of IL-8, MCP-1, and IL-6 in PIE cells challenged with ETEC by modulating the NF-kB and p38 MAPK pathways [12] [13] [16].

To dissect the mechanism(s) involved in the immunoregulatory effect of FMA5 we evaluated the expression of the negative TLR regulators in PIE cells. The expression of SIGIRR, Tollip, A20, Bcl-3, MKP-1, and IRAKM was studied, and it was found that SIGIRR, Tollip, A20 and IRAK-M expression was upregulated in PIE cells stimulated with FMA5.

It was shown *in vitro* that overexpression of SIGIRR inhibits TLR-induced NF-κB activation and attenuates the production of inflammatory cytokines [23]. The LPS-induced inflammatory response is enhanced in SIGIRR-deficient mice [24]. Notably, IRAK-M-deficient cells stimulated with TLR ligands or bacteria produce an increase in NF-κB and MAPK activation and elevated amounts of proinflammatory cytokines, such as IL-12, IL-6, and TNF-α [25]. IRAK-M expression is induced upon LPS stimulation, and endotoxin tolerance is diminished in IRAK-M-deficient cells; these observations indicate that IRAK-M plays a critical role in regulating

innate immunity through a negative feedback loop [26]. In addition, Tollip ensures a state of non-responsiveness in cultured enterocytes at re-exposure to LPS, due to down-regulation of TLR surface expression and decreased phosphorylation of IRAK-1. Tollip is known to be expressed at high levels in IECs, and to thereby contribute to the hyporesponsiveness of IECs to commensals [27]. Thus, the induction of these negative regulators by FMA5 in PIE cells may be important for establishing NF-kBand p38 MAPK-mediated tolerance of ETEC.

We also demonstrated previously that *L. jensenii* TL2937, *B. longum* BB536 and B. breve M-16V significantly upregulated A20 in PIE cells in a TLR2-dependent manner [12] [13]. A20 plays an essential role in the termination of NF-kB signaling in response to TNFand microbial products such as LPS [28]. Moreover, A20 deficiency in enterocytes renders mice sensitive to TNF-induced lethal inflammation, leading to disruption of the epithelial barrier and infiltration of commensal bacteria that initiate a systemic inflammatory response [29]. Therefore, we speculated that the strongest immunobiotic strains would induce anti-inflammatory effects in PIE cells through TLR2 and A20. We showed here that *B. thermophilum* present in FMA5 was able to upregulate A20 in PIE cells. Moreover, blocking experiments with anti-TLR2 antibodies significantly reduced the capacity of FMA5 to downregulate MCP-1 expression. Then, the present experiments confirm our previous results showing that A20 and TLR2 are important for the anti-inflammatory effect of immunobiotics.

In addition, TLR4 seems to have an important role in the immunoregulatory effect of FMA5 since blocking experiments with anti-TLR4 antibodies inhibit the reduction of IL-6 and IL-8 induced by FMA5. It was demonstrated that IECs shortly after birth exhibit a transient transcriptional activation induced by exposure to environmental endotoxin since it is almost completely abolished in TLR4-deficient mice [30] [31]. This transient intestinal epithelial activation precedes the acquisition of epithelial innate immune tolerance, a state of functional hyporesponsiveness to TLR4 stimuli. Epithelial tolerance is accompanied by a marked downregulation of the essential TLR signalling molecule interleukin 1 associated kinase (IRAK)-1 and a strong elevation of microRNA (miR)-146a expression [30] [31]. Moreover, epithelial tolerance appears to be maintained during the postnatal period by continuous TLR4 signalling. Further studies are necessary to establish the exact role of TLR4 in the immunoregulatory effect of FMA5. However, although the anti-inflammatory activities of probiotic microorganisms against ETEC, *Salmonella Typhimurium*, LPS or flagellin challenge have been reported previously using IECs cultures [20] [32] , this is the first report of the immunoregulatory effect of a FMA on porcine IECs that is able to induce its effect in a TLR4-dependent manner.

5. CONCLUSION

The scientific research into probiotic mode of actions has come to age and has shown how immunobiotics are able to induce beneficial changes in the host. The study of immunobiotics—intestinal cell interactions has unraveled several molecular mechanisms and cellular pathways, showing that these interactions play a crucial role in the regulation of several immunological functions in the gut. Our present research work allows us to give a view of the cellular and molecular mechanisms involved in the immunoregulatory effects of *Bifidobacterium thermophilum.* These results improve the scientific basis to use this feed material to modulate intestinal epithelial and immune cells and exert beneficial effects in pigs. This feed is expected to contribute to the healthy growth of young pigs without using antimicrobial feed materials.

ACKNOWLEDGEMENTS

This study was supported by a Grant-in-Aid for Scientific Research (B) (2) (No.21380164, 24380146) andChallenging Exploratory Research (No. 23658216) from the Japan Society for the Promotion of Science (JSPS)to Dr. H. Kitazawa. Dr. Paulraj Kanmani was supported by JSPS (Postdoctoral Fellowship for Foreign Researchers,Program No. 25-03397).

REFERENCES

1. Lalles, J.P., Boudry, G., Favier, C., Le Floc'h, N., Luron, I., Montagne, L., Oswald, I.P., Pié, S. and Piel Sève, B. (2004) Gut Function and Dysfunction in Young Pigs: Physiology. Animal Research, 53, 301-316. http://dx.doi.org/10.1051/animres:2004018
2. Williams, B.A., Verstegen, M.W.A. and Tamminga, S. (2001) Fermentation in the Large Intestine of Single-Stomached Animals and Its Relationship to Animal Health. Nutrition Research Reviews, 14, 207-227. http://dx.doi.org/10.1079/NRR200127
3. McDermott, P.F., Walker, R.D. and White, D.G. (2003) Antimicrobials: Modes of Action and Mechanisms of Resistance. International Journal of Toxicolology, 22, 135-143.
4. Kemper, N. (2008) Veterinary Antibiotics in the Aquatic and Terrestrial Environment. Ecological Indicators, 8, 1-13. http://dx.doi.org/10.1016/j.ecolind.2007.06.002
5. Qiao, S., Huang, C., Lifa, D., Piao, X. and Ren, J. (2004) Effects of Lactobacilli on the Performance, Diarrhoea Incidence, VFA Concentration and Gastrointestinal Microbial Flora of 29 Weaning Pigs. Asian-Australasian. Journal of Animal Science, 17, 401-409.
6. Herfel, T.M., Jacobi, S.K., Lin, X., Jouni, Z.E., Chichlowski, M., Stahl, C.H. and Odle, J. (2013) Dietary Supplementation of Bifidobacterium longum Strain AH1206 Increases Its Cecal Abundance and Elevates Intestinal Interleukin-10 Expression in the Neonatal Piglet. Food Chemical Toxicology, 60, 116-122. http://dx.doi.org/10.1016/j.fct.2013.07.020
7. Deng, J., Li, Y., Zhang, J. and Yang, Q. (2013) Co-Administration of Bacillus subtilis RJGP16 and Lactobacillus salivarius B1 Strongly Enhances the Intestinal Mucosal Immunity of

Piglets. Research in Veterinary Science, 94, 62-68. http://dx.doi.org/ 10.1016/ j.rvsc.2012.07.025
8. Lee, J.S., Awji, E.G., Lee, S.J., Tassew, D.D., Park, Y.B., Park, K.S., Kim, M.K., Kim, B. and Park, S.C. (2012) Effect of Lactobacillus plantarum CJLP243 on the Growth Performance and Cytokine Response of Weaning Pigs Challenged with Enterotoxigenic Escherichia coli. Journal of Animal Science, 90, 3709-3717. http://dx.doi.org/10.2527/jas.2011-4434
9. Li, X.Q., Zhu, Y.H., Zhang, H.F., Yue, Y., Cai, Z.X., Lu, Q.P., Zhang, L., Weng, X.G., Zhang, F.J., Zhou, D., Yang, J.C. and Wang, J.F. (2012) Risks Associated with High-Dose Lactobacillus rhamnosus in an Escherichia coli Model of Piglet Diarrhoea: Intestinal Microbiota and Immune Imbalances. PLoS ONE, 7, e40666. http://dx.doi.org/ 10.1371/ journal.pone.0040666
10. Sasaki, T., Maede, Y. and Namioka, S. (1987) Immunopotentiation of the Mucosa of the Small Intestine of Weaning Piglets by Peptidoglycan. Japanese Journal of Veterinary Science, 49, 235-243. http://dx.doi.org/10.1292/jvms1939.49.235
11. Fujie, H., Villena, J., Tohno, M., Morie, K., Shimazu, T., Aso, H., Suda, Y., Iwabuchi, N., Xiao, J., Iwatsuki, K., Kawai, Y., Saito, T. and Kitazawa, H. (2011) Toll-Like Receptor-2 Activating Bifidobacteria Strains Differentially Regulate Inflammatory Cytokines in Porcine Intestinal Epithelial Cell Culture System: Finding New Anti-Inflammatory Immunobiotics. FEMS Immunology and Medical Microbiology, 63, 129-139. http://dx.doi.org/ 10. 1111/j.1574-695X.2011.00837.x
12. Shimazu, T., Villena, J., Tohno, M., Fujie, H., Hosoya, S., Shimosato, T., Aso, H., Suda, Y., Kawai, Y., Saito, T., Makino, S., Ikegami, S., Itoh, H. and Kitazawa, H. (2012) Immunobiotic Lactobacillus jensenii Elicit Anti-Inflammatory Activity in Porcine Intestinal Epithelial Cells by Modulating Negative Regulators of the Toll-Like Receptor Signaling Pathway. Infection and Immunity, 80, 276-288. http://dx.doi.org/10.1128/IAI.05729-11
13. Tomosada, Y., Villena, J., Murata, K., Chiba, E., Shimazu, T., Aso, H., Iwabuchi, N., Xiao, J.Z., Saito, T. and Kitazawa, H. (2013) Immunoregulatory Effect of Bifidobacteria Strains in Porcine Intestinal Epithelial Cells through Modulation of Ubiquitin-Editing Enzyme A20 Expression. PLoS ONE, 8, e59259. http://dx.doi.org/10.1371/journal.pone.0059259
14. Villena, J., Suzuki, R., Fujie, H., Chiba, E., Takahashi, T., Shimazu, T., Aso, H., Ohwada, S., Suda, Y., Ikegami, S., Itoh, H., Alvarez, S., Saito, T. and Kitazawa, H. (2012) Immunobiotic Lactobacillus jensenii Modulates Toll-Like Receptor 4-Induced Inflammatory Response via Negative Regulation in Porcine Antigen Presenting Cells. Clinical and Vaccine Immunology, 19, 1038-1053. http://dx.doi.org/10.1128/CVI.00199-12
15. Murata, K., Tomosada, Y., Villena, J., Chiba, E., Shimazu, T., Aso, H., Iwabuchi, N., Xiao, J.Z., Saito, T. and Kitazawa, H. (2014) Bifidobacterium breve MCC-117 Induces Tolerance in Porcine Intestinal Epithelial Cells: Study of the Mechanisms Involved in the Immunoregulatory Effect. Bioscience of the Microbiota, Food and Health, 33, 1-10. http:// dx.doi.org/10.12938/bmfh.33.1
16. Moue, M., Tohno, M., Shimazu, T., Kido, T., Aso, H., Saito, T. and Kitazawa, H. (2008) Toll-Like Receptor 4 and Cytokine Expression Involved in Functional Immune Response in an Originally Established Porcine Intestinal Epitheliocyte Cell Line. Biochimeca and Biophysica Acta, 1780, 134-144.
17. Lalles, J.P., Bosia, P., Smidta, H. and Stokes, C.R. (2007) Nutritional Management of Gut Health in Pigs around Weaning. Proceedings of the Nutrition Society, 66, 260-268. http://dx.doi.org/10.1017/S0029665107005484
18. Salva, S., Villena, J. and Alvarez, S. (2010) Differential Immunomodulatory Activity of Lactobacillus rhamnosus Strains Isolated from Goat Milk: Impact on Intestinal and Respiratory Infections. International Journal of Food Microbiology, 141, 82-89.http://dx.doi.org/10.1016/j.ijfoodmicro.2010.03.013
19. Chiba, E., Tomosada, Y., Vizoso-Pinto, M.G., Takahashi, T., Tsukida, K., Kitazawa, H., Avarez, S. and Villena, J. (2013) Immunobiotic Lactobacillus rhamnosus Improves Resistance of Infant Mice against Respiratory Syncytial Virus Infection. International Immunopharmacolology, 17, 373-382. http://dx.doi.org/10.1016/j.intimp.2013.06.024
20. Roselli, M., Finamore, A., Britti, M.S. and Mengheri, E. (2006) Probiotic Bacteria Bifidobacterium animalis MB5 and Lactobacillus rhamnosus GG Protect Intestinal Caco-2

Cells from the Inflammation-Associated Response Induced by Enterotoxigenic Escherichia coli K88. British Journal of Nutrition, 95, 1177-1184. http://dx.doi.org/10.1079/BJN20051681

21. Roselli, M., Finamore, A., Britti, M.S., Konstantinov, S.R., Smidt, H., de Vos, W.M. and Mengheri, E. (2007) The Novel Porcine Lactobacillus sobrius Strain Protects Intestinal Cells from Enterotoxigenic Escherichia coli K88 Infection and Prevents Membrane Barrier Damage. Journal of Nutrition, 137, 2709-2716.
22. O'Hara, A.M., O'Regan, P., Fanning, A., O'Mahony, C., Macsharry, J., Lyons, A., Bienenstock, J., O'Mahony, L. and Shanahan, F. (2006) Functional Modulation of Human Intestinal Epithelial Cell Responses by Bifidobacterium infantis and Lactobacillus salivarius. Immunology, 118, 202-215. http://dx.doi.org/10.1111/j.1365-2567.2006.02358.x
23. Qin, J., Qian, Y., Yao, J., Grace, C. and Li, X. (2005) SIGIRR Inhibits Interleukin-1 Receptorand Toll-Like Receptor 4-Mediated Signaling through Different Mechanisms. Journal of Biological Chemistry, 280, 25233-25241. http://dx.doi.org/ 10.1074/jbc.M501363200
24. Lech Garlanda, M., Mantovani, C., Kirschning, A., Schlöndorff, C.J. and Anders, D. (2007) Different Roles of TiR8/Sigirr on Toll-Like Receptor Signaling in Intrarenal Antigen-Presenting Cells and Tubular Epithelial Cells. Kidney International, 72, 182-192. http://dx.doi.org/10.1038/sj.ki.5002293
25. Deng, J.C., Cheng, G., Newstead, M.W., Zeng, X., Kobayashi, K., Flavell, R.A. and Standiford, T.J. (2006) Sepsis-Induced Suppression of Lung Innate Immunity Is Mediated by IRAK-M. Journal of Clinical Investigation, 116, 2532-2542.
26. Escoll, P., del Fresno, C., García, L., Vallés, G., Lendínez, M.J., Arnalich, F. and López-Collazo, E. (2003) Rapid Up-Regulation of IRAK-M Expression Following a Second Endotoxin Challenge in Human Monocytes and in Monocytes Isolated from Septic Patients. Biochemical and Biophysical Research Communications, 311, 465-472. http://dx.doi.org/10.1016/j.bbrc.2003.10.019
27. Otte, J.M., Cario, E. and Podolsky, D.K. (2004) Mechanisms of Cross Hyporesponsiveness to Toll-Like Receptor Bacterial Ligands in Intestinal Epithelial Cells. Gastroenterology, 126, 1054-1070. http://dx.doi.org/10.1053/j.gastro.2004.01.007
28. Wang, J., Ouyang, Y., Guner, Y., Ford, H.R. and Grishin, A.V. (2009) Ubiquitin-Editing Enzyme A20 Promotes Tolerance to Lipopolysaccharide in Enterocytes. Journal of Immunology, 183, 1384-1392. http://dx.doi.org/10.4049/jimmunol.0803987
29. Lee, E.G., Boone, D.L., Chai, S., Libby, S.L. and Chien, M. (2000) Failure to Regulate TNF-Induced NF-kB and Cell Death Responses in A20-Deficient Mice. Science, 289, 2350-2354. http://dx.doi.org/10.1126/science.289.5488.2350
30. Lotz, M., Gutle, D., Walther, S., Menard, S., Bogdan, C. and Hornef, M.W. (2006) Postnatal Acquisition of Endotoxin Tolerance in Intestinal Epithelial Cells. The Journal of Experimental Medicine, 203, 973-984. http://dx.doi.org/10.1084/jem.20050625
31. Chassin, C., Kocur, M., Pott, J., Duerr, C.U., Gutle, D. and Lotz, M. (2010) miR-146a Mediates Protective Innate Immune Tolerance in the Neonate Intestine. Cell Host and Microbe, 8, 358-368. http://dx.doi.org/10.1016/j.chom.2010.09.005
32. Riedel, C.U., Foata, F., Philippe, D., Adolfsson, O., Eikmanns, B.J. and Blum, S. (2006) Anti-Inflammatory Effects of Bifidobacteria by Inhibition of LPS-Induced NF-kappaB Activation. World Journal of Gastroenterology, 12, 3729- 3735.

INDEX